Recommender Systems: The Textbook

Charu C. Aggarwal

Recommender Systems

The Textbook

 Springer

Charu C. Aggarwal
IBM T.J. Watson Research Center
Yorktown Heights, NY, USA

ISBN 978-3-319-80619-8   ISBN 978-3-319-29659-3   (eBook)
DOI 10.1007/978-3-319-29659-3

Springer Cham Heidelberg New York Dordrecht London
© Springer International Publishing Switzerland 2016
Softcover reprint of the hardcover 1st edition 2016

Printed on acid-free paper

Springer International Publishing AG Switzerland is part of Springer Science+Business Media (www.springer.com)

To my wife Lata, my daughter Sayani,
and my late parents Dr. Prem Sarup and Mrs. Pushplata Aggarwal.

Contents

1   **An Introduction to Recommender Systems**   **1**
   1.1   Introduction   1
   1.2   Goals of Recommender Systems   3
      1.2.1   The Spectrum of Recommendation Applications   7
   1.3   Basic Models of Recommender Systems   8
      1.3.1   Collaborative Filtering Models   8
         1.3.1.1   Types of Ratings   10
         1.3.1.2   Relationship with Missing Value Analysis   13
         1.3.1.3   Collaborative Filtering as a Generalization of Classification
                 and Regression Modeling   13
      1.3.2   Content-Based Recommender Systems   14
      1.3.3   Knowledge-Based Recommender Systems   15
         1.3.3.1   Utility-Based Recommender Systems   18
      1.3.4   Demographic Recommender Systems   19
      1.3.5   Hybrid and Ensemble-Based Recommender Systems   19
      1.3.6   Evaluation of Recommender Systems   20
   1.4   Domain-Specific Challenges in Recommender Systems   20
      1.4.1   Context-Based Recommender Systems   20
      1.4.2   Time-Sensitive Recommender Systems   21
      1.4.3   Location-Based Recommender Systems   21
      1.4.4   Social Recommender Systems   22
         1.4.4.1   Structural Recommendation of Nodes and Links   22
         1.4.4.2   Product and Content Recommendations with Social
                 Influence   23
         1.4.4.3   Trustworthy Recommender Systems   23
         1.4.4.4   Leveraging Social Tagging Feedback for
                 Recommendations   23
   1.5   Advanced Topics and Applications   23
      1.5.1   The Cold-Start Problem in Recommender Systems   24
      1.5.2   Attack-Resistant Recommender Systems   24
      1.5.3   Group Recommender Systems   24

	1.5.4	Multi-Criteria Recommender Systems	24
	1.5.5	Active Learning in Recommender Systems	25
	1.5.6	Privacy in Recommender Systems	25
	1.5.7	Application Domains	26
1.6	Summary		26
1.7	Bibliographic Notes		26
1.8	Exercises		28

2 Neighborhood-Based Collaborative Filtering   29

2.1	Introduction		29
2.2	Key Properties of Ratings Matrices		31
2.3	Predicting Ratings with Neighborhood-Based Methods		33
	2.3.1	User-Based Neighborhood Models	34
		2.3.1.1 Similarity Function Variants	37
		2.3.1.2 Variants of the Prediction Function	38
		2.3.1.3 Variations in Filtering Peer Groups	39
		2.3.1.4 Impact of the Long Tail	39
	2.3.2	Item-Based Neighborhood Models	40
	2.3.3	Efficient Implementation and Computational Complexity	41
	2.3.4	Comparing User-Based and Item-Based Methods	42
	2.3.5	Strengths and Weaknesses of Neighborhood-Based Methods	44
	2.3.6	A Unified View of User-Based and Item-Based Methods	44
2.4	Clustering and Neighborhood-Based Methods		45
2.5	Dimensionality Reduction and Neighborhood Methods		47
	2.5.1	Handling Problems with Bias	49
		2.5.1.1 Maximum Likelihood Estimation	49
		2.5.1.2 Direct Matrix Factorization of Incomplete Data	50
2.6	A Regression Modeling View of Neighborhood Methods		51
	2.6.1	User-Based Nearest Neighbor Regression	53
		2.6.1.1 Sparsity and Bias Issues	54
	2.6.2	Item-Based Nearest Neighbor Regression	55
	2.6.3	Combining User-Based and Item-Based Methods	57
	2.6.4	Joint Interpolation with Similarity Weighting	57
	2.6.5	Sparse Linear Models (SLIM)	58
2.7	Graph Models for Neighborhood-Based Methods		60
	2.7.1	User-Item Graphs	61
		2.7.1.1 Defining Neighborhoods with Random Walks	61
		2.7.1.2 Defining Neighborhoods with the Katz Measure	62
	2.7.2	User-User Graphs	63
	2.7.3	Item-Item Graphs	66
2.8	Summary		67
2.9	Bibliographic Notes		67
2.10	Exercises		69

3 Model-Based Collaborative Filtering   71

3.1	Introduction		71
3.2	Decision and Regression Trees		74
	3.2.1	Extending Decision Trees to Collaborative Filtering	76

3.3   Rule-Based Collaborative Filtering   77
        3.3.1   Leveraging Association Rules for Collaborative Filtering   79
        3.3.2   Item-Wise Models versus User-Wise Models   80
3.4   Naive Bayes Collaborative Filtering   82
        3.4.1   Handling Overfitting   84
        3.4.2   Example of the Bayes Method with Binary Ratings   85
3.5   Using an Arbitrary Classification Model as a Black-Box   86
        3.5.1   Example: Using a Neural Network as a Black-Box   87
3.6   Latent Factor Models   90
        3.6.1   Geometric Intuition for Latent Factor Models   91
        3.6.2   Low-Rank Intuition for Latent Factor Models   93
        3.6.3   Basic Matrix Factorization Principles   94
        3.6.4   Unconstrained Matrix Factorization   96
                3.6.4.1   Stochastic Gradient Descent   99
                3.6.4.2   Regularization   100
                3.6.4.3   Incremental Latent Component Training   103
                3.6.4.4   Alternating Least Squares and Coordinate Descent   105
                3.6.4.5   Incorporating User and Item Biases   106
                3.6.4.6   Incorporating Implicit Feedback   109
        3.6.5   Singular Value Decomposition   113
                3.6.5.1   A Simple Iterative Approach to SVD   114
                3.6.5.2   An Optimization-Based Approach   116
                3.6.5.3   Out-of-Sample Recommendations   116
                3.6.5.4   Example of Singular Value Decomposition   117
        3.6.6   Non-negative Matrix Factorization   119
                3.6.6.1   Interpretability Advantages   121
                3.6.6.2   Observations about Factorization with Implicit Feedback .   122
                3.6.6.3   Computational and Weighting Issues with Implicit
                          Feedback   124
                3.6.6.4   Ratings with Both Likes and Dislikes   124
        3.6.7   Understanding the Matrix Factorization Family   126
3.7   Integrating Factorization and Neighborhood Models   128
        3.7.1   Baseline Estimator: A Non-Personalized Bias-Centric Model   128
        3.7.2   Neighborhood Portion of Model   129
        3.7.3   Latent Factor Portion of Model   130
        3.7.4   Integrating the Neighborhood and Latent Factor Portions   131
        3.7.5   Solving the Optimization Model   131
        3.7.6   Observations about Accuracy   132
        3.7.7   Integrating Latent Factor Models with Arbitrary Models   133
3.8   Summary   134
3.9   Bibliographic Notes   134
3.10   Exercises   136

4   Content-Based Recommender Systems   139
    4.1   Introduction   139
    4.2   Basic Components of Content-Based Systems   141
    4.3   Preprocessing and Feature Extraction   142
            4.3.1   Feature Extraction   142
                    4.3.1.1   Example of Product Recommendation   143

    4.3.1.2   Example of Web Page Recommendation 143
    4.3.1.3   Example of Music Recommendation 144
  4.3.2   Feature Representation and Cleaning 145
  4.3.3   Collecting User Likes and Dislikes 146
  4.3.4   Supervised Feature Selection and Weighting 147
    4.3.4.1   Gini Index 147
    4.3.4.2   Entropy 148
    4.3.4.3 χ^2-Statistic 148
    4.3.4.4   Normalized Deviation 149
    4.3.4.5   Feature Weighting 150
4.4   Learning User Profiles and Filtering 150
  4.4.1   Nearest Neighbor Classification 151
  4.4.2   Connections with Case-Based Recommender Systems 152
  4.4.3   Bayes Classifier 153
    4.4.3.1   Estimating Intermediate Probabilities 154
    4.4.3.2   Example of Bayes Model 155
  4.4.4   Rule-based Classifiers 156
    4.4.4.1   Example of Rule-based Methods 157
  4.4.5   Regression-Based Models 158
  4.4.6   Other Learning Models and Comparative Overview 159
  4.4.7   Explanations in Content-Based Systems 160
4.5   Content-Based Versus Collaborative Recommendations 161
4.6   Using Content-Based Models for Collaborative Filtering 162
  4.6.1   Leveraging User Profiles 163
4.7   Summary 163
4.8   Bibliographic Notes 164
4.9   Exercises 165

5   **Knowledge-Based Recommender Systems**   **167**
5.1   Introduction 167
5.2   Constraint-Based Recommender Systems 172
  5.2.1   Returning Relevant Results 174
  5.2.2   Interaction Approach 176
  5.2.3   Ranking the Matched Items 178
  5.2.4   Handling Unacceptable Results or Empty Sets 179
  5.2.5   Adding Constraints 180
5.3   Case-Based Recommenders 181
  5.3.1   Similarity Metrics 183
    5.3.1.1   Incorporating Diversity in Similarity Computation 187
  5.3.2   Critiquing Methods 188
    5.3.2.1   Simple Critiques 188
    5.3.2.2   Compound Critiques 190
    5.3.2.3   Dynamic Critiques 192
  5.3.3   Explanation in Critiques 193
5.4   Persistent Personalization in Knowledge-Based Systems 194
5.5   Summary 195
5.6   Bibliographic Notes 195
5.7   Exercises 197

6 Ensemble-Based and Hybrid Recommender Systems   **199**
   6.1   Introduction 199
   6.2   Ensemble Methods from the Classification Perspective 204
   6.3   Weighted Hybrids 206
       6.3.1   Various Types of Model Combinations 208
       6.3.2   Adapting Bagging from Classification 209
       6.3.3   Randomness Injection 211
   6.4   Switching Hybrids 211
       6.4.1   Switching Mechanisms for Cold-Start Issues 212
       6.4.2   Bucket-of-Models 212
   6.5   Cascade Hybrids 213
       6.5.1   Successive Refinement of Recommendations 213
       6.5.2   Boosting 213
             6.5.2.1   Weighted Base Models 214
   6.6   Feature Augmentation Hybrids 215
   6.7   Meta-Level Hybrids 216
   6.8   Feature Combination Hybrids 217
       6.8.1   Regression and Matrix Factorization 218
       6.8.2   Meta-level Features 218
   6.9   Mixed Hybrids 220
   6.10   Summary 221
   6.11   Bibliographic Notes 222
   6.12   Exercises 224

7 Evaluating Recommender Systems   **225**
   7.1   Introduction 225
   7.2   Evaluation Paradigms 227
       7.2.1   User Studies 227
       7.2.2   Online Evaluation 227
       7.2.3   Offline Evaluation with Historical Data Sets 229
   7.3   General Goals of Evaluation Design 229
       7.3.1   Accuracy 229
       7.3.2   Coverage 231
       7.3.3   Confidence and Trust 232
       7.3.4   Novelty 233
       7.3.5   Serendipity 233
       7.3.6   Diversity 234
       7.3.7   Robustness and Stability 235
       7.3.8   Scalability 235
   7.4   Design Issues in Offline Recommender Evaluation 235
       7.4.1   Case Study of the Netflix Prize Data Set 236
       7.4.2   Segmenting the Ratings for Training and Testing 238
             7.4.2.1   Hold-Out 238
             7.4.2.2   Cross-Validation 239
       7.4.3   Comparison with Classification Design 239
   7.5   Accuracy Metrics in Offline Evaluation 240
       7.5.1   Measuring the Accuracy of Ratings Prediction 240
             7.5.1.1   RMSE versus MAE 241
             7.5.1.2   Impact of the Long Tail 241

    7.5.2   Evaluating Ranking via Correlation 242
    7.5.3   Evaluating Ranking via Utility 244
    7.5.4   Evaluating Ranking via Receiver Operating Characteristic 247
    7.5.5   Which Ranking Measure is Best? 250
  7.6   Limitations of Evaluation Measures 250
    7.6.1   Avoiding Evaluation Gaming 252
  7.7   Summary 252
  7.8   Bibliographic Notes 253
  7.9   Exercises 254

8   Context-Sensitive Recommender Systems   255
  8.1   Introduction 255
  8.2   The Multidimensional Approach 256
    8.2.1   The Importance of Hierarchies 259
  8.3   Contextual Pre-filtering: A Reduction-Based Approach 262
    8.3.1   Ensemble-Based Improvements 264
    8.3.2   Multi-level Estimation 265
  8.4   Post-Filtering Methods 266
  8.5   Contextual Modeling 268
    8.5.1   Neighborhood-Based Methods 268
    8.5.2   Latent Factor Models 269
      8.5.2.1   Factorization Machines 272
      8.5.2.2   A Generalized View of Second-Order Factorization
                       Machines 275
      8.5.2.3   Other Applications of Latent Parametrization 276
    8.5.3   Content-Based Models 277
  8.6   Summary 279
  8.7   Bibliographic Notes 280
  8.8   Exercises 281

9   Time- and Location-Sensitive Recommender Systems   283
  9.1   Introduction 283
  9.2   Temporal Collaborative Filtering 285
    9.2.1   Recency-Based Models 286
      9.2.1.1   Decay-Based Methods 286
      9.2.1.2   Window-Based Methods 288
    9.2.2   Handling Periodic Context 288
      9.2.2.1   Pre-Filtering and Post-Filtering 289
      9.2.2.2   Direct Incorporation of Temporal Context 290
    9.2.3   Modeling Ratings as a Function of Time 290
      9.2.3.1   The Time-SVD++ Model 291
  9.3   Discrete Temporal Models 295
    9.3.1   Markovian Models 295
      9.3.1.1   Selective Markov Models 298
      9.3.1.2   Other Markovian Alternatives 300
    9.3.2   Sequential Pattern Mining 300
  9.4   Location-Aware Recommender Systems 302
    9.4.1   Preference Locality 303
    9.4.2   Travel Locality 305
    9.4.3   Combined Preference and Travel Locality 305

9.5   Summary   305
9.6   Bibliographic Notes   306
9.7   Exercises   308

10 Structural Recommendations in Networks   **309**
10.1   Introduction 309
10.2   Ranking Algorithms   311
       10.2.1 PageRank   311
       10.2.2 Personalized PageRank   314
       10.2.3 Applications to Neighborhood-Based Methods   316
               10.2.3.1 Social Network Recommendations   317
               10.2.3.2 Personalization in Heterogeneous Social Media   317
               10.2.3.3 Traditional Collaborative Filtering   319
       10.2.4 SimRank   321
       10.2.5 The Relationship Between Search and Recommendation   322
10.3   Recommendations by Collective Classification   323
       10.3.1 Iterative Classification Algorithm   324
       10.3.2 Label Propagation with Random Walks   325
       10.3.3 Applicability to Collaborative Filtering in Social Networks   326
10.4   Recommending Friends: Link Prediction   326
       10.4.1 Neighborhood-Based Measures   327
       10.4.2 Katz Measure   328
       10.4.3 Random Walk-Based Measures   329
       10.4.4 Link Prediction as a Classification Problem   329
       10.4.5 Matrix Factorization for Link Prediction   330
               10.4.5.1 Symmetric Matrix Factorization   333
       10.4.6 Connections Between Link Prediction and Collaborative Filtering .   335
               10.4.6.1 Using Link Prediction Algorithms for Collaborative
                         Filtering   336
               10.4.6.2 Using Collaborative Filtering Algorithms for Link
                         Prediction   337
10.5   Social Influence Analysis and Viral Marketing   337
       10.5.1 Linear Threshold Model   339
       10.5.2 Independent Cascade Model   340
       10.5.3 Influence Function Evaluation   340
       10.5.4 Targeted Influence Analysis Models in Social Streams   341
10.6   Summary   342
10.7   Bibliographic Notes   343
10.8   Exercises   344

11 Social and Trust-Centric Recommender Systems   **345**
11.1   Introduction 345
11.2   Multidimensional Models for Social Context   347
11.3   Network-Centric and Trust-Centric Methods   349
       11.3.1 Collecting Data for Building Trust Networks   349
       11.3.2 Trust Propagation and Aggregation   351
       11.3.3 Simple Recommender with No Trust Propagation   353
       11.3.4 TidalTrust Algorithm   353

11.3.5   MoleTrust Algorithm 356
11.3.6   TrustWalker Algorithm 357
11.3.7   Link Prediction Methods   358
11.3.8   Matrix Factorization Methods 361
         11.3.8.1   Enhancements with Logistic Function 364
         11.3.8.2   Variations in the Social Trust Component   364
11.3.9   Merits of Social Recommender Systems 365
         11.3.9.1   Recommendations for Controversial Users and Items   . . . 365
         11.3.9.2   Usefulness for Cold-Start 366
         11.3.9.3   Attack Resistance 366
11.4   User Interaction in Social Recommenders 366
11.4.1   Representing Folksonomies 367
11.4.2   Collaborative Filtering in Social Tagging Systems 368
11.4.3   Selecting Valuable Tags 371
11.4.4   Social-Tagging Recommenders with No Ratings Matrix 372
         11.4.4.1   Multidimensional Methods for Context-Sensitive Systems . 372
         11.4.4.2   Ranking-Based Methods 373
         11.4.4.3   Content-Based Methods   374
11.4.5   Social-Tagging Recommenders with Ratings Matrix 377
         11.4.5.1   Neighborhood-Based Approach 378
         11.4.5.2   Linear Regression 379
         11.4.5.3   Matrix Factorization 380
         11.4.5.4   Content-Based Methods 382
11.5   Summary 382
11.6   Bibliographic Notes 382
11.7   Exercises 384

12 Attack-Resistant Recommender Systems   385
12.1   Introduction 385
12.2   Understanding the Trade-Offs in Attack Models 386
12.2.1   Quantifying Attack Impact 390
12.3   Types of Attacks   392
12.3.1   Random Attack 393
12.3.2   Average Attack 393
12.3.3   Bandwagon Attack 394
12.3.4   Popular Attack 395
12.3.5   Love/Hate Attack 395
12.3.6   Reverse Bandwagon Attack 396
12.3.7   Probe Attack   396
12.3.8   Segment Attack 396
12.3.9   Effect of Base Recommendation Algorithm 397
12.4   Detecting Attacks on Recommender Systems 398
12.4.1   Individual Attack Profile Detection 399
12.4.2   Group Attack Profile Detection 402
         12.4.2.1   Preprocessing Methods 402
         12.4.2.2   Online Methods 403
12.5   Strategies for Robust Recommender Design 403
12.5.1   Preventing Automated Attacks with CAPTCHAs 403
12.5.2   Using Social Trust 404

		12.5.3	Designing Robust Recommendation Algorithms	404
			12.5.3.1 Incorporating Clustering in Neighborhood Methods	405
			12.5.3.2 Fake Profile Detection during Recommendation Time	405
			12.5.3.3 Association-Based Algorithms	405
			12.5.3.4 Robust Matrix Factorization	405
	12.6	Summary		408
	12.7	Bibliographic Notes		408
	12.8	Exercises		410

13 Advanced Topics in Recommender Systems — **411**

	13.1	Introduction		411
	13.2	Learning to Rank		413
		13.2.1	Pairwise Rank Learning	415
		13.2.2	Listwise Rank Learning	416
		13.2.3	Comparison with Rank-Learning Methods in Other Domains	417
	13.3	Multi-Armed Bandit Algorithms		418
		13.3.1	Naive Algorithm	419
		13.3.2	ϵ-Greedy Algorithm	420
		13.3.3	Upper Bounding Methods	421
	13.4	Group Recommender Systems		423
		13.4.1	Collaborative and Content-Based Systems	424
		13.4.2	Knowledge-Based Systems	425
	13.5	Multi-Criteria Recommender Systems		426
		13.5.1	Neighborhood-Based Methods	427
		13.5.2	Ensemble-Based Methods	428
		13.5.3	Multi-Criteria Systems without Overall Ratings	429
	13.6	Active Learning in Recommender Systems		430
		13.6.1	Heterogeneity-Based Models	431
		13.6.2	Performance-Based Models	432
	13.7	Privacy in Recommender Systems		432
		13.7.1	Condensation-Based Privacy	434
		13.7.2	Challenges for High-Dimensional Data	434
	13.8	Some Interesting Application Domains		435
		13.8.1	Portal Content Personalization	435
			13.8.1.1 Dynamic Profiler	436
			13.8.1.2 Google News Personalization	436
		13.8.2	Computational Advertising versus Recommender Systems	438
			13.8.2.1 Importance of Multi-Armed Bandit Methods	442
		13.8.3	Reciprocal Recommender Systems	443
			13.8.3.1 Leveraging Hybrid Methods	444
			13.8.3.2 Leveraging Link Prediction Methods	445
	13.9	Summary		446
	13.10	Bibliographic Notes		446

Bibliography — **449**

Index — **493**

Preface

"Nature shows us only the tail of the lion. But I do not doubt that the lion belongs to it even though he cannot at once reveal himself because of his enormous size." – Albert Einstein

The topic of recommender systems gained increasing importance in the nineties, as the Web became an important medium for business and e-commerce transactions. It was recognized early on that the Web provided unprecedented opportunities for personalization, which were not available in other channels. In particular, the Web provided ease in data collection and a user interface that could be employed to recommend items in a non-intrusive way.

Recommender systems have grown significantly in terms of public awareness since then. An evidence of this fact is that many conferences and workshops are exclusively devoted to this topic. The *ACM Conference on Recommender Systems* is particularly notable because it regularly contributes many of the cutting-edge results in this topic. The topic of recommender systems is very diverse because it enables the ability to use various types of user-preference and user-requirement data to make recommendations. The most well-known methods in recommender systems include collaborative filtering methods, content-based methods, and knowledge-based methods. These three methods form the fundamental pillars of research in recommender systems. In recent years, specialized methods have been designed for various data domains and contexts, such as time, location and social information. Numerous advancements have been proposed for specialized scenarios, and the methods have been adapted to various application domains, such as query log mining, news recommendations, and computational advertising. The organization of the book reflects these important topics. The chapters of this book can be organized into three categories:

1. *Algorithms and evaluation:* These chapters discuss the fundamental algorithms in recommender systems, including collaborative filtering methods (Chapters 2 and 4), content-based methods (Chapter 4), and knowledge-based methods (Chapter 5). Techniques for hybridizing these methods are discussed in Chapter 6. The evaluation of recommender systems is discussed in Chapter 7.

2. *Recommendations in specific domains and contexts:* The context of a recommender system plays a critical role in providing effective recommendations. For example, a

user looking for a restaurant would want to use their location as additional *context*. The context of a recommendation can be viewed as important side information that affects the recommendation goals. Different types of domains such as temporal data, spatial data, and social data, provide different types of contexts. These methods are discussed in Chapters 8, 9, 10, and 11. Chapter 11 also discusses the issue of using social information to increase the trustworthiness of the recommendation process. Recent topics such as factorization machines and trustworthy recommender systems are also covered in these chapters.

3. *Advanced topics and applications:* In Chapter 12, we discuss various robustness aspects of recommender systems, such as shilling systems, attack models, and their defenses. In addition, recent topics, such as learning to rank, multi-armed bandits, group recommender systems, multi-criteria systems, and active learning systems, are discussed in Chapter 13. An important goal of this chapter is to introduce the reader to the basic ideas and principles underlying recent developments. Although it is impossible to discuss all the recent developments in detail in a single book, it is hoped that the material in the final chapter will play the role of "breaking the ice" for the reader in terms of advanced topics. This chapter also investigates some application settings in which recommendation technology is used, such as news recommendations, query recommendations, and computational advertising. The application section provides an idea of how the methods introduced in earlier chapters apply to these different domains.

Although this book is primarily written as a textbook, it is recognized that a large portion of the audience will comprise industrial practitioners and researchers. Therefore, we have taken pains to write the book in such a way that it is also useful from an applied and reference point of view. Numerous examples and exercises have been provided to enable its use as a textbook. As most courses on recommender systems will teach only the fundamental topics, the chapters on fundamental topics and algorithms are written with a particular emphasis on classroom teaching. On the other hand, advanced industrial practitioners might find the chapters on context-sensitive recommendation useful, because many real-life applications today arise in the domains where a significant amount of contextual side-information is available. The application portion of Chapter 13 is particularly written for industrial practitioners, although instructors might find it useful towards the end of a recommender course.

We conclude with a brief introduction to the notations used in this book. This book consistently uses an $m \times n$ ratings matrix denoted by R, where m is the number of users and n is the number of items. The matrix R is typically incomplete because only a subset of entries are observed. The (i, j)th entry of R indicates the rating of user i for item j, and it is denoted by r_{ij} when it is actually observed. When the entry (i, j) is *predicted* by a recommender algorithm (rather than being specified by a user), it is denoted by $\hat{r}_{ij}$, with a "hat" symbol (i.e., a circumflex) denoting that it is a predicted value. Vectors are denoted by an "overline," as in $\overline{X}$ or $\overline{y}$.

Acknowledgments

I would like to thank my wife and daughter for their love and support during the writing of this book. I also owe my late parents a debt of gratitude for instilling in me a love of education, which has played an important inspirational role in my book-writing efforts.

This book has been written with the direct and indirect support of many individuals to whom I am grateful. During the writing of this book, I received feedback from many colleagues. In particular, I received feedback from Xavier Amatriain, Kanishka Bhaduri, Robin Burke, Martin Ester, Bart Goethals, Huan Liu, Xia Ning, Saket Sathe, Jiliang Tang, Alexander Tuzhilin, Koen Versetrepen, and Jieping Ye. I would like to thank them for their constructive feedback and suggestions. Over the years, I have benefited from the insights of numerous collaborators. These insights have influenced this book directly or indirectly. I would first like to thank my long-term collaborator Philip S. Yu for my years of collaboration with him. Other researchers with whom I have had significant collaborations include Tarek F. Abdelzaher, Jing Gao, Quanquan Gu, Manish Gupta, Jiawei Han, Alexander Hinneburg, Thomas Huang, Nan Li, Huan Liu, Ruoming Jin, Daniel Keim, Arijit Khan, Latifur Khan, Mohammad M. Masud, Jian Pei, Magda Procopiuc, Guojun Qi, Chandan Reddy, Saket Sathe, Jaideep Srivastava, Karthik Subbian, Yizhou Sun, Jiliang Tang, Min-Hsuan Tsai, Haixun Wang, Jianyong Wang, Min Wang, Joel Wolf, Xifeng Yan, Mohammed Zaki, ChengXiang Zhai, and Peixiang Zhao. I would also like to thank my advisor James B. Orlin for his guidance during my early years as a researcher.

I would also like to thank my manager Nagui Halim for providing the tremendous support necessary for the writing of this book. His professional support has been instrumental for my many book efforts in the past and present.

Finally, I would like to thank Lata Aggarwal for helping me with some of the figures drawn using Microsoft Powerpoint.

Author Biography

Charu C. Aggarwal is a Distinguished Research Staff Member (DRSM) at the IBM T. J. Watson Research Center in Yorktown Heights, New York. He completed his B.S. from IIT Kanpur in 1993 and his Ph.D. from the Massachusetts Institute of Technology in 1996.

 He has worked extensively in the field of data mining. He has published more than 300 papers in refereed conferences and journals and authored over 80 patents. He is the author or editor of 15 books, including a textbook on data mining and a comprehensive book on outlier analysis. Because of the commercial value of his patents, he has thrice been designated a Master Inventor at IBM. He is a recipient of an IBM Corporate Award (2003) for his work on bio-terrorist threat detection in data streams, a recipient of the IBM Outstanding Innovation Award (2008) for his scientific contributions to privacy technology, a recipient of two IBM Outstanding Technical Achievement Award (2009, 2015) for his work on data streams and high-dimensional data, respectively. He received the EDBT 2014 Test of Time Award for his work on condensation-based privacy-preserving data mining. He is also a recipient of the IEEE ICDM Research Contributions Award (2015), which is one of the two highest awards for influential research contributions in the field of data mining.

He has served as the general co-chair of the IEEE Big Data Conference (2014), program co-chair of the ACM CIKM Conference (2015), IEEE ICDM Conference (2015), and the ACM KDD Conference (2016). He served as an associate editor of the IEEE Transactions on Knowledge and Data Engineering from 2004 to 2008. He is an associate editor of the ACM Transactions on Knowledge Discovery from Data, an associate editor of the IEEE Transactions on Big Data, an action editor of the Data Mining and Knowledge Discovery Journal, editor-in-chief of the ACM SIGKDD Explorations, and an associate editor of the Knowledge and Information Systems Journal. He serves on the advisory board of the Lecture Notes on Social Networks, a publication by Springer. He has served as the vice-president of the SIAM Activity Group on Data Mining. He is a fellow of the SIAM, ACM, and the IEEE, for "contributions to knowledge discovery and data mining algorithms."

Chapter 1

An Introduction to Recommender Systems

"Many receive advice, only the wise profit from it." – Harper Lee

1.1 Introduction

The increasing importance of the Web as a medium for electronic and business transactions has served as a driving force for the development of recommender systems technology. An important catalyst in this regard is the ease with which the Web enables users to provide feedback about their likes or dislikes. For example, consider a scenario of a content provider such as Netflix. In such cases, users are able to easily provide feedback with a simple click of a mouse. A typical methodology to provide feedback is in the form of *ratings*, in which users select numerical values from a specific evaluation system (e.g., five-star rating system) that specify their likes and dislikes of various items.

Other forms of feedback are not quite as explicit but are even easier to collect in the Web-centric paradigm. For example, the simple act of a user buying or browsing an item may be viewed as an endorsement for that item. Such forms of feedback are commonly used by online merchants such as Amazon.com, and the collection of this type of data is completely effortless in terms of the work required of a customer. The basic idea of recommender systems is to utilize these various sources of data to infer customer interests. The entity to which the recommendation is provided is referred to as the *user*, and the product being recommended is also referred to as an *item*. Therefore, recommendation analysis is often based on the previous interaction between users and items, because past interests and proclivities are often good indicators of future choices. A notable exception

© Springer International Publishing Switzerland 2016
C.C. Aggarwal, *Recommender Systems: The Textbook*,
DOI 10.1007/978-3-319-29659-3_1

is the case of *knowledge-based recommender systems*, in which the recommendations are suggested on the basis of user-specified *requirements* rather than the past history of the user.

So, what is the basic principle that underlies the working of recommendation algorithms? The basic principle of recommendations is that significant dependencies exist between user- and item-centric activity. For example, a user who is interested in a historical documentary is more likely to be interested in another historical documentary or an educational program, rather than in an action movie. In many cases, various categories of items may show significant correlations, which can be leveraged to make more accurate recommendations. Alternatively, the dependencies may be present at the finer granularity of individual items rather than categories. These dependencies can be *learned* in a data-driven manner from the ratings matrix, and the resulting model is used to make predictions for target users. The larger the number of rated items that are available for a user, the easier it is to make robust predictions about the future behavior of the user. Many different learning models can be used to accomplish this task. For example, the collective buying or rating behavior of various users can be leveraged to create cohorts of similar users that are interested in similar products. The interests and actions of these cohorts can be leveraged to make recommendations to individual members of these cohorts.

The aforementioned description is based on a very simple family of recommendation algorithms, referred to as *neighborhood models*. This family belongs to a broader class of models, referred to as *collaborative filtering*. The term "collaborative filtering" refers to the use of ratings from multiple users in a collaborative way to predict missing ratings. In practice, recommender systems can be more complex and data-rich, with a wide variety of auxiliary data types. For example, in content-based recommender systems, the content plays a primary role in the recommendation process, in which the ratings of users and the attribute descriptions of items are leveraged in order to make predictions. The basic idea is that user interests can be modeled on the basis of properties (or *attributes*) of the items they have rated or accessed in the past. A different framework is that of *knowledge-based systems*, in which users interactively specify their interests, and the user specification is combined with domain knowledge to provide recommendations. In advanced models, contextual data, such as temporal information, external knowledge, location information, social information, or network information, may be used.

This book will study all types of basic systems, including collaborative, content-based, and knowledge-based systems. We will also discuss both the basic and the enhanced models of recommender systems in different domains. We will study various aspects of the robustness of recommender systems, such as attack models, and the construction of trustworthy models. In addition, a variety of evaluation and hybridization models for recommender systems will be studied thoroughly. In this chapter, the goal is to provide an overview of the wide diversity of work in the field of recommender systems, and also relate the various topics to the individual chapters of this book.

This chapter is organized as follows. Section 1.2 discusses the main goals of recommender systems. Section 1.3 will introduce the basic models and evaluation methods used in recommender systems. The use of recommender systems in various data domains is discussed in section 1.4. Advanced models for recommender systems are discussed in section 1.5. Section 1.6 discusses the conclusions and summary.

1.2 Goals of Recommender Systems

Before discussing the goals of recommender systems, we introduce the various ways in which the recommendation problem may be formulated. The two primary models are as follows:

1. *Prediction version of problem:* The first approach is to predict the rating value for a user-item combination. It is assumed that *training* data is available, indicating user preferences for items. For m users and n items, this corresponds to an incomplete $m \times n$ matrix, where the specified (or *observed*) values are used for training. The missing (or *unobserved*) values are predicted using this training model. This problem is also referred to as the *matrix completion problem* because we have an incompletely specified matrix of values, and the remaining values are predicted by the learning algorithm.

2. *Ranking version of problem:* In practice, it is not necessary to predict the ratings of users for specific items in order to make recommendations to users. Rather, a merchant may wish to recommend the top-k items for a particular user, or determine the top-k users to target for a particular item. The determination of the top-k items is more common than the determination of top-k users, although the methods in the two cases are exactly analogous. Throughout this book, we will discuss only the determination of the top-k items, because it is the more common setting. This problem is also referred to as the *top-k recommendation problem*, and it is the ranking formulation of the recommendation problem.

In the second case, the absolute values of the predicted ratings are not important. The first formulation is more general, because the solutions to the second case can be derived by solving the first formulation for various user-item combinations and then ranking the predictions. However, in many cases, it is easier and more natural to design methods for solving the ranking version of the problem directly. Such methods will be discussed in Chapter 13.

Increasing product sales is the primary goal of a recommender system. Recommender systems are, after all, utilized by merchants to increase their profit. By recommending carefully selected items to users, recommender systems bring relevant items to the attention of users. This increases the sales volume and profits for the merchant. Although the primary goal of a recommendation system is to increase revenue for the merchant, this is often achieved in ways that are less obvious than might seem at first sight. In order to achieve the broader *business-centric* goal of increasing revenue, the common *operational* and *technical* goals of recommender systems are as follows:

1. *Relevance:* The most obvious operational goal of a recommender system is to recommend items that are relevant to the user at hand. Users are more likely to consume items they find interesting. Although relevance is the primary operational goal of a recommender system, it is not sufficient in isolation. Therefore, we discuss several secondary goals below, which are not quite as important as relevance but are nevertheless important enough to have a significant impact.

2. *Novelty:* Recommender systems are truly helpful when the recommended item is something that the user has not seen in the past. For example, popular movies of a preferred genre would rarely be novel to the user. Repeated recommendation of popular items can also lead to reduction in sales diversity [203].

3. *Serendipity:* A related notion is that of *serendipity* [229], wherein the items recommended are somewhat unexpected, and therefore there is a modest element of lucky

discovery, as opposed to obvious recommendations. Serendipity is different from novelty in that the recommendations are truly *surprising* to the user, rather than simply something they did not know about before. It may often be the case that a particular user may only be consuming items of a specific type, although a latent interest in items of other types may exist which the user might themselves find surprising. Unlike novelty, serendipitous methods focus on discovering such recommendations.

For example, if a new Indian restaurant opens in a neighborhood, then the recommendation of that restaurant to a user who normally eats Indian food is novel but not necessarily serendipitous. On the other hand, when the same user is recommended Ethiopian food, and it was unknown to the user that such food might appeal to her, then the recommendation is serendipitous. Serendipity has the beneficial side effect of increasing sales diversity or beginning a new trend of interest in the user. Increasing serendipity often has long-term and strategic benefits to the merchant because of the possibility of discovering entirely new areas of interest. On the other hand, algorithms that provide serendipitous recommendations often tend to recommend irrelevant items. In many cases, the longer term and strategic benefits of serendipitous methods outweigh these short-term disadvantages.

4. *Increasing recommendation diversity:* Recommender systems typically suggest a list of top-k items. When all these recommended items are very similar, it increases the risk that the user might not like *any* of these items. On the other hand, when the recommended list contains items of different types, there is a greater chance that the user might like at least one of these items. Diversity has the benefit of ensuring that the user does not get bored by repeated recommendation of similar items.

Aside from these concrete goals, a number of soft goals are also met by the recommendation process both from the perspective of the user and merchant. From the perspective of the user, recommendations can help improve overall user satisfaction with the Web site. For example, a user who repeatedly receives relevant recommendations from Amazon.com will be more satisfied with the experience and is more likely to use the site again. This can improve user loyalty and further increase the sales at the site. At the merchant end, the recommendation process can provide insights into the needs of the user and help customize the user experience further. Finally, providing the user an explanation for why a particular item is recommended is often useful. For example, in the case of Netflix, recommendations are provided along with previously watched movies. As we will see later, some recommendation algorithms are better suited to providing explanations than others.

There is a wide diversity in the types of products recommended by such systems. Some recommender systems, such as Facebook, do not directly recommend products. Rather they may recommend social connections, which have an indirect benefit to the site by increasing its usability and advertising profits. In order to understand the nature of these goals, we will discuss some popular examples of historical and current recommender systems. These examples will also showcase the broad diversity of recommender systems that were built either as research prototypes, or are available today as commercial systems in various problem settings.

GroupLens Recommender System

GroupLens was a pioneering recommender system, which was built as a research prototype for recommendation of Usenet news. The system collected ratings from Usenet readers and used them to predict whether or not other readers would like an article before they read it.

Some of the earliest automated collaborative filtering algorithms were developed in the GroupLens[1] setting. The general ideas developed by this group were also extended to other product settings such as books and movies. The corresponding recommender systems were referred to as *BookLens* and *MovieLens*, respectively. Aside from its pioneering contributions to collaborative filtering research, the GroupLens research team was notable for releasing several data sets during the early years of this field, when data sets were not easily available for benchmarking. Prominent examples include three data sets [688] from the MovieLens recommender system. These data sets are of successively increasing size, and they contain 10^5, 10^6, and 10^7 ratings, respectively.

Amazon.com Recommender System

Amazon.com [698] was also one of the pioneers in recommender systems, especially in the commercial setting. During the early years, it was one of the few retailers that had the foresight to realize the usefulness of this technology. Originally founded as a book e-retailer, the business expanded to virtually all forms of products. Consequently, Amazon.com now sells virtually all categories of products such as books, CDs, software, electronics, and so on. The recommendations in Amazon.com are provided on the basis of explicitly provided ratings, buying behavior, and browsing behavior. The ratings in Amazon.com are specified on a 5-point scale, with lowest rating being 1-star, and the highest rating being 5-star. The customer-specific buying and browsing data can be easily collected when users are logged in with an account authentication mechanism supported by Amazon. Recommendations are also provided to users on the main Web page of the site, whenever they log into their accounts. In many cases, explanations for recommendations are provided. For example, the relationship of a recommended item to previously purchased items may be included in the recommender system interface.

The purchase or browsing behavior of a user can be viewed as a type of *implicit rating*, as opposed to an *explicit rating*, which is specified by the user. Many commercial systems allow the flexibility of providing recommendations both on the basis of explicit and implicit feedback. In fact, several models have been designed (cf. section 3.6.4.6 of Chapter 3) to jointly account for explicit and implicit feedback in the recommendation process. Some of the algorithms used by early versions of the Amazon.com recommender system are discussed in [360].

Netflix Movie Recommender System

Netflix was founded as a mail-order digital video disc (DVD) rental company [690] of movies and television shows, which was eventually expanded to streaming delivery. At the present time, the primary business of Netflix is that of providing streaming delivery of movies and television shows on a subscription basis. Netflix provides users the ability to rate the movies and television shows on a 5-point scale. Furthermore, the user actions in terms of watching various items are also stored by Netflix. These ratings and actions are then used by Netflix to make recommendations. Netflix does an excellent job of providing *explanations* for the recommended items. It explicitly provides examples of recommendations based on specific items that were watched by the user. Such information provides the user with additional

---

[1] The term "GroupLens" currently refers to the academic group at the University of Minnesota [687] that developed these algorithms. This group continues to work in the area of recommender systems, and has made many pioneering contributions over the years.

information to decide whether or not to watch a specific movie. Presenting meaningful explanations is important to provide the user with an understanding of *why* they might find a particular movie interesting. This approach also makes it more likely for the user to act on the recommendation and truly improves the user experience. This type of interesting approach can also help improve customer loyalty and retention.

Netflix has contributed significantly to the research community as a result of the *Netflix Prize contest*. This contest was designed to provide a forum for competition among various collaborative filtering algorithms contributed by contestants. A data set of Netflix movie ratings was released, and the task was to predict ratings of particular user-item combinations. For this purpose, Netflix provided both a *training* data set, and a *qualifying* data set. The training data set contained 100,480,507 ratings that 480,189 users gave to 17,770 movies. The training set included a smaller *probe set* containing 1,408,395 ratings. The probe set was based on more recent ratings than the remaining training data, and it was statistically similar to the portion of the data set with hidden ratings. This portion of the data set was referred to as the *qualifying* data set, and it contained over 2,817,131 triplets of the form $\langle User, Movie, GradeDate \rangle$. Note that the triplet did not contain the actual rating, which was known only to the judges. Users needed to predict the ratings in the qualifying data set based on models of the training data. This prediction was scored by the judges (or an equivalent automated system), and the users were (continuously) informed of the prediction results on only half the qualifying data set on the *leader-board*. This half of the qualifying data set was referred to as the *quiz set*. The remaining half was used as the *test set* for computing the final score and determining the prize-winners. The scores of the remaining half were never revealed to the users until the very end. Furthermore, it was not revealed to the contestants which of the triplets in the qualifying set belonged to the quiz set, and which belonged to the test set. The reason for this unusual arrangement on the test set was to ensure that the users did not leverage the scores on the leader-board to overfit their algorithms to the test set. Issues related to overfitting will be described in Chapter 7 on evaluation algorithms. Indeed, Netflix's framework for handling the contestant entries is an excellent example of proper evaluation design of recommendation algorithms.

The probe set, quiz set, and test set were designed to have similar statistical characteristics. Prizes were given based on improvement of Netflix's own recommendation algorithm, known as *Cinematch*, or by improvement of the previous best score by a certain threshold. Many well-known recommendation algorithms, such as latent factor models, were popularized by the Netflix contest. The Netflix Prize contest is notable for its numerous contributions to recommendation [71, 373] research.

Google News Personalization System

The Google News personalization system [697] is able to recommend news to users based on their history of clicks. The clicks are associated with specific users based on identification mechanisms enabled by Gmail accounts. In this case, news articles are treated as items. The act of a user clicking on a news article can be viewed as a positive rating for that article. Such ratings can be viewed as *unary ratings*, in which a mechanism exists for a user to express their affinity for an item, but no mechanism exists for them to show their dislike. Furthermore, the ratings are *implicit*, because they are *inferred* from user actions rather than being explicitly specified by the user. Nevertheless, variations of the approach can also be applied to cases where ratings are explicitly specified. Collaborative recommendation algorithms are applied to the collected ratings, so that inferences can be made about the

Table 1.1: Examples of products recommended by various real-world recommender systems

System	Product Goal
Amazon.com [698]	Books and other products
Netflix [690]	DVDs, Streaming Video
Jester [689]	Jokes
GroupLens [687]	News
MovieLens [688]	Movies
last.fm [692]	Music
Google News [697]	News
Google Search [696]	Advertisements
Facebook [691]	Friends, Advertisements
Pandora [693]	Music
YouTube [694]	Online videos
Tripadvisor [695]	Travel products
IMDb [699]	Movies

personalized articles for specific users. A description of a collaborative filtering system for Google News is provided in [175]. More details of the Google News personalization engine are discussed in section 13.8.1.2 of Chapter 13.

Facebook Friend Recommendations

Social networking sites often recommend potential friends to users in order to increase the number of social connections at the site. Facebook [691] is one such example of a social networking Web site. This kind of recommendation has slightly different goals than a product recommendation. While a product recommendation directly increases the profit of the merchant by facilitating product sales, an increase in the number of social connections improves the experience of a user at a social network. This, in turn, encourages the growth of the social network. Social networks are heavily dependent on the growth of the network to increase their advertising revenues. Therefore, the recommendation of potential friends (or *links*) enables better growth and connectivity of the network. This problem is also referred to as *link prediction* in the field of social network analysis. Such forms of recommendations are based on *structural relationships* rather than ratings data. Therefore, the nature of the underlying algorithms is completely different. The link recommendation problem is explored in detail in Chapter 10. The relationship of computational advertising to recommender system technology is discussed in Chapter 13.

1.2.1 The Spectrum of Recommendation Applications

In the following, we will provide a brief overview of the application-specific goals accomplished by various implementations of recommender systems. A brief overview of the products suggested and the goals accomplished by various recommender systems are illustrated in Table 1.1. Many of these recommender systems are focused on traditional e-commerce applications for various products, including books, movies, videos, travel, and other goods and services. The broader applicability of recommender systems to e-commerce applications is discussed in [530]. However, recommender systems have expanded beyond the traditional domain of product recommendations. It is noteworthy that some of the systems in Table 1.1

may not recommend specific products. An example is the Google search application, which may advertise products along with their search results. This is the area of *computational advertising*, which is a distinct area in its own right, but it is nevertheless closely related to recommender systems. This area is discussed in detail in section 13.8.2 of Chapter 13. Similarly, Facebook recommends friends, and online recruitment sites recommend employers and job-seekers to one another. The last of these systems is also referred to as a *reciprocal recommender*. The models for some of these recommendation algorithms are quite different from those of traditional recommender systems. This book will study many of these variations in detail.

1.3 Basic Models of Recommender Systems

The basic models for recommender systems work with two kinds of data, which are (i) the user-item interactions, such as ratings or buying behavior, and (ii) the attribute information about the users and items such as textual profiles or relevant keywords. Methods that use the former are referred to as *collaborative filtering* methods, whereas methods that use the latter are referred to as *content-based recommender* methods. Note that content-based systems also use the ratings matrices in most cases, although the model is usually focused on the ratings of a single user rather than those of all users. In *knowledge-based* recommender systems, the recommendations are based on explicitly specified *user requirements*. Instead of using historical rating or buying data, external knowledge bases and constraints are used to create the recommendation. Some recommender systems combine these different aspects to create *hybrid* systems. Hybrid systems can combine the strengths of various types of recommender systems to create techniques that can perform more robustly in a wide variety of settings. In the following, we will discuss these basic models briefly, and also provide pointers to the relevant chapters in the book where they are discussed.

1.3.1 Collaborative Filtering Models

Collaborative filtering models use the collaborative power of the ratings provided by multiple users to make recommendations. The main challenge in designing collaborative filtering methods is that the underlying ratings matrices are *sparse*. Consider an example of a movie application in which users specify ratings indicating their like or dislike of specific movies. Most users would have viewed only a small fraction of the large universe of available movies. As a result, most of the ratings are unspecified. The specified ratings are also referred to as *observed ratings*. Throughout this book, the terms "specified" and "observed" will be used in an interchangeable way. The unspecified ratings will be referred to as "unobserved" or "missing."

The basic idea of collaborative filtering methods is that these unspecified ratings can be imputed because the observed ratings are often highly correlated across various users and items. For example, consider two users named Alice and Bob, who have very similar tastes. If the ratings, which both have specified, are very similar, then their similarity can be identified by the underlying algorithm. In such cases, it is very likely that the ratings in which only one of them has specified a value, are also likely to be similar. This similarity can be used to make inferences about incompletely specified values. Most of the models for collaborative filtering focus on leveraging either inter-item correlations or inter-user correlations for the prediction process. Some models use both types of correlations. Furthermore, some models use carefully designed optimization techniques to create a training model in much the same way a classifier creates a training model from the labeled data. This model is then used to impute the missing values in the matrix, in the same way that a classifier imputes the missing test labels. There are two types of methods that are commonly used in collaborative filtering, which are referred to as memory-based methods and model-based methods:

1. *Memory-based methods:* Memory-based methods are also referred to as *neighborhood-based collaborative filtering algorithms.* These were among the earliest collaborative filtering algorithms, in which the ratings of user-item combinations are predicted on the basis of their neighborhoods. These neighborhoods can be defined in one of two ways:

   - *User-based collaborative filtering:* In this case, the ratings provided by like-minded users of a *target* user A are used in order to make the recommendations for A. Thus, the basic idea is to determine users, who are similar to the target user A, and recommend ratings for the unobserved ratings of A by computing weighted averages of the ratings of this peer group. Therefore, if Alice and Bob have rated movies in a similar way in the past, then one can use Alice's observed ratings on the movie *Terminator* to predict Bob's unobserved ratings on this movie. In general, the k most similar users to Bob can be used to make rating predictions for Bob. Similarity functions are computed between the *rows* of the ratings matrix to discover similar users.

   - *Item-based collaborative filtering:* In order to make the rating predictions for target item B by user A, the first step is to determine a set S of items that are most similar to target item B. The ratings in item set S, which are specified by A, are used to predict whether the user A will like item B. Therefore, Bob's ratings on similar science fiction movies like *Alien* and *Predator* can be used to predict his rating on *Terminator*. Similarity functions are computed between the *columns* of the ratings matrix to discover similar items.

   The advantages of memory-based techniques are that they are simple to implement and the resulting recommendations are often easy to explain. On the other hand, memory-based algorithms do not work very well with sparse ratings matrices. For example, it might be difficult to find sufficiently similar users to Bob, who have rated *Gladiator*. In such cases, it is difficult to robustly predict Bob's rating of *Gladiator*. In other words, such methods might lack full *coverage* of rating predictions. Nevertheless, the lack of coverage is often not an issue, when only the top-k items are required. Memory-based methods are discussed in detail in Chapter 2.

2. *Model-based methods:* In model-based methods, machine learning and data mining methods are used in the context of predictive models. In cases where the model is parameterized, the parameters of this model are learned within the context of an optimization framework. Some examples of such model-based methods include decision trees, rule-based models, Bayesian methods and latent factor models. Many of these methods, such as latent factor models, have a high level of coverage even for sparse ratings matrices. Model-based collaborative filtering algorithms are discussed in Chapter 3.

Even though memory-based collaborative filtering algorithms are valued for their simplicity, they tend to be heuristic in nature, and they do not work well in all settings. However, the distinction between memory-based and model-based methods is somewhat artificial, because

Figure 1.1: Example of 5-point interval ratings

Overall Ratings

		Excellent	Very Good	Good	Fair	Poor	NA
1.	The quality of the course content	○	○	○	○	○	○
2.	The instructor's overall teaching	○	○	○	○	○	○

Figure 1.2: Example of ordinal ratings used in Stanford University course evaluations

memory-based methods can also be considered similarity-based models, albeit heuristic ones. In section 2.6 of Chapter 2, it will also be shown that some variations of neighborhood-based methods can be formally expressed as regression-based models. Latent factor models were popularized in later years as a result of the Netflix Prize contest, although similar algorithmswere proposed much earlier in the context of (generic) incomplete data sets [24]. Recently, it was shown that some combinations of memory-based and model-based methods [309] provide very accurate results.

1.3.1.1   Types of Ratings

The design of recommendation algorithms is influenced by the system used for tracking ratings. The ratings are often specified on a scale that indicates the specific level of like or dislike of the item at hand. It is possible for ratings to be continuous values, such as in the case of the Jester joke recommendation engine [228, 689], in which the ratings can take on any value between -10 and 10. This is, however, relatively rare. Usually, the ratings are interval-based, where a discrete set of ordered numbers are used to quantify like or dislike. Such ratings are referred to as *interval*-based ratings. For example, a 5-point rating scale might be drawn from the set $\{-2, -1, 0, 1, 2\}$, in which a rating of -2 indicates an extreme dislike, and a rating of 2 indicates a strong affinity to the item. Other systems might draw the ratings from the set $\{1, 2, 3, 4, 5\}$.

The number of possible ratings might vary with the system at hand. The use of 5-point, 7-point, and 10-point ratings is particularly common. The 5-star ratings system, illustrated in Figure 1.1, is an example of interval ratings. Along each of the possible ratings, we have indicated a semantic interpretation of the user's level of interest. This interpretation might vary slightly across different merchants, such as Amazon or Netflix. For example, Netflix uses a 5-star ratings system in which the 4-star point corresponds to *"really liked it,"* and the central 3-star point corresponds to *"liked it."* Therefore, there are three favorable ratings and two unfavorable ratings in Netflix, which leads to an *unbalanced rating scale*. In some cases, there may be an even number of possible ratings, and the neutral rating might be missing. This approach is referred to as a *forced choice rating system*.

One can also use ordered categorical values such as {Strongly Disagree, Disagree, Neutral, Agree, Strongly Agree} in order to achieve the same goals. In general, such ratings are referred to as *ordinal* ratings, and the term is derived from the concept of ordinal attributes. An example of ordinal ratings, used in Stanford University course evaluation forms, is illustrated in Figure 1.2. In *binary* ratings, the user may represent only a like or dislike for the item and nothing else. For example, the ratings may be 0, 1, or unspecified values. The unspecified values need to be predicted to 0-1 values. A special case of ratings is that of *unary* ratings, in which there is a mechanism for a user to specify a liking for an item but no mechanism to specify a dislike. Unary ratings are particularly common, especially in the case of *implicit feedback data sets* [259, 260, 457]. In these cases, customer preferences are derived from their activities rather than their explicitly specified ratings. For example, the buying behavior of a customer can be converted to unary ratings. When a customer buys an item, it can be viewed as a preference for the item. However, the act of not buying an item from a large universe of possibilities does not always indicate a dislike. Similarly, many social networks, such as Facebook, use *"like"* buttons, which provide the ability to express liking for an item. However, there is no mechanism to specify dislike for an item. The implicit feedback setting can be viewed as the matrix completion analog of the positive-unlabeled (PU) learning problem in data classification [259].

Examples of Explicit and Implicit Ratings

A quantitative example of explicit ratings is illustrated in Figure 1.3(a). In this case, there are 6 users, labeled $U_1 \ldots U_6$, and 6 movies with specified titles. Higher ratings indicate more positive feedback in Figure 1.3(a). The missing entries correspond to unspecified preferences. The example of this figure represents a small toy example. In general, the ratings could be represented as an $m \times n$ matrix, where m and n are typically very large and may range in the order of hundreds of thousands. Even though this particular example uses a 6×6 matrix, the values of m and n are typically not the same in real-world scenarios. A ratings matrix is sometimes referred to as a *utility matrix*, although the two may not always be the same. Strictly speaking, when the utility refers to the amount of profit, then the utility of a user-item combination refers to the amount of profit incurred by recommending that item to the particular user. While utility matrices are often set to be the same as the ratings matrices, it is possible for the application to explicitly transform the ratings to utility values based on domain-specific criteria. All collaborative filtering algorithms are then applied to the utility matrix rather than the ratings matrix. However, such an approach is rarely used in practice, and most collaborative filtering algorithms work directly with the ratings matrix.

An example of a unary ratings matrix is illustrated in Figure 1.3(b). For cases in which the ratings are unary, the matrix is referred to as a *positive preference utility matrix* because it allows only the specification of positive preferences. The two matrices in Figure 1.3 have the same set of observed entries, but they provide very different insights. For example, the users U_1 and U_3 are very different in Figure 1.3(a) because they have very different ratings for their mutually specified entries. On the other hand, these users would be considered very similar in Figure 1.3(b) because these users have expressed a positive preference for the same items. The ratings-based utility provides a way for users to express negative preferences for items. For example, user U_1 does not like the movie *Gladiator* in Figure 1.3(a). There is no mechanism to specify this in the positive-preference utility matrix of Figure 1.3(b) beyond

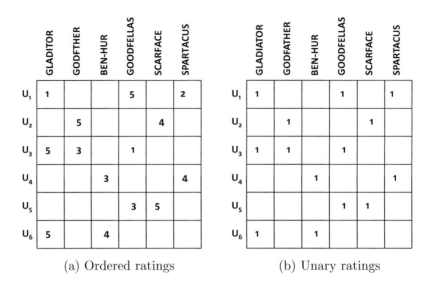

(a) Ordered ratings   (b) Unary ratings

Figure 1.3: Examples of utility matrices

a relatively ambiguous missing entry. In other words, the matrix in Figure 1.3(b) is less expressive. While Figure 1.3(b) provides an example of a binary matrix, it is possible for the nonzero entries to be arbitrary positive values. For example, they could correspond to the quantities of items bought by the different users. In general, unary matrices are created by user *actions* such as buying an item, and are therefore also referred to as implicit feedback matrices.

Unary ratings have a significant effect on the recommendation algorithm at hand, because no information is available about whether a user dislikes an item. In the case of unary matrices, it is often recommended [260] to perform the analysis in a simple way by treating the missing entries as 0s in the initial phase. However, the final predicted value by the learning algorithm might be much larger than 0, especially if the item matches user interests. The recommended items are therefore based on the entries with the largest positive prediction error over the initial "zero" assumption. In fact, if the missing entries are not substituted with 0s, significant overfitting is possible. This type of overfitting is an artifact of the fact that there is often not a sufficient level of discrimination between the various observed values of the ratings. In explicit feedback matrices, ratings correspond to (highly discriminated) *preferences*, whereas in implicit feedback matrices, ratings correspond to (less discriminated) *confidences*. In a later chapter, we will provide a specific example of overfitting with implicit feedback matrices when missing entries are not treated as zeros (cf. section 3.6.6.2 of Chapter 3).

Pre-substitution of missing ratings is not recommended in explicit ratings matrices. In explicit ratings matrices with both likes and dislikes, the substitution of missing entries with any value (such as 0 or the row/column/data mean) always leads to a significant amount of bias in the analysis. In the unary case, substituting missing entries with 0s also leads to some bias [457, 467, 468], although it is often small because the default assumption in implicit feedback data, such as buying data, is that the user will not buy most of the items. One is often willing to live with this bias in the unary case, because a significant amount of overfitting is reduced by the substitution. There are also some interesting computational effects of such choices. These trade-offs are discussed in Chapters 2 and 3.

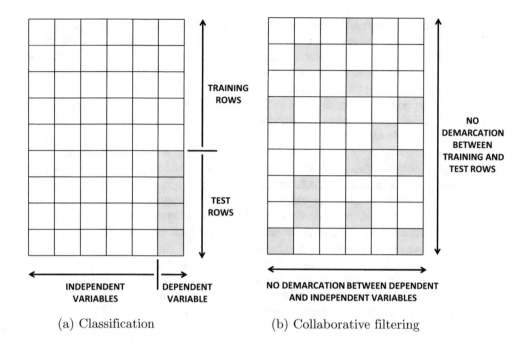

(a) Classification   (b) Collaborative filtering

Figure 1.4: Comparing the traditional classification problem with collaborative filtering. Shaded entries are missing and need to be predicted.

1.3.1.2   Relationship with Missing Value Analysis

Collaborative filtering models are closely related to missing value analysis. The traditional literature on missing value analysis studies the problem of imputation of entries in an incompletely specified data matrix. Collaborative filtering can be viewed as a (difficult)special case of this problem in which the underlying data matrix is very *large* and *sparse*. A detailed discussion of methods for missing value analysis in the statistical literature may be found in [362]. Many of these methods can also be used for recommender systems, although some of them might require specialized adaptations for very large and sparse matrices. In fact, some of the recent classes of models for recommender systems, such as latent factor models, were studied earlier in the context of missing value analysis [24]. Similar methods were independently proposed in the context of recommender systems [252, 309, 313, 500, 517, 525]. In general, many classical missing value estimation methods [362] can also be used for collaborative filtering.

1.3.1.3   Collaborative Filtering as a Generalization of Classification and Regression Modeling

Collaborative filtering methods can be viewed as generalizations of classification and regression modeling. In the classification and regression modeling problems, the class/dependent variable can be viewed as an attribute with missing values. Other columns are treated as features/independent variables. The collaborative filtering problem can be viewed as a generalization of this framework because any column is allowed to have missing values rather than (only) the class variable. In the recommendation problem, a clear distinction does not exist between class variables and feature variables because each feature plays the dual role of a dependent and independent variable. This distinction exists in the classification problem *only* because the missing entries are restricted to a special column. Furthermore, there is no distinction between training and test rows in collaborative filtering because any row might contain missing entries. Therefore, it is more meaningful to speak of training and test *entries* in collaborative filtering rather than training and test *rows*. Collaborative filtering is a generalization of classification/regression modeling in which the prediction is performed in entry-wise fashion rather than row-wise fashion. This relationship between classification/regression modeling and collaborative filtering is important to keep in mind because many principles of classification and regression modeling methods can be generalized to recommender systems. The relationship between the two problems is illustrated in Figure 1.4. This figure is particularly useful in relating collaborative filtering with classification, and it will be revisited multiple times in this book. wherever the similarities between these two problems are leveraged in some way for algorithmic or theoretical development.

The matrix completion problem also shares a number of characteristics with the *transductive* setting in classification and regression. In the transductive setting, the test instances are also included in the training process (typically with the use of a semisupervised algorithm), and it is often hard to make predictions for test instances that are not available at the time of training. On the other hand, models in which predictions can be easily made for new instances are referred to as *inductive*. For example, a naive Bayes model in classification is inherently inductive because one can easily use it to predict the label of a test instance for which the features were not known at the time of building the Bayes model.

The setting for matrix completion is inherently transductive because the training and test data are tightly integrated with one another in the $m \times n$ ratings matrix R, and many models cannot easily predict ratings for out-of-sample users and/or items. For example, if John is added to the ratings matrix (with many specified ratings) after the collaborative filtering model has already been constructed, many off-the-shelf methods will not be able to make predictions for John. This is especially true for model-based collaborative filtering methods. However, some recent matrix completion models have also been designed to be inductive in which ratings can be predicted for out-of-sample users and/or items.

1.3.2 Content-Based Recommender Systems

In content-based recommender systems, the descriptive attributes of items are used to make recommendations. The term "content" refers to these descriptions. In content-based methods, the ratings and buying behavior of users are combined with the content information available in the items. For example, consider a situation where John has rated the movie *Terminator* highly, but we do not have access to the ratings of other users. Therefore, collaborative filtering methods are ruled out. However, the item description of *Terminator* contains similar genre keywords as other science fiction movies, such as *Alien* and *Predator*. In such cases, these movies can be recommended to John.

In content-based methods, the item descriptions, which are labeled with ratings, are used as training data to create a user-specific classification or regression modeling problem. For each user, the training documents correspond to the descriptions of the items she has bought or rated. The class (or dependent) variable corresponds to the specified ratings or buying behavior. These training documents are used to create a classification or regression model, which is *specific* to the user at hand (or *active* user). This user-specific model is used to predict whether the corresponding individual will like an item for which her rating or buying behavior is unknown.

Content-based methods have some advantages in making recommendations for new items, when sufficient rating data are not available for that item. This is because other items with similar attributes might have been rated by the active user. Therefore, the supervised model will be able to leverage these ratings in conjunction with the item attributes to make recommendations even when there is no history of ratings for that item.

Content-based methods do have several disadvantages as well:

1. In many cases, content-based methods provide *obvious* recommendations because of the use of keywords or content. For example, if a user has never consumed an item with a particular set of keywords, such an item has no chance of being recommended. This is because the constructed model is specific to the user at hand, and the community knowledge from similar users is not leveraged. This phenomenon tends to reduce the diversity of the recommended items, which is undesirable.

2. Even though content-based methods are effective at providing recommendations for new *items*, they are not effective at providing recommendations for new *users*. This is because the training model for the target user needs to use the history of her ratings. In fact, it is usually important to have a large number of ratings available for the target user in order to make robust predictions without overfitting.

Therefore, content-based methods have different trade-offs from collaborative filtering systems.

Although the aforementioned description provides the conventional learning-based view of content-based methods, a broader view of these methods is sometimes used. For example, users can specify relevant keywords in their own profiles. These profiles can be matched with item descriptions in order to make recommendations. Such an approach does not use ratings in the recommendation process, and it is therefore useful in cold-start scenarios. However, such methods are often viewed as a distinct class of recommender systems, known as *knowledge-based systems*, because the similarity metrics are often based on domain knowledge. Knowledge-based recommender systems are often considered to be closely related to content-based recommender systems, and it is sometimes questioned whether a clear demarcation exists between the two classes of methods [558]. Methods for content-based recommender systems are discussed in Chapter 4.

1.3.3   Knowledge-Based Recommender Systems

Knowledge-based recommender systems are particularly useful in the context of items that are not purchased very often. Examples include items such as real estate, automobiles, tourism requests, financial services, or expensive luxury goods. In such cases, sufficient ratings may not be available for the recommendation process. As the items are bought rarely, and with different types of detailed options, it is difficult to obtain a sufficient number of ratings for a specific instantiation (i.e., combination of options) of the item at hand. This problem is also encountered in the context of the cold-start problem, when sufficient ratings are not available for the recommendation process. Furthermore, the nature of consumer preferences may evolve over time when dealing with such items. For example, the model of a car may evolve significantly over a few years, as a result of which the preferences may show a corresponding evolution. In other cases, it might be difficult to fully capture user interest with historical data such as ratings. A particular item may have attributes associated with it that correspond to its various properties, and a user may be interested only in items with specific properties. For example, cars may have several makes, models,

Table 1.2: The conceptual goals of various recommender systems

Approach	Conceptual Goal	Input
Collaborative	Give me recommendations based on a collaborative approach that leverages the ratings and actions of my peers/myself.	User ratings + community ratings
Content-based	Give me recommendations based on the content (attributes) I have favored in my past ratings and actions.	User ratings + item attributes
Knowledge-based	Give me recommendations based on my explicit specification of the kind of content (attributes) I want.	User specification + item attributes + domain knowledge

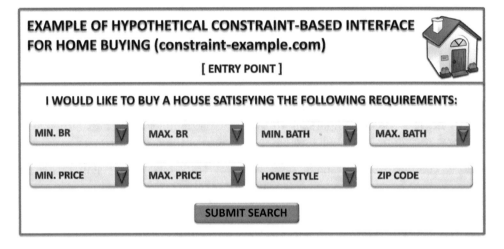

Figure 1.5: A hypothetical example of an initial user interface for a constraint-based recommender)

colors, engine options, and interior options, and user interests may be regulated by a very specific combination of these options. Thus, in these cases, the item domain tends to be *complex* in terms of its varied properties, and it is hard to associate sufficient ratings with the large number of combinations at hand.

Such cases can be addressed with knowledge-based recommender systems, in which ratings are not used for the purpose of recommendations. Rather, the recommendation process is performed on the basis of similarities between customer requirements and item descriptions, or the use of constraints specifying user requirements. The process is facilitated with the use of *knowledge bases*, which contain data about rules and similarity functions to use during the retrieval process. In fact, the knowledge bases are so important to the effective functioning of these methods that the approach takes its name from this fact. The explicit specification of requirements results in greater control of users over the recommendation process. In both collaborative and content-based systems, recommendations are decided entirely by either the user's past actions/ratings, the action/ratings of her peers, or a combination of the two. Knowledge-based systems are unique in that they allow the users to *explicitly specify what they want*. This difference is illustrated in Table 1.2.

Knowledge-based recommender systems can be classified on the basis of the type of the interface (and corresponding knowledge) used to achieve the aforementioned goals:

1. *Constraint-based recommender systems:* In constraint-based systems [196, 197], users typically specify requirements or constraints (e.g., lower or upper limits) on the item

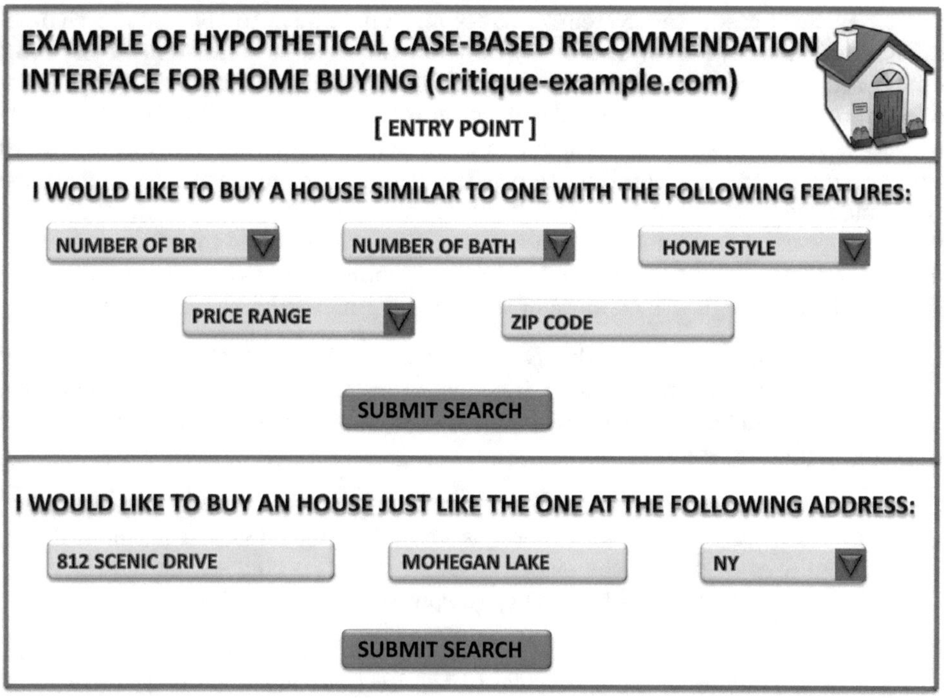

Figure 1.6: A hypothetical example of an initial user interface for a case-based recommender)

attributes. An example of such an interface is illustrated in Figure 1.5. Domain-specific rules are used to match the user requirements to item attributes. These rules represent the domain-specific knowledge used by the system. Such rules could take the form of domain-specific constraints on the item attributes (e.g., "*Cars before year 1970 do not have cruise control.*"). Furthermore, constraint-based systems often create rules relating user attributes to item attributes (e.g., "*Older investors do not invest in ultra high-risk products.*"). In such cases, user attributes may also be specified in the search process. Depending on the number and type of returned results, the user might have an opportunity to modify their original requirements. For example, they might relax some of their constraints when too few results are returned, or they might add more constraints. This search process is interactively repeated until the user arrives at her desired results.

2. *Case-based recommender systems:* In case-based recommender systems [102, 116, 377, 558], specific cases are specified by the user as targets or anchor points. Similarity metrics are defined on the item attributes to retrieve similar items to these cases. An example of such an interface is illustrated in Figure 1.6. The similarity metrics are often carefully defined in a domain-specific way. Therefore, the similarity metrics form the domain knowledge that is used in such systems. The returned results are often used as new target cases with some interactive modifications by the user. For example, when a user sees a returned result, which is almost similar to what they want, they might re-issue a query with that target, but with some of the attributes changed to the user's liking. This interactive process is used to guide the user towards items of interest.

Note that in both cases, the system provides an opportunity to the user to change their specified requirements. However, the way in which this is done is different in the two cases. In case-based systems, examples (or *cases*) are used as anchor points to guide the search in combination with similarity metrics. Critiquing interfaces are particularly popular for expressing feedback in such systems, where users iteratively modify one or more attributes of a preferred item in each iteration. In constraint-based systems, rules (or *constraints*) are used to guide the search. The form of the guidance may often take the form of search-based systems, where users specify their constraints with a search-based interface.

How is the interactivity in knowledge-based recommender systems achieved? This guidance takes place through one or more of the following methods:

1. *Conversational systems:* In this case, the user preferences are determined iteratively in the context of a feedback loop. The main reason for this is that the item domain is complex and the user preferences can be determined only in the context of an iterative conversational system.

2. *Search-based systems:* In search-based systems, user preferences are elicited by using a preset sequence of questions such as the following: "Do you prefer a house in a suburban area or within the city?" In some cases, specific search interfaces may be set up in order to provide the ability to specify user constraints.

3. *Navigation-based recommendation:* In navigation-based recommendation, the user specifies a number of change requests to the item being currently recommended. Through an iterative set of change requests, it is possible to arrive at a desirable item. An example of a change request specified by the user, when a specific house is being recommended is as follows: "I would like a similar house about 5 miles west of the currently recommended house." Such recommender systems are also referred to as *critiquing recommender systems* [417].

It is noteworthy that both knowledge-based and content-based systems depend significantly on the attributes of the items. Because of their use of content-attributes, knowledge-based systems inherit some of the same disadvantages as content-based systems. For example, just like content-based systems, the recommendations in knowledge-based systems can sometimes be obvious because the use of community (i.e., peer) ratings is not leveraged. In fact, knowledge-based systems are sometimes considered to be the "cousins" of content-based systems [558]. The main difference is that content-based systems learn from *past user behavior*, whereas knowledge-based recommendation systems recommend based on active user *specification of their needs and interests*. Therefore, in most of the recommendation literature, knowledge-based recommenders are considered to be a distinct category from content-based recommenders. These distinctions are based both on the goals of such systems and the kind of input data used (see Table 1.2). The different forms of knowledge-based recommender systems are discussed in Chapter 5.

1.3.3.1 Utility-Based Recommender Systems

In utility-based recommender systems, a utility function is defined on the product features in order to compute the probability of a user liking the item [239]. The central challenge in utility-based methods is in defining an appropriate utility function for the user at hand. It is noteworthy that all recommender schemes, whether collaborative, content-based, or knowledge-based methods, implicitly rank the recommended items on the basis of their perceived value (or *utility*) for the target user. In utility-based systems, this utility value is based on a function that is known *a priori*. In this sense, such functions can be viewed as a kind of external knowledge. Therefore, utility-based systems can be viewed as a specific case of knowledge-based recommender systems. In fact, it will be shown in Chapter 5 that utility functions are used frequently in various ways for ranking items in knowledge-based recommender systems.

1.3.4 Demographic Recommender Systems

In demographic recommender systems, the demographic information about the user is leveraged to learn classifiers that can map specific demographics to ratings or buying propensities. An early recommender system, referred to as *Grundy* [508], recommended books based on the library of manually assembled stereotypes. The characteristics of the user were collected with the use of an interactive dialogue. The work in [320] observed that the demographic groups from marketing research can be used to recommend items. Another work [475] makes Web page recommendations on the basis of the demographic characteristics of users that have rated a particular page highly. In many cases, demographic information can be combined with additional *context* to guide the recommendation process. This approach is related to the methodology of *context-sensitive recommender systems*. Some of these methods are discussed in section 8.5.3 of Chapter 8.

More recent techniques have focused on using classifiers for making recommendations. One of the interesting systems in this respect was a technique that extracted features from user home pages in order to predict their likelihood of liking certain restaurants. Rule-based classifiers [31, 32] are often used to relate the demographic profile to buying behavior in an interactive way. While the approach in [31, 32] was not specifically used to recommend specific items, it can easily be paired with a recommender system. Such recommender systems are not very different from the vanilla classification and regression modeling problem, in which feature variables correspond to the demographic profiles and the dependent variables correspond to the ratings or to the buying behavior. Although demographic recommender systems do not usually provide the best results on a stand-alone basis, they add significantly to the power of other recommender systems as a component of hybrid or ensemble models. Demographic techniques are sometimes combined with knowledge-based recommender systems to increase their robustness.

1.3.5 Hybrid and Ensemble-Based Recommender Systems

The three aforementioned systems exploit different sources of input, and they may work well in different scenarios. For example, collaborative filtering systems rely on community ratings, content-based methods rely on textual descriptions and the target user's own ratings, and knowledge-based systems rely on interactions with the user in the context of knowledge bases. Similarly, demographic systems use the demographic profiles of the users to make recommendations. It is noteworthy that these different systems use different types of input, and have different strengths and weaknesses. Some recommender systems, such as knowledge-based systems, are more effective in cold-start settings where a significant amount of data is not available. Other recommender systems, such as collaborative methods, are more effective when a lot of data is available.

In many cases where a wider variety of inputs is available, one has the flexibility of using different types of recommender systems for the same task. In such cases, many opportunities exist for hybridization, where the various aspects from different types of systems are combined to achieve the best of all worlds. Hybrid recommender systems are closely related to the field of ensemble analysis, in which the power of multiple types of machine learning algorithms is combined to create a more robust model. Ensemble-based recommender systems are able to combine not only the power of multiple data sources, but they are also able to improve the effectiveness of a particular class of recommender systems (e.g., collaborative systems) by combining multiple models of the same type. This scenario is not very different from that of ensemble analysis in the field of data classification. Chapter 6 studies various hybridization strategies for recommender systems.

1.3.6   Evaluation of Recommender Systems

Given a set of recommendation algorithms, how well do they perform? How can we evaluate their relative effectiveness? Recommender systems share several conceptual similarities with the classification and regression modeling problem. In classification and regression modeling, the missing class variable needs to be predicted from the feature variables. In recommender systems, any of the matrix entries may be missing and need to be predicted in a data-driven way from the observed entries in the remaining matrix. In this sense, the recommendation problem can be viewed as a generalization of the classification problem. Therefore, many of the models used for evaluation of classifiers can be used for evaluating recommender systems, albeit with some modifications. There are significant variations in the evaluation techniques used for different aspects of recommender systems, such as rating prediction or ranking. The former is closely related to classification and regression modeling, whereas the latter is closely related to the evaluation of retrieval effectiveness in search and information retrieval applications. Evaluation methods for recommender systems are discussed in detail in Chapter 7.

1.4   Domain-Specific Challenges in Recommender Systems

In different domains, such as temporal data, location-based data, and social data, the context of the recommendation plays a critical role. Therefore, the notion of *contextual recommender systems* was developed to address the additional side information that arises in these domains. This notion is used with different modifications for various types of data, such as temporal data, location data, or social data.

1.4.1   Context-Based Recommender Systems

Context-based or context-*aware* recommender systems take various types of contextual information into account, while making recommendations. Such contextual information could include time, location, or social data. For example, the types of clothes recommended by a retailer might depend both on the season and the location of the customer. Another example is the case in which a particular type of festival or holiday affects the underlying customer activity.

It has generally been observed that the use of such contextual information can greatly improve the effectiveness of the recommendation process. Context-based recommender systems are incredibly powerful because the underlying ideas are relevant to a wide variety of domain-specific settings. In fact, a recurring theme throughout the later chapters of the book, will be the use of a *multidimensional model* [7] for context-specific recommendations in different domain-specific settings. Context-aware recommender systems will be discussed in Chapter 8 in a general sense. However, individual aspects of the context, such as time, location, and social information, are studied in detail in other chapters. A general review of these different aspects is provided below.

1.4.2 Time-Sensitive Recommender Systems

In many settings, the recommendations for an item might evolve with time. For example, the recommendations for a movie may be very different at the time of release from the recommendations received several years later. In such cases, it is extremely important to incorporate temporal knowledge in the recommendation process. The temporal aspect in such recommender systems can be reflected in several ways:

1. The rating of an item might evolve with time, as community attitudes evolve and the interests of users change over time. User interests, likes, dislikes, and fashions inevitably evolve with time.

2. The rating of an item might be dependent on the specific time of day, day of week, month, or season. For example, it makes little sense to recommend winter clothing during the summer, or raincoats during the dry season.

The first type of recommender system is created by incorporating time as an explicit parameter in collaborative filtering systems. The second type can be viewed as a special case of context-based recommender systems. Temporal recommender systems are challenging because of the fact that the matrix of ratings is sparse, and the use of specific temporal context aggravates the sparsity problem. Therefore, it is particularly important to have access to large data sets in these settings.

Another common setting is that of implicit feedback data sets such as Web click-streams. The user activity on the Web and other internet platforms creates a lot of useful data that can be mined to make recommendations about future activity. In such cases, discrete sequential pattern mining and Markov models are helpful. The problem of time-sensitive recommendation is discussed in detail in Chapter 9.

1.4.3 Location-Based Recommender Systems

With the increasing popularity of GPS-enabled mobile phones, consumers are often interested in location-based recommendations. For example, a traveling user may wish to determine the closest restaurant based on her previous history of ratings for other restaurants. In general, the recommendation of *places* always has a location aspect built into it. An example of such a system is Foursquare[2], which recommends various types of places such as restaurants or nightlife venues. There are two types of spatial locality that are common to such systems:

1. *User-specific locality:* The geographical location of a user has an important role in her preferences. For example, a user from Wisconsin might not have the same movie preferences as a user from New York. This type of locality is referred to as *preference locality.*

---

[2]http://foursquare.com

2. *Item-specific locality:* The geographical location of an item (e.g., restaurant) might have an impact on the relevance of the item, depending on the current location of the user. Users are generally not willing to travel very far from their current location. This type of locality is referred to as *travel locality*.

The algorithms for preference locality and travel locality are quite different. The former are closer to context-sensitive systems, whereas the latter are usually designed as ad hoc heuristics. Location-based recommender systems have witnessed an increasing interest in recent years because of the increasing prevalence of mobile phones and other GPS-enabled devices. Location-based recommender systems are discussed in detail in Chapter 9.

1.4.4   Social Recommender Systems

Social recommender systems are based on network structures, social cues and tags, or a combination of these various network aspects. In general, the recommender systems that are based on social cues and tags are slightly different from those that are based purely on structural aspects. Recommender systems, which are based purely on structural aspects, are used to suggest nodes and links within the network itself. On the other hand, social recommender systems may be also be used to recommend various products with the use of social cues. Both these forms of recommender systems will be studied in this book. However, these forms of recommendation are sufficiently different that they will be studied in different chapters of this book. It is important to note that the utility of structural recommender systems extends beyond social networks, because such methods are applied to various types of Web-enabled networks.

1.4.4.1   Structural Recommendation of Nodes and Links

Various types of networks, including social networks, are composed of nodes and links. In many cases, it is desirable to recommend nodes and links. For example, a personalized Web search may require a recommendation of material which is related to a particular topic. Since the Web can be viewed as a graph, such methods can be viewed as a node recommendation problem. The problem of node recommendation is closely related to the problem of Web search. In fact, both problems require the use of various forms of ranking algorithms. A key component of these methods is the use of the *PageRank* algorithm, although the personalization of such algorithms is more closely related to recommendation algorithms. Therefore, such algorithms are also referred to as personalized *PageRank* algorithms. In cases where examples of nodes of interest are available, such nodes can be used as training data in order to determine other nodes of interest. This problem is referred to as *collective classification.* A closely related problem is that of the link recommendation or link prediction problem, where it is desirable to suggest friends (or potential links) for a user in a social network. The link prediction problem also has numerous applications beyond social networks. Interestingly, the problems of ranking, collective classification, and link recommendation are closely related. In fact, solutions to one problem are often used as subroutines for other problems. For example, ranking and link prediction methods are often used for traditional product recommendations in user-item graphs. In fact, these methods can be used to perform recommendations in many problem settings, which can be transformed into graphs. Methods for node and link recommendations are discussed in Chapter 10.

1.4.4.2 Product and Content Recommendations with Social Influence

Many forms of product and content recommendation are performed with the help of network connections and other social cues. This problem is also referred to as *viral marketing*. In viral marketing, products are recommended with the use of word-of-mouth systems. In order to achieve this goal, it is important to be able to determine influential and topically relevant entities in the network. This problem is referred to as *influence analysis* in social networks [297]. Many variations of this problem have been proposed, in which the influencers are found in a topically sensitive way, in the social stream scenario. For example, determining the influential users in a Twitter stream for specific topics may be very useful for viral marketing. In other cases, social cues are harvested from social networks in order to make recommendations. These methods for discussed in Chapter 10.

1.4.4.3 Trustworthy Recommender Systems

Many social media sites, such as Epinions [705] or Slashdot [706], allow users to express their trust and distrust in one another, either in a direct way, or through various feedback mechanisms. For example, users can express their trust or distrust in reviews of other users, or they may directly specify their trust or distrust relationships with other users. This trust information is very useful for making more robust recommendations. For example, it is evident that a user-based neighborhood method should be computed with the use of trustworthy peers to obtain robust recommendations. Recent research has shown [221, 588, 616] that the incorporation of trust information can lead to more robust recommendations. Trustworthy recommender systems are presented in Chapter 11.

1.4.4.4 Leveraging Social Tagging Feedback for Recommendations

Users have numerous methods for incorporating their feedback in recommender systems. The most common form of feedback is *social tagging*. Such forms of feedback are particularly common on content sharing sites on the Web, such as Flickr (photo sharing) [692], last.fm [692] (music sharing), and Bibsonomy [708] (scientific literature sharing). *Tags* are meta-data that users utilize to add short informative keywords to the content. For example, a user on a music site might tag Michael Jackson's *Thriller* album as "*rock.*" Such tags provide useful information about the interests of both the user and the content of the item because the tag is associated with both. The tags serve as useful context for performing the recommendations. Methods for context-sensitive recommendations can be directly used to incorporate this feedback into the recommendation process. Other specialized methods have also been developed for using social tagging feedback in the recommendation process. These methods are discussed in detail in Chapter 11.

1.5   Advanced Topics and Applications

This book will also introduce a number of advanced topics and applications. Most of the these topics are discussed in Chapters 12 and 13, although some of the topics are spread out over the book, where it is appropriate. In this section, we provide a brief introduction to these topics.

1.5.1 The Cold-Start Problem in Recommender Systems

One of the major problems in recommender systems is that the number of initially available ratings is relatively small. In such cases, it becomes more difficult to apply traditional collaborative filtering models. While content-based and knowledge-based methods are more robust than collaborative models in the presence of cold starts, such content or knowledge might not always be available. Therefore, a number of specific methods have been designed to ameliorate the problem of cold start in the context of recommender systems. The susceptibility of various models to the cold-start problem is also highlighted throughout this book, along with possible solutions.

1.5.2 Attack-Resistant Recommender Systems

The use of recommender systems has a significant impact on the sale of various products and services. As a result, the sellers of products and services have significant economic incentives to manipulate the output of recommender systems. One example of such a manipulation would be to submit inflated ratings of their own products to the recommender systems. A malicious rival might submit biased and negative reviews about the products of a competitor. Over the years, numerous sophisticated strategies have been developed for attacking recommender systems. Such attacks are highly undesirable because they reduce the overall effectiveness of the recommender system and reduce the quality of experience for legitimate users. Therefore, methods are needed that enable robust recommendations in the presence of such attacks. Attack methods, including the susceptibility of various types of algorithms to attacks, are discussed in detail in Chapter 12. In addition, Chapter 12 will provide a number of strategies for constructing robust recommender systems in the presence of such attacks.

1.5.3 Group Recommender Systems

An interesting extension of traditional recommender systems is the notion of *group recommender systems* [168]. In such cases, the recommendation system is tailored to recommend a particular activity to a group of users rather than a single user. Examples might include the watching of movie or television by a group [408, 653], the selection of music in a fitness center, or the travel recommendations to a group of tourists. The earliest systems, such as *PolyLens* [168], designed models that aggregated the preferences of individual users in order to create group recommendations. However, the consensus over the years has evolved into designing recommender systems, which are better than the sum of their parts and can take the interactions between the various users into account for designing recommendations [272, 413]. Simple averaging strategies do not work well when groups are heterogeneous and contain users with diverse tastes [653]. This is because users often have an impact on each other's tastes based on phenomena from social psychology, such as *emotional contagion* and *conformity*. Detailed surveys on the subject may be found in [45, 271, 407]. Group recommender systems are discussed in section 13.4 of Chapter 13.

1.5.4 Multi-Criteria Recommender Systems

In multi-criteria systems, ratings might be specified on the basis of different criteria by a single user. For example, a user might rate movies based on the plot, music, special effects, and so on. Such techniques often provide recommendations by modeling the user's utility for an item as a vector of ratings corresponding to various criteria. In multi-criteria recommender systems, one can often obtain misleading results by using only the overall rating in conjunction with a traditional recommender system. For example, if two users have the same overall rating for a movie, but their component ratings for the plot and music are very different, then the two users should not be considered similar from the perspective of a similarity-based collaborative filtering algorithm. In some of the multi-criteria systems, users may not specify an overall rating at all. In such cases, the problem is even more challenging because it is needed to present ranked lists of items to various users on the basis of multiple criteria. Excellent overviews of multi-criteria recommender systems may be found in [11, 398, 604] from various perspectives.

It has been shown [271, 410], that some of the methods for group recommender systems can also be adapted to multi-criteria recommender systems. However, the two topics are generally considered different because they emphasize different aspects of the recommendation process. Methods for multi-criteria recommender systems are discussed in section 13.5 of Chapter 13.

1.5.5 Active Learning in Recommender Systems

A major challenge in recommender systems is the acquisition of sufficient ratings in order to make robust predictions. The sparsity of the ratings matrix continues to be a significant impediment in effective functioning of recommender systems. The acquisition of sufficient ratings can reduce the sparsity problem. A variety of real-world recommender systems have mechanisms to encourage users to enter ratings in order to populate the system. For example, users might be provided incentives to rate certain items. In general, it is often difficult to obtain too many ratings from the single user because of the high cost of the acquisition process. Therefore, one must judiciously select the items to be rated by specific users. For example, if a user has already rated a lot of action movies, then asking the user to rate another action movie does not help much in predicting ratings of other action movies, and it helps even less in predicting ratings of movies belonging to unrelated genres. On the other hand, asking the user to rate movies belonging to less populated genres will help significantly in predicting ratings of movies belonging to that genre. Of course, if a user is asked to rate an unrelated movie, it is not necessary that she will be able to provide feedback because she might not have watched that movie at all. Therefore, there are many interesting trade-offs in the problem of active learning of recommender systems, that are not encountered in other problem domains like classification. A review of active learning methods for recommender systems may be found in [513]. Active learning methods are discussed in section 13.6 of Chapter 13.

1.5.6 Privacy in Recommender Systems

Recommender systems are based heavily on feedback from the users, which might be implicit or explicit. This feedback contains significant information about the interests of the user, and it might reveal information about their political opinions, sexual orientations, and personal preferences. In many cases, such information can be highly sensitive, which leads to privacy concerns. Such privacy concerns are significant in that they impede the release of data necessary for the advancement of recommendation algorithms. The availability of real data is crucial for algorithmic advances. For example, the contribution of the Netflix Prize data set to the recommender systems community is invaluable, in that it can be credited with motivating the development of many state-of-the-art algorithms [373]. In recent years, the topic of privacy has been explored in the context of a wide variety of data mining problems [20]. The recommendation domain is no exception, and numerous privacy-preserving algorithms have been developed [133, 484, 485]. The topic of privacy in recommender systems is discussed in detail in section 13.7 of Chapter 13.

1.5.7   Application Domains

Recommender systems are used in numerous application domains, such as retail, music, content, Web search, querying, and computational advertisements. Some of these domains require specialized methods for adapting recommender systems. In particular, Chapter 13 will study three specific domains corresponding to news recommendations, computational advertising, and reciprocal recommender systems. All these application domains are Web-centric in nature. An important aspect of recommender systems is that they assume the existence of strong user-identification mechanisms in order to track and identify long-term user interests. In many Web domains, mechanisms for strong user identification may not be available. In such cases, direct user of recommendation technology may not be feasible. Furthermore, since new items (advertisements) continually enter and leave the system, certain types of methods such as multi-armed bandits are particularly suitable. Therefore, Chapter 13 will discuss the scenarios in which recommendation technology can be used in these application domains. The specific changes that need to be made to off-the-shelf recommender systems will be discussed in this chapter together with advanced techniques such as multi-armed bandits.

1.6   Summary

This book will provide an overview of the most important classes of algorithms for recommender systems, their advantages and disadvantages, and the specific scenarios in which they are most effective. The recommendation problem will be studied in the context of different domain-specific scenarios and with different types of input information and knowledge bases. As this book will show, the recommendation problem is a rich one, and has many different manifestations depending on the nature of the input data and the scenario at hand. Furthermore, the relative effectiveness of different algorithms may vary with the specific problem setting. These trade-offs will also be explored by this book. In many cases, hybrid systems can be developed, which exploit these trade-offs effectively.

A number of advanced topics, such as attack models, group recommender systems, multi-criteria systems, active learning systems, will be studied in later chapters of this book. We will also explore a number of specific applications, such as news recommendations and computational advertising. It is hoped that this book will provide a comprehensive overview and understanding of the different scenarios that arise in the field of recommender systems.

1.7   Bibliographic Notes

Recommender systems became increasingly popular in the mid-nineties, as recommendation systems such as *GroupLens* [501] were developed. Since then, this topic has been explored extensively in the context of a wide variety of models such as collaborative systems, content-based systems, and knowledge-based systems. Detailed surveys and books on the topic may be found in [5, 46, 88, 275, 291, 307, 364, 378, 505, 529, 570]. Among these, the work in [5] is a very well written survey, which provides an excellent overview of the basic ideas. More recent surveys may be found in [88, 378, 570]. A survey of the use of non-traditional sources of information for recommendations, such as social, temporal, side information, or contextual data, is provided in [544]. A recent classification of various facets of recommender system research may be found in [462]. An excellent introductory book may be found in [275], whereas a detailed handbook [505] discusses various aspects of recommender systems in detail.

The problem of collaborative filtering with incomplete ratings matrices is closely related to the traditional literature on missing data analysis [362], although the two fields have often been studied independently. The earliest user-based collaborative filtering models were studied in [33, 98, 501, 540]. User-based methods utilize the ratings of *similar* users on the *same* item in order to make predictions. While such methods were initially quite popular, they are not easily scalable and sometimes inaccurate. Subsequently, item-based methods [181, 360, 524] were proposed, which compute predicted ratings as a function of the ratings of the *same* user on *similar* items. Another popular approach for making recommendations is the use of latent factor models. The earliest works in latent factor models independently appear in the contexts of recommendation [525] and missing value analysis [24]. Eventually, these methods were rediscovered as the most effective class of methods for performing recommendations [252, 309, 313, 500, 517]. Aside from their use in factor-based models, dimensionality reduction methods are also used to reduce the dimensionality of the ratings matrix to improve the *efficiency* of the user-to-user or item-to-item similarity in the reduced space [228, 525]. However, the work on missing data analysis is just as relevant to the recommendation literature. Other relevant models for collaborative filtering include the use of data mining models such as clustering [167, 360, 608], classification, or association pattern mining [524]. Sparsity is a major problem is such systems, and various graph-based systems have been designed to alleviate the problem of sparsity [33, 204, 647].

Content-based methods are closely related to the information retrieval literature [144, 364, 400], in which similarity retrieval methods are used in the recommendation process. Text classification methods are also particularly useful in the recommendation process. A detailed discussion on various text classification methods may be found in [22]. Some of the earliest works on content-based recommendations are found in [60, 69]. The general survey in [5] also discusses content-based recommendations quite extensively.

There are many cases in which collaborative and content-based methods are not useful in obtaining meaningful recommendations because of the high degree of complexity and constraints in the item space. In such cases, knowledge-based recommender systems [116] are particularly useful. Demographic recommender systems are discussed in [320, 475, 508], whereas utility-based recommender systems are discussed in [239]. An excellent survey on explanations in recommender systems is provided in [598].

Different recommender systems are more effective in different types of settings. The evaluation [246] of recommender systems is important in order to judge the effectiveness of different algorithms. A detailed discussion of evaluation methods may also be found in [538]. Hybrid systems [117] can combine various recommender systems to obtain more effective results. Furthermore, ensemble methods can also combine algorithms of the same type to obtain more effective results. The top entries of the Netflix Prize contest, such as *"The Ensemble"* [704] and *"Bellkor's Pragmatic Chaos,"* [311] were both ensemble methods.

Recommender systems require specialized methods to make them more effective in a wide variety of scenarios. A major problem in the effective use of such systems is the *cold-start* problem, in which a sufficient number of ratings is not available at the beginning of the recommendation process. Therefore, specialized methods are often used to address this problem [533]. In many cases, the context of the recommendation, such as the location, time, or social information, can significantly improve the recommendation process [7]. Each of these different types of context has also been studied individually as a separate area of recommender systems. Temporally-aware recommender systems have been studied in [310], whereas location-aware recommender systems have been discussed in [26]. The social context is particularly diverse because it allows for a wide variety of problem settings. One can either recommend nodes or links in social networks, or one can recommend products with the help of social cues. The first of these settings is closely related to the domain of social network analysis [656]. Each of the traditional problems of ranking, node classification, and link prediction [22, 656] can be viewed as a structural recommendation problem in social networks. Furthermore, these forms of recommendation are useful beyond the social network setting. Interestingly, methods such as link prediction can also be used for traditional recommendation by transforming the user-item interactions into a bipartite graph structure [261]. A different form of social recommendation is the case where social cues are used for performing recommendations [588]. The social network structure can also be directly used in the context of viral marketing applications [297].

Since recommender systems often help the sale of products, the sellers of those products or their competitors have significant motivations to attack recommender systems by manipulating the ratings. In such cases, the recommendations are unlikely to be of high quality, and therefore *untrustworthy*. In recent years, a significant amount of effort has been devoted to the design of trustworthy recommender systems [444]. Various group recommender systems are discussed in [45, 271, 272, 407, 408, 412, 413, 415, 653]. Multi-criteria recommender systems are discussed in [11, 398, 604]. Active learning methods are discussed in [513]. A general discussion of privacy-preservation methods may be found in [20]. The earliest studies on the topic of privacy-preserving recommendations were presented in [133, 451, 484, 485, 667]. Privacy continues to be a significant challenge to such systems because of the high dimensional nature of the data. It has been shown in [30, 451] how the dimensionality can be leveraged to make privacy attacks on different types of data sets.

1.8   Exercises

1. Explain why unary ratings are significantly different from other types of ratings in the design of recommender systems.

2. Discuss cases in which content-based recommendations will not perform as well as ratings-based collaborative filtering.

3. Suppose you set up a system, where a guided visual interface is used in order to determine the product of interest to a customer. What category of recommender system does this case fall into?

4. Discuss a scenario in which location plays an important role in the recommendation process.

5. The chapter mentions the fact that collaborative filtering can be viewed as a generalization of the classification problem. Discuss a simple method to generalize classification algorithms to collaborative filtering. Explain why it is difficult to use such methods in the context of sparse ratings matrices.

6. Suppose that you had a recommender system that could predict raw ratings. How would you use it to design a top-k recommender system? Discuss the computational complexity of such a system in terms of the number of applications of the base prediction algorithm. Under what circumstances would such an approach become impractical?

Chapter 2

Neighborhood-Based Collaborative Filtering

"When one neighbor helps another, we strengthen our communities." – Jennifer Pahlka

2.1   Introduction

Neighborhood-based collaborative filtering algorithms, also referred to as *memory-based algorithms*, were among the earliest algorithms developed for collaborative filtering. These algorithms are based on the fact that similar users display similar patterns of rating behavior and similar items receive similar ratings. There are two primary types of neighborhood-based algorithms:

1. *User-based collaborative filtering:* In this case, the ratings provided by similar users to a *target* user A are used to make recommendations for A. The predicted ratings of A are computed as the weighted average values of these "peer group" ratings for each item.

2. *Item-based collaborative filtering:* In order to make recommendations for *target* item B, the first step is to determine a set S of items, which are most similar to item B. Then, in order to predict the rating of any particular user A for item B, the ratings in set S, which are specified by A, are determined. The weighted average of these ratings is used to compute the predicted rating of user A for item B.

An important distinction between user-based collaborative filtering and item-based collaborative filtering algorithms is that the ratings in the former case are predicted using the ratings of neighboring *users*, whereas the ratings in the latter case are predicted using

© Springer International Publishing Switzerland 2016
C.C. Aggarwal, *Recommender Systems: The Textbook*,
DOI 10.1007/978-3-319-29659-3_2

the user's *own* ratings on neighboring (i.e., closely related) *items*. In the former case, neighborhoods are defined by similarities among users (rows of ratings matrix), whereas in the latter case, neighborhoods are defined by similarities among items (columns of ratings matrix). Thus, the two methods share a complementary relationship. Nevertheless, there are considerable differences in the types of recommendations that are achieved using these two methods.

For the purpose of subsequent discussion, we assume that the user-item ratings matrix is an incomplete $m \times n$ matrix $R = [r_{uj}]$ containing m users and n items. It is assumed that only a small subset of the ratings matrix is specified or observed. Like all other collaborative filtering algorithms, neighborhood-based collaborative filtering algorithms can be formulated in one of two ways:

1. *Predicting the rating value of a user-item combination:* This is the simplest and most primitive formulation of a recommender system. In this case, the missing rating r_{uj} of the user u for item j is predicted.

2. *Determining the top-k items or top-k users:* In most practical settings, the merchant is not necessarily looking for specific ratings values of user-item combinations. Rather, it is more interesting to learn the top-k most relevant items for a particular user, or the top-k most relevant users for a particular item. The problem of determining the top-k items is more common than that of finding the top-k users. This is because the former formulation is used to present lists of recommended items to users in Web-centric scenarios. In traditional recommender algorithms, the "top-k problem" almost always refers to the process of finding the top-k items, rather than the top-k users. However, the latter formulation is also useful to the merchant because it can be used to determine the best users to target with marketing efforts.

The two aforementioned problems are closely related. For example, in order to determine the top-k items for a particular user, one can predict the ratings of each item for that user. The top-k items can be selected on the basis of the predicted rating. In order to improve efficiency, neighborhood-based methods pre-compute some of the data needed for prediction in an offline phase. This pre-computed data can be used in order to perform the ranking in a more efficient way.

This chapter will discuss various neighborhood-based methods. We will study the impact of some properties of ratings matrices on collaborative filtering algorithms. In addition, we will study the impact of the ratings matrix on recommendation effectiveness and efficiency. We will discuss the use of clustering and graph-based representations for implementing neighborhood-based methods. We will also discuss the connections between neighborhood methods and regression modeling techniques. Regression methods provide an optimization framework for neighborhood-based methods. In particular, the neighborhood-based method can be shown to be a heuristic approximation of a least-squares regression model [72]. This approximate equivalence will be shown in section 2.6. Such an optimization framework also paves the way for the integration of neighborhood methods with other optimization models, such as latent factor models. The integrated approach is discussed in detail in section 3.7 of Chapter 3.

This chapter is organized as follows. Section 2.2 discusses a number of key properties of ratings matrices. Section 2.3 discusses the key algorithms for neighborhood-based collaborative filtering algorithms. Section 2.4 discusses how neighborhood-based algorithms can be made faster with the use of clustering methods. Section 2.5 discusses the use of dimensionality reduction methods for enhancing neighborhood-based collaborative filtering algorithms.

An optimization modeling view of neighborhood-based methods is discussed in section 2.6. A linear regression approach is used to simulate the neighborhood model within a learning and optimization framework. Section 2.7 discusses how graph-based representations can be used to alleviate the sparsity problem in neighborhood methods. The summary is provided in section 2.8.

2.2 Key Properties of Ratings Matrices

As discussed earlier, we assume that the ratings matrix is denoted by R, and it is an $m \times n$ matrix containing m users and n items. Therefore, the rating of user u for item j is denoted by r_{uj}. Only a small subset of the entries in the ratings matrix are typically specified. The specified entries of the matrix are referred to as the training data, whereas the unspecified entries of the matrix are referred to as the test data. This definition has a direct analog in classification, regression, and semisupervised learning algorithms [22]. In that case, all the unspecified entries belong to a special column, which is known as the class variable or dependent variable. Therefore, the recommendation problem can be viewed as a generalization of the problem of classification and regression.

Ratings can be defined in a variety of ways, depending on the application at hand:

1. *Continuous ratings:* The ratings are specified on a continuous scale, corresponding to the level of like or dislike of the item at hand. An example of such a system is the Jester joke recommendation engine [228, 689], in which the ratings can take on any value between -10 and 10. The drawback of this approach is that it creates a burden on the user of having to think of a real value from an infinite number of possibilities. Therefore, such an approach is relatively rare.

2. *Interval-based ratings:* In interval-based ratings, the ratings are often drawn from a 5-point or 7-point scale, although 10-point and 20-point scales are also possible. Examples of such ratings could be numerical integer values from 1 to 5, from -2 to 2, or from 1 to 7. An important assumption is that the numerical values explicitly define the distances between the ratings, and the rating values are typically equidistant.

3. *Ordinal ratings:* Ordinal ratings are much like interval-based ratings, except that ordered *categorical* values may be used. Examples of such ordered categorical values might be responses such as "Strongly Disagree," "Disagree," "Neutral," "Agree," and "Strongly Agree." A major difference from interval-based ratings is that it is not assumed that the difference between any pair of adjacent ratings values is the same. However, in practice, this difference is only theoretical, because these different ordered categorical values are often assigned to equally spaced utility values. For example, one might assign the "Strongly Disagree" response to a rating value of 1, and the "Strongly Agree" response to a rating value of 5. In such cases, ordinal ratings are almost equivalent to interval-based ratings. Generally, the numbers of positive and negative responses are equally balanced in order to avoid bias. In cases where an even number of responses are used, the "Neutral" option is not present. Such an approach is referred to as the *forced choice* method because the neutral option is not present.

4. *Binary ratings:* In the case of binary ratings, only two options are present, corresponding to positive or negative responses. Binary ratings can be considered a special case of both interval-based and ordinal ratings. For example, the Pandora Internet radio station provides users with the ability to either like or dislike a particular music track.

Binary ratings are an example of the case where forced choice is imposed on the user. In cases where the user is neutral, she will often not specify a rating at all.

5. *Unary ratings:* Such systems allow the user to specify a positive preference for an item, but there is no mechanism to specify a negative preference. This is often the case in many real-world settings, such as the use of a *"like"* button on Facebook. More often, unary ratings are derived from customer *actions.* For example, the act of a customer buying an item can be considered a positive vote for an item. On the other hand, if the customer has not bought the item, then it does not necessarily indicate a dislike for the item. Unary ratings are special because they simplify the development of specialized models in these settings.

It is noteworthy that the indirect derivation of unary ratings from customer actions is also referred to as *implicit feedback*, because the customer does not explicitly provide feedback. Rather, the feedback is inferred in an implicit way through the customer's actions. Such types of "ratings" are often easier to obtain because users are far more likely to interact with items on an online site than to explicitly rate them. The setting of implicit feedback (i.e., unary ratings) is inherently different, as it can be considered the matrix completion analog of the positive-unlabeled (PU) learning problem in classification and regression modeling.

The distribution of ratings among items often satisfies a property in real-world settings, which is referred to as the *long-tail* property. According to this property, only a small fraction of the items are rated frequently. Such items are referred to as *popular* items. The vast majority of items are rated rarely. This results in a highly skewed distribution of the underlying ratings. An example of a skewed rating distribution is illustrated in Figure 2.1. The X-axis shows the index of the item in order of decreasing frequency, and the Y-axis shows the frequency with which the item was rated. It is evident that most of the items are rated only a small number of times. Such a rating distribution has important implications for the recommendation process:

1. In many cases, the high-frequency items tend to be relatively competitive items with little profit for the merchant. On the other hand, the lower frequency items have larger profit margins. In such cases, it may be advantageous to the merchant to recommend lower frequency items. In fact, analysis suggests [49] that many companies, such as Amazon.com, make most of their profit by selling items in the long tail.

2. Because of the rarity of observed ratings in the long tail it is generally more difficult to provide robust rating predictions in the long tail. In fact, many recommendation algorithms have a tendency to suggest popular items rather than infrequent items [173]. This phenomenon also has a negative impact on diversity, and users may often become bored by receiving the same set of recommendations of popular items.

3. The long tailed distribution implies that the items, which are frequently rated by users, are fewer in number. This fact has important implications for neighborhood-based collaborative filtering algorithms because the neighborhoods are often defined on the basis of these frequently rated items. In many cases, the ratings of these high-frequency items are not representative of the low-frequency items because of the inherent differences in the rating patterns of the two classes of items. As a result, the prediction process may yield misleading results. As we will discuss in section 7.6 of Chapter 7, this phenomenon can also cause misleading *evaluations* of recommendation algorithms.

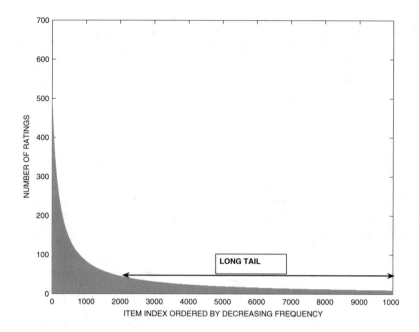

Figure 2.1: The long tail of rating frequencies

Important characteristics of ratings, such as sparsity and the long tail, need to be taken into account during the recommendation process. By adjusting the recommendation algorithms to take such real-world properties into account, it is possible to obtain more meaningful predictions [173, 463, 648].

2.3 Predicting Ratings with Neighborhood-Based Methods

The basic idea in neighborhood-based methods is to use either user-user similarity or item-item similarity to make recommendations from a ratings matrix. The concept of a *neighborhood* implies that we need to determine either similar users or similar items in order to make predictions. In the following, we will discuss how neighborhood-based methods can be used to predict the ratings of specific user-item combinations. There are two basic principles used in neighborhood-based models:

1. *User-based models:* Similar users have similar ratings on the same item. Therefore, if Alice and Bob have rated movies in a similar way in the past, then one can use Alice's observed ratings on the movie *Terminator* to predict Bob's unobserved ratings on this movie.

2. *Item-based models:* Similar items are rated in a similar way by the same user. Therefore, Bob's ratings on similar science fiction movies like *Alien* and *Predator* can be used to predict his rating on *Terminator*.

Since the collaborative filtering problem can be viewed as a generalization of the classification/regression modeling problem, neighborhood-based methods can be viewed as generalizations of nearest neighbor classifiers in the machine learning literature. Unlike

Table 2.1: User-user similarity computation between user 3 and other users

Item-Id ⇒	1	2	3	4	5	6	Mean Rating	Cosine(i, 3) (user-user)	Pearson(i, 3) (user-user)
User-Id ⇓									
1	7	6	7	4	5	4	5.5	0.956	0.894
2	6	7	?	4	3	4	4.8	0.981	0.939
3	?	3	3	1	1	?	2	1.0	1.0
4	1	2	2	3	3	4	2.5	0.789	-1.0
5	1	?	1	2	3	3	2	0.645	-0.817

Table 2.2: Ratings matrix of Table 2.1 with mean-centering for adjusted cosine similarity computation among items. The adjusted cosine similarities of items 1 and 6 with other items are shown in the last two rows.

Item-Id ⇒	1	2	3	4	5	6
User-Id ⇓						
1	1.5	0.5	1.5	-1.5	-0.5	-1.5
2	1.2	2.2	?	-0.8	-1.8	-0.8
3	?	1	1	-1	-1	?
4	-1.5	-0.5	-0.5	0.5	0.5	1.5
5	-1	?	-1	0	1	1
Cosine(1, j) (item-item)	1	0.735	0.912	-0.848	-0.813	-0.990
Cosine(6, j) (item-item)	-0.990	-0.622	-0.912	0.829	0.730	1

classification, where the nearest neighbors are always determined only on the basis of row similarity, it is possible to find the nearest neighbors in collaborative filtering on the basis of either rows or columns. This is because all missing entries are concentrated in a single column in classification, whereas the missing entries are spread out over the different rows and columns in collaborative filtering (cf. section 1.3.1.3 of Chapter 1). In the following discussion, we will discuss the details of both user-based and item-based neighborhood models, together with their natural variations.

2.3.1 User-Based Neighborhood Models

In this approach, user-based neighborhoods are defined in order to identify similar users to the *target* user for whom the rating predictions are being computed. In order to determine the neighborhood of the target user i, her similarity to all the other users is computed. Therefore, a similarity function needs to be defined between the ratings specified by users. Such a similarity computation is tricky because different users may have different scales of ratings. One user might be biased toward liking most items, whereas another user might be biased toward not liking most of the items. Furthermore, different users may have rated different items. Therefore, mechanisms need to be identified to address these issues.

For the $m \times n$ ratings matrix $R = [r_{uj}]$ with m users and n items, let I_u denote the set of item indices for which ratings have been specified by user (row) u. For example, if the ratings of the first, third, and fifth items (columns) of user (row) u are specified (observed) and the remaining are missing, then we have $I_u = \{1, 3, 5\}$. Therefore, the set of items rated by both users u and v is given by $I_u \cap I_v$. For example, if user v has rated the first four items, then $I_v = \{1, 2, 3, 4\}$, and $I_u \cap I_v = \{1, 3, 5\} \cap \{1, 2, 3, 4\} = \{1, 3\}$. It is possible (and quite common) for $I_u \cap I_v$ to be an empty set because ratings matrices are generally sparse. The set $I_u \cap I_v$ defines the mutually observed ratings, which are used to compute the similarity between the uth and vth users for neighborhood computation.

One measure that captures the similarity $\text{Sim}(u, v)$ between the rating vectors of two users u and v is the Pearson correlation coefficient. Because $I_u \cap I_v$ represents the set of item indices for which both user u and user v have specified ratings, the coefficient is computed only on this set of items. The first step is to compute the mean rating μ_u for each user u using her specified ratings:

$$\mu_u = \frac{\sum_{k \in I_u} r_{uk}}{|I_u|} \quad \forall u \in \{1 \ldots m\} \tag{2.1}$$

Then, the Pearson correlation coefficient between the rows (users) u and v is defined as follows:

$$\text{Sim}(u, v) = \text{Pearson}(u, v) = \frac{\sum_{k \in I_u \cap I_v} (r_{uk} - \mu_u) \cdot (r_{vk} - \mu_v)}{\sqrt{\sum_{k \in I_u \cap I_v} (r_{uk} - \mu_u)^2} \cdot \sqrt{\sum_{k \in I_u \cap I_v} (r_{vk} - \mu_v)^2}} \tag{2.2}$$

Strictly speaking, the traditional definition of $\text{Pearson}(u, v)$ mandates that the values of μ_u and μ_v should be computed *only* over the items that are rated *both* by users u and v. Unlike Equation 2.1, such an approach will lead to a different value of μ_u, depending on the choice of the other user v to which the Pearson similarity is being computed. However, it is quite common (and computationally simpler) to compute each μ_u just once for each user u, according to Equation 2.1. It is hard to make an argument that one of these two ways of computing μ_u always provides strictly better recommendations than the other. In extreme cases, where the two users have only one mutually specified rating, it can be argued that using Equation 2.1 for computing μ_u will provide more informative results, because the Pearson coefficient will be indeterminate over a single common item in the traditional definition. Therefore, we will work with the simpler assumption of using Equation 2.1 in this chapter. Nevertheless, it is important for the reader to keep in mind that many implementations of user-based methods compute μ_u and μ_v in pairwise fashion during the Pearson computation.

The Pearson coefficient is computed between the target user and all the other users. One way of defining the peer group of the target user would be to use the set of k users with the highest Pearson coefficient with the target. However, since the number of observed ratings in the top-k peer group of a target user may vary significantly with the item at hand, the closest k users are found for the target user separately for each predicted item, such that each of these k users have specified ratings for that item. The weighted average of these ratings can be returned as the predicted rating for that item. Here, each rating is weighted with the Pearson correlation coefficient of its owner to the target user.

The main problem with this approach is that different users may provide ratings on different scales. One user might rate all items highly, whereas another user might rate all items negatively. The raw ratings, therefore, need to be mean-centered in row-wise fashion, before determining the (weighted) average rating of the peer group. The mean-centered rating s_{uj} of a user u for item j is defined by subtracting her mean rating from the raw rating r_{uj}.

$$s_{uj} = r_{uj} - \mu_u \quad \forall u \in \{1 \ldots m\} \tag{2.3}$$

As before, the weighted average of the mean-centered rating of an item in the top-k peer group of target user u is used to provide a *mean-centered* prediction. The mean rating of the target user is then added back to this prediction to provide a *raw* rating prediction $\hat{r}_{uj}$ of target user u for item j. The hat notation "ˆ" on top of r_{uj} indicates a *predicted* rating, as opposed to one that was already observed in the original ratings matrix. Let $P_u(j)$ be the set[1] of k closest users to target user u, who have specified ratings for item j. Users with very low or negative correlations with target user u are sometimes filtered from $P_u(j)$ as a heuristic enhancement. Then, the overall neighborhood-based *prediction function* is as follows:

$$\hat{r}_{uj} = \mu_u + \frac{\sum_{v \in P_u(j)} \text{Sim}(u,v) \cdot s_{vj}}{\sum_{v \in P_u(j)} |\text{Sim}(u,v)|} = \mu_u + \frac{\sum_{v \in P_u(j)} \text{Sim}(u,v) \cdot (r_{vj} - \mu_v)}{\sum_{v \in P_u(j)} |\text{Sim}(u,v)|} \qquad (2.4)$$

This broader approach allows for a number of different variations in terms of how the similarity or prediction function is computed or in terms of which items are filtered out during the prediction process.

Example of User-Based Algorithm

Consider the example of Table 2.1. In this case, the ratings of five users $1 \ldots 5$ are indicated for six items denoted by $1 \ldots 6$. Each rating is drawn from the range $\{1 \ldots 7\}$. Consider the case where the target user index is 3, and we want to make item predictions on the basis of the ratings in Table 2.1. We need to compute the predictions $\hat{r}_{31}$ and $\hat{r}_{36}$ of user 3 for items 1 and 6 in order to determine the top recommended item.

The first step is to compute the similarity between user 3 and all the other users. We have shown two possible ways of computing similarity in the last two columns of the same table. The second-last column shows the similarity based on the raw cosine between the ratings and the last column shows the similarity based on the Pearson correlation coefficient. For example, the values of Cosine$(1,3)$ and Pearson$(1,3)$ are computed as follows:

$$\text{Cosine}(1,3) = \frac{6*3 + 7*3 + 4*1 + 5*1}{\sqrt{6^2 + 7^2 + 4^2 + 5^2} \cdot \sqrt{3^2 + 3^2 + 1^2 + 1^2}} = 0.956$$

$$\text{Pearson}(1,3) =$$

$$= \frac{(6-5.5)*(3-2) + (7-5.5)*(3-2) + (4-5.5)*(1-2) + (5-5.5)*(1-2)}{\sqrt{1.5^2 + 1.5^2 + (-1.5)^2 + (-0.5)^2} \cdot \sqrt{1^2 + 1^2 + (-1)^2 + (-1)^2}}$$

$$= 0.894$$

The Pearson and raw cosine similarities of user 3 with all other users are illustrated in the final two columns of Table 2.1. Note that the Pearson correlation coefficient is much more discriminative and the sign of the coefficient provides information about similarity and dissimilarity. The top-2 closest users to user 3 are users 1 and 2 according to both measures. By using the Pearson-weighted average of the *raw* ratings of users 1 and 2, the following predictions are obtained for user 3 with respect to her unrated items 1 and 6:

$$\hat{r}_{31} = \frac{7*0.894 + 6*0.939}{0.894 + 0.939} \approx 6.49$$

$$\hat{r}_{36} = \frac{4*0.894 + 4*0.939}{0.894 + 0.939} = 4$$

---

[1] In many cases, k valid peers of target user u with observed ratings for item j might not exist. This scenario is particularly common in sparse ratings matrices, such as the case where user u has less than k observed ratings. In such cases, the set $P_u(j)$ will have cardinality less than k.

Thus, item 1 should be prioritized over item 6 as a recommendation to user 3. Furthermore, the prediction suggests that user 3 is likely to be interested in *both* movies 1 and 6 to a greater degree than *any* of the movies she has already rated. This is, however, a result of the bias caused by the fact that the peer group $\{1, 2\}$ of user indices is a far more optimistic group with positive ratings, as compared to the target user 3. Let us now examine the impact of mean-centered ratings on the prediction. The mean-centered ratings are illustrated in Table 2.2. The corresponding predictions with mean-centered Equation 2.4 are as follows:

$$\hat{r}_{31} = 2 + \frac{1.5 * 0.894 + 1.2 * 0.939}{0.894 + 0.939} \approx 3.35$$

$$\hat{r}_{36} = 2 + \frac{-1.5 * 0.894 - 0.8 * 0.939}{0.894 + 0.939} \approx 0.86$$

Thus, the mean-centered computation also provides the prediction that item 1 should be prioritized over item 6 as a recommendation to user 3. There is, however, one crucial difference from the previous recommendation. In this case, the predicted rating of item 6 is only 0.86, which is *less* than all the other items that user 3 has rated. This is a drastically different result than in the previous case, where the predicted rating for item 6 was greater than all the other items that user 3 had rated. Upon visually inspecting Table 2.1 (or Table 2.2), it is indeed evident that item 6 ought to be rated very low by user 3 (compared to her other items), because her closest peers (users 1 and 2) have also rated it lower than their other items. Thus, the mean-centering process enables a much better *relative* prediction with respect to the ratings that have already been observed. In many cases, it can also affect the relative order of the predicted items. The only weakness in this result is that the predicted rating of item 6 is 0.85, which is outside the range of allowed ratings. Such ratings can always be used for ranking, and the predicted value can be corrected to the closest value in the allowed range.

2.3.1.1   Similarity Function Variants

Several other variants of the similarity function are used in practice. One variant is to use the cosine function on the *raw* ratings rather than the mean-centered ratings:

$$\text{RawCosine}(u, v) = \frac{\sum_{k \in I_u \cap I_v} r_{uk} \cdot r_{vk}}{\sqrt{\sum_{k \in I_u \cap I_v} r_{uk}^2} \cdot \sqrt{\sum_{k \in I_u \cap I_v} r_{vk}^2}} \qquad (2.5)$$

In some implementations of the raw cosine, the normalization factors in the denominator are based on all the specified items and not the mutually rated items.

$$\text{RawCosine}(u, v) = \frac{\sum_{k \in I_u \cap I_v} r_{uk} \cdot r_{vk}}{\sqrt{\sum_{k \in I_u} r_{uk}^2} \cdot \sqrt{\sum_{k \in I_v} r_{vk}^2}} \qquad (2.6)$$

In general, the Pearson correlation coefficient is preferable to the raw cosine because of the *bias adjustment* effect of mean-centering. This adjustment accounts for the fact that different users exhibit different levels of generosity in their global rating patterns.

The reliability of the similarity function $\text{Sim}(u, v)$ is often affected by the number of common ratings $|I_u \cap I_v|$ between users u and v. When the two users have only a small number of ratings in common, the similarity function should be reduced with a discount factor to de-emphasize the importance of that user pair. This method is referred to as *significance weighting*. The discount factor kicks in when the number of common ratings between the two users is less than a particular threshold β. The value of the discount factor is given by $\frac{\min\{|I_u \cap I_v|, \beta\}}{\beta}$, and it always lies in the range $[0, 1]$. Therefore, the discounted similarity DiscountedSim(u, v) is given by the following:

$$\text{DiscountedSim}(u, v) = \text{Sim}(u, v) \cdot \frac{\min\{|I_u \cap I_v|, \beta\}}{\beta} \tag{2.7}$$

The discounted similarity is used both for the process of determining the peer group and for computing the prediction according to Equation 2.4.

2.3.1.2   Variants of the Prediction Function

There are many variants of the prediction function used in Equation 2.4. For example, instead of mean-centering the raw rating r_{uj} to the centered value s_{uj}, one might use the Z-score z_{uj}, which further divides s_{uj} with the standard deviation σ_u of the observed ratings of user u. The standard deviation is defined as follows:

$$\sigma_u = \sqrt{\frac{\sum_{j \in I_u} (r_{uj} - \mu_u)^2}{|I_u| - 1}} \quad \forall u \in \{1 \ldots m\} \tag{2.8}$$

Then, the standardized rating is computed as follows:

$$z_{uj} = \frac{r_{uj} - \mu_u}{\sigma_u} = \frac{s_{uj}}{\sigma_u} \tag{2.9}$$

Let $P_u(j)$ denote the set of the top-k similar users of target user u, for which the ratings of item j have been observed. In this case, the predicted rating $\hat{r}_{uj}$ of target user u for item j is as follows:

$$\hat{r}_{uj} = \mu_u + \sigma_u \frac{\sum_{v \in P_u(j)} \text{Sim}(u, v) \cdot z_{vj}}{\sum_{v \in P_u(j)} |\text{Sim}(u, v)|} \tag{2.10}$$

Note that the weighted average needs to be *multiplied* with σ_u in this case. In general, if a function $g(\cdot)$ is applied during ratings normalization, then its inverse needs to be applied during the final prediction process. Although it is generally accepted that normalization improves the prediction, there seem to be conflicting conclusions in various studies on whether mean-centering or the Z-score provides higher-quality results [245, 258]. One problem with the Z-score is that the predicted ratings might frequently be outside the range of the permissible ratings. Nevertheless, even when the predicted values are outside the range of permissible ratings, they can be used to *rank* the items in order of desirability for a particular user.

A second issue in the prediction is that of the weighting of the various ratings in Equation 2.4. Each mean-centered rating s_{vj} of user v for item j is weighted with the similarity Sim(u, v) of user v to the target user u. While the value of Sim(u, v) was chosen to be the Pearson correlation coefficient, a commonly used practice is to *amplify* it by exponentiating it to the power of α. In other words, we have:

$$\text{Sim}(u, v) = \text{Pearson}(u, v)^{\alpha} \tag{2.11}$$

By choosing $\alpha > 1$, it is possible to amplify the importance of the similarity in the weighting of Equation 2.4.

As discussed earlier, neighborhood-based collaborative filtering methods are generalizations of nearest neighbor classification/regression methods. The aforementioned discussion is closer to nearest neighbor regression modeling, rather than nearest neighbor classification, because the predicted value is treated as a continuous variable throughout the prediction process. It is also possible to create a prediction function which is closer to a classification method by treating ratings as categorical values and ignoring the ordering among the ratings. Once the peer group of the target user u has been identified, the number of *votes* for each possible rating value (e.g., Agree, Neutral, Disagree) within the peer group is determined. The rating with the largest number of votes is predicted as the relevant one. This approach has the advantage of providing the most *likely* rating rather than the average rating. Such an approach is generally more effective in cases where the number of distinct ratings is small. It is also useful in the case of ordinal ratings, where the exact distances between pairs of rating values are not defined. In cases where the granularity of ratings is high, such an approach is less robust and loses a lot of ordering information among the ratings.

2.3.1.3 Variations in Filtering Peer Groups

The peer group for a target user may be defined and filtered in a wide variety of ways. The simplest approach is to use the top-k most similar users to the target user as her peer group. However, such an approach might include users that are weakly or negatively correlated with the target. Weakly correlated users might add to the error in the prediction. Furthermore, negatively correlated ratings often do not have as much predictive value in terms of potential inversion of the ratings. Although the prediction function technically allows the use of weak or negative ratings, their use is not consistent with the broader principle of neighborhood methods. Therefore, ratings with weak or negative correlations are often filtered out.

2.3.1.4 Impact of the Long Tail

As discussed in section 2.2, the distribution of ratings typically shows a long-tail distribution in many real scenarios. Some movies may be very popular and they may repeatedly occur as commonly rated items by different users. Such ratings can sometimes worsen the quality of the recommendations because they tend to be less discriminative across different users. The negative impact of these recommendations can be experienced both during the peer group computation and also during the prediction computation (cf. Equation 2.4). This notion is similar in principle to the deterioration in retrieval quality caused by popular and noninformative words (e.g., "a," "an," "the") in document retrieval applications. Therefore, the proposed solutions used in collaborative filtering are also similar to those used in the information retrieval literature. Just as the notion of *Inverse Document Frequency* (idf) exists in the information retrieval literature [400], one can use the notion of *Inverse User Frequency* in this case. If m_j is the number of ratings of item j, and m is the total number of users, then the weight w_j of the item j is set to the following:

$$w_j = \log\left(\frac{m}{m_j}\right) \quad \forall j \in \{1 \ldots n\} \tag{2.12}$$

Each item j is weighted by w_j both during the similarity computation and during the recommendation process. For example, the Pearson correlation coefficient can be modified to include the weights as follows:

$$\text{Pearson}(u, v) = \frac{\sum_{k \in I_u \cap I_v} w_k \cdot (r_{uk} - \mu_u) \cdot (r_{vk} - \mu_v)}{\sqrt{\sum_{k \in I_u \cap I_v} w_k \cdot (r_{uk} - \mu_u)^2} \cdot \sqrt{\sum_{k \in I_u \cap I_v} w_k \cdot (r_{vk} - \mu_v)^2}} \tag{2.13}$$

Item weighting can also be incorporated in other collaborative filtering methods. For example, the final prediction step of item-based collaborative filtering algorithms can be modified to use weights, even though the adjusted cosine similarity between two items remains unchanged by the weights.

2.3.2 Item-Based Neighborhood Models

In item-based models, peer groups are constructed in terms of *items* rather than *users*. Therefore, similarities need to be computed between items (or columns in the ratings matrix). Before computing the similarities between the columns, each row of the ratings matrix is centered to a mean of zero. As in the case of user-based ratings, the average rating of each item in the ratings matrix is subtracted from each rating to create a mean-centered matrix. This process is identical to that discussed earlier (see Equation 2.3), which results in the computation of mean-centered ratings s_{uj}. Let U_i be the indices of the set of users who have specified ratings for item i. Therefore, if the first, third, and fourth users have specified ratings for item i, then we have $U_i = \{1, 3, 4\}$.

Then, the *adjusted* cosine similarity between the items (columns) i and j is defined as follows:

$$\text{AdjustedCosine}(i, j) = \frac{\sum_{u \in U_i \cap U_j} s_{ui} \cdot s_{uj}}{\sqrt{\sum_{u \in U_i \cap U_j} s_{ui}^2} \cdot \sqrt{\sum_{u \in U_i \cap U_j} s_{uj}^2}} \qquad (2.14)$$

This similarity is referred to as the adjusted cosine similarity because the ratings are mean-centered before computing the similarity value. Although the Pearson correlation can also be used on the columns in the case of the item-based method, the adjusted cosine generally provides superior results.

Consider the case in which the rating of target item t for user u needs to be determined. The first step is to determine the top-k most similar *items* to *item* t based on the aforementioned adjusted cosine similarity. Let the top-k matching items to item t, for which the user u has specified ratings, be denoted by $Q_t(u)$. The *weighted* average value of these (raw) ratings is reported as the predicted value. The weight of item j in this average is equal to the adjusted cosine similarity between item j and the target item t. Therefore, the predicted rating $\hat{r}_{ut}$ of user u for target item t is as follows:

$$\hat{r}_{ut} = \frac{\sum_{j \in Q_t(u)} \text{AdjustedCosine}(j, t) \cdot r_{uj}}{\sum_{j \in Q_t(u)} |\text{AdjustedCosine}(j, t)|} \qquad (2.15)$$

The basic idea is to leverage the user's *own* ratings on similar items in the final step of making the prediction. For example, in a movie recommendation system, the item peer group will typically be movies of a similar genre. The ratings history of the *same* user on such movies is a very reliable predictor of the interests of that user.

The previous section discussed a number of variants of the basic approach for user-based collaborative filtering. Because item-based algorithms are very similar to user-based algorithms, similar variants of the similarity function and the prediction function can be designed for item-based methods.

Example of Item-Based Algorithm

In order to illustrate the item-based algorithm, we will use the same example of Table 2.1, which was leveraged to demonstrate the user-based algorithm. The missing ratings of user

3 are predicted with the item-based algorithm. Because the ratings of items 1 and 6 are missing for user 3, the similarity of the columns for items 1 and 6 needs to be computed with respect to the other columns (items).

First, the similarity between items are computed after adjusting for mean-centering. The mean-centered ratings matrix is illustrated in Table 2.2. The corresponding adjusted cosine similarities of each item to 1 and 6, respectively, are indicated in the final two rows of the table. For example, the value of the adjusted cosine between items 1 and 3, denoted by $\text{AdjustedCosine}(1, 3)$, is as follows:

$$\text{AdjustedCosine}(1, 3) = \frac{1.5 * 1.5 + (-1.5) * (-0.5) + (-1) * (-1)}{\sqrt{1.5^2 + (-1.5)^2 + (-1)^2} \cdot \sqrt{1.5^2 + (-0.5)^2 + (-1)^2}} = 0.912$$

Other item-item similarities are computed in an exactly analogous way, and are illustrated in the final two rows of Table 2.2. It is evident that items 2 and 3 are most similar to item 1, whereas items 4 and 5 are most similar to item 6. Therefore, the weighted average of the *raw* ratings of user 3 for items 2 and 3 is used to predict the rating $\hat{r}_{31}$ of item 1, whereas the weighted average of the raw ratings of user 3 for items 4 and 5 is used to predict the rating $\hat{r}_{36}$ of item 6:

$$\hat{r}_{31} = \frac{3 * 0.735 + 3 * 0.912}{0.735 + 0.912} = 3$$
$$\hat{r}_{36} = \frac{1 * 0.829 + 1 * 0.730}{0.829 + 0.730} = 1$$

Thus, the item-based method also suggests that item 1 is more likely to be preferred by user 3 than item 6. However, in this case, because the ratings are predicted using the ratings of user 3 herself, the predicted ratings tend to be much more consistent with the other ratings of this user. As a specific example, it is noteworthy that the predicted rating of item 6 is no longer outside the range of allowed ratings, as in the case of the user-based method. The greater prediction accuracy of the item-based method is its main advantage. In some cases, the item-based method might provide a different set of top-k recommendations, even though the recommended lists will generally be roughly similar.

2.3.3   Efficient Implementation and Computational Complexity

Neighborhood-based methods are always used to determine the best item recommendations for a target user or the best user recommendations for a target item. The aforementioned discussion only shows how to predict the ratings for a particular user-item *combination*, but it does not discuss the actual ranking process. A straightforward approach is to compute all possible rating predictions for the relevant user-item pairs (e.g., all items for a particular user) and then rank them. While this is the basic approach used in current recommender systems, it is important to observe that the prediction process for many user-item combinations reuses many intermediate quantities. Therefore, it is advisable to have an offline phase to store these intermediate computations and then leverage them in the ranking process.

Neighborhood-based methods are always partitioned into an *offline* phase and an *online* phase. In the offline phase, the user-user (or item-item) similarity values and peer groups of the users (or items) are computed. For each user (or item), the relevant peer group is prestored on the basis of this computation. In the online phase, these similarity values and peer groups are leveraged to make predictions with the use of relationships such as Equation 2.4. Let $n' \ll n$ be the maximum number of specified ratings of a user (row), and

$m' \ll m$ be the maximum number of specified ratings of an item (column). Note that n' is the maximum running time for computing the similarity between a pair of users (rows), and m' is the maximum running time for computing the similarity between a pair of items (columns). In the case of user-based methods, the process of determining the peer group of a target user may require $O(m \cdot n')$ time. Therefore, the offline running time for computing the peer groups of all users is given by $O(m^2 \cdot n')$. For item-based methods, the corresponding offline running time is given by $O(n^2 \cdot m')$.

In order to be able to use the approach for varying values of k, one might end up having to store all pairs of nonzero similarities between pairs of users (or items). Therefore, the space requirements of user-based methods are $O(m^2)$, whereas the space requirements of item-based methods are $O(n^2)$. Because the number of users is typically greater than the number of items, the space requirements of user-based methods are generally greater than those of item-based methods.

The online computation of the predicted value according to Equation 2.4 requires $O(k)$ time for both user-based and item-based methods, where k is the size of the user/item neighborhood used for prediction. Furthermore, if this prediction needs to be executed over all items in order to rank them for a target *user*, then the running time is $O(k \cdot n)$ for both user-based and item-based methods. On the other hand, a merchant may occasionally wish to determine the top-r users to be targeted for a specific item. In this case, the prediction needs to be executed over all users in order to rank them for a target *item*, and the running time is $O(k \cdot m)$ for both user-based and item-based methods. It is noteworthy that the primary computational complexity of neighborhood-based methods resides in the offline phase, which needs to be executed occasionally. As a result, neighborhood-based methods tend to be efficient when they are used for online prediction. After all, one can afford to be generous in allocating significantly more computational time to the offline phase.

2.3.4   Comparing User-Based and Item-Based Methods

Item-based methods often provide more relevant recommendations because of the fact that a user's *own* ratings are used to perform the recommendation. In item-based methods, similar *items* are identified to a target item, and the user's own ratings on those items are used to extrapolate the ratings of the target. For example, similar items to a target historical movie might be a set of other historical movies. In such cases, the user's own recommendations for the similar set might be highly indicative of her preference for the target. This is not the case for user-based methods in which the ratings are extrapolated from other users, who might have overlapping but different interests. As a result, item-based methods often exhibit better accuracy.

Although item-based recommendations are often more likely to be accurate, the relative accuracy between item-based and user-based methods also depends on the data set at hand. As you will learn in Chapter 12, item-based methods are also more robust to *shilling attacks* in recommender systems. On the other hand, it is precisely these differences that can lead to greater diversity in the recommendation process for user-based methods over item-based methods. Diversity refers to the fact that the items in the ranked list tend to be somewhat different. If the items are not diverse, then if the user does not like the first item, she might not also like any of the other items in the list. Greater diversity also encourages serendipity, through which somewhat surprising and interesting items are discovered. Item-based methods might sometimes recommend obvious items, or items which are not *novel* from previous user experiences. The notions of novelty, diversity, and serendipity are discussed in detail in Chapter 7. Without sufficient novelty, diversity, and serendipity, users might become bored with very similar recommendations to what they have already watched.

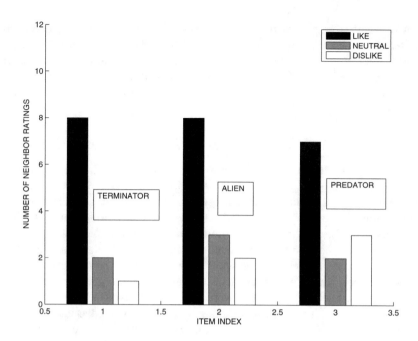

Figure 2.2: Explaining Alice's top recommendations with her neighbor rating histogram

Item-based methods can also provide a concrete reason for the recommendation. For example, Netflix often provides recommendations with statements such as the following:

Because you watched "Secrets of the Wings," [the recommendations are] ⟨List⟩ .

Such explanations can be concretely addressed with item-based methods[2] by using the item neighborhoods. On the other hand, these explanations are harder to address withuser-based methods, because the peer group is simply a set of anonymous users and not directly usable in the recommendation process.

User-based methods provide different types of explanations. For example, consider a scenario where the movies *Terminator, Alien,* and *Predator,* are recommended to Alice. Then, a histogram of her neighbor's ratings for these movies can be shown to her. An example of such a histogram is shown in Figure 2.2. This histogram can be used by Alice to obtain an idea of how much she might like this movie. Nevertheless, the power of this type of explanation is somewhat limited because it does not give Alice an idea of how these movies relate to her *own* tastes or to those of friends she actually knows and trusts. Note that the identity of her neighbors is usually not available to Alice because of privacy concerns.

Finally, item-based methods are more stable with changes to the ratings. This is because of two reasons. First, the number of users is generally much larger than the number of items. In such cases, two users may have a very small number of mutually rated items, but two items are more likely to have a larger number of users who have co-rated them. In the case of user-based methods, the addition of a few ratings can change the similarity values drastically. This is not the case for item-based methods, which are more stable to changes in the values of the ratings. Second, new users are likely to be added more frequently in

---

[2]The precise method used by Netflix is proprietary and therefore not known. However, item-based methods do provide a viable methodology to achieve similar goals.

commercial systems than new items. In such cases, the computation of neighborhood items can be done only occasionally because item neighborhoods are unlikely to change drastically with the addition of new users. On the other hand, the computation of user neighborhoods needs to be performed more frequently with the addition of new users. In this context, incremental maintenance of the recommendation model is more challenging in the case of user-based methods.

2.3.5   Strengths and Weaknesses of Neighborhood-Based Methods

Neighborhood methods have several advantages related to their simplicity and intuitive approach. Because of the simple and intuitive approach of these methods, they are easy to implement and debug. It is often easy to justify why a specific item is recommended, and the interpretability of item-based methods is particularly notable. Such justifications are often not easily available in many of the model-based methods discussed in later chapters. Furthermore, the recommendations are relatively stable with the addition of new items and users. It is also possible to create incremental approximations of these methods.

The main disadvantage of these methods is that the offline phase can sometimes be impractical in large-scale settings. The offline phase of the user-based method requires at least $O(m^2)$ time and space. This might sometimes be too slow or space-intensive with desktop hardware, when m is of the order of tens of millions. Nevertheless, the online phase of neighborhood methods is always efficient. The other main disadvantage of these methods is their limited coverage because of sparsity. For example, if none of John's nearest neighbors have rated *Terminator*, it is not possible to provide a rating prediction of *Terminator* for John. On the other hand, we care only about the top-k items of John in most recommendation settings. If none of John's nearest neighbors have rated *Terminator*, then it might be evidence that this movie is not a good recommendation for John. Sparsity also creates challenges for robust similarity computation when the number of mutually rated items between two users is small.

2.3.6   A Unified View of User-Based and Item-Based Methods

The respective weaknesses of user-based and item-based methods arise out of the fact that the former ignores the similarity between the columns of the ratings matrix, whereas the latter ignores the similarity between the rows while determining the most similar entries. A natural question arises whether we can determine the most similar *entries* to a target entry by unifying the two methods. By doing so, one does not need to ignore the similarity along either rows or columns. Rather, one can *combine* the similarity information between rows and columns.

In order to achieve this goal, it is crucial to understand that the user-based and item-based methods are almost identical (with some minor differences), once the rows have been mean-centered. We can assume without loss of generality that the rows of the ratings matrix are mean-centered because the mean of each row can be added back to each entry after the prediction. It is also noteworthy that if the rows are mean-centered then the Pearson correlation coefficient between rows is identical[3] to the cosine coefficient. Based on this

---

[3]There can be some minor differences depending on how the mean is computed for each row within the Pearson coefficient. If the mean for each row is computed using all the observed entries of that row (rather than only the mutually specified entries), then the Pearson correlation coefficient is identical to the cosine coefficient for row-wise mean-centered matrices.

assumption, the user-based and item-based methods can be described in a unified way to predict the entry r_{uj} in the ratings matrix R:

1. For a target entry (u, j) determine the most similar rows/columns of the ratings matrix with the use of the cosine coefficient between rows/columns. For user-based methods rows are used, whereas for item-based methods, columns are used.

2. Predict the target entry (u, j) using a weighted combination of the ratings in the most similar rows/columns determined in the first step.

Note that the aforementioned description ignores *either* the rows or the columns in each step. One can, of course, propose a generalized description of the aforementioned steps in which the similarity and prediction information along rows and columns are *combined*:

1. For a target entry (u, j) determine the most similar *entries* of the ratings matrix with the use of a combination function of the similarity between rows and columns. For example, one can use the sum of the cosine similarity between rows and between columns to determine the most similar entries in the ratings matrix to (u, j).

2. Predict the target entry (u, j) using a weighted combination of the ratings in the most similar *entries* determined in the first step. The weights are based on the similarities computed in the first step.

We have highlighted the steps, which are different in the generalized method. This approach fuses the similarities along rows and columns with the use of a combination function. One can experiment with the use of various combination functions to obtain the most effective results. Detailed descriptions of such unified methods may be found in [613, 622]. This basic principle is also used in the multidimensional model of context-sensitive recommender systems, in which the similarities along users, items, and other contextual dimensions are unified into a single framework (cf. section 8.5.1 of Chapter 8).

2.4   Clustering and Neighborhood-Based Methods

The main problem with neighborhood-based methods is the complexity of the offline phase, which can be quite significant when the number of users or the number of items is very large. For example, when the number of users m is of the order of a few hundred million, the $O(m^2 \cdot n')$ running time of a user-based method will become impractical even for occasional offline computation. Consider the case where $m = 10^8$ and $n' = 100$. In such a case, $O(m^2 \cdot n') = O(10^{18})$ operations will be required. If we make the conservative assumption that each operation requires an elementary machine cycle, a 10GHz computer will require 10^8 seconds, which is approximately 115.74 days. Clearly, such an approach will not be very practical from a scalability point of view.

   The main idea of clustering-based methods is to replace the offline nearest-neighbor computation phase with an offline clustering phase. Just as the offline nearest-neighbor phase creates a large number of peer groups, which are centered *at each possible target*, the clustering process creates a smaller number of peer groups which are not necessarily centered at each possible target. The process of clustering is much more efficient than the $O(m^2 \cdot n')$ time required for construction of the peer groups of every possible target. Once the clusters have been constructed, the process of predicting ratings is similar to the approach used in Equation 2.4. The main difference is that the top-k closest peers within the same cluster are used to perform the prediction. It is noteworthy that the pairwise similarity computation needs to be performed only within the same cluster and therefore, the approach can be significantly more efficient. This efficiency does result in some loss of accuracy because the set of closest neighbors to each target within a cluster is of lower quality than that over the entire data. Furthermore, the clustering granularity regulates the trade-off between accuracy and efficiency. When the clusters are fine-grained, the efficiency improves, but the accuracy is reduced. In many cases, very large gains in efficiency can be obtained for small reductions in accuracy. When the ratings matrices are very large, this approach provides a very practical alternative at a small cost.

One challenge with the use of this approach is the fact that the ratings matrix is incomplete. Therefore, clustering methods need to be adapted to work with massively incomplete data sets. In this context, k-means methods can be easily adapted to incomplete data. The basic idea of a k-means approach is to work with k central points (or "means"), which serve as the representatives of k different clusters. In k-means methods, the solution to a clustering can be fully represented by the specification of these k representatives. Given a set of k representatives $\overline{Y_1} \ldots \overline{Y_k}$, each data point is assigned to its closest representative with the use of a similarity or distance function. Therefore, the data partitioning can be uniquely defined by the set of representatives. For an $m \times n$ data set, each representative $\overline{Y_i}$ is an n-dimensional data point, which is a central point of the ith cluster. Ideally, we would like the central representative to be the mean of the cluster.

Therefore, the clusters are dependent on the representatives and vice versa. Such an interdependency is achieved with an iterative approach. We start with a set of representatives $\overline{Y_1} \ldots \overline{Y_k}$, which might be randomly chosen points generated in the range of the data space. We iteratively compute the cluster partitions using the representatives, and then recompute the representatives as the centroids of the resulting clusters. While computing the centroids, care must be taken to use only the observed values in each dimension. This two-step iterative approach is executed to convergence. The two-step approach is summarized as follows:

1. Determine the clusters $\mathcal{C}_1 \ldots \mathcal{C}_k$ by assigning each row in the $m \times n$ matrix to its closest representative from $\overline{Y_1} \ldots \overline{Y_k}$. Typically, the Euclidean distance or the Manhattan distance is used for similarity computation.

2. For each $i \in \{1 \ldots k\}$, reset $\overline{Y_i}$ to the centroid of the current set of points in $\mathcal{C}_i$.

The main problem with the use of this approach is that the $m \times n$ ratings matrix is incomplete. Therefore, the computation of the mean and the distance values becomes undefined. However, it is relatively easy to compute the means using only the observed values within a cluster. In some cases, the centroid itself might not be fully specified, when no rating is specified for one or more items in the cluster. The distance values are computed using only the subset of dimensions, which are specified both for the data point and cluster representative. The distance is also divided by the number of dimensions used in the computation. This is done in order to adjust for the fact that different numbers of dimensions are used for computing the distance of a data point to various centroids, when all the centroids are not fully specified. In this context, the Manhattan distance yields better adjustments than the Euclidean distance, and the normalized value can be interpreted more easily as an average distance along each observed value.

The aforementioned approach clusters the rows for user-based collaborative filtering. In item-based methods, it would be necessary to cluster the columns. The approach is exactly similar except that it is applied to the columns rather than the rows. A number of clustering methods for efficient collaborative filtering are discussed in [146, 167, 528, 643,

644, 647]. Some of these methods are user-based methods, whereas others are item-based methods. A number of co-clustering methods [643] can be used to cluster rows and columns simultaneously.

2.5 Dimensionality Reduction and Neighborhood Methods

Dimensionality reduction methods can be used to improve neighborhood-based methods both in terms of quality and in terms of efficiency. In particular, even though pairwise similarities are hard to robustly compute in sparse rating matrices, dimensionality reduction provides a dense low-dimensional representation in terms of latent factors. Therefore, such models are also referred to as *latent factor models*. Even when two users have very few items rated in common, a distance can be computed between their low-dimensional latent vectors. Furthermore, it is more efficient to determine the peer groups with low-dimensional latent vectors. Before discussing the details of dimensionality reduction methods, we make some comments about two distinct ways in which latent factor models are used in recommender systems:

1. A reduced representation of the data can be created in terms of *either* row-wise latent factors or in terms of column-wise latent factors. In other words, the reduced representation will either compress the item dimensionality or the user dimensionality into latent factors. This reduced representation can be used to alleviate the sparsity problem for neighborhood-based models. Depending on which dimension has been compressed into latent factors, the reduced representation can be used for either user-based neighborhood algorithms or item-based neighborhood algorithms.

2. The latent representations of *both* the row space and the column space are determined simultaneously. These latent representations are used to reconstruct the entire ratings matrix in one shot without the use of neighborhood-based methods.

Because the second class of methods is not directly related to neighborhood-based methods, it will not be discussed in this chapter. A detailed discussion of the second class of methods will be provided in Chapter 3. In this chapter, we will focus only on the first class of methods.

For ease of discussion, we will first describe only the user-based collaborative filtering method. In user-based collaborative filtering methods, the basic idea is to transform the $m \times n$ ratings matrix R into a lower-dimensional space by using principal component analysis. The resulting matrix R' is of size $m \times d$, where $d \ll n$. Thus, each of the (sparse) n-dimensional vector of ratings corresponding to a user is transformed into a reduced d-dimensional space. Furthermore, unlike the original rating vector, each of the d dimensions is fully specified. After this d-dimensional representation of each user is determined, the similarity is computed from the target user to each user using the reduced representation. The similarity computations in the reduced representation are more robust because the new low-dimensional vector is fully specified. Furthermore, the similarity computations are more efficient because of the low dimensionality of the latent representation. A simple cosine or dot product on the reduced vectors is sufficient to compute the similarity in this reduced space.

It remains to be described how the low-dimensional representation of each data point is computed. The low-dimensional representation can be computed using either SVD-like methods or PCA-like methods. In the following, we describe an SVD-like method.

Table 2.3: Example of bias in estimating covariances

User Index	Godfather	Gladiator	Nero
1	1	1	1
2	7	7	7
3	3	1	1
4	5	7	7
5	3	1	?
6	5	7	?
7	3	1	?
8	5	7	?
9	3	1	?
10	5	7	?
11	3	1	?
12	5	7	?

The first step is to augment the $m \times n$ incomplete ratings matrix R to fill in the missing entries. The missing entry is estimated to be equal to the mean of the corresponding row in the matrix (i.e., the mean rating of the corresponding user). An alternative approach is to estimate the missing entry as the mean of the corresponding column in the matrix (i.e., the mean rating of the corresponding item). Let the resulting matrix be denoted by R_f. Then, we compute the $n \times n$ similarity matrix between pairs of items, which is given by $S = R_f^T R_f$. This matrix is positive semi-definite. In order to determine the dominant basis vectors of R_f for SVD, we perform the diagonalization of the similarity matrix S as follows:

$$S = P \Delta P^T \tag{2.16}$$

Here, P is an $n \times n$ matrix, whose columns contain the orthonormal eigenvectors of S. Δ is a diagonal matrix containing the non-negative eigenvalues of S along its diagonal. Let P_d be the $n \times d$ matrix containing only the columns of P corresponding to the largest d eigenvectors. Then, the low-dimensional representation of R_f is given by the matrix product $R_f P_d$. Note that the dimensions of the reduced representation $R_f P_d$ are $m \times d$, because R_f is an $m \times n$ matrix and P_d is an $n \times d$ matrix. Therefore, each of the m users is now represented in a d-dimensional space. This representation is then used to determine the peer group of each user. Once the peers have been determined, the rating prediction can be easily performed with Equation 2.4. Such an approach can also be used for item-based collaborative filtering by applying the entire dimensionality reduction method to the transpose of R_f instead of R_f.

The aforementioned methodology can be viewed as a *singular value decomposition (SVD)* of the ratings matrix R_f. A number of other methods [24, 472] use *principal component analysis (PCA)* instead of SVD, but the overall result is very similar. In the PCA method, the covariance matrix of R_f is used instead of the similarity matrix $R_f^T R_f$. For data, which is mean-centered along columns, the two methods are identical. Therefore, one can subtract the mean of each column from its entries, and then apply the aforementioned approach to obtain a transformed representation of the data. This transformed representation is used to determine the peers of each user. Mean-centering has benefits in terms of reducing *bias* (see next section). An alternative approach is to first mean center along each row and then mean-center along each column. SVD can be applied to the transformed representation. This type of approach generally provides the most robust results.

2.5.1 Handling Problems with Bias

It is noteworthy that the matrix R_f is derived from the incomplete matrix R by filling in the unspecified entries with average values along either the rows or the columns. Such an approach is likely to cause considerable *bias*. To understand the nature of this bias, consider the example in Table 2.3 of ratings given by 12 users to the three movies *Godfather*, *Gladiator*, and *Nero*. Let us assume that PCA is used for dimensionality reduction, and therefore the covariance matrix needs to be estimated. Let us assume that missing values are replaced with the averages along the columns.

In this case, the ratings are drawn on a scale from 1 to 7 by a set of 4 users for 3 movies. It is visually evident that the correlations between the ratings of the movies *Gladiator* and *Nero* are extremely high because the ratings are very similar in the four cases in which they are specified. The correlation between *Godfather* and *Gladiator* seems to be less significant. However, many users have not specified their ratings for *Nero*. Because the mean rating of *Nero* is $(1 + 7 + 1 + 7)/4 = 4$, these unspecified ratings are replaced with the mean value of 4. The addition of these new entries significantly reduces the estimated covariance between *Gladiator* and *Nero*. However, the addition of the new entries has no impact on the covariance between *Godfather* and *Gladiator*. After filling in the missing ratings, the pairwise covariances between the three movies can be estimated as follows:

	Godfather	Gladiator	Nero
Godfather	2.55	4.36	2.18
Gladiator	4.36	9.82	3.27
Nero	2.18	3.27	3.27

According to the aforementioned estimation, the covariance between *Godfather* and *Gladiator* is larger than that between *Gladiator* and *Nero*. This does not seem to be correct because the ratings in Table 2.3 for *Gladiator* and *Nero* are identical for the case where both are specified. Therefore, the correlation between *Gladiator* and *Nero* ought to be higher. This error is a result of the bias caused by filling in the unspecified entries with the mean of that column. This kind of bias can be very significant in sparse matrices because most of the entries are unspecified. Therefore, methods need to be designed to reduce the bias caused by using the mean ratings in place of the unspecified entries. In the following, we explore two possible solutions to this problem.

2.5.1.1 Maximum Likelihood Estimation

The conceptual reconstruction method [24, 472] proposes the use of probabilistic techniques, such as the EM-algorithm, in order to estimate the covariance matrix. A generative model is assumed for the data and the specified entries are viewed as the outcomes of the generative model. The covariance matrix can be estimated as part of the process of estimating the parameters of this generative model. In the following, we provide a simplification of this approach. In this simplified approach, the maximum likelihood estimate of the covariance matrix is computed. The maximum likelihood estimate of the covariance between each pair of items is estimated as the covariance between only the specified entries. In other words, only the users that have specified ratings for a particular pair of items are used to estimate the covariance. In the event that there are no users in common between a pair of items, the covariance is estimated to be 0. By using this approach, the following covariance matrix is estimated for the data in Table 2.3.

	Godfather	Gladiator	Nero
Godfather	2.55	4.36	8
Gladiator	4.36	9.82	12
Nero	8	12	12

In this case, it becomes immediately evident that the covariance between *Gladiator* and *Nero* is almost three times that between *Godfather* and *Gladiator*. Furthermore, the movie *Nero* has more than three times as much variance than was originally estimated and has the largest variance in ratings among all movies. While the pairwise covariance between *Godfather* and *Gladiator* was the largest compared to all other pairwise covariances using the mean-filling technique, this same pair now shows the least of all pairwise covariances. This example suggests that the bias corrections can be very significant in some situations. The greater the proportion of unspecified entries in the matrix, the greater the bias of the mean-filling technique. Therefore, the modified technique of leveraging only the specified entries is used for computing the covariance matrix. While such a technique is not always effective, it is superior to the mean-filling technique. The reduced $n \times d$ basis matrix P_d is computed by selecting the top-d eigenvectors of the resulting covariance matrix.

In order to further reduce the bias in representation, the incomplete matrix R can be directly projected on the reduced matrix P_d, rather than projecting the filled matrix R_f on P_d. The idea is to compute the contribution of each observed rating to the projection on each latent vector of P_d, and then average the contribution over the number of such ratings. This averaged contribution is computed as follows. Let $\overline{e_i}$ be the ith column (eigenvector) of P_d, for which the jth entry is e_{ji}. Let r_{uj} be the observed rating of user u for item j in matrix R. Then, the contribution of user u to the projection on latent vector $\overline{e_i}$ is given by $r_{uj}e_{ji}$. Then, if the set I_u represents the indices of the specified item ratings of user u, the averaged contribution a_{ui} of user u on the ith latent vector is as follows:

$$a_{ui} = \frac{\sum_{j \in I_u} r_{uj}e_{ji}}{|I_u|} \tag{2.17}$$

This type of averaged normalization is particularly useful in cases where the different users have specified different numbers of ratings. The resulting $m \times d$ matrix $A = [a_{ui}]_{m \times d}$ is used as the reduced representation of the underlying ratings matrix. This reduced matrix is used to compute the neighborhood of the target user efficiently for user-based collaborative filtering. It is also possible to apply the approach to the transpose of the matrix R and reduce the dimensionality along the user dimension, rather than the item dimension. Such an approach is useful for computing the neighborhood of a target item in item-based collaborative filtering. This approach of using the reduced representation for missing value imputation is discussed in [24, 472].

2.5.1.2   Direct Matrix Factorization of Incomplete Data

Although the aforementioned methodology can correct for the bias in covariance estimation to some extent, it is not completely effective when the sparsity level of the ratings is high. This is because the covariance matrix estimation requires a sufficient number of observed ratings for each pair of items for robust estimation. When the matrix is sparse, the covariance estimates will be statistically unreliable.

A more direct approach is to use matrix factorization methods. Methods such as singular value decomposition are essentially matrix factorization methods. For a moment, assume that the $m \times n$ ratings matrix R is fully specified. It is a well-known fact of linear algebra [568] that any (fully specified) matrix R can be factorized as follows:

$$R = Q\Sigma P^T \tag{2.18}$$

Here, Q is an $m \times m$ matrix with columns containing the m orthonormal eigenvectors of RR^T. The matrix P is an $n \times n$ matrix with columns containing the n orthonormal eigenvectors of $R^T R$. Σ is an $m \times n$ diagonal matrix in which only diagonal entries[4] are nonzero and they contain the square-root of the nonzero eigenvalues of $R^T R$ (or equivalently, RR^T). It is noteworthy that the eigenvectors of $R^T R$ and RR^T are not the same and will have different dimensionality when $m \neq n$. However, they will always have the same number of (nonzero) eigenvalues, which are identical in value. The values on the diagonal of Σ are also referred to as *singular values*.

Furthermore, one can *approximately* factorize the matrix by using *truncated* SVD, where only the eigenvectors corresponding to the $d \leq \min\{m, n\}$ largest singular values are used. Truncated SVD is computed as follows:

$$R \approx Q_d \Sigma_d P_d^T \tag{2.19}$$

Here, Q_d, Σ_d, and P_d are $m \times d$, $d \times d$, and $n \times d$ matrices, respectively. The matrices Q_d and P_d, respectively, contain the d largest eigenvectors of RR^T and $R^T R$, whereas the matrix Σ_d contains the square-roots of the d largest eigenvalues of either matrix along its diagonal. It is noteworthy that the matrix P_d contains the top eigenvectors of $R^T R$, which is the *reduced* basis representation required for dimensionality reduction. Furthermore, the matrix $Q_d \Sigma_d$ contains the transformed and reduced $m \times d$ representation of the original ratings matrix in the basis corresponding to P_d. It can be shown that such an approximate factorization has the least mean-squared error of the approximated entries as compared to any other rank-d factorization. Therefore, if we can approximately factorize the ratings matrix R in the form corresponding to Equation 2.19, it provides us with the reduced basis as well as the representation of the ratings in the reduced basis. The main problem of using such an approach is that the ratings matrix is not fully specified. As a result, this factorization is undefined. Nevertheless, it is possible to recast the formulation as an optimization problem, in which the squared error of factorization is optimized *only over the observed entries* of the ratings matrix. It is also possible to explicitly solve this modified formulation using nonlinear optimization techniques. This results in a robust and unbiased lower dimensional representation. Furthermore, such an approach can be used to directly estimate the ratings matrix by using Equation 2.19, once the reduced factor matrices have been determined. In other words, such methods have a direct utility beyond neighborhood-based methods. More details of these latent factor models and nonlinear optimization techniques will be discussed in section 3.6 of Chapter 3. The reader should consult this section to learn how the reduced representation may be computed by using modified optimization formulations.

2.6 A Regression Modeling View of Neighborhood Methods

An important observation about both user-based and item-based methods is that they predict ratings as *linear functions* of either the ratings of the same item by neighboring users, or of the same user on neighboring items. In order to understand this point, we

---

[4]Diagonal matrices are usually square. Although this matrix is not square, only entries with equal indices are nonzero. This is a generalized definition of a diagonal matrix.

replicate the prediction function of user-based neighborhood methods (cf. Equation 2.4) below:

$$\hat{r}_{uj} = \mu_u + \frac{\sum_{v \in P_u(j)} \text{Sim}(u,v) \cdot (r_{vj} - \mu_v)}{\sum_{v \in P_u(j)} |\text{Sim}(u,v)|} \qquad (2.20)$$

Note that the predicted rating is a *weighted* linear combination of other ratings of the same item. The linear combination has been restricted only to the ratings of item j belonging to users with sufficiently similar tastes to target user u. This restriction is enabled with the use of the peer rating set $P_u(j)$. Recall from the discussion earlier in this chapter that $P_u(j)$ is the set of k nearest users to target user u, who have also rated item j. Note that if we allowed the set $P_u(j)$ to contain all ratings of item j (and not just specific peer users), then the prediction function becomes similar[5] to that of linear regression [22]. In linear regression, the ratings are also predicted as weighted combinations of other ratings, and the weights (coefficients) are determined with the use of an optimization model. In the neighborhood-based approach, the coefficients of the linear function are chosen in a heuristic way with the user-user similarities, rather than with the use of an optimization model.

A similar observation applies to the case of item-based neighborhood methods, where the prediction function (cf. Equation 2.15) is as follows:

$$\hat{r}_{ut} = \frac{\sum_{j \in Q_t(u)} \text{AdjustedCosine}(j,t) \cdot r_{uj}}{\sum_{j \in Q_t(u)} |\text{AdjustedCosine}(j,t)|} \qquad (2.21)$$

The set $Q_t(u)$ represents the set of the k closest items to target item t that have also been rated by user u. In this case, the rating of a user u for a target item t is expressed as a linear combination of her *own* ratings. As in the case of user-based methods, the coefficients of the linear combination are heuristically defined with similarity values. Therefore, a user-based model expresses a predicted rating as a linear combination of ratings in the same *column*, whereas an item-based model expresses a predicted rating as a linear combination of ratings in the same *row*. From this point of view, *neighborhood-based models are heuristic variants of linear regression models*, in which the regression coefficients are heuristically set to similarity values for related (neighboring) items/users and to 0 for unrelated items/users.

It is noteworthy that the use of similarity values as combination weights is rather heuristic and arbitrary. Furthermore, the coefficients do not account for interdependencies among items. For example, if a user has rated certain sets of correlated items in a very similar way, then the coefficients associated with these items will be interdependent as well. The use of similarities as heuristic weights does not account for such interdependencies.

A question arises as to whether one can do better by *learning* the weights with the use of an optimization formulation. It turns out that one can derive analogous regression-based models to the user-based and item-based models. Several different optimization formulations have been proposed in the literature, which can leverage user-based models, item-based models, or a combination of the two. These models can be viewed as theoretical generalizations of the heuristic nearest neighbor model. The advantage of such models is that they are mathematically better founded in the context of a crisp optimization formulation, and the weights for combining the ratings can be better justified because of their optimality from a *modeling* perspective. In the following, we discuss an optimization-based neighborhood model, which is a simplification of the work in [309]. This also sets the stage for combining the power of this model with other optimization models, such as matrix factorization, in section 3.7 of Chapter 3.

---

[5]A discussion of linear regression is provided in section 4.4.5 of Chapter 4, but in the context of content-based systems.

2.6.1 User-Based Nearest Neighbor Regression

Consider the user-based prediction of Equation 2.20. One can replace the (normalized) similarity coefficient with the unknown parameter w_{vu}^{user} to *model* the predicted rating $\hat{r}_{uj}$ of target user u for item j as follows:

$$\hat{r}_{uj} = \mu_u + \sum_{v \in P_u(j)} w_{vu}^{user} \cdot (r_{vj} - \mu_v) \tag{2.22}$$

As in the case of neighborhood models, one can use the Pearson correlation coefficient to define $P_u(j)$. There is, however, a subtle but important difference in terms of how $P_u(j)$ is defined in this case. In neighborhood-based models, $P_u(j)$ is the set of k closest users to target user u, who have specified ratings for item j. Therefore, the size of $P_u(j)$ is often exactly k, when at least k users have rated item j. In the case of regression methods, the set $P_u(j)$ is defined by first determining the k closest peers for each user, and then retaining only those for which ratings are observed. Therefore, the size of set $P_u(j)$ is often *significantly* less than k. Note that the parameter k needs to be set to much larger values in the regression framework as compared to that in neighborhood models because of its different interpretation.

Intuitively, the unknown coefficient w_{vu}^{user} controls the portion of the prediction of ratings given by user u, which comes from her similarity to user v, because this portion is given by $w_{vu}^{user} \cdot (r_{vj} - \mu_v)$. It is possible for w_{vu}^{user} to be different from w_{uv}^{user}. It is also noteworthy that w_{vu}^{user} is only defined for the k different values of v (user indices) that are closest to user u on the basis of the Pearson coefficient. The other values of w_{vu}^{user} are not needed by the prediction function of Equation 2.22, and they therefore do not need to be learned. This has the beneficial effect of reducing the number of regression coefficients.

One can use the aggregate squared difference between the predicted ratings $\hat{r}_{uj}$ (according to Equation 2.22) and the observed ratings r_{uj} to create an objective function that estimates the quality of a particular set of coefficients. Therefore, one can use the observed ratings in the matrix to set up a least-squares optimization problem over the unknown values of w_{vu}^{user} in order to minimize the overall error. The idea is to predict each (observed) rating of user u with her nearest k users in a formal regression model, and then measure the error of the prediction. The squared errors can be added over all items rated by user u to create a least-squares formulation. Therefore, the optimization problem is set up for each target user u. Let I_u be the set of items that have been rated by the target user u. The least-squares objective function for the uth user can be stated as the sum of the squares of the errors in predicting each item in I_u with the k nearest neighbors of the user in a formal regression model:

$$\text{Minimize } J_u = \sum_{j \in I_u} (r_{uj} - \hat{r}_{uj})^2$$

$$= \sum_{j \in I_u} \left(r_{uj} - \left[\mu_u + \sum_{v \in P_u(j)} w_{vu}^{user} \cdot (r_{vj} - \mu_v) \right] \right)^2$$

The second relationship is obtained by substituting the expression in Equation 2.22 for $\hat{r}_{uj}$. Note that this optimization problem is formulated separately for each target user u. However, one can add up the objective function values J_u over different target users $u \in \{1 \ldots m\}$ with no difference to the optimal solution. This is because the various values of J_u are expressed in terms of mutually disjoint sets of optimization variables w_{vu}^{user}. Therefore, the consolidated optimization problem is expressed as follows:

$$\text{Minimize} \sum_{u=1}^{m} J_u = \sum_{u=1}^{m} \sum_{j \in I_u} \left(r_{uj} - \left[\mu_u + \sum_{v \in P_u(j)} w_{vu}^{user} \cdot (r_{vj} - \mu_v) \right] \right)^2 \tag{2.23}$$

One can solve each of the smaller optimization problems (i.e., objective function J_u) in their decomposed form more efficiently without affecting the overall solution. However, the consolidated formulation has the advantage that it can be combined with other optimization models such as matrix factorization methods (cf. section 3.7 of Chapter 3) in which such a decomposition is not possible. Nevertheless, if linear regression is to be used on a standalone basis, it makes sense to solve these problems in their decomposed form.

Both the consolidated and decomposed versions of the optimization models are least-squares optimization problems. These methods can be solved with the use of any off-the-shelf optimization solver. Refer to section 4.4.5 of Chapter 4 for a discussion of closed form solutions to linear regression problems. A desirable property of most of these solvers is that they usually have *regularization* built in them, and they can therefore avoid overfitting to some extent. The basic idea in regularization is to reduce model complexity by adding the term $\lambda \sum_{j \in I_u} \sum_{v \in P_u(j)} (w_{vu}^{user})^2$ to each (decomposed) objective function J_u, where $\lambda > 0$ is a user-defined parameter regulating the weight of the regularization term. The term $\lambda \sum_{j \in I_u} \sum_{v \in P_u(j)} (w_{vu}^{user})^2$ penalizes large coefficients, and it therefore shrinks the absolute values of the coefficients. Smaller coefficients result in simpler models and reduce overfitting. However, as discussed below, it is sometimes not sufficient to use regularization alone to reduce overfitting.

2.6.1.1 Sparsity and Bias Issues

One problem with this regression approach is that the size of the $P_u(j)$ can be vastly different for the same user u and varying item indices (denoted by j). This is because of the extraordinary level of sparsity inherent in ratings matrices. As a result, the regression coefficients become heavily dependent on the *number* of peer users that have rated a particular item j along with user u. For example, consider a scenario where the target user u has rated both *Gladiator* and *Nero*. Out of the k nearest neighbors of the target u, only one user might rate the movie *Gladiator*, whereas all k might have rated *Nero*. As a result, the regression coefficient w_{vu}^{user} of the peer user v who rated *Gladiator* will be heavily influenced by the fact that she is the only user who has rated *Gladiator*. This will result in overfitting because this (statistically unreliable) regression coefficient might add noise to the rating predictions of other movies.

The basic idea is to change the prediction function and assume that the regression for item j predicts only a fraction $\frac{|P_u(j)|}{k}$ of the rating of target user u for item j. The implicit assumption is that the regression coefficients are based on *all* the peers of the target user, and one must interpolate incomplete information as a fraction. Therefore, this approach changes the interpretation of the regression coefficients. In this case, the prediction function of Equation 2.22 is modified as follows:

$$\hat{r}_{uj} \cdot \frac{|P_u(j)|}{k} = \mu_u + \sum_{v \in P_u(j)} w_{vu}^{user} \cdot (r_{vj} - \mu_v) \tag{2.24}$$

A number of other heuristic adjustments are sometimes used. For example, along the lines of the ideas in [312], one can use a heuristic adjustment factor of $\sqrt{|P_u(j)|/k}$. This factor can often be simplified to $\sqrt{|P_u(j)|}$ because constant factors are absorbed by the optimization variables. A related enhancement is that the *constant* offset μ_v is replaced with a bias *variable* b_u, which is learned in the optimization process. The corresponding prediction model, including heuristic adjustment factors, is as follows:

$$\hat{r}_{uj} = b_u^{user} + \frac{\sum_{v \in P_u(j)} w_{vu}^{user} \cdot (r_{vj} - b_v^{user})}{\sqrt{|P_u(j)|}} \tag{2.25}$$

Note that this model is no longer linear because of the multiplicative term $w_{vu}^{user} \cdot b_v^{user}$ between two optimization variables. Nevertheless, it is relatively easy to use the same least-squares formulation, as in the previous case. In addition to user biases, one can also incorporate *item* biases. In such a case, the model becomes the following:

$$\hat{r}_{uj} = b_u^{user} + b_j^{item} + \frac{\sum_{v \in P_u(j)} w_{vu}^{user} \cdot (r_{vj} - b_v^{user} - b_j^{item})}{\sqrt{|P_u(j)|}} \tag{2.26}$$

Furthermore, it is recommended to center the entire ratings matrix around its global mean by subtracting the mean of all the observed entries from it. The global mean needs to be added back to the predictions. The main problem with this model is computational. One must pre-compute and store all user-user relations, which is computationally expensive and requires $O(m^2)$ space over m users. This problem is similar to that encountered in traditional neighborhood-based models. Such models are suitable in settings in which the item space changes rapidly, but the users are relatively stable over time [312]. An example is the case of news recommender systems.

2.6.2 Item-Based Nearest Neighbor Regression

The item-based approach is similar to the user-based approach, except that the regression learns and leverages item-item correlations rather than user-user correlations. Consider the item-based prediction of Equation 2.21. One can replace the (normalized) similarity coefficient AdjustedCosine(j, t) with the unknown parameter w_{jt}^{item} to model the rating prediction of user u for target item t:

$$\hat{r}_{ut} = \sum_{j \in Q_t(u)} w_{jt}^{item} \cdot r_{uj} \tag{2.27}$$

The nearest items in $Q_t(u)$ can be determined using the adjusted cosine, as in item-based neighborhood methods. The set $Q_t(u)$ represents the subset of the k nearest neighbors of the target item t, for which user u has provided ratings. This way of defining $Q_t(u)$ is subtly different from that of traditional neighborhood-based methods, because the size of set $Q_t(u)$ might be significantly less than k. In traditional neighborhood methods, one determines the closest k items to target item t, for which the user u has specified ratings, and therefore the size of the neighborhood set is often exactly k. This change is required to be able to effectively implement the regression-based method.

Intuitively, the unknown coefficient w_{jt}^{item} controls the portion of the rating of item t, which comes from its similarity to item j, because this portion is given by $w_{jt}^{item} \cdot r_{uj}$. The prediction error of Equation 2.27 should be minimized to ensure the most robust predictive model. One can use the known ratings in the matrix to set up a least-squares optimization problem over the unknown values of w_{jt}^{item} in order to minimize the overall error. The idea is to predict each (observed) rating of target item t with its nearest k items and then, create an expression for the least-squares error. The optimization problem is set up for each target item t. Let U_t be the set of users who have rated the target item t. The least-squares objective function for the tth item can be stated as the sum of the squares of the errors in predicting each specified rating in U_t:

$$\text{Minimize } J_t = \sum_{u \in U_t} (r_{ut} - \hat{r}_{ut})^2$$

$$= \sum_{u \in U_t} \left(r_{ut} - \sum_{j \in Q_t(u)} w_{jt}^{item} \cdot r_{uj} \right)^2$$

Note that this optimization problem is formulated separately for each target item t. However, one can add up the terms over various values of the target item t with no difference to the optimization solution, because the unknown coefficients w_{jt}^{item} in the various objective functions are non-overlapping over different values of the target item $t \in \{1 \ldots n\}$. Therefore, we have the following consolidated formulation:

$$\text{Minimize } \sum_{t=1}^{n} \sum_{u \in U_t} \left(r_{ut} - \sum_{j \in Q_t(u)} w_{jt}^{item} \cdot r_{uj} \right)^2 \tag{2.28}$$

This is a least-squares regression problem and it can be solved with the use of any off-the-shelf solver. Furthermore, one can also solve each of the smaller optimization problems (i.e., objective function J_t) in its decomposed form more efficiently without affecting the overall solution. However, the consolidated formulation has the advantage that it can be combined with other optimization models, such as matrix factorization methods (cf. section 3.7 of Chapter 3). As in the case of user-based methods, significant challenges are associated with the problem of overfitting. One can add the regularization term $\lambda \sum_{u \in U_t} \sum_{j \in Q_t(u)} (w_{jt}^{item})^2$ to the objective function J_t.

As discussed in section 2.6.1.1 for the case of the user-based model, one can incorporate adjustment factors and bias variables to improve performance. For example, the user-based prediction model of Equation 2.26 takes on the following form in the item-wise model:

$$\hat{r}_{ut} = b_u^{user} + b_t^{item} + \frac{\sum_{j \in Q_t(u)} w_{jt}^{item} \cdot (r_{uj} - b_u^{user} - b_j^{item})}{\sqrt{|Q_t(u)|}} \tag{2.29}$$

Furthermore, it is assumed that the ratings are centered around the global mean of the entire ratings matrix. Therefore, the global mean is subtracted from each of the ratings before building the model. All predictions are performed on the centered ratings, and then the global mean is added back to each prediction. In some variations of the model, the bias terms $b_u^{user} + b_j^{item}$ within brackets are replaced with a consolidated *constant* term B_{uj}. This constant term is derived using a non-personalized approach described in section 3.7.1 of Chapter 3. The resulting prediction model is as follows:

$$\hat{r}_{ut} = b_u^{user} + b_t^{item} + \frac{\sum_{j \in Q_t(u)} w_{jt}^{item} \cdot (r_{uj} - B_{uj})}{\sqrt{|Q_t(u)|}} \tag{2.30}$$

A least-squares optimization model is formulated, and a gradient descent approach is used to solve for the optimization parameters. This is precisely the model used in [309]. The resulting gradient-descent steps are discussed in section 3.7.2 of Chapter 3. The user-user model is known to perform slightly better than the item-item model [312]. However, the item-based model is far more computationally and space-efficient in settings where the number of items is much smaller than the number of users.

2.6.3 Combining User-Based and Item-Based Methods

It is natural to combine the user and item-based models in a unified regression frame-work [312]. Therefore, a rating is predicted based on its relationship with similar users as well as similar items. This is achieved by combining the ideas in Equations 2.26 and 2.30 as follows:

$$\hat{r}_{uj} = b_u^{user} + b_j^{item} + \frac{\sum_{v \in P_u(j)} w_{vu}^{user} \cdot (r_{vj} - B_{vj})}{\sqrt{|P_u(j)|}} + \frac{\sum_{j \in Q_t(u)} w_{jt}^{item} \cdot (r_{uj} - B_{uj})}{\sqrt{|Q_t(u)|}} \quad (2.31)$$

As in previous cases, it is assumed that the ratings matrix is centered around its global mean. A similar least-squares optimization formulation can be used in which the squared error over all the observed entries is minimized. In this case, it is no longer possible to decompose the optimization problem into independent subproblems. Therefore, a single least-squares optimization model is constructed over all the observed entries in the ratings matrix. As in the previous cases, the gradient-descent approach can be used. It was reported in [312] that the fusion of the user-based and item-based models generally performs better than the individual models.

2.6.4 Joint Interpolation with Similarity Weighting

The method in [72] uses a different idea to set up the joint neighborhood-based model. The basic idea is to predict each rating of target user u with the user-based model of Equation 2.22. Then, instead of comparing it with the observed value of the *same* item, we compare it with the observed ratings of *other* items of that user.

Let S be the set of all pairs of user-item combinations in the ratings matrix, which have been observed:

$$S = \{(u, t) : r_{ut} \text{ is observed}\} \quad (2.32)$$

We set up an objective function which is penalized when the predicted rating $\hat{r}_{uj}$ of an item j is far away from the observed rating given to a similar item s by the same target user u. In other words, the objective function for target user u is defined as follows:

$$\text{Minimize} \sum_{s:(u,s) \in S} \sum_{j:j \neq s} AdjustedCosine(j,s) \cdot (r_{us} - \hat{r}_{uj})^2$$

$$= \sum_{s:(u,s) \in S} \sum_{j:j \neq s} AdjustedCosine(j,s) \cdot \left(r_{us} - \left[\mu_u + \sum_{v \in P_u(j)} w_{vu}^{user} \cdot (r_{vj} - \mu_v) \right] \right)^2$$

Regularization can be added to the objective function to reduce overfitting. Here, $P_u(j)$ is defined as the k closest users to target user u, who have also rated item j. Therefore, the conventional definition of $P_u(j)$ as used in neighborhood-based models is leveraged in this case.

By using the adjusted cosine as a multiplicative factor of each individual term in the objective function, the approach forces the target user's ratings of similar items to be more similar as well. It is noteworthy that both user and item similarities are used in this approach, but in different ways:

1. The item-item similarities are used as multiplicative factors of the terms in the objective function to force predicted ratings to be more similar to observed ratings of similar items.

2. The user-user similarities are used for predicting the ratings by restricting the regression coefficients to the relevant peer group $P_u(j)$ of the target user u.

Although it is also possible, in principle, to switch the roles of users and items to set up a different model, it is stated in [72] that the resulting model is not as effective as the one discussed above. This model can be solved with any off-the-shelf least-squares solver. A number of methods are also discussed in [72] for handling sparsity.

2.6.5 Sparse Linear Models (SLIM)

An interesting method, based on the item-item regression in section 2.6.2, is proposed in [455]. This family of models is referred to as *sparse linear models* because they encourage sparsity in the regression coefficients with the use of regularization methods. Unlike the methods in [72, 309], these methods work with non-negative rating values. Therefore, unlike the techniques in the previous sections, it will not be assumed that the ratings matrix is mean-centered. This is because mean-centering will automatically create negative ratings, corresponding to dislikes. However, the approach is designed to work with non-negative ratings, in which there is no mechanism to specify dislikes. From a practical point of view, the approach is most appropriate[6] for implicit feedback matrices (e.g., click-through data or sales data), where only positive preferences are expressed through user actions. Furthermore, as is common in implicit feedback settings, missing values are treated as 0s for the purposes of training in the optimization formulation. However, the optimization model might eventually predict some of these values to be highly positive, and such user-item combinations are excellent candidates for recommendation. Therefore, the approach ranks items on the basis of prediction errors on the training entries that have been set to 0.

Unlike the technique in section 2.6.2, these methods do not restrict the regression coefficients to only the neighborhood of the target item t. Then, the prediction function in *SLIM* is expressed as follows:

$$\hat{r}_{ut} = \sum_{j=1}^{n} w_{jt}^{item} \cdot r_{uj} \quad \forall u \in \{1 \ldots m\}, \ \forall t \in \{1 \ldots n\} \tag{2.33}$$

Note the relationship with Equation 2.27 in which only the neighborhood of the target item is used to construct the regression. It is important to exclude the target item itself on the right-hand side to prevent overfitting. This can be achieved by requiring the constraint that $w_{tt}^{item} = 0$. Let $\hat{R} = [\hat{r}_{uj}]$ represent the predicted ratings matrix and let $W^{item} = [w_{jt}^{item}]$ represent the item-item regression matrix. Therefore, if we assume that the diagonal elements of W^{item} are constrained to be 0, then we can stack up the instantiations of Equation 2.33 over different users and target items to create the following matrix-based prediction function:

$$\hat{R} = RW^{item}$$

$$\text{Diagonal}(W^{item}) = 0$$

Therefore, the main goal is to minimize the Frobenius norm $||R - RW^{item}||^2$ along with some regularization terms. This objective function is disjoint over different columns of W (i.e., target items in regression). Therefore, one can solve each optimization problem (for

---

[6]The approach can be adapted to arbitrary rating matrices. However, the main advantages of the approach are realized for non-negative ratings matrices.

a given value of the target item t) independently, while setting w_{tt}^{item} to 0. In order to create a more interpretable sum-of-parts regression, the weight vectors are constrained to be non-negative. Therefore, the objective function for target item t may be expressed as follows:

$$\text{Minimize } J_t^s = \sum_{u=1}^{m} (r_{ut} - \hat{r}_{ut})^2 + \lambda \cdot \sum_{j=1}^{n} (w_{jt}^{item})^2 + \lambda_1 \cdot \sum_{j=1}^{n} |w_{jt}^{item}|$$

$$= \sum_{u=1}^{m} \left(r_{ut} - \sum_{j=1}^{n} w_{jt}^{item} \cdot r_{uj} \right)^2 + \lambda \cdot \sum_{j=1}^{n} (w_{jt}^{item})^2 + \lambda_1 \cdot \sum_{j=1}^{n} |w_{jt}^{item}|$$

subject to:

$$w_{jt}^{item} \geq 0 \ \ \forall j \in \{1 \ldots n\}$$

$$w_{tt}^{item} = 0$$

The last two terms in the objective function correspond to the *elastic-net regularizer*, which combines L_1- and L_2-regularization. It can be shown [242] that the L_1-regularization component leads to sparse solutions for the weights w_{jt}, which means that most of the coefficients w_{jt} have zero values. The sparsity ensures that each predicted rating can be expressed as a more interpretable linear combination of the ratings of a small number of other related items. Furthermore, since the weights are non-negative, the corresponding items are positively related in a highly interpretable way in terms of the specific level of impact of each rating in the regression. The optimization problem is solved using the coordinate descent method, although any off-the-shelf solver can be used in principle. A number of faster techniques are discussed in [347]. The technique can also be hybridized [456] with side-information (cf. section 6.8.1 of Chapter 6).

It is evident that this model is closely related to the neighborhood-based regression models discussed in the previous sections. The main differences of the *SLIM* model from the linear regression model in [309] are as follows:

1. The method in [309] restricts the nonzero coefficients for each target to at most the k most similar items. The *SLIM* method can use as many as $|U_t|$ nonzero coefficients. For example, if an item is rated by all users, then all coefficients will be used. However, the value of w_{tt}^{item} is set to 0 to avoid overfitting. Furthermore, the *SLIM* method forces sparsity by using the elastic-net regularizer, whereas the method in [309] preselects the weights on the basis of explicit neighborhood computation. In other words, the work in [309] uses a heuristic approach for feature selection, whereas the *SLIM* approach uses a learning (regularization) approach for feature selection.

2. The *SLIM* method is primarily designed for implicit feedback data sets (e.g., buying an item or customer clicks), rather than explicit ratings. In such cases, ratings are typically unary, in which customer actions are indications of positive preference, but the act of not buying or clicking on an item does not necessarily indicate a negative preference. The approach can also be used for cases in which the "ratings" are arbitrary values indicating only positive preferences (e.g., amount of product bought). Note that such scenarios are generally conducive to regression methods that impose non-negativity in the coefficients of the model. As you will learn in Chapter 3, this observation is also true for other models, such as matrix factorization. For example, non-negative matrix factorization is primarily useful for implicit feedback data sets, but it is not quite as useful for arbitrary ratings. This is, in part, because the non-negative, sum-of-parts decomposition loses its interpretability when a rating indicates either a like or a dislike. For example, two "dislike" ratings do not add up to a "like" rating.

3. The regression coefficients in [309] can be either positive or negative. On the other hand, the coefficients in *SLIM* are constrained to be non-negative. This is because the *SLIM* method is primarily designed for the implicit feedback setting. Non-negativity is often more intuitive in these settings and the results are more interpretable. In fact, in some cases, imposing non-negativity might improve[7] the accuracy. However, some limited experimental results have been presented [347], which suggest that removing non-negativity constraints provides superior performance.

4. Although the *SLIM* method also proposes a prediction model for the ratings (according to Equation 2.33), the final *use* of the predicted values is for *ranking* the items in order of the predicted value. Note that the approach is generally used for data sets with unary ratings and therefore, it makes sense to use the predicted values to rank the items, rather than predict ratings. An alternative way of interpreting the predicted values is that each of them can be viewed as the *error* of replacing a non-negative rating with 0 in the ratings matrix. The larger the error is, the greater the predicted value of the rating will be. Therefore, the items can be ranked in the order of the predicted value.

5. Unlike the work in [309], the *SLIM* method does not explicitly adjust for the varying number of specified ratings with heuristic adjustment factors. For example, the right-hand side of Equation 2.29 uses an adjustment factor of $\sqrt{|Q_t(u)|}$ in the denominator. On the other hand, no such adjustment factor is used in the *SLIM* method. The adjustment issue is less pressing for the case of unary data sets, in which the presence of an item is usually the only information available. In such cases, replacing missing values with 0s is a common practice, and the bias of doing so is much lower than in the case where ratings indicate varying levels of likes or dislikes.

Therefore, the models share a number of conceptual similarities, although there are some differences at the detailed level.

2.7 Graph Models for Neighborhood-Based Methods

The sparsity of observed ratings causes a major problem in the computation of similarity in neighborhood-based methods. A number of graph models are used in order to define similarity in neighborhood-based methods, with the use of either structural transitivity or ranking techniques. Graphs are a powerful abstraction that enable many algorithmic tools from the network domain. The graphs provide a structural representation of the relationships among various users and/or items. The graphs can be constructed on the users, on the items, or on both. These different types of graphs result in a wide variety of algorithms, which use

---

[7] It is noteworthy that imposing an additional constraint, such as non-negativity, always reduces the quality of the optimal solution on the *observed* entries. On the other hand, imposing constraints increases the model bias and reduces model variance, which might reduce overfitting on the *unobserved* entries. In fact, when two closely related models have contradicting relative performances on the observed and unobserved entries, respectively, it is almost always a result of differential levels of overfitting in the two cases. You will learn more about the bias-variance trade-off in Chapter 6. In general, it is more reliable to predict item ratings with positive item-item relationships rather than negative relationships. The non-negativity constraint is based on this observation. The incorporation of model biases in the form of such natural constraints is particularly useful for smaller data sets.

either random-walk or shortest-path methods for recommendation. In the following, we will describe the algorithms used for performing recommendations with various types of graph representations of ratings matrices.

2.7.1   User-Item Graphs

It is possible to use structural measures on the *user-item graph*, rather than the Pearson correlation coefficient, for defining neighborhoods. Such an approach is more effective for sparse ratings matrices because one can use structural transitivity of edges for the recommendation process.

The user-item graph is defined as an undirected and bipartite graph $G = (N_u \cup N_i, A)$, where N_u is the set of nodes representing users, and N_i is the set of nodes representing items. All edges in the graph exist only between users and items. An undirected edge exists in A between a user i and an item j, if and only if user i has rated item j. Therefore, the number of edges is equal to the number of observed entries in the utility matrix. For example, the user-item graph for the ratings matrix of Figure 2.3(a) is illustrated in Figure 2.3(b). The main advantage of graph-based methods is that two users do not need to have rated many of the same items to be considered neighbors as long as many short paths exist between the two users. Therefore, this definition allows the construction of neighborhoods with the notion of *indirect* connectivity between nodes. Of course, if two users have rated many common items, then such a definition will also consider them close neighbors. Therefore, the graph-based approach provides a different way of defining neighborhoods, which can be useful in sparse settings.

The notion of indirect connectivity is achieved with the use of path- or walk-based definitions. Some common methods for achieving this goal include the use of random-walk measures or the *Katz measure*, which is discussed in section 2.7.1.2. Both these measures are closely related to the problem of *link prediction* in social network analysis (cf. section 10.4 of Chapter 10), and they demonstrate the fact that graphical models of recommender systems connect the link-prediction problem to the vanilla recommendation problem. In the following, we discuss different ways of defining neighborhoods on the graph representation.

2.7.1.1   Defining Neighborhoods with Random Walks

The neighborhood of a user is defined by the set of users that are encountered frequently in a random walk starting at that user. How can the expected frequency of such random walks be measured? The answer to this problem is closely related to the random-walk methods, which are used frequently in Web-ranking applications. One can use either the personalized *PageRank* or the *SimRank* method (cf. Chapter 10) to determine the k most similar users to a given user for user-based collaborative filtering. Similarly, one can use this method to determine the k most similar items to a given item by starting the random walk at a given *item*. This approach is useful for item-based collaborative filtering. The other steps of user-based collaborative filtering and item-based collaborative filtering remain the same.

Why is this approach more effective for sparse matrices? In the case of the Pearson's correlation coefficient, two users need to be connected *directly* to a set of common items for the neighborhood to be defined meaningfully. In sparse user-item graphs, such direct connectivity may not exist for many nodes. On the other hand, a random-walk method also considers *indirect* connectivity, because a walk from one node to another may use any number of steps. Therefore, as long as large portions of the user-item graphs are connected,

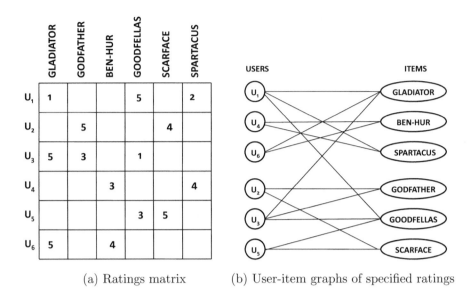

(a) Ratings matrix   (b) User-item graphs of specified ratings

Figure 2.3: A ratings matrix and corresponding user-item graph

it is always possible to meaningfully define neighborhoods. Such user-item graphs can also be used to directly predict ratings with the use of a variety of models. Such related methods will be discussed in section 10.2.3.3 of Chapter 10.

2.7.1.2 Defining Neighborhoods with the Katz Measure

Rather than using a probabilistic measure, such as random walks, it is possible to use the weighted number of walks between a pair of nodes in order to determine the affinity between them. The weight of each walk is a discount factor in $(0, 1)$, which is typically a decreasing function of its length. The weighted number of walks between a pair of nodes is referred to as the *Katz* measure. The weighted number of walks between a pair of nodes is often used as a link-prediction measure. The intuition is that if two users belong to the same neighborhood based on walk-based connectivity, then there is a propensity for a link to be formed between them in the user-item graph. The specific level of propensity is measured with the number of (discounted) walks between them.

Definition 2.7.1 (Katz Measure) *Let $n_{ij}^{(t)}$ be the number of walks of length t between nodes i and j. Then, for a user-defined parameter $\beta < 1$, the Katz measure between nodes i and j is defined as follows:*

$$Katz(i, j) = \sum_{t=1}^{\infty} \beta^t \cdot n_{ij}^{(t)} \qquad (2.34)$$

The value of β is a discount factor that de-emphasizes walks of longer lengths. For small enough values of β, the infinite summation of Equation 2.34 will converge.

Let K be the $m \times m$ matrix of Katz coefficients between pairs of users. If A is the symmetric adjacency matrix of an undirected network, then the pairwise Katz coefficient matrix K can be computed as follows:

$$K = \sum_{i=1}^{\infty} (\beta A)^i = (I - \beta A)^{-1} - I \qquad (2.35)$$

The value of β should always be selected to be smaller than the inverse of the largest eigenvalue of A to ensure convergence of the infinite summation. The Katz measure is closely rated to diffusion kernels in graphs. In fact, several collaborative recommendation methods directly use diffusion kernels to make recommendations [205].

A weighted version of the measure can be computed by replacing A with the weight matrix of the graph. This can be useful in cases where one wishes to weight the edges in the user-item graph with the corresponding rating. The top-k nodes with the largest Katz measures to the target node are isolated as its neighborhood. Once the neighborhood has been determined, it is used to perform the prediction according to Equation 2.4. Many variations of this basic principle are used to make recommendations:

1. It is possible to use a threshold on the maximum path length in Equation 2.34. This is because longer path lengths generally become noisy for the prediction process. Nevertheless, because of the use of the discount factor β, the impact of long paths on the measure is generally limited.

2. In the aforementioned discussion, the Katz measure is used only to determine the neighborhoods of users. Therefore, the Katz measure is used to compute the affinity between pairs of *users*. After the neighborhood of a user has been determined, it is used to make predictions in the same way as any other neighborhood-based method.

   However, a different way of *directly* performing the prediction, without using neighborhood methods, would be to measure the affinity between users and *items*. The Katz measure can be used to compute these affinities. In such cases, the links are weighted with ratings, and the problem is reduced to that of predicting links between users and items. These methods will be discussed in more detail in section 10.4.6 of Chapter 10.

The bibliographic notes contain a number of references to various path-based methods.

2.7.2 User-User Graphs

In user-item graphs, the user-user connectivity is defined by an even number of hops in the user-item graph. Instead of constructing user-item graphs, one might instead directly create user-user graphs based on 2-hop connectivity between users. The advantage of user-user graphs over user-item graphs is that the edges of the graph are more informative in the former. This is because the 2-hop connectivity can directly take the number and similarity of common items between the two users into account, while creating the edges. These notions, referred to as *horting* and *predictability*, will be discussed slightly later. The algorithm uses the notion of horting to quantify the number of mutually specified ratings between two users (nodes), whereas it uses the notion of predictability to quantify the level of similarity among these common ratings.

The user-user graph is constructed as follows. Each node u corresponds to one of the m users in the $m \times n$ user-item matrix. Let I_u be the set of items for which ratings have been specified by user u, and let I_v be the set of items for which ratings have been specified by user v. Edges are defined in this graph with the notion of *horting*. Horting is an asymmetric relationship between users, which is defined on the basis of their having rated similar items.

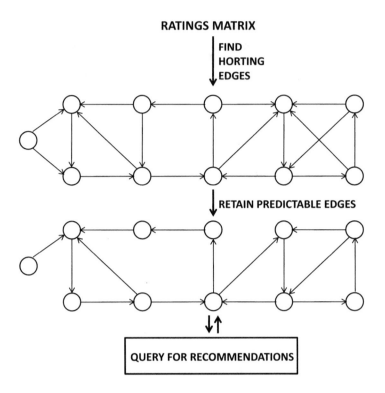

Figure 2.4: The user-user predictability approach

Definition 2.7.2 (Horting) *A user u is said to hort user v at level (F, G), if either of the following are true:*

$$|I_u \cap I_v| \geq F$$
$$|I_u \cap I_v| / |I_u| \geq G$$

Here, F and G are algorithm parameters. Note that it is sufficient for one of the two aforementioned conditions to hold for user u to hort user v. The notion of horting is used to further define predictability.

Definition 2.7.3 (Predictability) *The user v predicts user u, if u horts v and there exists a linear transformation function $f(\cdot)$ such that the following is true:*

$$\frac{\sum_{k \in I_u \cap I_v} |r_{uk} - f(r_{vk})|}{|I_u \cap I_v|} \leq U$$

Here, U is another algorithm parameter. It is noteworthy that the distance $\frac{\sum_{k \in I_u \cap I_v} |r_{uk} - f(r_{vk})|}{|I_u \cap I_v|}$ between the ratings of user u and the transformed ratings of user v is a variant of the Manhattan distance on their common specified ratings. The main difference from the Manhattan distance is that the distance is normalized by the number of mutually specified ratings between the two users. This distance is also referred to as the *Manhattan segmental distance*.

The directions of horting and predictability are opposite one another. In other words, for user v to predict user u, u must hort v. A directed graph G is defined, in which an edge exists from u to v, if v predicts u. This graph is referred to as the *user-user predictability graph*. Each edge in this graph corresponds to a linear transformation, as discussed in Definition 2.7.3. The linear transformation defines a prediction, where the rating at the head of the edge can be used to predict the rating at the tail of the edge. Furthermore, it is assumed that one can apply these linear transformations in a transitive way over a directed path in order to predict the rating of the source of the path from the rating at the destination of the path.

Then, the rating of a target user u for an item k is computed by determining all the directed shortest paths from user u to all other users who have rated item k. Consider a directed path of length r from user u to a user v who has rated item k. Let $f_1 \ldots f_r$ represent the sequence of linear transformations along the directed path starting from node u to this user v. Then, the rating prediction $\hat{r}_{uk}^{(v)}$ of the rating of target user u for item k (based only on user v) is given by applying the composition of the r linear mappings along this path from user u to v, to the rating r_{vk} of user v on item k:

$$\hat{r}_{uk}^{(v)} = (f_1 \circ f_2 \ldots \circ f_r)(r_{vk}) \tag{2.36}$$

The rating prediction $\hat{r}_{uk}^{(v)}$ contains the superscript v because it is based only on the rating of user v. Therefore, the final rating prediction $\hat{r}_{uk}$ is computing by averaging the value of $\hat{r}_{uk}^{(v)}$ over all users v that have rated item k, within a threshold distance D of the target user u.

Given a target user (node) u, one only needs to determine directed *paths* from this user to other users, who have rated the item at hand. The shortest path can be determined with the use of a breadth-first algorithm, which is quite efficient. Another important detail is that a threshold is imposed on the maximum path length that is usable for prediction. If no user, who has rated item k is found within a threshold length D of the target node u, then the algorithm terminates with failure. In other words, the rating of the target user u for item k simply cannot be determined robustly with the available ratings matrix. It is important to impose such thresholds to improve efficiency and also because the linear transformation along very long path lengths might lead to increasing distortion in the rating prediction. The overall approach is illustrated in Figure 2.4. Note that a directed edge exists from u to v in the horting graph if u horts v. On the other hand, an edge exists in the predictability graph if u horts v *and* v predicts u. Therefore, the predictability graph is obtained from the horting graph by dropping a few edges. This graph is set up in an offline phase and it is repeatedly queried for recommendations. In addition, a number of index data structures are set up from the ratings matrix during the offline setup phase. These data structures are used along with the predictability graph in order to resolve the queries efficiently. More details on the horting approach may be found in [33].

This approach can work for very sparse matrices because it uses transitivity to predict ratings. An important challenge in neighborhood methods is the lack of coverage of rating prediction. For example, if none of John's immediate neighbors have rated *Terminator*, it is impossible to provide a rating prediction for John. However, structural transitivity allows us to check whether the *indirect* neighbors of John have rated *Terminator*. Therefore, the main advantage of this approach is that it has better coverage compared to competing methods.

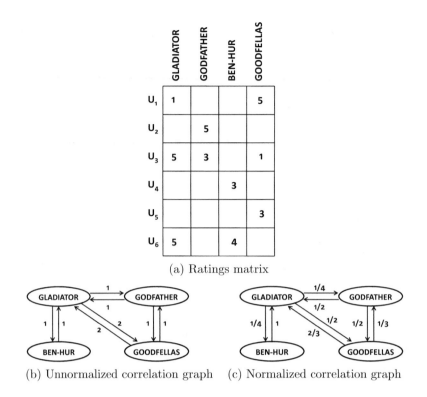

(a) Ratings matrix

(b) Unnormalized correlation graph   (c) Normalized correlation graph

Figure 2.5: A ratings matrix and its correlation graphs

2.7.3 Item-Item Graphs

It is also possible to leverage item-item graphs to perform the recommendations. Such a graph is also referred to as the *correlation graph* [232]. In this case, a weighted and directed network $G = (N, A)$ is constructed, in which each node in N corresponds to an item, and each edge in A corresponds to a relationship between items. The weight w_{ij} is associated with each edge (i, j). If items i and j have both been rated by at least one common user, then both the directed edges (i, j) and (j, i) exist in the network. Otherwise, no edges exist between nodes i and j. The directed network is, however, asymmetric because the weight of edge (i, j) is not necessarily the same as that of edge (j, i). Let U_i be the set of users that have specified ratings for item i and U_j be the set of users that have specified ratings for item j. Then, the weight of the edge (i, j) is computed using the following simple algorithm.

First, we initialize the weight w_{ij} of each edge (i, j) to $|U_i \cap U_j|$. At this point, the edge weights are symmetric because $w_{ij} = w_{ji}$. Then, the weights of the edges are normalized, so that the sum of the weights of the outgoing edges of a node is equal to 1. This normalization is achieved by dividing w_{ij} with the sum of the outgoing weights from node i. The normalization step results in asymmetric weights, because each of the weights w_{ij} and w_{ji} are divided by different quantities. This results in a graph in which the weights on edges correspond to *random-walk probabilities*. An example of the correlation graph for a ratings matrix is illustrated in Figure 2.5. It is clear that the weights on the *normalized* correlation graph are not symmetric because of the scaling of the weights to transition probabilities. Furthermore, it is noteworthy that the rating values are not used in the construction of the correlation graph. Only the *number* of observed ratings in common between two items is used. This is sometimes not desirable. It is, of course, possible to define the correlation graph in other ways, such as with the use of the cosine function between the *rating* vectors of the two items.

As discussed in Chapter 10, random-walk methods can be used to determine the neighborhood of a given item. The resulting neighborhood can be used for item-based collaborative filtering methods. Furthermore, personalized *PageRank* methods can be used to directly determine the ratings on the item-item graph. This method is referred to as *ItemRank*, and it is discussed in section 10.2.3.3 of Chapter 10.

2.8 Summary

Because collaborative filtering can be viewed as a generalization of classification and regression problems, the methodologies for the latter classes of problems can also be applied to the former. Neighborhood-based methods derive their inspiration from nearest neighbor classification and regression methods. In user-based methods, the first step is to determine the neighborhood of the target user. In order to compute the neighborhood, a variety of similarity functions, such as the Pearson correlation coefficient or the cosine, are used. The neighborhood is used in order to extrapolate the unknown ratings of a record. In item-based methods, the most similar items are computed with respect to a target item. Then, the user's own ratings on these similar items are used in order to make a rating prediction. Item-based methods are likely to have more relevant recommendations, but they are less likely to yield diverse recommendations. In order to speed up neighborhood-based methods, clustering is often used.

Neighborhood-based methods can be viewed as linear models, in which the weights are chosen in a heuristic way with the use of similarity values. One can also learn these weights with the use of linear regression models. Such methods have the advantage that they can be combined with other optimization models, such as matrix factorization, for better prediction. Such methods are discussed in the next chapter.

Neighborhood-based methods face numerous challenges because of data sparsity. Users often specify only a small number of ratings. As a result, a pair of users may often have specified only a small number of ratings. Such situations can be addressed effectively with the use of both dimensionality reduction and graph-based models. While dimensionality reduction methods are often used as standalone methods for collaborative filtering, they can also be combined with neighborhood-based methods to improve the effectiveness and efficiency of collaborative filtering. Various types of graphs can be extracted from rating patterns, such as user-item graphs, user-user graphs, or item-item graphs. Typically, random-walk or shortest-path methods are used in these cases.

2.9 Bibliographic Notes

Neighborhood-based methods were among the earliest techniques used in the field of recommender systems. The earliest user-based collaborative filtering models were studied in [33, 98, 501, 540]. A comprehensive survey of neighborhood-based recommender systems may be found in [183]. Sparsity is a major problem in such systems, and various graph-based systems have been designed to alleviate the problem of sparsity [33, 204, 647]. Methods that are specifically designed for the long tail in recommender algorithms are discussed in [173, 463, 648].

User-based methods utilize the ratings of *similar* users on the *same* item in order to make predictions. While such methods were initially quite popular, they are not easily scalable and sometimes inaccurate. Subsequently, item-based methods [181, 360, 524] were proposed, which compute predicted ratings as a function of the ratings of the *same* user on *similar* items. Item-based methods provide more accurate but less diverse recommendations.

The notion of mean-centering for improving recommendation algorithms was proposed in [98, 501]. A comparison of the use of the Z-score with mean-centering is studied in [245, 258], and these two studies provide somewhat conflicting results. A number of methods which do not use the absolute ratings, but instead focus on ordering the ratings in terms of preference weights, are discussed in [163, 281, 282]. The significance-weighting methods of de-emphasizing the neighbors who have too few common ratings with a given neighbor are discussed in [71, 245, 247, 380]. Many different variants of the similarity function are used for computing the neighbor. Two such examples are the *mean-squared distance* [540] and the *Spearman rank correlation* [299]. The specific advantage of these distance measures is not quite clear because conflicting results have been presented in the literature [247, 258]. Nevertheless, the consensus seems to be that the Pearson rank correlation provides the most accurate results [247]. Techniques for adjusting for the impact of very popular items are discussed in [98, 280]. The use of exponentiated amplification for prediction in neighborhood-based methods is discussed in [98]. A discussion of the use of voting techniques in nearest neighbor methods may be found in [183]. Voting methods can be viewed as a direct generalization of the nearest neighbor classifier, as opposed to a generalization of nearest neighbor regression modeling.

Methods for item-based collaborative filtering were proposed in [181, 524, 526]. A detailed study of different variations of item-based collaborative filtering algorithms is provided in [526], along with a comparison with respect to user-based methods. The item-based method in [360] is notable because it describes one of Amazon.com's collaborative filtering methods. The user-based and item-based collaborative filtering methods have also been unified with the notion of similarity fusion [622]. A more generic unification framework may be found in [613]. Clustering methods are used frequently to improve the efficiency of neighborhood-based collaborative filtering. A number of clustering methods are described in [146, 167, 528, 643, 644, 647]. The extension of neighborhood methods to very large-scale data sets has been studied in [51].

Dimensionality reduction techniques have a rich history of being used in missing-value estimation [24, 472] and recommender systems [71, 72, 228, 252, 309, 313, 500, 517, 525]. In fact, most of these techniques directly use such latent models to predict the ratings without relying on neighborhood models. However, some of these dimensionality reduction techniques [71, 72, 309, 525] are specifically designed to improve the effectiveness and efficiency of neighborhood-based techniques. A key contribution of [72] is to provide an insight about the relationship between neighborhood methods and regression-based methods. This relationship is important because it shows how one can formulate neighborhood-based methods as model-based methods with a crisp optimization formulation. Note that many other model-based methods, such as latent factor models, can also be expressed as optimization formulations. This observation paves the way for combining neighborhood methods with latent factor models in a unified framework [309] because one can now combine the two objective functions. Other regression-based models for recommender systems, such as *slope-one predictors* and ordinary least-squares methods, are proposed in [342, 620]. Methods for learning pairwise preferences over itemsets are discussed in [469]. Item-item regression models have also been studied in the context of *Sparse Linear Models (SLIM)* [455], where an elastic-net regularizer is used on the linear model without restricting the coefficients to the neighborhood of the item. Higher-order sparse learning methods, which model the effects of using combinations of items, are discussed in [159]. Efficient methods for training linear models and tuning regularization parameters are discussed in [347]. Constrained linear regression methods are discussed in [430].

A general examination of linear classifiers, such as least-squares regression and support vector machines, is provided in [669]. However, the approach is designed for implicit feedback data sets in which only positive preferences are specified. It was observed that collaborative filtering, in such cases, is similar to text categorization. However, because of the noise in the data and the imbalanced nature of the class distribution, a direct use of SVM methods is sometimes not effective. Changes to the loss function are suggested in [669] in order to provide more accurate results.

Many graph-based methods have been proposed for improving collaborative filtering algorithms. Most of these methods are based on either user-item graphs, but a few are also based on user-user graphs. An important observation from the perspective of graph-based methods is that they show an interesting relationship between the problems of ranking, recommendation, and link-prediction. The use of random walks for determining the neighborhood in recommendation systems is discussed in [204, 647]. A method, which uses the *number* of discounted paths between a pair of nodes in a user-item graph for recommendations, was proposed in [262]. This approach is equivalent to using the Katz measure between user-user pairs in order to determine whether they reside in each other's neighborhoods. This approach is related to link-prediction [354], because the Katz measure is often used to determine the linkage affinity between a pair of nodes. A survey on link prediction methods may be found in [17]. Some graph-based methods do not directly use neighborhoods. For example, the *ItemRank* method proposed in [232] shows how to use ranking directly to make predictions, and the method in [261] shows how to use link-prediction methods directly for collaborative filtering. These methods are also discussed in Chapter 10 of this book. Techniques for leveraging user-user graphs are discussed in [33]. These methods have the advantage that they directly encode the user-user similarity relationships in the edges of the graph. As a result, the approach provides better coverage than competing methods.

2.10 Exercises

1. Consider the ratings matrix of Table 2.1. Predict the absolute rating of item 3 for user 2 using:

   (a) User-based collaborative filtering with Pearson correlation and mean-centering
   (b) Item-based collaborative filtering with adjusted cosine similarity

   Use a neighborhood of size 2 in each case.

2. Consider the following ratings table between five users and six items:

Item-Id $\Rightarrow$	1	2	3	4	5	6
1	5	6	7	4	3	?
2	4	?	3	?	5	4
3	?	3	4	1	1	?
4	7	4	3	6	?	4
5	1	?	3	2	2	5

(a) Predict the values of unspecified ratings of user 2 using user-based collaborative filtering algorithms. Use the Pearson correlation with mean-centering.

(b) Predict the values of unspecified ratings of user 2 using item-based collaborative filtering algorithms. Use the adjusted cosine similarity.

Assume that a peer group of size at most 2 is used in each case, and negative correlations are filtered out.

3. Discuss the similarity between a k-nearest neighbor classifier in traditional machine learning and the user-based collaborative filtering algorithm. Describe an analogous classifier to item-based collaborative filtering.

4. Consider an algorithm that performs clustering of users based on their ratings matrix and reports the average ratings within a cluster as the predicted items ratings for every user within a cluster. Discuss the effectiveness and efficiency trade-offs of such an approach compared to a neighborhood model.

5. Propose an algorithm that uses random walks on a user-user graph to perform neighborhood-based collaborative filtering. [This question requires a background in ranking methods.]

6. Discuss various ways in which graph clustering algorithms can be used to perform neighborhood-based collaborative filtering.

7. Implement the user-based and item-based collaborative filtering algorithms.

8. Suppose you had content-based profiles associated with users indicating their interests and profiles associated with items corresponding to their descriptions. At the same time, you had a ratings matrix between users and items. Discuss how you can incorporate the content-based information within the framework of graph-based algorithms.

9. Suppose that you had a unary ratings matrix. Show how collaborative filtering algorithms can be solved using content-based methods by treating the ratings of an item as its features. Refer to Chapter 1 for a description of content-based methods. What type of a content-based classifier does an item-based collaborative filtering algorithm correspond to?

Chapter 3

Model-Based Collaborative Filtering

"Do not quench your inspiration and your imagination; do not become the slave of your model." – Vincent van Gogh

3.1 Introduction

The neighborhood-based methods of the previous chapter can be viewed as generalizations of k-nearest neighbor classifiers, which are commonly used in machine learning. These methods are instance-based methods, whereby a model is not specifically created up front for prediction other than an optional preprocessing[1] phase, which is required to ensure efficient implementation. Neighborhood-based methods are generalizations of *instance-based learning methods* or *lazy learning methods* in which the prediction approach is *specific to the instance being predicted*. For example, in user-based neighborhood methods, the peers of the *target* user are determined in order to perform the prediction.

In model-based methods, a summarized model of the data is created up front, as with supervised or unsupervised machine learning methods. Therefore, the training (or *model-building* phase) is clearly separated from the prediction phase. Examples of such methods in traditional machine learning include decision trees, rule-based methods, Bayes classifiers, regression models, support vector machines, and neural networks [22]. Interestingly, almost all these models can be generalized to the collaborative filtering scenario, just as k-nearest neighbor classifiers can be generalized to neighborhood-based models for collaborative filtering. This is because the traditional classification and regression problems are special cases of the matrix completion (or collaborative filtering) problem.

In the data classification problem, we have an $m \times n$ matrix, in which the first $(n-1)$ columns are feature variables (or independent variables), and the last (i.e., nth) column is

---

[1]From a practical point of view, preprocessing is essential for efficiency. However, one could implement the neighborhood method without a preprocessing phase, albeit with larger latencies at query time.

© Springer International Publishing Switzerland 2016
C.C. Aggarwal, *Recommender Systems: The Textbook*,
DOI 10.1007/978-3-319-29659-3_3

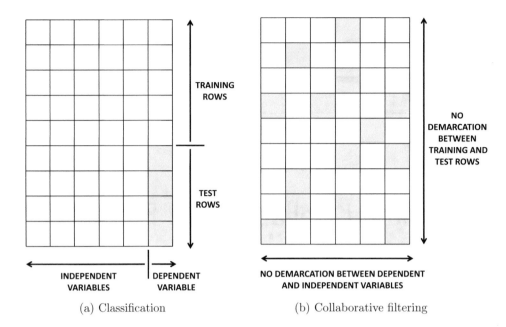

(a) Classification   (b) Collaborative filtering

Figure 3.1: Revisiting Figure 1.4 of Chapter 1. Comparing the traditional classification problem with collaborative filtering. Shaded entries are missing and need to be predicted.

the class variable (or dependent variable). All entries in the first $(n-1)$ columns are fully specified, whereas only a subset of the entries in the nth column is specified. Therefore, a subset of the rows in the matrix is fully specified, and these rows are referred to as the *training data*. The remaining rows are referred to as the *test data*. The values of the missing entries need to be learned for the test data. This scenario is illustrated in Figure 3.1(a), where the shaded values represent missing entries in the matrix.

Unlike data classification, any entry in the ratings matrix may be missing, as illustrated by the shaded entries in Figure 3.1(b). Thus, it can be clearly seen that the matrix completion problem is a generalization of the classification (or regression modeling) problem. Therefore, the crucial differences between these two problems may be summarized as follows:

1. In the data classification problem, there is a clear separation between feature (independent) variables and class (dependent) variables. In the matrix completion problem, this clear separation does not exist. Each column is both a dependent and independent variable, depending on which entries are being considered for predictive modeling at a given point.

2. In the data classification problem, there is a clear separation between the training and test data. In the matrix completion problem, this clear demarcation does not exist among the *rows* of the matrix. At best, one can consider the specified (observed) *entries* to be the training data, and the unspecified (missing) entries to be the test data.

3. In data classification, columns represent features, and rows represent data instances. However, in collaborative filtering, it is possible to apply the same approach to either the ratings matrix or to its transpose because of how the missing entries are distributed. For example, user-based neighborhood models can be viewed as direct

generalizations of nearest neighbor classifiers. When such methods are applied to the transpose of the ratings matrix, they are referred to as *item*-based neighborhood models. In general, many classes of collaborative filtering algorithms have both user-wise and item-wise versions.

These differences between data classification and collaborative filtering are illustrated in Figure 3.1. The greater generality of the collaborative filtering problem leads to a richer number of algorithmic *possibilities* in collaborative filtering, as compared to data classification.

The similarity between the collaborative filtering problem and the data classification problem is useful to keep in mind when designing learning algorithms for the former. This is because data classification is a relatively well-studied field, and the various types of solutions to classification also provide important hints for the design of collaborative filtering algorithms. In fact, most machine learning and classification algorithms have direct analogs in the collaborative filtering literature. Understanding recommender systems in a similar way to classification models enables the application of a significant number of meta-algorithms from the classification literature. For example, classical meta-algorithms from the classification literature, such as bagging, boosting or model combination, can be extended to collaborative filtering. Interestingly, much of the theory developed for ensemble methods in classification continues to apply to recommender systems. In fact, the ensemble-based methods [311, 704] were among the best performing methods in the Netflix challenge. Ensemble methods are discussed in detail in Chapter 6.

It is not always easy, however, to generalize data classification models directly to the matrix completion problem, especially when the vast majority of the entries are missing. Furthermore, the relative effectiveness of the various models are different in different settings. For example, a number of recent collaborative filtering models, such as latent factor models, are particularly well suited to collaborative filtering. Such models are, however, not considered competitive models in the context of data classification.

Model-based recommender systems often have a number of advantages over neighborhood-based methods:

1. *Space-efficiency:* Typically, the size of the learned model is much smaller than the original ratings matrix. Thus, the space requirements are often quite low. On the other hand, a user-based neighborhood method might have $O(m^2)$ space complexity, where m is the number of users. An item-based method will have $O(n^2)$ space complexity.

2. *Training speed and prediction speed:* One problem with neighborhood-based methods is that the pre-processing stage is quadratic in either the number of users or the number of items. Model-based systems are usually much faster in the preprocessing phase of constructing the trained model. In most cases, the compact and summarized model can be used to make predictions efficiently.

3. *Avoiding overfitting:* Overfitting is a serious problem in many machine learning algorithms, in which the prediction is overly influenced by random artifacts in the data. This problem is also encountered in classification and regression models. The summarization approach of model-based methods can often help in avoiding overfitting. Furthermore, *regularization* methods can be used to make these models robust.

Even though neighborhood-based methods were among the earliest collaborative filtering methods and were also among the most popular because of their simplicity, they are not necessarily the most accurate models available today. In fact, some of the most accurate methods are based on model-based techniques in general, and on latent factor models in particular.

This chapter is organized as follows. Section 3.2 discusses the use of decision and regression trees for recommender systems. Rule-based collaborative filtering methods are discussed in section 3.3. The use of the naive Bayes model for recommender systems is discussed in section 3.4. A general discussion of how other classification methods are extended to collaborative filtering is provided in section 3.5. Latent factor models are discussed in section 3.6. The integration of latent factor models with neighborhood models is discussed in section 3.7. A summary is given in section 3.8.

3.2   Decision and Regression Trees

Decision and regression trees are frequently used in data classification. Decision trees are designed for those cases in which the dependent variable is categorical, whereas regression trees are designed for those cases in which the dependent variable is numerical. Before discussing the generalization of decision trees to collaborative filtering, we will first discuss the application of decision trees to classification.

Consider the case in which we have an $m \times n$ matrix R. Without loss of generality, assume that the first $(n - 1)$ columns are the independent variables, and the final column is the dependent variable. For ease in discussion, assume that all variables are binary. Therefore, we will discuss the creation of a decision tree rather than a regression tree. Later, we will discuss how to generalize this approach to other types of variables.

The decision tree is a hierarchical partitioning of the data space with the use of a set of hierarchical decisions, known as the *split criteria* in the independent variables. In a *univariate decision tree*, a single feature is used at one time in order to perform a split. For example, in a binary matrix R, in which the feature values are either 0 or 1, all the data records in which a carefully chosen feature variable takes on the value of 0 will lie in one branch, whereas all the data records in which the feature variable takes on the value of 1 will lie in the other branch. When the feature variable is chosen in such a way, so that it is correlated with the class variable, the data records within each branch will tend to be *purer*. In other words, most of the records belonging to the different classes will be separated out. In other words, one of the two branches will predominantly contain one class, whereas the other branch will predominantly contain the other class. When each node in a decision tree has two children, the resulting decision tree is said to be a binary decision tree.

The quality of the split can be evaluated by using the weighted average *Gini index* of the child nodes created from a split. If $p_1 \ldots p_r$ are the fractions of data records belonging to r different classes in a node S, then the Gini index $G(S)$ of the node is defined as follows:

$$G(S) = 1 - \sum_{i=1}^{r} p_i^2 \tag{3.1}$$

The Gini index lies between 0 and 1, with smaller values being more indicative of greater discriminative power. The overall Gini index of a split is equal to the weighted average of the Gini index of the children nodes. Here, the weight of a node is defined by the number of data points in it. Therefore, if S_1 and S_2 are the two children of node S in a binary decision tree, with n_1 and n_2 data records, respectively, then the Gini index of the split $S \Rightarrow (S_1, S_2)$ may be evaluated as follows:

$$\text{Gini}\,(S \Rightarrow [S_1, S_2]) = \frac{n_1 \cdot G(S_1) + n_2 \cdot G(S_2)}{n_1 + n_2} \tag{3.2}$$

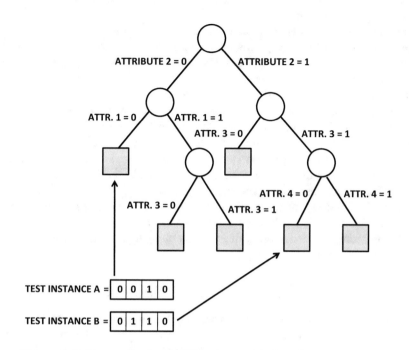

Figure 3.2: Example of a decision tree with four binary attributes

The Gini index is used for selecting the appropriate attribute to use for performing the split at a given level of the tree. One can test each attribute to evaluate the Gini index of its split according to Equation 3.2. The attribute with the smallest Gini index is selected for performing the split. The approach is executed hierarchically, in top-down fashion, until each node contains only data records belonging to a particular class. It is also possible to stop the tree growth early, when a minimum fraction of the records in the node belong to a particular class. Such a node is referred to as a leaf node, and it is labeled with the dominant class of the records in that node. To classify a test instance with an unknown value of the dependent variable, its independent variables are used to map a path in the decision tree from the root to the leaf. Because the decision tree is a hierarchical partitioning of the data space, the test instance will follow exactly one path from the root to the leaf. The label of the leaf is reported as the relevant one for the test instance. An example of a decision tree, constructed on four binary attributes, is illustrated in Figure 3.2. The leaf nodes of the the tree are shaded in the figure. Note that all attributes are not necessarily used for splits by the decision tree. For example, the leftmost path uses attributes 1 and 2, but it does not use attributes 3 and 4. Furthermore, different paths in the decision tree may use different sequences of attributes. This situation is particularly common with high-dimensional data. Examples of the mappings of test instances A= 0010 and B= 0110 to respective leaf nodes are illustrated in Figure 3.2. Each of these test instances is mapped to a unique leaf node because of the hierarchical nature of the data partitioning.

The approach can be extended to numerical dependent and independent variables with minor modifications. To handle numerical independent (feature) variables, the attribute values can be divided into intervals in order to perform the splits. Note that this approach might result in a multi-way split, where each branch of the split corresponds to a different interval. The split is then performed by choosing the attribute on the basis of the Gini index criterion. Such an approach also applies to categorical feature variables, wherein each value of the categorical attribute corresponds to a branch of the split.

To handle numeric *dependent* variables, the split criterion is changed from the Gini index to a measure better suited to numeric attributes. Specifically, the variance of the numeric dependent variable is used instead of the Gini index. A lower variances is more desirable because it means that the node contains training instances that are discriminatively mapped in the locality of the dependent variable. Either the average value in the leaf node, or a linear regression model, is used at the leaf node to perform the prediction [22].

In many cases, the tree is pruned to reduce overfitting. In this case, a portion of the training data is not used during the tree construction phase. Then, the effect of pruning the node is tested on the portion of the training data that is held out. If the removal of the node improves the accuracy of the decision tree prediction on the held out data, then the node is pruned. Additionally, other variations of the split criteria, such as error rates and entropy, are commonly used. Detailed discussions of various design choices in decision tree construction may be found in [18, 22].

3.2.1   Extending Decision Trees to Collaborative Filtering

The main challenge in extending decision trees to collaborative filtering is that the predicted entries and the observed entries are not clearly separated in column-wise fashion as feature and class variables. Furthermore, the ratings matrix is very sparsely populated, where the majority of entries are missing. This creates challenges in hierarchically partitioning the training data during the tree-building phase. Furthermore, since the dependent and independent variables (items) are not clearly demarcated in collaborative filtering, what item should be predicted by the decision tree?

The latter issue is relatively easy to address by constructing separate decision trees to predict the rating of each item. Consider an $m \times n$ ratings matrix R with m users and n items. A separate decision tree needs to be constructed by fixing each attribute (item) to be dependent and the remaining attributes as independent. Therefore, the number of decision trees constructed is exactly equal to the number n of attributes (items). While predicting the rating of a particular item for a user, the decision tree corresponding to the relevant item is used for prediction.

On the other hand, the issue of missing independent features is more difficult to address. Consider the case, where a particular item (say, a particular movie) is used as a splitting attribute. All users whose rating is less than a threshold are assigned to one branch of the tree, whereas the users whose ratings are larger than the threshold are assigned to the other branch. Because ratings matrices are sparse, most users will not have specified ratings for this item. Which branch should such users be assigned to? Logic dictates that such users should be assigned to *both* branches. However, in such a case, the decision tree no longer remains a strict partitioning of the training data. Furthermore, according to this approach, test instances will map to multiple paths in the decision tree, and the possibly conflicting predictions from the various paths will need to combined into a single prediction.

A second (and more reasonable) approach is to create a lower-dimensional representation of the data using the dimensionality reduction methods discussed in section 2.5.1.1 of Chapter 2. Consider the scenario, where the rating of the jth item needs to be predicted. At the very beginning, the $m \times (n-1)$ ratings matrix, excluding the jth column, is converted into a lower-dimensional $m \times d$ representation, in which $d \ll n-1$ and all attributes are fully specified. The covariance between each pair of items in the $m \times (n-1)$ ratings matrix is estimated using the methods discussed in section 2.5.1.1 of Chapter 2. The top-d

eigenvectors $\overline{e_1} \ldots \overline{e_d}$ of the estimated $(n-1) \times (n-1)$ covariance matrix are determined. Note that each eigenvector is a vector containing $(n-1)$ elements. Equation 2.17 is used for projecting the ratings of each user on the eigenvectors, except that the jth item is not included on the right-hand side of Equation 2.17. This results in a d-dimensional vector of ratings for each user, which is completely specified. This reduced representation is used to construct the decision tree for the jth item by treating the problem as a standard classification or regression modeling problem. This approach is repeated by varying the value of j from 1 to n, in order to construct a total of n decision trees. Therefore, the jth decision tree is useful only for predicting the ratings of the jth item. Both the eigenvectors and the trees for each of the n cases are stored as a part of the model.

To predict the rating of item j for a user i, the ith row of the $m \times d$ matrix is used as the test instance, and the jth decision/regression tree is used as the model to predict the value of the corresponding rating. The first step is to use the remaining $n-1$ items (except for the jth item) in order to create the reduced d-dimensional representation of the test instance according to Equation 2.17. Note that the jth set of eigenvectors are used for the projection and reduction process. This representation is then used with the corresponding decision or regression tree for the jth item to perform the prediction. It is noteworthy that this broader approach of combining dimensionality reduction with a classification model is not restricted to decision trees. It is relatively easy to use this approach in conjunction with virtually any classification model. Furthermore, dimensionality reduction methods are also used in isolation to predict ratings in recommender systems. Both these issues are discussed later in this chapter.

3.3 Rule-Based Collaborative Filtering

The relationship between *association rules* [23] and collaborative filtering is a natural one because the association rule problem was first proposed in the context of discovering relationships between supermarket data. Association rules are naturally defined over binary data, although the approach can be extended to categorical and numerical data by converting these data types to binary data. For the purpose of this discussion, we will assume the simplified case of unary data, which are common in supermarket transactions and in implicit feedback data sets.

Consider a transaction database $\mathcal{T} = \{T_1 \ldots T_m\}$, containing m transactions, which are defined on n items I. Therefore, I is the universal set of items, and each transaction T_i is a subset of the items in I. The key in association rule mining is to determine sets of items that are closely correlated in the transaction database. This is achieved with the notions of *support* and *confidence*. These measures quantify the relationships between sets of items.

Definition 3.3.1 (Support) *The support of an itemset $X \subseteq I$ is the fraction of transactions in $\mathcal{T}$, of which X is a subset.*

If the support of an itemset is at least equal to a predefined threshold s, then the itemset is said to be frequent. This threshold is referred to as the *minimum support*. These itemsets are referred to as *frequent itemsets* or *frequent patterns*. Frequent itemsets can provide important insights about correlations in customer buying behavior.

For example, consider the data set illustrated in Table 3.1. In this table, the rows correspond to customers and columns correspond to items. The 1s correspond to cases in which a particular customer has bought an item. Although this data set is unary, and the 0s correspond to missing values, a common practice in such implicit feedback data sets is to

Table 3.1: Example of market basket data

Item ⇒ Customer ⇓	Bread	Butter	Milk	Fish	Beef	Ham
Jack	1	1	1	0	0	0
Mary	0	1	1	0	1	0
Jane	1	1	0	0	0	0
Sayani	1	1	1	1	1	1
John	0	0	0	1	0	1
Tom	0	0	0	1	1	1
Peter	0	1	0	1	1	0

approximate missing values with 0s. It is evident that the columns of the table can be partitioned into two sets of closely related items. One of these sets is $\{Bread, Butter, Milk\}$, and the other set is $\{Fish, Beef, Ham\}$. These are the only itemsets with at least 3 items, which also have a support of at least 0.2. Therefore, both of these itemsets are frequent itemsets or frequent patterns. Finding such patterns with high support is useful to the merchant, because she can use them to make recommendations and other target marketing decisions. For example, it is reasonable to conclude that Mary is likely to eventually buy *Bread*, because she has already bought $\{Butter, Milk\}$. Similarly, John is likely to buy *Beef* because he has also bought $\{Fish, Ham\}$. Such inferences are very useful from the point of view of a recommendation system.

A further level of insight may be obtained in terms of the *directions* of these correlations by using the notion of *association rules* and *confidence*. An association rule is denoted in the form $X \Rightarrow Y$, where the "$\Rightarrow$" is intended to give a direction to the nature of the correlation between the set of items X and Y. For example, a rule such as $\{Butter, Milk\} \Rightarrow \{Bread\}$ would be very useful to recommend *Bread* to Mary, because it is already known that she has bought *Milk* and *Butter*. The strength of such a rule is measured by its *confidence*.

Definition 3.3.2 (Confidence) *The confidence of the rule $X \Rightarrow Y$ is the conditional probability that a transaction in $\mathcal{T}$ contains Y, given that it also contains X. Therefore, the confidence is obtained by dividing the support of $X \cup Y$ with the support of X.*

Note that the support of $X \cup Y$ will always be less than the support of X. This is because if a transaction contains $X \cup Y$, then it will always contain X. However, the reverse might not be true. Therefore, the confidence of a rule must always lie in the range $(0, 1)$. Higher values of the confidence are always indicative of greater strength of the rule. For example, if a rule $X \Rightarrow Y$ is true, then a merchant, who knows that a specific set of customers has bought the set of items X, can also target these customers with the set of items Y. An association rule is defined on the basis of a minimum support s and minimum confidence c:

Definition 3.3.3 (Association Rules) *A rule $X \Rightarrow Y$ is said to be an association rule at a minimum support of s and minimum confidence of c, if the following two conditions are satisfied:*

1. The support of $X \cup Y$ is at least s.

2. The confidence of $X \Rightarrow Y$ is at least c.

The process of finding association rules is a two-phase algorithm. In the first phase, all the itemsets that satisfy a minimum support threshold s are determined. From each of these itemsets Z, all possible 2-way partitions $(X, Z - X)$ are used to create a potential rule $X \Rightarrow Z - X$. Those rules satisfying the minimum confidence are retained. The first phase of determining the frequent itemsets is the computationally intensive one, especially when the underlying transaction database is very large. Numerous computationally efficient algorithms have been devoted to the problem of efficient frequent itemset discovery. The discussion of these algorithms is beyond the scope of this book, because it is a distinct field of data mining in its own right. Interested readers are referred to [23] for a detailed discussion of frequent pattern mining. In this book, we will show how to use these algorithms as tools for collaborative filtering.

3.3.1 Leveraging Association Rules for Collaborative Filtering

Association rules are particularly useful for performing recommendations in the context of *unary* ratings matrices. As discussed in Chapters 1 and 2, unary ratings matrices are created by customer activity (e.g., buying behavior), wherein there is a natural mechanism for the customer to specify a liking for an item, but no mechanism to specify a dislike. In these cases, the items bought by a customer are set to 1, whereas the missing items are set to 0 as an approximation. Setting missing values to 0 is not common for most types of ratings matrices because doing so would cause bias in the predictions. However, it is generally considered an acceptable practice in sparse unary matrices because the most common value of an attribute is usually 0 in these cases. As a result, the effect of bias is relatively small, and one can now treat the matrix as a binary data set.

The first step of rule-based collaborative filtering is to discover all the association rules at a pre-specified level of minimum support and minimum confidence. The minimum support and minimum confidence can be viewed as parameters, which are tuned[2] to maximize predictive accuracy. Only those rules are retained in which the consequent contains exactly one item. This set of rules is the model, which can be used to perform recommendations for specific users. Consider a given customer A to which it desired to recommend relevant items. The first step is to determine all association rules that have been *fired* by customer A. An association rule is said to be fired by a customer A, if the itemset in the antecedent of the rule is a subset of the items preferred by that customer. All of the fired rules are then sorted in order of reducing confidence. The first k items discovered in the consequents of these sorted rules are recommended as the top-k items to customer A. The approach described here is a simplification of the algorithm described in [524]. Numerous other variations of this basic approach are used in the recommender systems literature. For example, sparsity can be addressed using dimensionality reduction methods [524].

The aforementioned association rules are based on unary ratings matrices, which allow the ability to specify likes, but they do not allow the ability to specify dislikes. However, numeric ratings can be easily handled by using variations of this basic methodology. When the number of possible ratings is small, each value of the rating-item combination can be treated as a pseudo-item. An example of such as pseudo-item is $(Item = Bread, Rating = Dislike)$. A new set of transactions is created in terms of these pseudo-items. The rules are then constructed in terms of these pseudo-items by using the approach as discussed earlier.

---

[2]Parameter-tuning methods, such as hold-out and cross-validation, are discussed in Chapter 7.

Therefore, such rules could appear as follows:

$$(Item = Bread, Rating = Like) \Rightarrow (Item = Eggs, Rating = Like)$$

$$(Item = Bread, Rating = Like) \text{ AND } (Item = Fish, Rating = Dislike)$$
$$\Rightarrow (Item = Eggs, Rating = Dislike)$$

For a given customer, the set of fired rules is determined by identifying the rules whose antecedents contain a subset of the pseudo-items for that user. The rules are sorted in decreasing order of confidence. These sorted rules can be used to predict ratings for items by selecting the top-k pseudo-items in the consequents of these rules. An additional step that might be required in this case is to resolve the conflicts between the various rules because different pseudo-items in the rules fired by a customer might be conflicting. For example, the pseudo-items $(Item = Bread, Rating = Like)$ and $(Item = Bread, Rating = Dislike)$ are conflicting pseudo-items. Such conflicts can be resolved by finding a way of aggregating the ratings in the consequents in order to create the final sorted list of recommendations. It is also possible to numerically aggregate the ratings in the consequents by using a variety of heuristics. For example, one can first determine all the fired rules in which the consequents correspond to an item of interest. The item ratings in the consequents of these fired rules are voted on in a weighted way in order to make a prediction for that user-item combination. One can weight the ratings in the fired rules by the corresponding confidence in the averaging process. For example, if two rules contain the rating "*like*" in the consequent (for a particular item), with confidences of 0.9 and 0.8, respectively, then the total number of votes for "*like*" for that item is $0.9 + 0.8 = 1.7$. The votes can be used to predict an average value of the rating for that item. Such predicted values can be determined for all items in the consequents of the fired rules. The resulting values can be used to sort the items in reducing order of priority. The voting approach is more appropriate when the granularity of the rating scale is very limited (e.g., *like* or *dislike*). In the case of interval-based ratings with high granularity, it is possible to discretize the ratings into a smaller number of intervals, and then use the same approach discussed above. Other heuristic methods for aggregating the predictions from rule-based methods are discussed in [18]. In many cases, it has been shown that the most effective results are not necessarily obtained by using the same support level for each item. Rather, it is often desirable to make the support level specific to the item whose rating is being predicted [358, 359, 365].

3.3.2   Item-Wise Models versus User-Wise Models

The dual relationship between user-wise and item-wise models is a recurrent theme in collaborative filtering. The neighborhood models of Chapter 2 provide the most well-known example of this duality. In general, every user-wise model can be converted to an item-wise model by applying it to the transpose of the rating matrix, and vice versa. Minor adjustments might sometimes be required to account for the varying semantic interpretations in the two cases. For example, one uses the adjusted cosine for similarity computation in item-based neighborhood models rather than the Pearson correlation coefficient.

The aforementioned discussion focuses on item-wise models for rule-based collaborative filtering. It is also possible to create user-wise models. These methods leverage user associations rather than item associations [358, 359]. In these cases, the rules associate the user tastes with one another rather than associating the item tastes with one another. Therefore, one works with *pseudo-users* corresponding to user-rating combinations. Examples of such rules are as follows:

$$(User = Alice, Rating = Like) \Rightarrow (User = Bob, Rating = Disike)$$

$$(User = Alice, Rating = Like) \text{ AND } (User = Peter, Rating = Dislike)$$
$$\Rightarrow (User = John, Rating = Like)$$

The first rule implies that Bob is likely to dislike items that Alice likes. The second rule implies that John is likely to like items that Alice likes and Peter dislikes. Such rules can be mined by applying exactly the same approach as the previous case on the *transpose* of the transaction matrix constructed from the pseudo-users. In other words, each list of pseudo-users for an item is now treated as a "transaction." Association rules are mined from this database at the required level of minimum support and confidence. In order to predict the rating of a user-item combination, the pseudo-user-based "transaction" for the relevant item is determined. Rules are fired by this transaction when the antecedent of this rule contains a subset of the pseudo-users in the transaction. All the fired rules are determined. Among these fired rules, all those in which the consequents correspond to the user of interest are determined. The ratings in the consequents of the fired rules may be averaged or voted on to make a prediction. The averaging process can be weighted with the confidence of the corresponding rule to provide a more robust prediction. Thus, the user-based approach is exactly analogous to the item-based approach. It is noteworthy that the two ways of performing collaborative filtering with association rules share a complimentary relationship, which is reminiscent of user-based and item-based neighborhood algorithms.

The association rule approach is useful not only for collaborative filtering, but also for content-based recommender systems, in which customer profiles are matched to specific items. These rules are referred to as *profile association rules*, and are used popularly for profile-based recommendations. It has been shown in [31, 32] how an efficient interactive interface can be constructed for performing profile-based recommendations for a variety of different types of queries.

Association rule-based recommender systems can be viewed as generalizations of rule-based systems that are used commonly for the classification problem [18]. The main difference is that consequents of the generated rules in the classification problem always contain the class variable. However, in the case of recommender systems, the consequents of the generated rules might contain[3] any item. Furthermore, the heuristics for sorting the fired rules and combining the possibly conflicting results from the rules are also similar in collaborative filtering and classification. This natural relationship between these methods is a direct result of the relationship between the classification and collaborative filtering problems. The main distinction between the two cases is that there is no clear demarcation between the feature variables and the class variables in collaborative filtering. This is why any association rule can be generated, rather than simply rules that contain the class variable in the consequent.

A number of comparative studies have shown [358, 359] that association rule systems can provide accurate results in certain types of settings. This is particularly true of unary data, which is commonly encountered in Web recommender systems. Association rule-based systems have found significant applications in Web-based personalization and recommender systems [441, 552]. The approach is naturally suited to Web personalization systems because it is specifically designed for sparse transaction data, which is commonly encountered in Web click behavior. Such methods can even be extended to include temporal information by using *sequential pattern mining models* [23].

---

[3]In the case of user-based associations, the consequents might contain any *user*.

3.4 Naive Bayes Collaborative Filtering

In the following, we will assume that there are a small number of distinct ratings, each of which can be treated as a categorical value. Therefore, the orderings among the ratings will be ignored in the following discussion. For example, three ratings, such as *Like*, *Neutral*, and *Dislike*, will be treated as unordered discrete values. In the case where the number of distinct ratings is small, such an approximation can be reasonably used without significant loss in accuracy.

Assume that there are l distinct values of the ratings, which are denoted by $v_1 \ldots v_l$. As in the case of the other models discussed in this chapter, we assume that we have an $m \times n$ matrix R containing the ratings of m users for n items. The (u, j)th entry of the matrix is denoted by r_{uj}.

The naive Bayes model is a generative model, which is commonly used for classification. One can treat the items as features and users as instances in order to infer the missing entries with a classification model. The main challenge in using this approach for collaborative filtering is that any feature (item) can be the target class in collaborative filtering, and one also has to work with incomplete feature variables. These differences can be handled with minor modifications to the basic methodology of the naive Bayes model.

Consider the uth user, who has specified ratings for the set of items I_u. In other words, if the uth row has specified ratings for the first, third, and fifth columns, then we have $I_i = \{1, 3, 5\}$. Consider the case where the Bayes classifier needs to predict the unobserved rating r_{uj} of user u for item j. Note that r_{uj} can take on any one of the discrete possibilities in $\{v_1 \ldots v_l\}$. Therefore, we would like to determine the probability that r_{uj} takes on any of these values *conditional on the observed ratings in* I_u. Therefore, for each value of $s \in \{1 \ldots l\}$, we would like to determine the probability $P(r_{uj} = v_s | Observed\ ratings\ in\ I_u)$. This expression appears in the form $P(A|B)$, where A and B are events corresponding to the value of r_{uj}, and the values of the observed ratings in I_u, respectively. The expression can be simplified using the well-known Bayes rule in probability theory:

$$P(A|B) = \frac{P(A) \cdot P(B|A)}{P(B)} \tag{3.3}$$

Therefore, for each value of $s \in \{1 \ldots l\}$, we have the following:

$$P(r_{uj} = v_s | Observed\ ratings\ in\ I_u) = \frac{P(r_{uj} = v_s) \cdot P(Observed\ ratings\ in\ I_u | r_{uj} = v_s)}{P(Observed\ ratings\ in\ I_u)} \tag{3.4}$$

We need to determine the value of s in the aforementioned expression for which the value of $P(r_{uj} = v_s | Observed\ ratings\ in\ I_u)$ on the left-hand side is as large as possible. It is noteworthy that the denominator on the right-hand side of Equation 3.4 is independent of the value of s. Therefore, in order to determine the value of s at which the right-hand side takes on the maximum value, one can ignore the denominator and express the aforementioned equation in terms of a constant of proportionality:

$$P(r_{uj} = v_s | Observed\ ratings\ in\ I_u) \propto P(r_{uj} = v_s) \cdot P(Observed\ ratings\ in\ I_u | r_{uj} = v_s) \tag{3.5}$$

If desired, the constant of proportionality can be derived by ensuring that all the resulting probability values $P(r_{uj} = v_s | Observed\ ratings\ in\ I_u)$ for $s \in \{1 \ldots l\}$ sum to 1. A key observation is that all the expressions on the right-hand side of Equation 3.5 can be estimated easily in a data-driven manner. The value of $P(r_{uj} = v_s)$, which is also referred to as the *prior probability* of rating r_{uj}, is estimated to the fraction of the users that have specified the rating v_s for the jth item. Note that the fraction is computed only out of those users that have rated item j, and the other users are ignored. The expression $P(Observed\ ratings\ in\ I_u | r_{uj} = v_s)$ is estimated with the use of the *naive assumption*. The naive assumption is based on *conditional independence* between the ratings. The conditional independence assumption says that the ratings of user u for various items in I_u are independent of one another, *conditional* of the fact that the value of r_{uj} was observed to be v_s. This condition may be mathematically expressed as follows:

$$P(Observed\ ratings\ in\ I_u | r_{uj} = v_s) = \prod_{k \in I_u} P(r_{uk} | r_{uj} = v_s) \qquad (3.6)$$

The value of $P(r_{uk} | r_{uj} = v_s)$ is estimated as the fraction of users that have specified the rating of r_{uk} for the kth item, given that they have specified the rating of their jth item to v_s. By plugging in the estimation of the prior probability $P(r_{uj} = v_s)$ and that of Equation 3.6 into Equation 3.5, it is possible to obtain an estimate of the posterior probability of the rating of item j for user u as follows:

$$P(r_{uj} = v_s | Observed\ ratings\ in\ I_u) \propto P(r_{uj} = v_s) \cdot \prod_{k \in I_u} P(r_{uk} | r_{uj} = v_s) \qquad (3.7)$$

This estimate of the posterior probability of the rating r_{uj} can be used to estimate its value in one of the following two ways:

1. By computing each of the expressions on the right-hand side of Equation 3.7 for each $s \in \{1 \ldots l\}$, and determining the value of s at which it is the largest, one can determine the most likely value $\hat{r}_{uj}$ of the missing rating r_{uj}. In other words, we have:

$$\hat{r}_{uj} = \text{argmax}_{v_s} P(r_{uj} = v_s | Observed\ ratings\ in\ I_u)$$
$$= \text{argmax}_{v_s} P(r_{uj} = v_s) \cdot \prod_{k \in I_u} P(r_{uk} | r_{uj} = v_s)$$

   Such an approach, however, treats a rating purely as a categorical value and ignores all ordering among the various ratings. When the number of possible ratings is small, this is a reasonable approach to use.

2. Rather than determining the rating that takes on the maximum probability, one can estimate the predicted value as the weighted average of all the ratings, where the weight of a rating is its probability. In other words, the weight of the rating v_s is proportional to the value of $P(r_{uj} = v_s | Observed\ ratings\ in\ I_u)$, as computed in Equation 3.7. Note that the constant of proportionality in the equation is irrelevant for computing the weighted average. Therefore, the estimated value $\hat{r}_{uj}$ of the missing rating r_{uj} in the matrix R is as follows:

$$\hat{r}_{uj} = \frac{\sum_{s=1}^{l} v_s \cdot P(r_{uj} = v_s | Observed\ ratings\ in\ I_u)}{\sum_{s=1}^{l} P(r_{uj} = v_s | Observed\ ratings\ in\ I_u)}$$
$$= \frac{\sum_{s=1}^{l} v_s \cdot P(r_{uj} = v_s) \cdot P(Observed\ ratings\ in\ I_u | r_{uj} = v_s)}{\sum_{s=1}^{l} P(r_{uj} = v_s) \cdot P(Observed\ ratings\ in\ I_u | r_{uj} = v_s)}$$
$$= \frac{\sum_{s=1}^{l} v_s \cdot P(r_{uj} = v_s) \cdot \prod_{k \in I_u} P(r_{uk} | r_{uj} = v_s)}{\sum_{s=1}^{l} P(r_{uj} = v_s) \cdot \prod_{k \in I_u} P(r_{uk} | r_{uj} = v_s)}$$

   This approach is preferable when the granularity of the ratings distribution is greater.

For a given user u, all her unobserved ratings are estimated using this approach. The items with the top-k estimated values of the ratings are reported.

It is noteworthy that this approach computes the conditional probability of a rating, based on the ratings of the other *items* (or dimensions). Therefore, this approach is an *item*-based Bayes approach. This approach is a straightforward adaptation of traditional classification methods, except that the predicted (class) dimension is fixed in traditional classification, whereas the predicted dimension varies in collaborative filtering. This difference occurs because collaborative filtering is a generalization of classification (cf. Figure 3.1). In the particular case of collaborative filtering, it is also possible to compute the probability of a rating based on the ratings of the other *users* for the same item (see Exercise 4). Such an approach can be viewed as a *user*-based Bayes approach. It is even possible to combine the predictions from the user-based and item-based Bayes methods. In virtually all forms of collaborative filtering, such as neighborhood-based and rule-based methods, it is possible to provide a solution from the user-based perspective, the item-based perspective, or a combination of the two methods.

3.4.1 Handling Overfitting

A problem arises when the underlying ratings matrix is sparse and the number of observed ratings is small. In such cases, the data-driven estimations may not remain robust. For example, the estimation of the prior probability $P(r_{uj} = v_s)$ is unlikely to be robust if a small number of users have specified ratings for the jth item. For example, if no user has specified a rating for the jth item, the estimation is of the form $0/0$, which is indeterminate. Furthermore, the estimation of each value $P(r_{uk}|r_{uj} = v_s)$ on the right-hand side of Equation 3.6 is likely to be even less robust than the estimation of the prior probability. This is because only a small portion of the ratings matrix will be conditional on the event $r_{uj} = v_s$. In this case, the portion of the ratings matrix that needs to be analyzed is only those users that have specified the rating v_s for item j. If the number of such users is small, the estimation will be inaccurate and the multiplicative terms in Equation 3.6 will produce a large error. For example, for any value of $k \in I_u$, if no user has specified the rating r_{uk} in cases where the rating of the jth item is set to v_s, the entire expression of Equation 3.6 will be set to 0 because of its multiplicative nature. This is, of course, an erroneous and overfitting result, which is obtained because of the estimation of the model parameters from a small amount of data.

In order to handle this problem, the method of Laplacian smoothing is commonly used. For example, let $q_1 \ldots q_l$ be the number of users that have respectively specified the ratings $v_1 \ldots v_l$ for the jth item. Then, instead of estimating $P(r_{uj} = v_s)$ in a straightforward way to $q_s / \sum_{t=1}^{l} q_t$, it is smoothed with a Laplacian smoothing parameter α:

$$P(r_{uj} = v_s) = \frac{q_s + \alpha}{\sum_{t=1}^{l} q_t + l \cdot \alpha} \tag{3.8}$$

Note that if no ratings are specified for the jth item, then such an approach will set the prior probability of each possible rating to $1/l$. The value of α controls the level of smoothing. Larger values of α will lead to more smoothing, but the results will become insensitive to the underlying data. An exactly similar approach can be used to smooth the estimation of $P(r_{uk}|r_{uj} = v_s)$, by adding α and $l \cdot \alpha$ to the numerator and denominator, respectively.

Table 3.2: Illustration of the Bayes method with a binary ratings matrix

Item-Id ⇒	1	2	3	4	5	6
User-Id ⇓						
1	1	-1	1	-1	1	-1
2	1	1	?	-1	-1	-1
3	?	1	1	-1	-1	?
4	-1	-1	-1	1	1	1
5	-1	?	-1	1	1	1

3.4.2 Example of the Bayes Method with Binary Ratings

In this section, we will illustrate the Bayes method with a binary ratings matrix on 5 users and 6 items. The ratings are drawn from $\{v_1, v_2\} = \{-1, 1\}$. This matrix is shown in Table 3.2. For ease in discussion, we will not use Laplacian smoothing although it is essential to do so in practice. Consider the case in which we wish to predict the ratings of the two unspecified items of user 3. Therefore, we need to compute the probabilities of the unspecified ratings r_{31} and r_{36} taking on each of the values from $\{-1, 1\}$, conditional on the observed values of the other ratings of user 3. By using Equation 3.7, we obtain the following posterior probability for the rating of item 1 by user 3:

$$P(r_{31} = 1 | r_{32}, r_{33}, r_{34}, r_{35}) \propto P(r_{31} = 1) \cdot P(r_{32} = 1 | r_{31} = 1) \cdot P(r_{33} = 1 | r_{31} = 1) \cdot$$
$$\cdot P(r_{34} = -1 | r_{31} = 1) \cdot P(r_{35} = -1 | r_{31} = 1)$$

The values of the individual terms on the right-hand side of the aforementioned equation are estimated using the data in Table 3.2 as follows:

$$P(r_{31} = 1) = 2/4 = 0.5$$
$$P(r_{32} = 1 | r_{31} = 1) = 1/2 = 0.5$$
$$P(r_{33} = 1 | r_{31} = 1) = 1/1 = 1$$
$$P(r_{34} = -1 | r_{31} = 1) = 2/2 = 1$$
$$P(r_{35} = -1 | r_{31} = 1) = 1/2 = 0.5$$

Upon substituting these values in the aforementioned equation, we obtain the following:

$$P(r_{31} = 1 | r_{32}, r_{33}, r_{34}, r_{35}) \propto (0.5)(0.5)(1)(1)(0.5) = 0.125$$

Upon performing the same steps for the probability of r_{31} taking on the value of -1, we obtain:

$$P(r_{31} = -1 | r_{32}, r_{33}, r_{34}, r_{35}) \propto (0.5) \left(\frac{0}{1}\right) \left(\frac{0}{2}\right) \left(\frac{0}{2}\right) \left(\frac{0}{2}\right) = 0$$

Therefore, the rating r_{31} has a higher probability of taking on the value of 1, as compared to -1, and its predicted value is set to 1. One can use a similar argument to show that the predicted value of the rating r_{36} is -1. Therefore, in a top-1 recommendation scenario, item 1 should be prioritized over item 6 in a recommendation to user 3.

3.5 Using an Arbitrary Classification Model as a Black-Box

Many other classification (or regression modeling) methods can be extended to the case of collaborative filtering. The main challenge in these methods is the incomplete nature of the underlying data. In the case of some classifiers, it is more difficult to adjust the model to handle the case of missing attribute values. An exception is the case of *unary* data, in which missing values are often estimated to be 0, and the specified entries are set to 1. Therefore, the underlying matrix resembles sparse binary data of high dimensionality. In such cases, the data can be treated as a complete data set and any classifiers that are designed for sparse and high dimensional data can be used. Fortunately, many forms of data, including customer transaction data, Web click data, or other activity data, can be formulated as a unary matrix. It is noteworthy that text data is also sparse and high-dimensional; as a result, many of the classification algorithms used in text mining can be directly adapted to these data sets. In fact, it has been shown in [669] that one can directly leverage the success of support vector machines in text data to (unary) collaborative filtering, albeit with a squared form of the loss function. The squared form of the loss function makes the model more akin to regularized linear regression. It has also been suggested in [669] that the use of rare class learning methods can be effective in collaborative filtering due to the imbalanced nature of the class distribution. For example, one might use different loss functions for the majority and minority classes while adapting the support vector machine to the collaborative filtering scenario. Numerous ad hoc methods have also been proposed to extend various classification and regression methods to collaborative filtering. For example, smoothing support vector machines [638] have been used to estimate the missing values in the user-item matrix in an iterative way.

For cases in which the ratings matrix is not unary, it is no longer possible to fill in the missing entries of the matrix with 0s without causing significant bias. This issue is discussed in detail in section 2.5 of Chapter 2. Nevertheless, as discussed in the same section, several dimensionality reduction methods can be used to create a low-dimensional representation of the data, which is fully specified. In such cases, any known classification method can be used effectively by treating the low-dimensional representation as the feature variables of the training data. Any column that needs to be completed is treated as the class variable. The main problem with this approach is a loss of interpretability in the classification process. When the reduced representation represents a linear combination of the original columns, it is difficult to provide any type of explanation of the predictions.

In order to work in the original feature space, it is possible to use classification methods as meta-algorithms in conjunction with iterative methods. In other words, an off-the-shelf classification algorithm is used as a *black-box* to predict the ratings of one of the items with the ratings of other items. How does one overcome the problem that the training columns haven been incompletely specified? The trick is to iteratively fill in the missing values of the training columns with successive refinement. This successive refinement is achieved with the use of our black-box, which is an off-the-shelf classification (or regression modeling) algorithm.

Consider an arbitrary classification/regression modeling algorithm $\mathcal{A}$, which is designed to work with a completely specified matrix. The first step is to initialize the missing entries in the matrix with row averages, column averages, or with any simple collaborative filtering algorithm. For example, one might use a simple user-based algorithm for the initialization process. As an optional enhancement, one might center each row of the ratings matrix as a preprocessing step to remove user bias. In this case, the bias of each user needs to be added back to the predicted values in a post-processing phase. Removing user bias during pre-processing often makes[4] the approach more robust. If the user bias is removed, then the missing entries are always filled in with row averages, which are 0.

These simple initializations and bias removal methods will still lead to prediction bias, when one attempts to use the artificially filled in values as training data. Then, the bias in the predicted entries can be iteratively reduced by using the following two-step iterative approach:

1. **(Iterative step 1):** Use algorithm $\mathcal{A}$ to estimate the missing entries of each column by setting it as the target variable and the remaining columns as the feature variables. For the remaining columns, use the current set of filled in values to create a complete matrix of feature variables. The observed ratings in the target column are used for training, and the missing ratings are predicted.

2. **(Iterative step 2):** Update all the missing entries based on the prediction of algorithm $\mathcal{A}$ on each target column.

These two steps are iteratively executed to convergence. The approach can be sensitive to the quality of the initialization and the algorithm $\mathcal{A}$. Nevertheless, the merit of the approach is that it is a simple method that can easily be implemented with any off-the-shelf classification or regression model. Numerical ratings can also be handled with a linear regression model. The work in [571] uses a similar approach in which the ratings matrix is imputed with artificial entries predicted by an ensemble of different classifiers.

3.5.1 Example: Using a Neural Network as a Black-Box

In this section, we will provide a simple example of the aforementioned approach, when neural networks are used as black-boxes to implement the approach. For the purpose of the following discussion, we will assume that the reader is already familiar with the basics of neural networks [87]. Nevertheless, we will introduce them very briefly to ensure continuity of discussion.

Neural networks simulate the human brain with the use of *neurons*, which are connected to one another via *synaptic connections*. In biological systems, learning is performed by changing the strength of synaptic connections in response to external stimuli. In artificial neural networks, the basic computation unit is also referred to as a neuron, and the strengths of the synaptic connections correspond to *weights*. These weights define the parameters

---

[4]It is also possible to use more sophisticated ways of removing bias for better performance. For example, the bias B_{ij}, which is specific to user i and item j, can be computed using the approach discussed in section 3.7.1. This bias is subtracted from observed entries and all missing entries are initialized to 0s during pre-processing. After computing the predictions, the biases B_{ij} are added back to the predicted values during postprocessing.

used by the learning algorithm. The most basic architecture of the neural network is the *perceptron*, which contains a set of input nodes and an output node. An example of a perceptron is shown in Figure 3.3(a). For a data set containing d different dimensions, there are d different input units. The output node is associated with a set of weights W, which is used to compute a function $f(\cdot)$ of the d inputs. A typical example of such a function is the signed linear function, which would work well for binary output:

$$z_i = \text{sign}\{\overline{W} \cdot \overline{X_i} + b\} \tag{3.9}$$

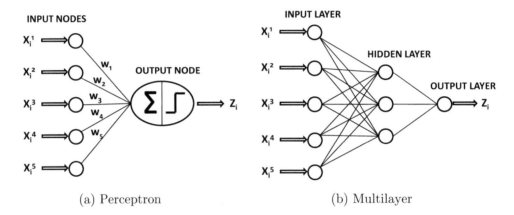

(a) Perceptron   (b) Multilayer

Figure 3.3: Single and multilayer neural networks

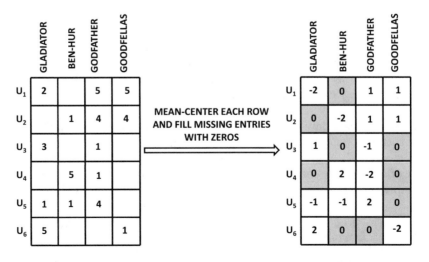

Figure 3.4: Pre-processing the ratings matrix. Shaded entries are iteratively updated.

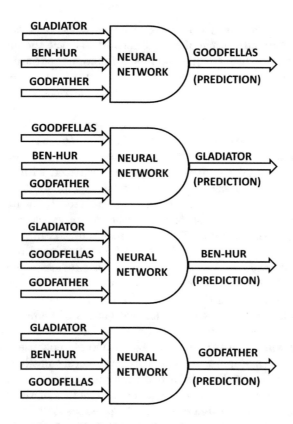

Figure 3.5: Neural networks for predicting and updating missing entries. Shaded entries of Figure 3.4 are iteratively updated by the neural networks.

Here, $\overline{X_i}$ is a d-dimensional row vector defining the d inputs of the ith training instance, and $\overline{W}$ is the coefficient vector. In the context of collaborative filtering, the d inputs correspond to the $(n-1)$ items, which are used to predict the rating of the remaining item. Assume that the label of the ith instance is y_i. In the context of collaborative filtering, y_i represents the observed ratings of the items being predicted. The parameter b denotes the bias. One can already notice the similarity of this approach with linear regression although the prediction function is slightly different. The value of z_i is the predicted output, and the error $(z_i - y_i)^2$ of this predicted output is used to update the weights in $\overline{W}$ in a manner similar to linear regression. This update is similar to the updates in gradient descent, which are made for least-squares optimization. In the case of neural networks, the update function is as follows:

$$\overline{W}^{t+1} = \overline{W}^t + \alpha(y_i - z_i)\overline{X_i} \tag{3.10}$$

Here, $\alpha > 0$ denotes the learning rate and $\overline{W}^t$ is the value of the weight vector in the tth iteration. It is not difficult to show that the incremental update vector, is the negative gradient of $(y_i - z_i)^2$ with respect to $\overline{W}$. We iterate through all the observed ratings in the item being predicted in order to make these updates. Since it was assumed that y_i is binary, this approach is designed for binary ratings matrices. One can also design neural networks in which the output need not be binary, and the prediction function need not be linear.

In general, a neural network can have multiple layers, and the intermediate nodes can compute nonlinear functions. An example of such a multi-layer neural network is illustrated in Figure 3.3(b). Of course, such a network would have a larger number of learning parameters. The corresponding learning algorithm is referred to as the *back-propagation algorithm* [87]. The main advantage of neural networks is that the multi-layer architecture provides the ability to compute complex nonlinear functions that are not easily computable with other classification methods. Therefore, neural networks are also referred to as *universal function approximators*. For noisy data like ratings matrices, regularization can be used to reduce the impact of noise.

Consider a ratings matrix with four items, illustrated on the left-hand side of Figure 3.4. In this example, the items correspond to movies. The first step is to mean-center each row, in order to remove user biases. The resulting mean-centered matrix is shown on the right-hand side of Figure 3.4. Note that the missing values are replaced with the corresponding row average, which is 0 after mean-centering. Since there are four items, there are four possible neural network models, whereby each model is constructed by using the ratings input of the other three items as training columns, and the fourth as the test column. These four neural networks are shown in Figure 3.5. The completed matrix of Figure 3.4 is used to train each of these neural networks in the first iteration. For each column of the ratings matrix, the relevant neural network in Figure 3.5 is used for prediction purposes. The resulting predictions made by the neural networks are then used to create a new matrix in which the missing entries are updated with the predicted values. In other words, the neural networks are used only to update the values in the shaded entries of Figure 3.4 with the use of an off-the-shelf neural network training and prediction procedure. After the update, the shaded entries of Figure 3.4 will no longer be zeros. This matrix is now used to predict the entries for the next iteration. This approach is repeated iteratively until convergence. Note that each iteration requires the application of n training procedures, where n is the number of items. However, one does not need to learn the parameters of the neural networks from scratch in each iteration. The parameters from the previous iteration can be used as a good starting point. It is important to use regularization because of the high dimensionality of the underlying data [220].

This model can be considered an *item-wise* model, in which the inputs represent the ratings of various items. It is also possible to create a *user-wise* model [679], in which the inputs correspond to the ratings of various users. The main challenge with such an approach is that the number of inputs to the neural network becomes very large. Therefore, it is recommended in [679] that not all users should be used as input nodes. Rather, only users who have rated at least a minimum threshold number of items are used. Furthermore, the users should not all be highly similar to one another. Therefore, heuristics are proposed in [679] to preselect mutually diverse users in the initial phase. This approach can be considered a type of feature selection for neural networks, and it can also be used in the item-wise model.

3.6 Latent Factor Models

In section 2.5 of Chapter 2, we discussed some dimensionality reduction methods to create a new fully specified representation of an incomplete data set. In Chapter 2, a number of heuristic methods were discussed, which create a full dimensional representation for enabling the use of neighborhood algorithms [525]. Such data reduction techniques are also used to enable other model-based methods, which use classification algorithms as a subroutine. Therefore, in all the methods previously discussed, dimensionality reduction only plays an *enabling* role of creating a more convenient data representation for other model-based methods. In this chapter, more sophisticated methods will be discussed, because the goal is to use dimensionality reduction methods to directly estimate the data matrix in one shot.

The earliest discussions on the use of latent factor models as a direct method for matrix completion may be found in [24, 525]. The basic idea is to exploit the fact that significant portions of the rows and columns of data matrices are highly correlated. As a result, the data has built-in redundancies and the resulting data matrix is often approximated quite well by a *low-rank* matrix. Because of the inherent redundancies in the data, the *fully specified* low-rank approximation can be determined even with a small subset of the entries in the original matrix. This fully-specified low rank approximation often provides a robust estimation of the missing entries. The approach in [24] combines the expectation-maximization (EM) technique with dimensionality reduction to reconstruct the entries of the incomplete data matrix.

Latent factor models are considered to be state-of-the-art in recommender systems. These models leverage well-known dimensionality reduction methods to fill in the missing entries. Dimensionality reduction methods are used commonly in other areas of data analytics to represent the underlying data in a small number of dimensions. The basic idea of dimensionality reduction methods is to rotate the axis system, so that pairwise correlations between dimensions are removed. The key idea in dimensionality reduction methods is that the reduced, rotated, and completely specified representation can be robustly estimated from an incomplete data matrix. Once the completely specified representation has been obtained, one can rotate it back to the original axis system in order to obtain the fully specified representation [24]. Under the covers, dimensionality reduction methods leverage the row and column correlations to create the fully specified and reduced representation. The use of such correlations is, after all, fundamental to all collaborative filtering methods, whether they are neighborhood methods or model-based methods. For example, user-based neighborhood methods leverage user-wise correlations, whereas item-based neighborhood methods leverage item-wise correlations. Matrix factorization methods provide a neat way to leverage all row and column correlations in one shot to estimate the entire data matrix. This sophistication of the approach is one of the reasons that latent factor models have become the state-of-the-art in collaborative filtering. In order to understand why latent factor models are effective, we will provide two pieces of intuition, one of which is geometric and the other elucidates the semantic interpretation directly. Both these intuitions show how data redundancies in highly correlated data can be exploited to create a low-rank approximation.

3.6.1 Geometric Intuition for Latent Factor Models

We will first provide a geometric intuition for latent factor models, based on a discussion provided in [24]. In order to understand the intuition of how the notions of low-rank, redundancy, and correlation are related, consider a ratings matrix with three items, in which all three items are positively correlated. Assume a movie rating scenario, in which the three items correspond to *Nero*, *Gladiator*, and *Spartacus*. For ease of discussion, assume that the ratings are continuous values, which lie in the range $[-1, 1]$. If the ratings are positively correlated, then the 3-dimensional scatterplot of the ratings might be roughly arranged along a 1-dimensional line, as shown in Figure 3.6. Since the data is mostly arranged along a 1-dimensional line, it means that the original data matrix has a rank of approximately 1 after removing the noisy variations. For example, the rank-1 approximation of Figure 3.6 would be the 1-dimensional line (or *latent vector*) that passes through the center of the data and aligned with the elongated data distribution. Note that dimensionality reduction methods such as Principal Component Analysis (PCA) and (mean-centered) Singular Value Decomposition (SVD) typically represent the projection of the data along this line as an approximation. When the $m \times n$ ratings matrix has a rank of $p \ll \min\{m, n\}$ (after

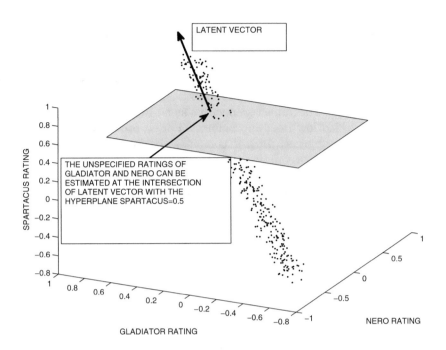

Figure 3.6: Leveraging correlation-based redundancies in missing data estimation for a user whose only specified rating is a value of 0.5 for the movie *Spartacus*

removing noisy variations), the data can be approximately represented on a p-dimensional hyperplane. In such cases, the missing ratings of a user can often be robustly estimated with as few as p specified entries as long as the p-dimensional hyperplane is known. For example, in the case of Figure 3.6, only one rating needs to be specified in order to determine the other two ratings, because the rank of the ratings matrix is only 1 after noise removal. For example, if the rating of *Spartacus* is fixed at 0.5, then the ratings of *Nero* and *Gladiator* can be estimated[5] as the intersection of the 1-dimensional latent vector with the axis-parallel hyperplane, in which the rating of *Spartacus* is fixed to 0.5. This hyperplane is illustrated in Figure 3.6. Therefore, dimensionality reduction methods such as SVD leverage the inter-attribute correlations and redundancies in order to infer unspecified entries.

In this case, it was assumed that a specified data matrix was available to estimate the relevant latent vector. In practice, the data matrix does not need to be fully specified in order to estimate the *dominant* latent vectors, such as the line aligned with the elongated shape of the data distribution in Figure 3.6. The ability to estimate such latent vectors with missing data is the key to the success of the latent factor approach. The basic idea in all these methods is to find a set of latent vectors, in which the average squared distance of the data points (representing individual user ratings) from the hyperplane defined by these latent vectors is as small as possible. Therefore, *we must use a partially specified data set to recover the low-dimensional hyperplane on which the data approximately lies*. By doing so, we can implicitly capture the underlying redundancies in the correlation structure of the data and reconstruct all the missing values in one shot. It is the knowledge of these implicit redundancies that helps us to predict the missing entries in the matrix. It is noteworthy that if the data does not have any correlations or redundancies, then a latent factor model will simply not work.

---

[5] A detailed description of the method used for performing this estimation in various scenarios is discussed in section 3.6.5.3.

3.6.2 Low-Rank Intuition for Latent Factor Models

The *geometric* intuition of the previous section is helpful in understanding the impact of latent vectors when they are mutually orthogonal. However, latent vectors are not always mutually orthogonal. In such cases, it is helpful to obtain some intuition from linear algebra. One way of understanding the effectiveness of latent factor models is by examining the role that *factorization* plays in such matrices. Factorization is, in fact, a more general way of approximating a matrix when it is prone to dimensionality reduction because of correlations between columns (or rows). Most dimensionality reduction methods can also be expressed as matrix factorizations.

First, let us consider the simple case in which all entries in the ratings matrix R are observed. The key idea is that any $m \times n$ matrix R of rank $k \ll \min\{m, n\}$ can always be expressed in the following product form of rank-k factors:

$$R = UV^T \tag{3.11}$$

Here, U is an $m \times k$ matrix, and V is an $n \times k$ matrix. Note that the rank of both the row space[6] and the column space of R is k. Each column of U be viewed as one of the k basis vectors of the k-dimensional column space of R, and the jth row of V contains the corresponding coefficients to combine these basis vectors into the jth column of R. Alternatively, one can view the columns of V as the basis vectors of the row space of R, and the rows of U as the corresponding coefficients. The ability to factorize any rank-k matrix in this form is a fundamental fact of linear algebra [568], and there are an infinite number of such factorizations corresponding to various sets of basis vectors. SVD is one example of such a factorization in which the basis vectors represented by the columns of U (and the columns of V) are orthogonal to one another.

Even when the matrix R has rank larger than k, it can often be *approximately* expressed as the product of rank-k factors:

$$R \approx UV^T \tag{3.12}$$

As before, U is an $m \times k$ matrix, and V is an $n \times k$ matrix. The error of this approximation is equal to $||R - UV^T||^2$, where $|| \cdot ||^2$ represents the sum of the squares of the entries in the resulting *residual matrix* $(R - UV^T)$. This quantity is also referred to as the (squared) *Frobenius norm* of the residual matrix. The residual matrix typically represents the noise in the underlying ratings matrix, which cannot be modeled by the low-rank factors. For simplicity in discussion, let us consider the straightforward case in which R is fully observed. We will first examine the intuition behind the factorization process, and then we will discuss the implication of this intuition in the context of matrices with missing entries.

What is the implication of the factorization process, and its impact on a matrix with highly correlated rows and columns? In order to understand this point, consider the ratings matrix illustrated in Figure 3.7. In this figure, a 7×6 ratings matrix with 7 users and 6 items is illustrated. All ratings are drawn from $\{1, -1, 0\}$, which correspond to like, dislike, and neutrality. The items are movies, and they belong to the romance and history genres, respectively. One of the movies, titled *Cleopatra*, belongs to both genres. Because of the nature of the genres of the underlying movies, users also show clear trends in their ratings. For example, users 1 to 3 typically like historical movies, but they are neutral to the romance genre. User 4 likes movies of both genres. Users 5 to 7 like movies belonging to the romance genre, but they explicitly dislike historical movies. Note that this matrix has a significant

---

[6]The row space of a matrix is defined by all possible linear combinations of the rows of the matrix. The column space of a matrix is defined by all possible linear combinations of the columns of the matrix.

number of correlations among the users and items, although the ratings of movies belonging to the two distinct genres seem to be relatively independent. As a result, this matrix can be approximately factorized into rank-2 factors, as shown in Figure 3.7(a). The matrix U is a 7×2 matrix, which shows the proclivity of users towards the two genres, whereas the matrix V is a 6×2 matrix, which shows the membership of the movies in the two genres. In other words, the matrix U provides the basis for the column space, whereas the matrix V provides the basis for the row space. For example, the matrix U shows that user 1 likes history movies, whereas user 4 likes both genres. A similar inference can be made using the rows of V. The columns of V correspond to the latent vectors, such as those shown in Figure 3.6. Unlike SVD, however, the latent vectors in this case are not mutually orthogonal.

The corresponding residual matrix for the factorization is shown in Figure 3.7(b). The residual matrix typically corresponds to the ratings of users for *Cleopatra*, which do not follow the set pattern. It needs to be pointed out that in real-world applications, the matrix entries in the factors are typically real numbers (rather than integral). An example with integral factors is shown here for visual simplicity. Furthermore, a neat semantic interpretation of the factors in terms of genres or categories is sometimes not possible, especially when the factors contain both positive and negative values. For example, if we multiply both U and V with -1 in Figure 3.7, the factorization is still valid, but the interpretation becomes more difficult. Nevertheless, the k columns of U and V do represent key correlations among the users and items, respectively, and they can be viewed abstractly as *latent concepts*, whether or not they are semantically interpretable. In some forms of factorization, such as non-negative matrix factorization, the interpretability of these concepts is retained to a greater degree.

In this example, the matrix R was fully specified, and therefore the factorization is not particularly helpful from the perspective of missing value estimation. The key usefulness of the approach arises when the matrix R is not fully specified, but one can still robustly estimate *all* entries of the latent factors U and V, respectively. *For low values of the rank,* this is still possible from sparsely specified data. This is because one does not need too many observed entries to estimate the latent factors from inherently redundant data. Once the matrices U and V have been estimated, the entire ratings matrix can be estimated as UV^T in one shot, which provides all the missing ratings.

3.6.3 Basic Matrix Factorization Principles

In the basic matrix factorization model, the $m \times n$ ratings matrix R is approximately factorized into an $m \times k$ matrix U and an $n \times k$ matrix V, as follows:

$$R \approx UV^T \tag{3.13}$$

Each column of U (or V) is referred to as a latent *vector* or latent *component*, whereas each row of U (or V) is referred to as a latent *factor*. The ith row $\overline{u_i}$ of U is referred to as a *user factor*, and it contains k entries corresponding to the affinity of user i towards the k concepts in the ratings matrix. For example, in the case of Figure 3.7, $\overline{u_i}$ is a 2-dimensional vector containing the affinity of user i towards the history and romance genres in the ratings matrix. Similarly, each row $\overline{v_i}$ of V is referred to as an *item factor*, and it represents the affinity of the ith item towards these k concepts. In Figure 3.7, the item factor contains the affinity of the item towards the two categories of movies.

From Equation 3.13, it follows that each rating r_{ij} in R can be approximately expressed as a dot product of the ith user factor and jth item factor:

$$r_{ij} \approx \overline{u_i} \cdot \overline{v_j} \tag{3.14}$$

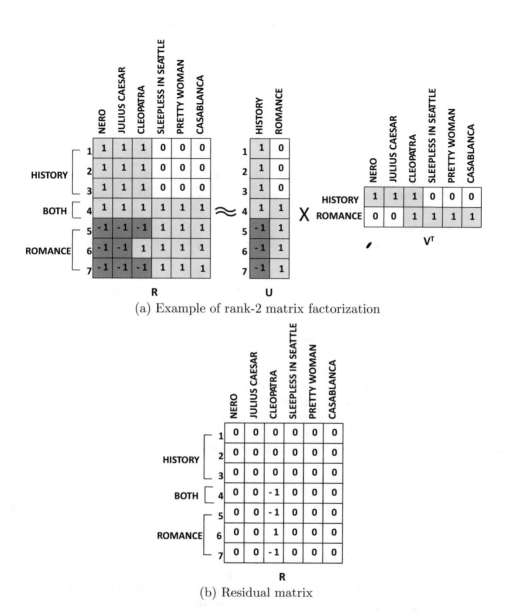

(a) Example of rank-2 matrix factorization

(b) Residual matrix

Figure 3.7: Example of a matrix factorization and its residual matrix

Since the latent factors $\overline{u_i} = (u_{i1} \ldots u_{ik})$ and $\overline{v_j} = (v_{j1} \ldots v_{jk})$ can be viewed as the affinities of the users for k different concepts, an intuitive way of expressing Equation 3.14 would be as follows:

$$r_{ij} \approx \sum_{s=1}^{k} u_{is} \cdot v_{js}$$

$$= \sum_{s=1}^{k} (\textit{Affinity of user i to concept s}) \times (\textit{Affinity of item j to concept s})$$

In the case of Figure 3.7, the two concepts in the aforementioned summation correspond to the romance and historical genres. Therefore, the summation may be expressed as follows:

$$r_{ij} \approx (\textit{Affinity of user i to history}) \times (\textit{Affinity of item j to history})$$
$$+ (\textit{Affinity of user i to romance}) \times (\textit{Affinity of item j to romance})$$

It needs to be pointed out that the notion of concepts is often not semantically interpretable, as illustrated in Figure 3.7. A latent vector may often be an arbitrary vector of positive and negative values and it becomes difficult to give it a semantic interpretation. However, it does represent a dominant correlation pattern in the ratings matrix, just as the latent vector of Figure 3.6 represents a geometric correlation pattern. As we will see later, some forms of factorization, such as non-negative matrix factorization, are explicitly designed to achieve greater interpretability in the latent vectors.

The key differences among various matrix factorization methods arise in terms of the constraints imposed on U and V (e.g., orthogonality or non-negativity of the latent vectors) and the nature of the objective function (e.g., minimizing the Frobenius norm or maximizing the likelihood estimation in a generative model). These differences play a key role in the usability of the matrix factorization model in various real-world scenarios.

3.6.4 Unconstrained Matrix Factorization

The most fundamental form of matrix factorization is the unconstrained case, in which no constraints are imposed on the factor matrices U and V. Much of the recommendation literature refers to unconstrained matrix factorization as singular value decomposition (SVD). Strictly speaking, this is technically incorrect; in SVD, the columns of U and V must be orthogonal. However, the use of the term "SVD" to refer to unconstrained matrix factorization[7] is rather widespread in the recommendation literature, which causes some confusion to practitioners from outside the field. In this chapter, we will deviate from this incorrect practice and treat unconstrained matrix factorization and SVD in a distinct way. This section will discuss unconstrained matrix factorization, and the following section will discuss SVD.

Before discussing the factorization of incomplete matrices, let us first visit the problem of factorizing fully specified matrices. How can one determine the factor matrices U and V,

---

[7]In SVD [568], the basis vectors are also referred to as *singular* vectors, which, by definition, must be mutually orthonormal.

so that the fully specified matrix R matches UV^T as closely as possible? One can formulate an optimization problem with respect to the matrices U and V in order to achieve this goal:

$$\text{Minimize } J = \frac{1}{2}||R - UV^T||^2$$

subject to:

No constraints on U and V

Here, $||.||^2$ represents the squared Frobenius norm of the matrix, which is equal to the sum of the squares of the matrix entries. Thus, the objective function is equal to the sum of the squares of the entries in the residual matrix $(R - UV^T)$. The smaller the objective function is, the better the quality of the factorization $R \approx UV^T$ will be. This objective function can be viewed as a *quadratic loss* function, which quantifies the loss of accuracy in estimating the matrix R with the use of low-rank factorization. A variety of gradient descent methods can be used to provide an optimal solution to this factorization.

However, in the context of a matrix with *missing entries*, only a subset of the entries of R are known. Therefore, the objective function, as written above, is undefined as well. After all, one cannot compute the Frobenius norm of a matrix in which some of the entries are missing! The objective function, therefore, needs to be rewritten only in terms of the observed entries in order to learn U and V. The nice part about this process is that once the latent factors U and V are learned, *the entire ratings matrix can be reconstructed as* UV^T *in one shot.*

Let the set of all user-item pairs (i, j), which are observed in R, be denoted by S. Here, $i \in \{1 \ldots m\}$ is the index of a user, and $j \in \{1 \ldots n\}$ is the index of an item. Therefore, the set S of observed user-item pairs is defined as follows:

$$S = \{(i, j) : r_{ij} \text{ is observed}\} \tag{3.15}$$

If we can somehow factorize the incomplete matrix R as the approximate product UV^T of fully specified matrices $U = [u_{is}]_{m \times k}$ and $V = [v_{js}]_{n \times k}$, then all the entries in R can be predicted as well. Specifically, the (i, j)th entry of matrix R can be predicted as follows:

$$\hat{r}_{ij} = \sum_{s=1}^{k} u_{is} \cdot v_{js} \tag{3.16}$$

Note the "hat" symbol (i.e., circumflex) on the rating on the left-hand side to indicate that it is a predicted value rather than an observed value. The difference between the observed and predicted value of a specified entry (i, j) is given by $e_{ij} = (r_{ij} - \hat{r}_{ij}) = (r_{ij} - \sum_{s=1}^{k} u_{is} \cdot v_{js})$. Then, the modified objective function, which works with incomplete matrices, is computed only over the observed entries in S as follows:

$$\text{Minimize } J = \frac{1}{2} \sum_{(i,j) \in S} e_{ij}^2 = \frac{1}{2} \sum_{(i,j) \in S} \left(r_{ij} - \sum_{s=1}^{k} u_{is} \cdot v_{js} \right)^2$$

subject to:

No constraints on U and V

Note that the aforementioned objective function sums up the error *only over the observed entries in S.* Furthermore, each of the terms $(r_{ij} - \sum_{s=1}^{k} u_{is} \cdot v_{js})^2$ is the squared error e_{ij}^2 between the observed and predicted values of the entry (i, j). Here, u_{is} and v_{js} are the

Algorithm GD(Ratings Matrix: R, Learning Rate: α)
begin
    Randomly initialize matrices U and V;
    $S = \{(i,j) : r_{ij} \text{ is observed}\}$;
    while not(convergence) **do**
    begin
        Compute each error $e_{ij} \in S$ as the observed entries of $R - UV^T$;
        for each user-component pair (i,q) **do** $u_{iq}^+ \Leftarrow u_{iq} + \alpha \cdot \sum_{j:(i,j)\in S} e_{ij} \cdot v_{jq}$;
        for each item-component pair (j,q) **do** $v_{jq}^+ \Leftarrow v_{jq} + \alpha \cdot \sum_{i:(i,j)\in S} e_{ij} \cdot u_{iq}$;
        for each user-component pair (i,q) **do** $u_{iq} \Leftarrow u_{iq}^+$;
        for each item-component pair (j,q) **do** $v_{jq} \Leftarrow v_{jq}^+$;
        Check convergence condition;
    end
end

Figure 3.8: Gradient descent

unknown variables, which need to be learned to minimize the objective function. This can be achieved simply with gradient descent methods. Therefore, one needs to compute the partial derivative of J with respect to the decision variables u_{iq} and v_{jq}:

$$\frac{\partial J}{\partial u_{iq}} = \sum_{j:(i,j)\in S} \left(r_{ij} - \sum_{s=1}^{k} u_{is} \cdot v_{js} \right)(-v_{jq}) \quad \forall i \in \{1 \ldots m\}, q \in \{1 \ldots k\}$$

$$= \sum_{j:(i,j)\in S} (e_{ij})(-v_{jq}) \quad \forall i \in \{1 \ldots m\}, q \in \{1 \ldots k\}$$

$$\frac{\partial J}{\partial v_{jq}} = \sum_{i:(i,j)\in S} \left(r_{ij} - \sum_{s=1}^{k} u_{is} \cdot v_{js} \right)(-u_{iq}) \quad \forall j \in \{1 \ldots n\}, q \in \{1 \ldots k\}$$

$$= \sum_{i:(i,j)\in S} (e_{ij})(-u_{iq}) \quad \forall j \in \{1 \ldots n\}, q \in \{1 \ldots k\}$$

Note that the entire vector of partial derivatives provides us with the gradient with respect to the vector of $(m \cdot k + n \cdot k)$ decision variables in the matrices U and V. Let this gradient vector be denoted by $\overline{\nabla J}$. Let the vector of $(m \cdot k + n \cdot k)$ decision variables corresponding to the entries in U and V be denoted by $\overline{VAR}$. Then, one can update the entire vector of decision variables as $\overline{VAR} \Leftarrow \overline{VAR} - \alpha \cdot \overline{\nabla J}$. Here, $\alpha > 0$ is the step size, which can be chosen using standard numerical methods in nonlinear programming [76]. In many cases, the step sizes are set to small constant values. The iterations are executed to convergence. This approach is referred to as *gradient descent*. The algorithmic framework for gradient-descent is illustrated in Figure 3.8. It is noteworthy that the intermediate variables u_{iq}^+ and v_{jq}^+ are used to ensure that all updates to the entries in U and V are performed simultaneously.

One can also perform the updates in Figure 3.8 using a matrix representation. The first step is to compute an error matrix $E = R - UV^T$ in which the unobserved entries of E (i.e., entries not in S) are set to 0. Note that E is a very sparse matrix, and it makes sense to compute the value of e_{ij} for only the observed entries $(i,j) \in S$ and store the matrix using a sparse data structure. Subsequently, the updates can be computed as follows:

$$U \Leftarrow U + \alpha E V$$
$$V \Leftarrow V + \alpha E^T U$$

These updates can be executed to convergence, while taking care to update all entries in both matrices simultaneously with the use of intermediate variables (as in Figure 3.8).

3.6.4.1 Stochastic Gradient Descent

The aforementioned method is referred to as the *batch update method*. An important observation is that the updates are linear functions of the errors in the observed entries of the ratings matrix. The update can be executed in other ways by *decomposing* it into smaller components associated with the errors in *individual* observed entries rather than all entries. This update can be *stochastically approximated* in terms of the error in a (randomly chosen) observed entry (i, j) as follows:

$$\cdot\; u_{iq} \Leftarrow u_{iq} - \alpha \cdot \left[\frac{\partial J}{\partial u_{iq}}\right]_{\text{Portion contributed by } (i,j)} \quad \forall q \in \{1 \ldots k\}$$

$$v_{jq} \Leftarrow v_{jq} - \alpha \cdot \left[\frac{\partial J}{\partial v_{jq}}\right]_{\text{Portion contributed by } (i,j)} \quad \forall q \in \{1 \ldots k\}$$

One can cycle through the observed entries in R one at a time (in random order) and update only the relevant set of $2 \cdot k$ entries in the factor matrices rather than all $(m \cdot k + n \cdot k)$ entries in the factor matrices. In such a case, the $2 \cdot k$ updates *specific to the observed entry* $(i, j) \in S$, are as follows:

$$u_{iq} \Leftarrow u_{iq} + \alpha \cdot e_{ij} \cdot v_{jq} \quad \forall q \in \{1 \ldots k\}$$
$$v_{jq} \Leftarrow v_{jq} + \alpha \cdot e_{ij} \cdot u_{iq} \quad \forall q \in \{1 \ldots k\}$$

For each observed rating r_{ij}, the error e_{ij} is used to update the k entries in row i of U and the k entries in the row j of V. Note that $e_{ij} \cdot v_{jq}$ is the component of partial derivative of J with respect to u_{iq}, that *is specific to* a single observed entry (i, j). For better efficiency, each of these k entries can be updated simultaneously in vectorized form. Let $\overline{u_i}$ be the ith row of U and $\overline{v_j}$ be the jth row of V. Then, the aforementioned updates can be rewritten in k-dimensional vectorized form as follows:

$$\overline{u_i} \Leftarrow \overline{u_i} + \alpha\, e_{ij}\, \overline{v_j}$$
$$\overline{v_j} \Leftarrow \overline{v_j} + \alpha\, e_{ij}\, \overline{u_i}$$

We cycle through all the observed entries multiple times (i.e., use multiple iterations) until convergence is reached. This approach is referred to as *stochastic gradient descent* in which the gradient is approximated by that computed on the basis of the error of a single randomly chosen entry in the matrix. The pseudo-code for the stochastic gradient descent method is illustrated in Figure 3.9. It is noteworthy that temporary variables u_{iq}^+ and v_{jq}^+ are used to store intermediate results during an update, so that the $2 \cdot k$ updates do not affect each other. This is a general approach that should be used in all group-wise updates discussed in this book, although we might not state it explicitly.

In practice, faster convergence is achieved by the stochastic gradient descent method as compared to the batch method, although the convergence is much smoother in the latter.

Algorithm SGD(Ratings Matrix: R, Learning Rate: α)
begin
 Randomly initialize matrices U and V;
 $S = \{(i,j) : r_{ij} \text{ is observed}\}$;
 while not(convergence) **do**
 begin
  Randomly shuffle observed entries in S;
  for each $(i,j) \in S$ in shuffled order **do**
  begin
   $e_{ij} \Leftarrow r_{ij} - \sum_{s=1}^{k} u_{is}v_{js}$;
   for each $q \in \{1 \ldots k\}$ **do** $u_{iq}^{+} \Leftarrow u_{iq} + \alpha \cdot e_{ij} \cdot v_{jq}$;
   for each $q \in \{1 \ldots k\}$ **do** $v_{jq}^{+} \Leftarrow v_{jq} + \alpha \cdot e_{ij} \cdot u_{iq}$;
   for each $q \in \{1 \ldots k\}$ **do** $u_{iq} = u_{iq}^{+}$ and $v_{jq} = v_{jq}^{+}$;
  end
  Check convergence condition;
 end
end

Figure 3.9: Stochastic gradient descent

This is because the entries of U and V are updated simultaneously in the latter case with the use of all observed entries, rather than a single randomly chosen observed entry. This noisy approximation of stochastic gradient descent can sometimes impact solution quality and smoothness of convergence. In general, stochastic gradient descent is preferable when the data size is very large and computational time is the primary bottleneck. In other "compromise" methods, mini-batches are used in which a subset of observed entries is used to construct the update. These different methods provide different trade-offs between solution quality and computational efficiency.

As one repeatedly cycles through the observed entries in the matrix to update the factor matrices, convergence will eventually be reached. In general, the global method is known to have guaranteed convergence, even though it is generally slower than the local method. A typical value of the step size (or *learning rate*) is a small constant value such as $\alpha = 0.005$. A more effective approach to avoid local minima and speed up convergence is to use the *bold driver algorithm* [58, 217] to select α adaptively in each iteration. It is also possible, in principle, to use different step sizes for different factors [586]. An interesting observation about some of these models is that executing them until convergence for too many iterations can sometimes lead to slight worsening of the solution quality on the unobserved entries. Therefore, it is sometimes advisable not to set the convergence criteria too strictly.

Another issue with these latent factor models is that of *initialization*. For example, one can initialize the factor matrices to small numbers in $(-1, 1)$. However, the choice of initialization can affect the final solution quality. It is possible to use a number of heuristics to improve quality. For example, one can use some simple SVD-based heuristics, discussed later in this section, to create an approximate initialization.

3.6.4.2 Regularization

One of the main problems with this approach arises when the ratings matrix R is sparse and relatively few entries are observed. This is almost always the case in real settings.

In such cases, the observed set S of ratings is small, which can cause overfitting. Note that overfitting is also a common problem in classification when training data are limited. A common approach for addressing this problem is to use *regularization*. Regularization reduces the tendency of the model to overfit at the expense of introducing a *bias*[8] in the model.

In regularization, the idea is to discourage very large values of the coefficients in U and V in order to encourage stability. Therefore, a regularization term, $\frac{\lambda}{2}(||U||^2 + ||V||^2)$, is added to the objective function, where $\lambda > 0$ is the regularization parameter. Here, $||\cdot||^2$ denotes the (squared) Frobenius norm of the matrix. The basic idea is to create a bias in favor of simpler solutions by penalizing large coefficients. This is a standard approach, which is used in many forms of classification and regression, and also leveraged by collaborative filtering. The parameter λ is always non-negative and it controls the weight of the regularization term. The method for choosing λ is discussed later in this section.

As in the previous case, assume that $e_{ij} = (r_{ij} - \sum_{s=1}^{k} u_{is} \cdot v_{js})$ represents the difference between the observed value and predicted value of specified entry $(i, j) \in S$. The regularized objective function is as follows:

$$
\text{Minimize } J = \frac{1}{2} \sum_{(i,j) \in S} e_{ij}^2 + \frac{\lambda}{2} \sum_{i=1}^{m} \sum_{s=1}^{k} u_{is}^2 + \frac{\lambda}{2} \sum_{j=1}^{n} \sum_{s=1}^{k} v_{js}^2
$$

$$
= \frac{1}{2} \sum_{(i,j) \in S} \left(r_{ij} - \sum_{s=1}^{k} u_{is} \cdot v_{js} \right)^2 + \frac{\lambda}{2} \sum_{i=1}^{m} \sum_{s=1}^{k} u_{is}^2 + \frac{\lambda}{2} \sum_{j=1}^{n} \sum_{s=1}^{k} v_{js}^2
$$

Upon taking the partial derivative of J with respect to each of the decision variables, one obtains almost the same result as the unregularized case, except that the terms λu_{iq} and λv_{jq}, respectively, are added to the corresponding gradients in the two cases.

$$
\frac{\partial J}{\partial u_{iq}} = \sum_{j:(i,j) \in S} \left(r_{ij} - \sum_{s=1}^{k} u_{is} \cdot v_{js} \right)(-v_{jq}) + \lambda u_{iq} \quad \forall i \in \{1 \ldots m\}, q \in \{1 \ldots k\}
$$

$$
= \sum_{j:(i,j) \in S} (e_{ij})(-v_{jq}) + \lambda u_{iq} \quad \forall i \in \{1 \ldots m\}, q \in \{1 \ldots k\}
$$

$$
\frac{\partial J}{\partial v_{jq}} = \sum_{i:(i,j) \in S} \left(r_{ij} - \sum_{s=1}^{k} u_{is} \cdot v_{js} \right)(-u_{iq}) + \lambda v_{jq} \quad \forall j \in \{1 \ldots n\}, q \in \{1 \ldots k\}
$$

$$
= \sum_{i:(i,j) \in S} (e_{ij})(-u_{iq}) + \lambda v_{jq} \quad \forall j \in \{1 \ldots n\}, q \in \{1 \ldots k\}
$$

The steps for performing the gradient descent remain similar to those discussed in the case without regularization. Either the batch or the local methods may be used. For example, consider the global update method. Let the vector of $(m \cdot k + n \cdot k)$ decision variables corresponding to the entries in U and V be denoted by $\overline{VAR}$ and let the corresponding gradient vector be denoted by $\overline{\nabla J}$. Then, one can update the entire vector of decision variables as $\overline{VAR} \Leftarrow \overline{VAR} - \alpha \cdot \overline{\nabla J}$. This can be effectively achieved by modifying the

---

[8]Refer to Chapter 6 for a discussion of the bias-variance trade-off.

(unregularized) updates in Figure 3.8 to include regularization terms. The modified updates may be written as follows:

$$u_{iq} \Leftarrow u_{iq} + \alpha \left(\sum_{j:(i,j) \in S} e_{ij} \cdot v_{jq} - \lambda \cdot u_{iq} \right) \quad \forall q \in \{1 \ldots k\}$$

$$v_{jq} \Leftarrow v_{jq} + \alpha \left(\sum_{i:(i,j) \in S} e_{ij} \cdot u_{iq} - \lambda \cdot v_{jq} \right) \quad \forall q \in \{1 \ldots k\}$$

The updates can be executed to convergence. One can also write these updates in terms of the $m \times n$ error matrix $E = [e_{ij}]$ in which unobserved entries of E are set to 0:

$$U \Leftarrow U(1 - \alpha \cdot \lambda) + \alpha E V$$
$$V \Leftarrow V(1 - \alpha \cdot \lambda) + \alpha E^T U$$

Note that the multiplicative term $(1 - \alpha \cdot \lambda)$ shrinks the parameters in each step, which is a result of regularization. If the matrix form is to be used for updates, care must be taken to compute and use sparse representations of E. It makes sense to compute the value of e_{ij} for only the observed entries $(i, j) \in S$ and store E using a sparse data structure.

In the case of local updates (i.e., stochastic gradient descent), the partial derivatives are computed with respect to the error in a randomly chosen observed entry (i, j) rather than all the entries. The following $2 \cdot k$ updates may be executed for each observed entry $(i, j) \in S$, which are processed in random order:

$$u_{iq} \Leftarrow u_{iq} + \alpha(e_{ij} \cdot v_{jq} - \lambda \cdot u_{iq}) \quad \forall q \in \{1 \ldots k\}$$
$$v_{jq} \Leftarrow v_{jq} + \alpha(e_{ij} \cdot u_{iq} - \lambda \cdot v_{jq}) \quad \forall q \in \{1 \ldots k\}$$

For better efficiency, these updates are executed in vectorized form over the k-dimensional factor vectors of user i and item j as follows:

$$\overline{u_i} \Leftarrow \overline{u_i} + \alpha(e_{ij}\overline{v_j} - \lambda\overline{u_i})$$
$$\overline{v_j} \Leftarrow \overline{v_j} + \alpha(e_{ij}\overline{u_i} - \lambda\overline{v_j})$$

These updates are used within the framework of the algorithm described in Figure 3.9. It is noteworthy that the local updates are not exactly equivalent[9] to the vectorized global updates in terms of how the regularization term is treated. This is because the regularization components of the updates, which are $-\lambda u_{iq}$ and $-\lambda v_{jq}$, are used multiple times in a cycle of local updates through *all* the observed entries; updates are executed to u_{iq} for each observed entry in row i and updates are executed to v_{jq} for each observed entry in column j. Furthermore, different rows and columns may have different numbers of observed entries, which can further affect the relative level of regularization of various user and

---

[9]A more precise update should be $\overline{u_i} \Leftarrow \overline{u_i} + \alpha(e_{ij}\overline{v_j} - \lambda\overline{u_i}/n_i^{user})$ and $\overline{v_j} \Leftarrow \overline{v_j} + \alpha(e_{ij}\overline{u_i} - \lambda\overline{v_j}/n_j^{item})$. Here, n_i^{user} represents the number of observed ratings for user i and n_j^{item} represents the number of observed ratings for item j. Here, the regularization terms for various user/item factors are divided equally among the corresponding observed entries for various users/items. In practice, the (simpler) heuristic update rules discussed in the chapter are often used. We have chosen to use these (simpler) rules throughout this chapter to be consistent with the research literature on recommender systems. With proper parameter tuning, λ will automatically adjust to a smaller value in the case of the simpler update rules.

item factors. In the vectorized global method, the regularization is done more gently and uniformly because each entry u_{iq} and v_{jq} is updated only once. Nevertheless, since λ is chosen adaptively during parameter tuning, the local update method will automatically select smaller values of λ than the global method. From a heuristic point of view, the two methods provide roughly similar results, but with different trade-offs between quality and efficiency.

As before, $\alpha > 0$ represents the step size, and $\lambda > 0$ is the regularization parameter. For example, a small constant value of α, such as 0.005, is known to work reasonably well in the case of the Netflix Prize data set. Alternatively, one might use the bold driver algorithm [58, 217] to select α adaptively in each iteration in order to avoid local optima and speed up convergence. It remains to discuss how the regularization parameter λ is selected. The simplest method is to hold out a fraction of the observed entries in the ratings matrix and not use them to train the model. The prediction accuracy of the model is tested over this subset of held out entries. Different values of λ are tested, and the value of λ that provides the highest accuracy is used. If desired, the model can be retrained on the entire set of specified entries (with no hold outs), once the value of λ is selected. This method of parameter tuning is referred to as the *hold out* method. A more sophisticated approach is to use a method referred to as *cross-validation*. This method is discussed in Chapter 7 on evaluating recommender systems. For better results, different regularization parameters λ_1 and λ_2 may be used for the user factors and item factors.

Often, it can be expensive to try different values of λ on the hold-out set in order to determine the optimal value. This restricts the ability to try many choices of λ. As a result, the values of λ are often not well-optimized. One approach, proposed in [518], is to treat the entries of matrices U and V as parameters, and the regularization parameters as hyper-parameters, which are optimized *jointly* with a probabilistic approach. A Gibbs sampling approach is proposed in [518] to jointly learn the parameters and hyper-parameters.

3.6.4.3 Incremental Latent Component Training

One variant of these training methods is to train the latent components incrementally. In other words, we first perform the updates $u_{iq} \Leftarrow u_{iq} + \alpha(e_{ij} \cdot v_{jq} - \lambda \cdot u_{iq})$ and $v_{jq} \Leftarrow v_{jq} + \alpha(e_{ij} \cdot u_{iq} - \lambda \cdot v_{jq})$ only for $q = 1$. The approach repeatedly cycles through all the observed entries in S while performing these updates for $q = 1$ until convergence is reached. Therefore, we can learn the first pair of columns, $\overline{U_1}$ and $\overline{V_1}$, of U and V, respectively. Then, the $m \times n$ *outer-product*[10] matrix $\overline{U_1}\,\overline{V_1}^T$ is subtracted from R (for observed entries). Subsequently, the updates are performed for $q = 2$ with the (residual) ratings matrix to learn the second pair of columns, $\overline{U_2}$ and $\overline{V_2}$, of U and V, respectively. Then, $\overline{U_2}\,\overline{V_2}^T$ is subtracted from R. This process is repeated each time with the residual matrix until $q = k$. The resulting approach provides the required matrix factorization because the overall rank-k factorization can be expressed as the sum of k rank-1 factorizations:

$$R \approx UV^T = \sum_{q=1}^{k} \overline{U_q}\,\overline{V_q}^T \tag{3.17}$$

---

[10]The inner-product of two column-vectors $\overline{x}$ and $\overline{y}$ is given by the scalar $\overline{x}^T\overline{y}$, whereas the outer-product is given by the rank-1 matrix $\overline{x}\,\overline{y}^T$. Furthermore, $\overline{x}$ and $\overline{y}$ need not be of the same size in order to compute an outer-product.

Algorithm *ComponentWise-SGD*(Ratings Matrix: R, Learning Rate: α)
begin
   Randomly initialize matrices U and V;
   $S = \{(i,j) : r_{ij} \text{ is observed}\}$;
   for $q = 1$ to k **do**
   begin
     while not(convergence) **do**
     begin
       Randomly shuffle observed entries in S;
       for each $(i,j) \in S$ in shuffled order **do**
       begin
         $e_{ij} \Leftarrow r_{ij} - u_{iq}v_{jq}$;
         $u_{iq}^{+} \Leftarrow u_{iq} + \alpha \cdot (e_{ij} \cdot v_{jq} - \lambda \cdot u_{iq})$;
         $v_{jq}^{+} \Leftarrow v_{jq} + \alpha \cdot (e_{ij} \cdot u_{iq} - \lambda \cdot v_{jq})$;
         $u_{iq} = u_{iq}^{+}; v_{jq} = v_{jq}^{+}$;
       end
       Check convergence condition;
     end
     { Element-wise implementation of $R \Leftarrow R - \overline{U_q}\,\overline{V_q}^{T}$ }
     for each $(i,j) \in S$ **do** $r_{ij} \Leftarrow r_{ij} - u_{iq}v_{jq}$;
   end
end

Figure 3.10: Component-wise implementation of stochastic gradient descent

A description of this procedure is illustrated in Figure 3.10. The differences of this approach from the version discussed earlier can be understood in terms of the differences in their nested loop structures. Incremental component training loops through various values of q in the outermost loops and cycles through the observed entries repeatedly in the inner loops to reach convergence for each value of q (cf. Figure 3.10). The earlier method loops through the observed entries repeatedly to reach convergence in the outer loops and cycles though various values of q in the inner loop (cf. Figure 3.9). Furthermore, the incremental method needs to adjust the ratings matrix between two executions of the outer loop. This approach leads to faster and more stable convergence in each component because a smaller number of variables is optimized at one time.

It is noteworthy that different strategies for gradient descent will lead to solutions with different properties. This particular form of incremental training will lead to the earlier latent components being the dominant ones, which provides a similar flavor to that of SVD. However, the resulting columns in U (or V) might not be mutually orthogonal. It is also possible to force mutual orthogonality of the columns of U (and V) by using *projected* gradient descent for $q > 1$. Specifically, the gradient vector with respect to the variables in column $\overline{U_q}$ (or $\overline{V_q}$) is projected in an orthogonal direction to the $(q-1)$ columns of U (or V) found so far.

3.6.4.4 Alternating Least Squares and Coordinate Descent

The stochastic gradient method is an efficient methodology for optimization. On the other hand, it is rather sensitive, both to the initialization and the way in which the step sizes are chosen. Other methods for optimization include the use of *alternating least squares (ALS)* [268, 677], which is generally more stable. The basic idea of this approach to use the following iterative approach, starting with an initial set of matrices U and V:

1. Keeping U fixed, we solve for each of the n rows of V by treating the problem as a least-squares regression problem. Only the observed ratings in S can be used for building the least-squares model in each case. Let $\overline{v_j}$ be the jth row of V. In order to determine the optimal vector $\overline{v_j}$, we wish to minimize $\sum_{i:(i,j)\in S}(r_{ij} - \sum_{s=1}^{k} u_{is}v_{js})^2$, which is a least-squares regression problem in $v_{j1}\ldots v_{jk}$. The terms $u_{i1}\ldots u_{ik}$ are treated as constant values, whereas $v_{j1}\ldots v_{jk}$ are treated as optimization variables. Therefore, the k latent factor components in $\overline{v_j}$ for the jth item are determined with least-squares regression. A total of n such least-squares problems need to be executed, and each least-squares problem has k variables. Because the least-squares problem for each item is independent, this step can be parallelized easily.

2. Keeping V fixed, solve for each of the m rows of U by treating the problem as a least-squares regression problem. Only the specified ratings in S can be used for building the least-squares model in each case. Let $\overline{u_i}$ be the ith row of U. In order to determine the optimal vector $\overline{u_i}$, we wish to minimize $\sum_{j:(i,j)\in S}(r_{ij} - \sum_{s=1}^{k} u_{is}v_{js})^2$, which is a least-squares regression problem in $u_{i1}\ldots u_{ik}$. The terms $v_{j1}\ldots v_{jk}$ are treated as constant values, whereas $u_{i1}\ldots u_{ik}$ are treated as optimization variables. Therefore, the k latent factor components for the ith user are determined with least-squares regression. A total of m such least-squares problems need to be executed, and each least-squares problem has k variables. Because the least-squares problem for each user is independent, this step can be parallelized easily.

These two steps are iterated to convergence. When regularization is used in the objective function, it amounts to using Tikhonov regularization [22] in the least-squares approach. The value of the regularization parameter $\lambda > 0$ can be fixed across all the independent least-squares problems, or it can be chosen differently. In either case, one might need to determine the optimal value of λ by using a hold-out or cross-validation methodology. A brief discussion of linear regression with Tikhonov regularization is provided in section 4.4.5 of Chapter 4. Although the linear regression discussion in Chapter 4 is provided in the context of content-based models, the basic regression methodology is invariant across the different scenarios in which it is used.

Interestingly, a weighted version *ALS* is particularly well-suited to implicit feedback settings in which the matrix is assumed to be fully specified with many zero values. Furthermore, the nonzero entries are often weighted more heavily in these settings. In such cases, stochastic gradient descent becomes too expensive. When most of the entries are zeros, some tricks can be used to make weighted *ALS* an efficient option. The reader is referred to [260].

The drawback of *ALS* is that it is not quite as efficient as stochastic-gradient descent in large-scale settings with explicit ratings. Other methods such as coordinate descent can effectively address the trade-off between efficiency and stability [650]. In coordinate-descent, the approach of fixing a subset of variables (as in *ALS*) is pushed to the extreme. Here, all entries in U and V are fixed except for a single entry (or *coordinate*) in one of the two matrices, which is optimized using the objective function of section 3.6.4.2. The resulting optimization solution can be shown to have closed form because it is a quadratic objective function in a single variable. The corresponding value of u_{iq} (or v_{jq}) can be determined efficiently according to one of the following two updates:

$$u_{iq} \Leftarrow \frac{\sum_{j:(i,j)\in S}(e_{ij} + u_{iq}v_{jq})v_{jq}}{\lambda + \sum_{j:(i,j)\in S} v_{jq}^2}$$

$$v_{jq} \Leftarrow \frac{\sum_{i:(i,j)\in S}(e_{ij} + u_{iq}v_{jq})u_{iq}}{\lambda + \sum_{i:(i,j)\in S} u_{iq}^2}$$

Here, S denotes the set of observed entries in the ratings matrix and $e_{ij} = r_{ij} - \hat{r}_{ij}$ is the prediction error of entry (i, j). One cycles through the $(m + n) \cdot k$ parameters in U and V with these updates until convergence is reached. It is also possible to combine coordinate descent with incremental latent component training just as stochastic gradient descent is combined with increment component training (cf. section 3.6.4.3).

3.6.4.5   Incorporating User and Item Biases

A variation on the unconstrained model was introduced by Paterek [473] to incorporate variables that can learn user and item biases. Assume for the purpose of discussion that the ratings matrix is mean-centered by subtracting the *global* mean μ of the *entire* ratings matrix from all the entries as a preprocessing step. After predicting the entries with the latent factor model, the value μ is added back to the predicted values as a postprocessing step. Therefore, in this section, we will simply assume that the ratings matrix R has already been centered in this way, and ignore the preprocessing and postprocessing steps.

Associated with each user i, we have a variable o_i, which indicates the general bias of users to rate items. For example, if user i is a generous person, who tends to rate all items highly, then the variable o_i will be a positive quantity. On the other hand, the value of o_i will be negative for a curmudgeon who rates most items negatively. Similarly, the variable p_j denotes the bias in the ratings of item j. Highly liked items (e.g., a box-office hit) will tend to have larger (positive) values of p_j, whereas globally disliked items will have negative values of p_j. It is the job of the factor model to learn the values of o_i and p_j in a data-driven manner. The main change to the original latent factor model is that a part of the (i, j)th rating is explained by $o_i + p_j$ and the remainder by the (i, j)th entry of the product UV^T of the latent factor matrices. Therefore, the predicted value of the rating of entry (i, j) is given by the following:

$$\hat{r}_{ij} = o_i + p_j + \sum_{s=1}^{k} u_{is} \cdot v_{js} \qquad (3.18)$$

Thus, the error e_{ij} of an observed entry $(i, j) \in S$ is given by the following:

$$e_{ij} = r_{ij} - \hat{r}_{ij} = r_{ij} - o_i - p_j - \sum_{s=1}^{k} u_{is} \cdot v_{js} \qquad (3.19)$$

Note that the values o_i and p_j are also variables that need to be learned in a data-driven manner along with the latent factor matrices U and V. Then, the minimization objective function J may be formulated by aggregating the squared errors over the observed entries of the ratings matrix (i.e., set S) as follows:

$$J = \frac{1}{2} \sum_{(i,j) \in S} e_{ij}^2 + \frac{\lambda}{2} \sum_{i=1}^{m} \sum_{s=1}^{k} u_{is}^2 + \frac{\lambda}{2} \sum_{j=1}^{n} \sum_{s=1}^{k} v_{js}^2 + \frac{\lambda}{2} \sum_{i=1}^{m} o_i^2 + \frac{\lambda}{2} \sum_{j=1}^{n} p_j^2$$

$$= \frac{1}{2} \sum_{(i,j) \in S} \left(r_{ij} - o_i - p_j - \sum_{s=1}^{k} u_{is} \cdot v_{js} \right)^2 + \frac{\lambda}{2} \left(\sum_{i=1}^{m} \sum_{s=1}^{k} u_{is}^2 + \sum_{j=1}^{n} \sum_{s=1}^{k} v_{js}^2 + \sum_{i=1}^{m} o_i^2 + \sum_{j=1}^{n} p_j^2 \right)$$

It turns out that this problem is different from unconstrained matrix factorization to only a minor degree. Instead of having separate bias variables o_i and p_j for users and items, we can increase the size of the factor matrices to incorporate these bias variables. We need to add two additional columns to each factor matrix U and V, to create larger factor matrices of size $m \times (k+2)$ and $n \times (k+2)$, respectively. The last two columns of each factor matrix are special, because they correspond to the bias components. Specifically, we have:

$$u_{i,k+1} = o_i \quad \forall i \in \{1 \dots m\}$$
$$u_{i,k+2} = 1 \quad \forall i \in \{1 \dots m\}$$
$$v_{j,k+1} = 1 \quad \forall j \in \{1 \dots n\}$$
$$v_{j,k+2} = p_j \quad \forall j \in \{1 \dots n\}$$

Note that the conditions $u_{i,k+2} = 1$ and $v_{j,k+1} = 1$ are constraints on the factor matrices. *In other words, we need to constrain the last column of the user-factor matrix to all 1s, and the second last column of the item-factor matrix to all 1s.* This scenario is pictorially shown in Figure 3.11. Then, the modified optimization problem with these enlarged factor matrices is as follows:

$$\text{Minimize } J = \frac{1}{2} \sum_{(i,j) \in S} \left(r_{ij} - \sum_{s=1}^{k+2} u_{is} \cdot v_{js} \right)^2 + \frac{\lambda}{2} \sum_{s=1}^{k+2} \left(\sum_{i=1}^{m} u_{is}^2 + \sum_{j=1}^{n} v_{js}^2 \right)$$

subject to:

$(k+2)$th column of U contains only 1s

$(k+1)$th column of V contains only 1s

It is noteworthy that the summations in the objective are up to $(k+2)$ rather than k. Note that this problem is virtually identical to the unconstrained case except for the minor

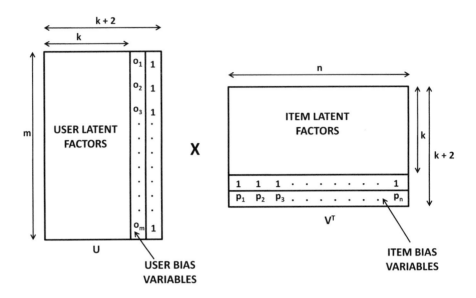

Figure 3.11: Incorporating user and item biases in the latent factor model

constraints on the factors. The other change is the increase in the sizes of the factor matrices to incorporate the user and item bias variables. Because of the minor change in the problem formulation, one only needs to make corresponding changes to the gradient descent method. For initialization, the $(k + 1)$th column of V and the $(k + 2)$th column of U are set to 1s. Exactly the same (local) update rules are used as in the unconstrained case, except that the two perturbed entries in the $(k + 1)$th column of V and the $(k + 2)$th column of U are reset to their fixed values after each update (or simply not updated). The following updates may be executed by cycling over each specified entry $(i, j) \in S$:

$$u_{iq} \Leftarrow u_{iq} + \alpha(e_{ij} \cdot v_{jq} - \lambda \cdot u_{iq}) \quad \forall q \in \{1 \ldots k + 2\}$$
$$v_{jq} \Leftarrow v_{jq} + \alpha(e_{ij} \cdot u_{iq} - \lambda \cdot v_{jq}) \quad \forall q \in \{1 \ldots k + 2\}$$
Reset perturbed entries in $(k + 2)$th column of U and $(k + 1)$th column of V to 1s

This group of updates is performed simultaneously as a group. It is also possible to use the alternating least-squares method with minor variations (see Exercise 11). The aforementioned discussion uses the same regularization parameters and learning rates for each type of variable. It is sometimes recommended to use different regularization parameters and learning rates for the user biases, item biases, and factor variables [586]. This can be achieved with minor modifications of the aforementioned updates.

A natural question that arises is why this formulation should perform better than unconstrained matrix factorization. The addition of constraints on the last two columns of the factor matrices should only reduce the global solution quality, because one is now optimizing over a smaller space of solutions. However, in many cases, adding such constraints biases the solution while reducing overfitting. In other words, the addition of such intuitive constraints can often improve the generalizability of the learning algorithm to *unseen* entries, even though the error over the *specified* entries may be higher. This is particularly helpful when the number of observed ratings for a user or for an item is small [473]. Bias variables add a component to the ratings that are global to either the users or the items. Such global properties are useful when limited data is available. As a specific example, consider the case in which a user has provided ratings for only a small number (1 or 2) items. In such cases, many recommendation algorithms, such as neighborhood-based methods, will not give reliable predictions for the user. On the other hand, the (non-personalized) predictions of the item bias variables will be able to give reasonable predictions. After all, if a particular movie is a box-office hit on a global basis, then the relevant user is also more likely to appreciate it. The bias variables will also reflect this fact and incorporate it into the learning algorithm.

In fact, it has been shown [73, 310, 312] that using *only* the bias variables (i.e., $k = 0$) can often provide reasonably good rating predictions. This point was emphasized as one of the practical lessons learned from the Netflix Prize contest [73]:

> "Of the numerous new algorithmic contributions, I would like to highlight one – those humble baseline predictors (or biases), which capture main effects in the data. While the literature mostly concentrates on the more sophisticated algorithmic aspects, we have learned that an accurate treatment of main effects is probably at least as significant as coming up with modeling breakthroughs."

This means that a significant part of the ratings can be explained by user generosity and item popularity, rather than any specific *personalized* preferences of users for items. Such a non-personalized model is discussed in section 3.7.1, which is equivalent to setting $k = 0$ in the aforementioned model. As a result, only the biases of users and items are learned, and a *baseline* rating B_{ij} is predicted for user i and item j by summing their biases. One can use such a baseline rating to enhance any off-the-shelf collaborative filtering model. To do so, one can simply subtract each B_{ij} from the (i, j)th (observed) entry of the ratings matrix before applying collaborative filtering. These values are added back in a postprocessing phase to the predicted values. Such an approach is especially useful for models in which one cannot easily parameterize bias variables. For example, (traditional) neighborhood models accomplish these bias-correction goals with row-wise mean-centering, although the use of B_{ij} to correct the matrix entries would be a more sophisticated approach because it adjusts for both user and item biases.

3.6.4.6 Incorporating Implicit Feedback

Generally, implicit feedback scenarios correspond to the use of unary ratings matrices in which users express their interests by buying items. However, even in cases in which users explicitly rate items, the *identity of the items they rate* can be viewed as an implicit feedback. In other words, a significant predictive value is captured by the identity of the items that users rate, *irrespective of the actual values of the ratings*. A recent paper [184] describes this phenomenon elegantly in the context of the music domain:

> "Intuitively, a simple process could explain the results [showing the predictive value of implicit feedback]: users chose to rate songs they listen to, and listen to music they expect to like, while avoiding genres they dislike. Therefore, most of the songs that would get a bad rating are not voluntarily rated by the users. Since people rarely listen to random songs, or rarely watch random movies, we should expect to observe in many areas a difference between the distribution of ratings for random items and the corresponding distribution for the items selected by the users."

Various frameworks such as *asymmetric factor models* and *SVD++* have been proposed to incorporate implicit feedback. These algorithms use two different item factor matrices V

and Y, corresponding to explicit and implicit feedback, respectively. The user latent factors are either wholly or partially derived using a linear combination of those rows of the (implicit) item latent factor matrix Y that correspond to rated items of the user. The idea is that user factors correspond to user preferences, and user preferences should therefore be influenced by the items they have chosen to rate. In the simplest version of asymmetric factor models, a linear combination of the (implicit) factor vectors of the rated items is used to create the user factors. This results in an asymmetric approach in which we no longer have independent variables for user factors. Instead, we have *two* sets of independent *item* factors (i.e., explicit and implicit), and user factors are derived as a linear combination of the implicit item factors. Many variants [311] of this methodology are discussed in the literature, although the original idea is credited to Paterek [473]. The SVD++ model further combines this asymmetric approach with (explicit) user factors and a traditional factorization framework. The asymmetric approach can, therefore, be viewed as a simplified precursor to SVD++. For clarity in exposition, we will first discuss the asymmetric model briefly.

Asymmetric Factor Models: To capture the implicit feedback information, we first derive an *implicit feedback matrix* from the explicit ratings matrix. For an $m \times n$ ratings matrix R, the $m \times n$ implicit feedback matrix $F = [f_{ij}]$ is defined by setting it to 1, if the value r_{ij} is observed, and 0, if it is missing. The feedback matrix F is subsequently normalized so that the L_2-norm of each row is 1. Therefore, if I_i is the set of indices of the items rated by user i, then each nonzero entry in the ith row is $1/\sqrt{|I_i|}$. An example of a ratings matrix R together with its corresponding implicit feedback matrix F is illustrated below:

$$
\underbrace{\begin{pmatrix} 1 & -1 & 1 & ? & 1 & 2 \\ ? & ? & -2 & ? & -1 & ? \\ 0 & ? & ? & ? & ? & ? \\ -1 & 2 & -2 & ? & ? & ? \end{pmatrix}}_{R} \Rightarrow \underbrace{\begin{pmatrix} 1/\sqrt{5} & 1/\sqrt{5} & 1/\sqrt{5} & 0 & 1/\sqrt{5} & 1/\sqrt{5} \\ 0 & 0 & 1/\sqrt{2} & 0 & 1/\sqrt{2} & 0 \\ 1/\sqrt{1} & 0 & 0 & 0 & 0 & 0 \\ 1/\sqrt{3} & 1/\sqrt{3} & 1/\sqrt{3} & 0 & 0 & 0 \end{pmatrix}}_{F}
$$

An $n \times k$ matrix $Y = [y_{ij}]$ is used as the implicit item-factor matrix and the matrix F provides the linear combination coefficients to create a user-factor matrix from it. The variables in Y encode the propensity of each factor-item combination to contribute to implicit feedback. For example, if $|y_{ij}|$ is large, then it means that simply the *act of rating* item i contains significant information about the affinity of that *action* for the jth latent component, no matter what the actual value of the rating might be. In the simplified asymmetric model, user factors are encoded as linear combinations of the implicit item factors of rated items; the basic idea is that linear combinations of user *actions* are used to define their preferences (factors). Specifically, the matrix product FY is an $m \times k$ user-factor matrix, and each (user-specific) row in it is a (user-specific) linear combination of implicit item factors depending on the items rated by the user. The matrix FY is used in lieu of the user-factor matrix U, and the ratings matrix is factorized as $R \approx [FY]V^T$, where V is the $n \times k$ *explicit* item-factor matrix. If desired, bias variables can be incorporated in the model by mean-centering the ratings matrix and appending two additional columns to each of Y and V, as discussed in section 3.6.4.5 (see Exercise 13).

This simple approach often provides excellent[11] results because it reduces the redundancy in user factors by deriving them as linear combinations of item factors. The basic

---

[11]In many cases, this approach can outperform SVD++, especially when the number of observed ratings is small.

idea here is that two users will have similar user factors if they have rated similar items, *irrespective of the values of the ratings*. Note that the $n \times k$ matrix Y contains fewer parameters than an $m \times k$ user-factor matrix U because $n \ll m$. Another advantage of this approach is that it is possible to incorporate other types of *independent* implicit feedback (such as buying or browsing behavior) by incorporating it in the implicit feedback matrix F. In such cases, the approach can usually do better than most other forms of matrix factorization (with explicit ratings) because of its ability to use *both* explicit and implicit ratings. Nevertheless, even in cases where no independent implicit feedback is available, this model seems to be perform better than straightforward variations of matrix factorization for very sparse matrices with a large number of users (compared to the number of items). An additional advantage of this model is that no user parameterizations are needed; therefore, the model can work well for out-of-sample users, although it cannot be used for out-of-sample items. In other words, the model is at least partially inductive unlike most matrix factorization methods. We omit discussing the gradient-descent steps of this model, because the generalization of this model is discussed in the next section. The corresponding steps are, nevertheless, enumerated in the problem statement of Exercise 13.

The item-based parametrization of asymmetric factor models also provides it the merit of *explainability*. Note that one can re-write the factorization $[FY]V^T$ as $F[YV^T]$. The matrix YV^T can be viewed as an $n \times n$ item-to-item prediction matrix in which $[YV^T]_{ij}$ tells us how much the act of rating item i contributes to the predicted rating of item j. The matrix F provides the corresponding $m \times n$ user-to-item coefficients and, therefore, multiplying F with $[YV^T]$ provides user-to-item predictions. Therefore, one can now explain, which items previously consumed/rated by the user contribute most to the prediction in $F[YV^T]$. This type of explainability is inherent to item-centric models.

SVD++: The derivation of user factors *purely* on the basis of the identities of rated items seems like a rather extreme use of implicit feedback in asymmetric factor models. This is because such an approach does not discriminate *at all* between pairs of users who have rated exactly the same set of items but have very different observed values of the ratings. Two such users will receive exactly the same rating prediction for an item that is not rated by both.

In SVD++, a more nuanced approach is used. The implicit user-factor matrix FY is used only to *adjust* the explicit user-factor matrix U rather than to create it. Therefore, FY needs to be added to U before multiplying with V^T. Then, the reconstructed $m \times n$ ratings matrix R is given by $(U + FY)V^T$, and the implicit feedback component of the predicted rating is given by $(FY)V^T$. The price for the additional modeling flexibility in SVD++ is that the number of parameters is increased, which can cause overfitting in very sparse ratings matrices. The implicit feedback matrix can be derived from the ratings matrix (as in asymmetric factor models), although other forms of implicit feedback (e.g., buying or browsing behavior) can also be included.

The user and item biases are included in this model in a manner similar to section 3.6.4.5. We can assume, without loss[12] of generality, that the ratings matrix is mean-centered around the global mean μ of all the entries. Therefore, we will work with $m \times (k+2)$ and $n \times (k+2)$ factor matrices U and V, respectively, in which the last two columns contain either 1s or bias variables according to section 3.6.4.5. We also assume[13] that Y is an $n \times (k+2)$ matrix,

---

[12]For matrices, which are not mean-centered, the global mean can be subtracted during preprocessing and then added back at prediction time.

[13]We use a slightly different notation than the original paper [309], although the approach described here is equivalent. This presentation simplifies the notation by introducing fewer variables and viewing bias

and the last two columns of Y contain 0s. This is because the bias component is already addressed by the last two columns of U, but we need the last two dummy columns in Y to ensure that we can add U and FY as matrices of the same dimensions. Therefore, the predicted rating $\hat{r}_{ij}$ can be expressed in terms of these variables as follows:

$$\hat{r}_{ij} = \sum_{s=1}^{k+2} (u_{is} + [FY]_{is}) \cdot v_{js} \tag{3.20}$$

$$= \sum_{s=1}^{k+2} \left(u_{is} + \sum_{h \in I_i} \frac{y_{hs}}{\sqrt{|I_i|}} \right) \cdot v_{js} \tag{3.21}$$

The first term $\sum_{s=1}^{k+2} u_{is} v_{js}$ on the right-hand side of the aforementioned equation is the (i,j)th term of UV^T, and the second term $\sum_{s=1}^{k+2} \sum_{h \in I_i} \frac{y_{hs}}{\sqrt{|I_i|}} v_{js}$ is the (i,j)th term of $[FY]V^T$. Note that the (i,s)th entry of $[FY]$ is given by $\sum_{h \in I_i} \frac{y_{hs}}{\sqrt{|I_i|}}$. One can view this model as a combination of the unconstrained matrix factorization model (with biases) and the asymmetric factorization model discussed in the previous section. Therefore, it combines the strengths of both models.

The corresponding optimization problem, which minimizes the aggregate squared error $e_{ij}^2 = (r_{ij} - \hat{r}_{ij})^2$ over all observed entries (denoted by set S) in the ratings matrix, may be stated as follows:

$$\text{Min. } J = \frac{1}{2} \sum_{(i,j) \in S} \left(r_{ij} - \sum_{s=1}^{k+2} \left[u_{is} + \sum_{h \in I_i} \frac{y_{hs}}{\sqrt{|I_i|}} \right] \cdot v_{js} \right)^2 + \frac{\lambda}{2} \sum_{s=1}^{k+2} \left(\sum_{i=1}^{m} u_{is}^2 + \sum_{j=1}^{n} v_{js}^2 + \sum_{j=1}^{n} y_{js}^2 \right)$$

subject to:

$(k+2)$th column of U contains only 1s

$(k+1)$th column of V contains only 1s

Last two columns of Y contain only 0s

Note that this optimization formulation is different from that in the previous section in terms of its having an implicit feedback term together with its regularizer. One can use the partial derivative of this objective function to derive the update rules for matrices U and V, as well as the variables in Y. The update rules are then expressed in terms of the error values $e_{ij} = r_{ij} - \hat{r}_{ij}$ of the observed entries. The following updates[14] may be used for each

---

variables as constraints on the factorization process.

[14] The literature often describes these updates in vectorized form. These updates may be applied to the rows of U, V, and Y as follows:

$$\overline{u_i} \Leftarrow \overline{u_i} + \alpha(e_{ij}\overline{v_j} - \lambda \overline{u_i})$$

$$\overline{v_j} \Leftarrow \overline{v_j} + \alpha \left(e_{ij} \cdot \left[\overline{u_i} + \sum_{h \in I_i} \frac{\overline{y_h}}{\sqrt{|I_i|}} \right] - \lambda \cdot \overline{v_j} \right)$$

$$\overline{y_h} \Leftarrow \overline{y_h} + \alpha \left(\frac{e_{ij} \cdot \overline{v_j}}{\sqrt{|I_i|}} - \lambda \cdot \overline{y_h} \right) \quad \forall h \in I_i$$

Reset perturbed entries in fixed columns of U, V, and Y

observed entry $(i, j) \in S$ in the ratings matrix:

$$u_{iq} \Leftarrow u_{iq} + \alpha(e_{ij} \cdot v_{jq} - \lambda \cdot u_{iq}) \quad \forall q \in \{1 \ldots k+2\}$$

$$v_{jq} \Leftarrow v_{jq} + \alpha \left(e_{ij} \cdot \left[u_{iq} + \sum_{h \in I_i} \frac{y_{hq}}{\sqrt{|I_i|}} \right] - \lambda \cdot v_{jq} \right) \quad \forall q \in \{1 \ldots k+2\}$$

$$y_{hq} \Leftarrow y_{hq} + \alpha \left(\frac{e_{ij} \cdot v_{jq}}{\sqrt{|I_i|}} - \lambda \cdot y_{hq} \right) \quad \forall q \in \{1 \ldots k+2\}, \forall h \in I_i$$

Reset perturbed entries in fixed columns of U, V, and Y

The updates are executed by repeatedly looping over all the observed ratings in S. The perturbed entries in the fixed columns of U, V, and Y are reset by these rules to either 1s and 0s. A more efficient (and practical) alternative would be to simply not update the fixed entries by keeping track of them during the update. Furthermore, these columns are always initialized to fixed values that respect the constraints of the optimization model. The nested loop structure of stochastic-gradient descent is similar across the family of matrix factorization methods. Therefore, the basic framework described in Figure 3.9 may be used, although the updates are based on the aforementioned discussion. Better results may be obtained by using different regularization parameters for different factor matrices. A fast variation of stochastic gradient descent is described in [151]. It is also possible to develop an alternating least-squares approach to solve the aforementioned problem (see Exercise 12). Although this model is referred to as SVD++ [309], the name is slightly misleading because the basis vectors of the factorized matrices are not orthogonal. Indeed, the term "SVD" is often loosely applied in the literature on latent factor models. In the next section, we will discuss the use of singular value decomposition with orthogonal vectors.

3.6.5 Singular Value Decomposition

Singular value decomposition (SVD) is a form of matrix factorization in which the columns of U and V are constrained to be mutually orthogonal. Mutual orthogonality has the advantage that the concepts can be completely independent of one another, and they can be geometrically interpreted in scatterplots. However, the semantic interpretation of such a decomposition is generally more difficult, because these latent vectors contain both positive and negative quantities, and are constrained by their orthogonality to other concepts. For a *fully specified matrix*, it is relatively easy to perform SVD with the use of eigen-decomposition methods. We will first briefly recap the discussion on singular value decomposition in section 2.5.1.2 of Chapter 2.

Consider the case in which the ratings matrix is fully specified. One can *approximately* factorize the ratings matrix R by using *truncated* SVD of rank $k \ll \min\{m, n\}$. Truncated SVD is computed as follows:

$$R \approx Q_k \Sigma_k P_k^T \tag{3.22}$$

Here, Q_k, Σ_k, and P_k are matrices of size $m \times k$, $k \times k$, and $n \times k$, respectively. The matrices Q_k and P_k respectively contain the k largest eigenvectors of RR^T and R^TR, whereas the (diagonal) matrix Σ_k contains the (non-negative) square roots of the k largest eigenvalues of either matrix along its diagonal. It is noteworthy that the nonzero eigenvalues of RR^T and R^TR are the same, even though they will have a different number of zero eigenvalues when $m \neq n$. The matrix P_k contains the top eigenvectors of R^TR, which is the *reduced* basis representation required for dimensionality reduction of the row space. These eigenvectors contain information about the directions of item-item correlations among ratings, and therefore they provide the ability to represent each user in a reduced number of dimensions in a rotated axis system. For example, in Figure 3.6, the top eigenvector corresponds to the latent vector representing the dominant directions of item-item correlations. Furthermore, the matrix $Q_k \Sigma_k$ contains the transformed and reduced $m \times k$ representation of the original ratings matrix in the basis corresponding to P_k. Therefore, in Figure 3.6, the matrix $Q_k \Sigma_k$ would be a 1-dimensional column vector containing the coordinates of the ratings along the dominant latent vector.

It is easy to see from Equation 3.22 that SVD is inherently defined as a matrix factorization. Of course, the factorization here is into *three* matrices rather than *two*. However, the diagonal matrix Σ_k can be absorbed in either the user factors Q_k or the item factors P_k. By convention, the user factors and item factors are defined as follows:

$$U = Q_k \Sigma_k$$
$$V = P_k$$

As before, the factorization of the ratings matrix R is defined as $R = UV^T$. As long as the user and item factor matrices have orthogonal columns, it is easy to convert the resulting factorization into a form that is compliant with SVD (see Exercise 9). Therefore, the goal of the factorization process is to discover matrices U and V with orthogonal columns. Therefore, SVD can be formulated as the following optimization problem over the matrices U and V:

$$\text{Minimize } J = \frac{1}{2}||R - UV^T||^2$$

$$\text{subject to:}$$

$$\text{Columns of } U \text{ are mutually orthogonal}$$

$$\text{Columns of } V \text{ are mutually orthogonal}$$

It is easy to see that the only difference from the case of unconstrained factorization is the presence of orthogonality constraints. In other words, the same objective function is being optimized over a smaller space of solutions compared to unconstrained matrix factorization. Although one would expect that the presence of constraints would increase the error J of the approximation, it turns out that the optimal value of J is identical in the case of SVD and unconstrained matrix factorization, if the matrix R is fully specified and regularization is not used. Therefore, for fully specified matrices, the optimal solution to SVD is one of the alternate optima of unconstrained matrix factorization. This is not necessarily true in the cases in which R is not fully specified, and the objective function $J = \frac{1}{2}||R - UV^T||^2$ is computed *only over the observed entries*. In such cases, unconstrained matrix factorization will typically provide lower error on the observed entries. However, the relative performance on the unobserved entries can be unpredictable because of varying levels of *generalizability* of different models.

3.6.5.1   A Simple Iterative Approach to SVD

In this section, we discuss how to solve the optimization problem when the matrix R is incompletely specified. The first step is to mean-center each row of R by subtracting the average rating μ_i of the user i from it. These row-wise averages are stored because they will eventually be needed to reconstruct the raw ratings of the missing entries. Let the centered matrix be denoted by R_c. Then, the missing entries of R_c are set to 0. This approach effectively sets the missing entries to the average rating of the corresponding user, because the missing entries of the centered matrix are set to 0. SVD is then applied to R_c to obtain the decomposition $R_c = Q_k \Sigma_k P_k^T$. The resulting user factors and item factors are given by $U = Q_k \Sigma_k$ and $V = P_k$. Let the ith row of U be the k-dimensional vector denoted by $\overline{u_i}$ and the jth row of V be the k-dimensional vector denoted by $\overline{v_j}$. Then, the rating $\hat{r}_{ij}$ of user i for item j is estimated as the following adjusted dot product of $\overline{u_i}$ and $\overline{v_j}$:

$$\hat{r}_{ij} = \overline{u_i} \cdot \overline{v_j} + \mu_i \tag{3.23}$$

Note that the mean μ_i of user i needs to added to the estimated rating to account for the mean-centering applied to R in the first step.

The main problem with this approach is that the substitution of missing entries with row-wise means can lead to considerable bias. A specific example of how column-wise mean substitution leads to bias is provided in section 2.5.1 of Chapter 2. The argument for row-wise substitution is exactly similar. There are several ways of reducing this bias. One of the methods is to use maximum-likelihood estimation [24, 472], which is discussed in section 2.5.1.1 of Chapter 2. Another approach is to use a method, which reduces the bias iteratively by improving the estimation of the missing entries. The approach uses the following steps:

1. **Initialization:** Initialize the missing entries in the ith row of R to be the mean μ_i of that row to create R_f.

2. **Iterative step 1:** Perform rank-k SVD of R_f in the form $Q_k \Sigma_k P_k^T$.

3. **Iterative step 2:** Readjust only the (originally) missing entries of R_f to the corresponding values in $Q_k \Sigma_k P_k^T$. Go to iterative step 1.

The iterative steps 1 and 2 are executed to convergence. In this method, although the initialization step causes bias in the early SVD iterations, later iterations tend to provide more robust estimates. This is because the matrix $Q_k \Sigma_k P_k^T$ will differ from R to a greater degree in the biased entries. The final ratings matrix is then given by $Q_k \Sigma_k P_k^T$ at convergence.

The approach can become stuck in a local optimum when the number of missing entries is large. In particular, the local optimum at convergence can be sensitive to the choice of initialization. It is also possible to use the baseline predictor discussed in section 3.7.1 to perform more robust initialization. The idea is to compute an initial predicted value B_{ij} for user i and item j with the use of *learned* user and item biases. This approach is equivalent to applying the method in section 3.6.4.5 at $k = 0$, and then adding the bias of user i to that of item j to derive B_{ij}. The value of B_{ij} is subtracted from each observed entry (i, j) in the ratings matrix, and missing entries are set to 0 at initialization. The aforementioned iterative approach is applied to this adjusted matrix. The value of B_{ij} is added back to entry (i, j) at prediction time. Such an approach tends to be more robust because of better initialization.

Regularization can be used in conjunction with the aforementioned iterative method. The idea is to perform regularized SVD of R_f in each iteration rather than using only vanilla SVD. Because the matrix R_f is fully specified in each iteration, it is relatively easy to apply regularized SVD methods to these intermediate matrices. Regularized singular value decomposition methods for complete matrices are discussed in [541]. The optimal values of the regularization parameters λ_1 and λ_2 are chosen adaptively by using either the hold-out or the cross-validation methods.

3.6.5.2   An Optimization-Based Approach

The iterative approach is quite expensive because it works with fully specified matrices. It is simple to implement for smaller matrices but does not scale well in large-scale settings. A more efficient approach is to add orthogonality constraints to the optimization model of the previous sections. A variety of gradient-descent methods can be used for solve the model. Let S be the set of specified entries in the ratings matrix. The optimization problem (with regularization) is stated as follows:

$$\text{Minimize } J = \frac{1}{2} \sum_{(i,j) \in S} \left(r_{ij} - \sum_{s=1}^{k} u_{is} \cdot v_{js} \right)^2 + \frac{\lambda_1}{2} \sum_{i=1}^{m} \sum_{s=1}^{k} u_{is}^2 + \frac{\lambda_2}{2} \sum_{j=1}^{n} \sum_{s=1}^{k} v_{js}^2$$

$$\text{subject to:}$$

$$\text{Columns of } U \text{ are mutually orthogonal}$$

$$\text{Columns of } V \text{ are mutually orthogonal}$$

The primary difference of this model from unconstrained matrix factorization is the addition of orthogonality constraints, which makes the problem more difficult. For example, if one tries to directly use the update equations of the previous section on unconstrained matrix factorization, the orthogonality constraints will be violated. However, a variety of modified update methods exist to handle this case. For example, one can use a *projected gradient descent* [76] method, wherein all components of a particular column of U or V are updated at one time. In projected gradient descent, the descent direction for the pth column of U (or V), as indicated by the equations of the previous section, is projected in a direction that is orthogonal to the first $(p-1)$ columns of U (or V). For example, the implementation of section 3.6.4.3 can be adapted to learn orthogonal factors by projecting each factor in a direction orthogonal to those learned so far at each step. One can easily incorporate user and item biases by computing the baseline predictions B_{ij} (discussed in the previous section) and subtracting them from the observed entries in the ratings matrix before modeling. Subsequently, the baseline values can be added back to the predicted values as a postprocessing step.

3.6.5.3   Out-of-Sample Recommendations

Many matrix completion methods like matrix factorization are inherently *transductive*, in which predictions can be made only for users and items already included in the ratings matrix at the time of training. It is often not easy to make predictions for new users and items from the factors U and V, if they were not included in the original ratings matrix R at factorization time. One advantage of orthogonal basis vectors is that they can be leveraged more easily to perform out-of-sample recommendations for new users and items. This problem is also referred to as *inductive matrix completion*.

The geometric interpretation provided in Figure 3.6 is helpful in understanding why orthogonal basis vectors are helpful in predicting missing ratings. Once the latent vectors have been obtained, one can project the information in the specified ratings on the corresponding latent vectors; this is much easier when the vectors are mutually orthogonal. Consider a situation where SVD has obtained latent factors U and V, respectively. The columns of V define a k-dimensional hyperplane, $\mathcal{H}_1$, passing through the origin. In Figure 3.6, the number of latent factors is 1, and therefore the single latent vector (i.e., 1-dimensional hyperplane) is shown. If two factors had been used, it would have been a plane.

Now imagine a new user whose ratings have been added into the system. Note that this new user is not represented in the latent factors in U or V. Consider the scenario in which the new user has specified a total of h ratings. The space of possibilities of ratings for this user is an $(n - h)$-dimensional hyperplane in which h values are fixed. An example is illustrated in Figure 3.6, where one rating for *Spartacus* is fixed, and the hyperplane is defined on the other two dimensions. Let this hyperplane be denoted by $\mathcal{H}_2$. The goal is then to determine the point on $\mathcal{H}_2$, which is as close to $\mathcal{H}_1$ as possible. That point on $\mathcal{H}_2$ yields the values of all the other ratings. Three possibilities arise:

1. $\mathcal{H}_1$ *and* $\mathcal{H}_2$ *do not intersect:* The point on $\mathcal{H}_2$ that is closest to $\mathcal{H}_1$ is returned. The smallest distance between a pair of hyperplanes can be formulated as a simple sum-of-squares optimization problem.

2. $\mathcal{H}_1$ *and* $\mathcal{H}_2$ *intersect at a unique point:* This case is similar to that of Figure 3.6. In that case, the values of the ratings of the intersection point can be used.

3. $\mathcal{H}_1$ *and* $\mathcal{H}_2$ *intersect on an t-dimensional hyperplane, where* $t \geq 1$: All ratings, which are as close as possible to the t-dimensional hyperplane, should be found. The average values of the ratings of the corresponding users are returned. Note that this approach combines latent factor and neighborhood methods. The main difference from neighborhood methods is that the neighborhood is discovered in a more refined way with the use of feedback from latent factor models.

Orthogonality has significant advantages in terms of geometric interpretability. The ability to discover out-of-sample recommendations is one example of such an advantage.

3.6.5.4 Example of Singular Value Decomposition

In order to illustrate the use of singular value decomposition, let us apply this approach to the example of Table 3.2. We will use the iterative approach of estimating the missing entries repeatedly. The first step is to fill in the missing entries with the average of each row. As a result, the filled in ratings matrix R_f becomes:

$$
R_f = \begin{pmatrix}
1 & -1 & 1 & -1 & 1 & -1 \\
1 & 1 & -0.2 & -1 & -1 & -1 \\
0 & 1 & 1 & -1 & -1 & 0 \\
-1 & -1 & -1 & 1 & 1 & 1 \\
-1 & 0.2 & -1 & 1 & 1 & 1
\end{pmatrix}
$$

Upon applying rank-2 truncated SVD to the matrix, and absorbing the diagonal matrix within the user factors, we obtain the following:

$$
R_f \approx \begin{pmatrix}
1.129 & -2.152 \\
1.937 & 0.640 \\
1.539 & 0.873 \\
-2.400 & -0.341 \\
-2.105 & 0.461
\end{pmatrix}
\begin{pmatrix}
0.431 & 0.246 & 0.386 & -0.518 & -0.390 & -0.431 \\
-0.266 & 0.668 & -0.249 & 0.124 & -0.578 & 0.266
\end{pmatrix}
$$

$$
= \begin{pmatrix}
1.0592 & -1.1604 & 0.9716 & -0.8515 & 0.8040 & -1.0592 \\
0.6636 & 0.9039 & 0.5881 & -0.9242 & -1.1244 & -0.6636 \\
0.4300 & 0.9623 & 0.3764 & -0.6891 & -1.1045 & -0.4300 \\
-0.9425 & -0.8181 & -0.8412 & 1.2010 & 1.1320 & 0.9425 \\
-1.0290 & -0.2095 & -0.9270 & 1.1475 & 0.5535 & 1.0290
\end{pmatrix}
$$

Note that even after the first iteration, a reasonable estimate is obtained of the missing entries. In particular, the estimated values are $\hat{r}_{23} \approx 0.5581$, $\hat{r}_{31} \approx 0.43$, $\hat{r}_{36} \approx -0.43$, and $\hat{r}_{52} \approx -0.2095$. Of course, these entries are biased by the fact that the initial filled-in entries were based on the row averages, and thus did not accurately reflect the correct values. Therefore, in the next iteration, we fill in these four missing values in the original matrix to obtain the following matrix:

$$R_f = \begin{pmatrix} 1 & -1 & 1 & -1 & 1 & -1 \\ 1 & 1 & 0.5581 & -1 & -1 & -1 \\ 0.43 & 1 & 1 & -1 & -1 & -0.43 \\ -1 & -1 & -1 & 1 & 1 & 1 \\ -1 & -0.2095 & -1 & 1 & 1 & 1 \end{pmatrix}$$

This matrix is still biased, but it is better than filling in missing entries with row averages. In the next iteration, we apply SVD with this new matrix, which is clearly a better starting point. Upon applying the entire process of rank-2 SVD again, we obtain the following matrix in the next iteration:

$$R_f = \begin{pmatrix} 1 & -1 & 1 & -1 & 1 & -1 \\ 1 & 1 & 0.9274 & -1 & -1 & -1 \\ 0.6694 & 1 & 1 & -1 & -1 & -0.6694 \\ -1 & -1 & -1 & 1 & 1 & 1 \\ -1 & -0.5088 & -1 & 1 & 1 & 1 \end{pmatrix}$$

Note that the newly estimated entries have further changed in the next iteration. The new estimated values are $\hat{r}_{23} \approx 0.9274$, $\hat{r}_{31} \approx 0.6694$, $\hat{r}_{36} \approx -0.6694$, and $\hat{r}_{52} \approx -0.5088$. Furthermore, the entries have changed to a smaller degree than in the first iteration. Upon applying the process for one more iteration to the latest value of R_f, we obtain the following:

$$R_f = \begin{pmatrix} 1 & -1 & 1 & -1 & 1 & -1 \\ 1 & 1 & 0.9373 & -1 & -1 & -1 \\ 0.7993 & 1 & 1 & -1 & -1 & -0.7993 \\ -1 & -1 & -1 & 1 & 1 & 1 \\ -1 & -0.6994 & -1 & 1 & 1 & 1 \end{pmatrix}$$

The estimated values are now $\hat{r}_{23} \approx 0.9373$, $\hat{r}_{31} \approx 0.7993$, $\hat{r}_{36} \approx -0.7993$, and $\hat{r}_{52} \approx -0.6994$. Note that the change is even smaller than in the previous iteration. In fact, the change in entry $\hat{r}_{23}$ is very small. Over successive iterations, the changes in the entries tend to become smaller and smaller, until convergence is reached. The resulting entries can be used as the predicted values. A large number of iterations are typically not required in the process. In fact, for *ranking* the items for a given user, only 5 to 10 iterations might be sufficient. In this particular example, one can correctly rank the two missing ratings for user 3 after the very first iteration. The approach can also be applied after mean-centering the rows or columns, or both. This approach has the effect of removing user and item biases before the estimation process. Applying such bias correction methods often has a positive effect on prediction.

The approach is not guaranteed to converge to a global optimum, especially if poor initialization points have been used. This is especially true when a large fraction of the entries in the matrix are missing. In these cases, the initial bias can be significant enough to affect the quality of the final solution. Therefore, it is sometimes advisable to use a simple heuristic, such as a neighborhood model, in order to obtain a first estimate of the missing entries. Choosing such a robust estimate as a starting point will speed up the convergence, and it will also lead to more accurate results. Furthermore, one could easily apply this entire process with regularized singular value decomposition of the filled-in matrices. The main difference is that each iteration uses regularized singular value decomposition of the current matrix, which is filled in with the estimated values. The work in [541] may be used as the relevant subroutine for regularized singular value decomposition.

3.6.6 Non-negative Matrix Factorization

Non-negative matrix factorization (NMF) may be used for ratings matrices that are non-negative. The major advantage of this approach is not necessarily one of accuracy, but that of the high level of interpretability it provides in understanding the user-item interactions. The main difference from other forms of matrix factorization is that the factors U and V must be non-negative. Therefore, the optimization formulation in non-negative matrix factorization is stated as follows:

$$\text{Minimize } J = \frac{1}{2}||R - UV^T||^2$$
$$\text{subject to:}$$
$$U \geq 0$$
$$V \geq 0$$

Although non-negative matrix factorization can be used for any non-negative ratings matrix (e.g., ratings from 1 to 5), its greatest interpretability advantages arise in cases in which users have a mechanism to specify a liking for an item, but no mechanism to specify a dislike. Such matrices include unary ratings matrices or matrices in which the non-negative entries correspond to the activity frequency. These data sets are also referred to as *implicit feedback data sets* [260, 457]. Some examples of such matrices are as follows:

1. In customer transaction data, the purchase of an item corresponds to expressing a liking for an item. However, not buying an item does not necessarily imply a dislike because the user might have purchased the item elsewhere or they may not be aware of the item. When amounts are associated with transactions, the matrix R may contain arbitrary non-negative numbers. However, all these numbers specify the degree of liking for an item, but they do not indicate dislike. In other words, the numerical quantities in implicit feedback indicate *confidence*, whereas the numerical quantities in explicit feedback indicate *preference*.

2. Similar to the case of purchasing an item, the browsing of an item may be indicative of a like. In some cases, the frequency of the buying or browsing behavior can be quantified as a non-negative value.

3. In Web click data, the selection of an item corresponds to a unary rating of liking an item.

4. A "*like*" button on Facebook can be considered a mechanism to provide a unary rating for an item.

The implicit feedback setting can be considered the matrix completion analog to the positive-unlabeled (PU) learning problem in classification and regression modeling. In classification and regression modeling, reasonable results can often be obtained by treating the unlabeled entries as belonging to the negative class when the positive class is already known to be a very small minority class. Similarly, a helpful aspect of such matrices and problem settings is that it is often reasonably possible to set the unspecified entries to 0, rather than treat them as missing values. For example, consider a customer transaction data set, in which values indicate quantities purchased by a customer. In such a case, it is reasonable to set a value to 0, when that item has not been bought by the customer. Therefore, in this case, one only has to perform non-negative matrix factorization of a fully specified matrix, which is a standard problem in the machine learning literature. This problem is also referred to as *one class collaborative filtering*. Although some recent works argue that the missing values should not be set to 0 in such cases [260, 457, 467, 468] to reduce bias, a considerable amount of work in the literature shows that reasonably robust solutions can be obtained by treating the missing entries as 0 in the modeling process. This is especially the case when the prior probability of an entry to be 0 is very large. For instance, in the supermarket scenario, a customer would typically never buy the vast majority of items in the store. In such cases, setting the missing values to 0 (in the initial matrix for factorization purposes but not in the final prediction) would result in a small amount of bias, but explicitly treating the entries as unspecified in the initial matrix will lead to greater solution complexity. Unnecessary complexity always leads to overfitting. These effects are especially significant[15] in smaller data sets.

Note that the optimization formulation of non-negative matrix factorization is a constrained optimization formulation, which can be solved using standard methods such as Lagrangian relaxation. Although a detailed derivation of the algorithm used for non-negative matrix factorization is beyond the scope of this book, we refer the reader to [22] for details. Here, we present only a brief discussion of how non-negative matrix factorization is performed.

An iterative approach is used to update the matrices U and V. Let u_{ij} and v_{ij}, respectively, be the (i, j)th entries of the matrices U and V. The following multiplicative update rules for u_{ij} and v_{ij} are used:

$$u_{ij} \Leftarrow \frac{(RV)_{ij} u_{ij}}{(UV^T V)_{ij} + \epsilon} \quad \forall i \in \{1 \ldots m\}, \forall j \in \{1 \ldots k\} \tag{3.24}$$

$$v_{ij} \Leftarrow \frac{(R^T U)_{ij} v_{ij}}{(VU^T U)_{ij} + \epsilon} \quad \forall i \in \{1 \ldots n\}, \forall j \in \{1 \ldots k\} \tag{3.25}$$

Here, ϵ is a small value such as 10^{-9} to increase numerical stability. All entries in U and V on the right-hand side of the update equations are fixed to the values obtained at the end of the previous iteration during the course of a particular iteration. In other words, all entries in U and V are updated "simultaneously." Small values are sometimes added to the denominator of the update equations to prevent division by 0. The entries in U and V are initialized to random values in $(0, 1)$, and the iterations are executed to convergence. It is possible to obtain better solutions by performing the initialization in a more judicious way [331, 629].

As in the case of other types of matrix factorization, regularization can be used to improve the quality of the underlying solution. The basic idea is to add the penalties $\frac{\lambda_1 \|U\|^2}{2} + \frac{\lambda_2 \|V\|^2}{2}$ to the objective function. Here $\lambda_1 > 0$ and $\lambda_2 > 0$ are the regularization

---

[15]These effects are best understood in terms of the bias-variance trade-off in machine learning [22]. Setting the unspecified values to 0 increases bias, but it reduces variance. When a large number of entries are unspecified, and the prior probability of a missing entry to be 0 is very high, the variance effects can dominate.

parameters. This results in a modification [474] of the update equations as follows:

$$u_{ij} \Leftarrow \max \left\{ \left[\frac{(RV)_{ij} - \lambda_1 u_{ij}}{(UV^T V)_{ij} + \epsilon} \right] u_{ij}, 0 \right\} \quad \forall i \in \{1 \ldots m\}, \forall j \in \{1 \ldots k\} \tag{3.26}$$

$$v_{ij} \Leftarrow \max \left\{ \left[\frac{(R^T U)_{ij} - \lambda_2 v_{ij}}{(VU^T U)_{ij} + \epsilon} \right] v_{ij}, 0 \right\} \quad \forall i \in \{1 \ldots n\}, \forall j \in \{1 \ldots k\} \tag{3.27}$$

The maximization function is used to impose non-negativity and the small additive term $\epsilon \approx 10^{-9}$ in the denominator is used to ensure numerical stability. The parameters λ_1 and λ_2 can be determined using the same approach as described earlier. Instead of using the gradient-descent method, one can also use alternating least-squares methods in which non-negative linear regression is used. Tikhonov regularization can be used within the regression model to prevent overfitting. Details of the alternating least-squares method for non-negative matrix factorization may be found in [161, 301]. The main challenges with these off-the-shelf methods is their lack of computational efficiency with large ratings matrices, because all entries are treated as observed. In section 3.6.6.3, we will discuss how these issues can be addressed.

3.6.6.1 Interpretability Advantages

The main advantage of non-negative matrix factorization is that a high degree of interpretability is achieved in the solution. It is always useful to pair recommender systems with explanations for the recommendations, and this is provided by non-negative matrix factorization. In order to better understand this point, consider a situation in which the preference matrix contains quantities of items bought by customers. An example of a toy 6×6 matrix with 6 items and 6 customers is illustrated in Figure 3.12. It is clear that there are two classes of products corresponding to dairy products and drinks, respectively. It is clear that the customer buying behavior is highly correlated on the basis of the classes of items although all customers seem to like juice. These classes of items are referred to as *aspects*. The corresponding factor matrices also provide a clear interpretability about the affinity of customers and items to these aspects. For example, customers 1 to 4 like dairy products, whereas customers 4 to 6 like drinks. This is clearly reflected in the 6×2 user-factor matrix U. In this simplified example, we have shown all the factored values in U and V to be integral for visual simplicity. In practice, the optimal values are almost always real numbers. The magnitude of the entry of a user in each of the two columns quantifies her level of interest in the relevant aspect. Similarly, the factor matrix V shows how the items are related to the various aspects. Therefore, in this case, the condition $r_{ij} \approx \sum_{s=1}^{k} u_{is} \cdot v_{js}$ can be semantically interpreted in terms of the $k = 2$ aspects:

$$r_{ij} \approx (\text{Affinity of user } i \text{ to dairy aspect}) \times (\text{Affinity of item } j \text{ to dairy aspect})$$
$$+ (\text{Affinity of user } i \text{ to drinks aspect}) \times (\text{Affinity of item } j \text{ to drinks aspect})$$

This way of predicting the value of r_{ij} shows a "sum-of-parts" decomposition of the matrix. Each of these parts can also be viewed as a user-item co-cluster. This is also one of the reasons that non-negative matrix factorization is often used in clustering. In practical applications, it is often possible to look at each of these clusters and semantically interpret the associations between users and items. When semantic labels can be manually attached to the various clusters, the factorization process provides a neat explanation of the ratings in terms of the contributions of various semantic "genres" of items.

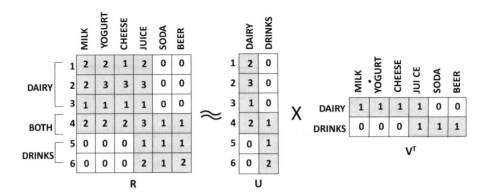

Figure 3.12: An example of non-negative matrix factorization

This sum-of-parts decomposition can be represented mathematically as follows. The rank-k matrix factorization UV^T can be decomposed into k components by expressing the matrix product in terms of the k columns $\overline{U_i}$ and $\overline{V_i}$, respectively, of U and V:

$$UV^T = \sum_{i=1}^{k} \overline{U_i}\, \overline{V_i}^T \tag{3.28}$$

Each $m \times n$ matrix $\overline{U_i}\, \overline{V_i}^T$ is a rank-1 matrix that corresponds to an aspect in the data. Because of the interpretable nature of non-negative decomposition, it is easy to map these aspects to clusters. For example, the two latent components of the aforementioned example corresponding to dairy products and drinks, respectively, are illustrated in Figure 3.13. Note that Equation 3.28 decomposes the factorization in terms of the *columns* of U and V, whereas Equation 3.14 is a different way of understanding the factorization in terms of the *rows* of U and V. For a given user-item combination, the rating prediction is given by the sum of the contributions of these aspects, and one can even gain a better understanding of why a rating is predicted in a certain way by the approach.

3.6.6.2   Observations about Factorization with Implicit Feedback

Non-negative matrix factorization is particularly well suited to implicit feedback matrices in which ratings indicate positive preferences. Unlike explicit feedback data sets, it is not possible to ignore the missing entries in the optimization model because of the lack of negative feedback in such data. It is noteworthy that the non-negative matrix factorization model treats missing entries as negative feedback by setting them to 0s. Not doing so would grossly increase the error on the unobserved entries. In order to understand this point, consider a unary ratings matrix in which likes are specified by 1s. The factorization shown in Figure 3.14 will provide 100% accuracy on an *arbitrary* unary matrix when computed only over observed entries. This is because the multiplication of U and V^T in Figure 3.14 leads to a matrix containing only 1s and no 0s. Of course, such a factorization will have very high error on the unobserved entries because many unobserved entries may correspond to negative preferences. This example is a manifestation of overfitting caused by lack of negative feedback data. Therefore, for ratings matrices in which negative preferences are missing and it is known that negative preferences vastly outnumber positive preferences, it is important to treat missing entries as 0s. For example, in a customer transaction data set,

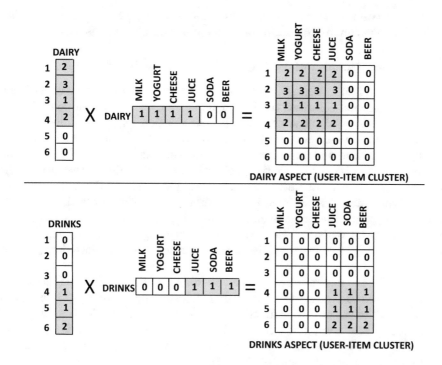

Figure 3.13: The sum-of-parts interpretation of NMF

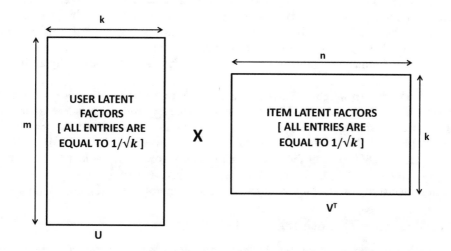

Figure 3.14: Overfitting caused by ignoring missing entries in a unary matrix

if the values indicate the amounts bought by various users and most items are not bought by default, then one can approximate the value of a missing entry as 0.

3.6.6.3 Computational and Weighting Issues with Implicit Feedback

The treatment of missing entries as 0s leads to computational challenges with large matrices. There are several solutions to this dilemma. For example, a sample of the missing entries can be treated as 0s. The gradient-descent solution for the sampled case is similar to that discussed in the next section. It is possible to further improve the accuracy with an ensemble approach. The matrix is factorized multiple times with a different sample of 0s, and each factorization is used to predict (a slightly different) value of the rating. The different predictions of a particular rating are then averaged to create the final result. By using samples of varying sizes, it is also possible to weight the negative feedback entries differently from the positive feedback entries. Such an approach can be important in cost-sensitive settings where false positives and false negatives are weighted differently. Typically, the zero entries should be weighted less than the nonzero entries, and therefore down-sampling the zero entries is useful.

It is also possible to incorporate such weights directly in the objective function and treat all missing entries as 0s. The errors on the zero entries should be weighted less than those on the nonzero entries in the objective function to prevent the zero entries from dominating the optimization. The relative weights can be determined using cross-validation with respect to a particular accuracy measure. Alternatively, the work in [260] suggests the following heuristic to select the weight w_{ij} of entry (i, j):

$$w_{ij} = 1 + \theta \cdot r_{ij} \qquad (3.29)$$

Note that all missing values of r_{ij} are treated as 0s in Equation 3.29, and a typical value of θ is 40. This approach also works in settings, where the ratings r_{ij} represent quantities that are bought, rather than binary indicators. In such cases, the weights w_{ij} are computed by treating these quantities as the ratings in Equation 3.29, but the factorized matrix is a derivative binary indicator matrix R_I of the quantity matrix $R = [r_{ij}]$. This indicator matrix R_I is derived from R by copying the zero entries and substituting the nonzero entries with 1s. This approach of weighted factorization of the indicator matrix is therefore slightly different from the example of Figure 3.12, which was presented purely for illustrative purposes.

When working with weighted entries, it is possible to modify stochastic gradient descent methods with weights (cf. section 6.5.2.1 of Chapter 6). However, the problem is that implicit feedback matrices are fully specified, and many of the gradient-descent methods no longer remain computationally viable in large-scale settings. An efficient (weighted) ALS method was proposed in [260] for the factorization process in order to avoid the computational challenge of handing the large number of zero entries. Although this approach does not impose non-negativity on the factors, it can be easily generalized to the non-negative setting.

3.6.6.4 Ratings with Both Likes and Dislikes

Our discussion of non-negative matrix factorization so far has focussed only on implicit feedback matrices in which there is a mechanism to specify a liking for an item but no mechanism to specify a dislike. As a result, the underlying "ratings" matrices are always non-negative. Although one can use non-negative matrix factorization for *nominally* non-negative ratings (e.g., from 1 to 5), which explicitly specify both likes and dislikes, there are no special interpretability advantages of using non-negative matrix factorization in such cases. For example, the rating scale might be from 1 to 5, wherein a value of 1 indicates extreme dislike. In this case, one cannot treat the unspecified entries as 0, and one must work only with the set of observed entries. As before, we denote the set of observed entries in the ratings matrix $R = [r_{ij}]$ by S:

$$S = \{(i, j) : r_{ij} \text{ is observed}\} \qquad (3.30)$$

The optimization problem (with regularization) is stated in terms of these observed entries as follows:

$$\text{Minimize } J = \frac{1}{2} \sum_{(i,j) \in S} \left(r_{ij} - \sum_{s=1}^{k} u_{is} \cdot v_{js} \right)^2 + \frac{\lambda}{2} \sum_{i=1}^{m} \sum_{s=1}^{k} u_{is}^2 + \frac{\lambda}{2} \sum_{j=1}^{n} \sum_{s=1}^{k} v_{js}^2$$

subject to:

$$U \geq 0$$
$$V \geq 0$$

This formulation is similar to the regularized formulation in unconstrained matrix factorization. The only difference is the addition of the non-negativity constraints. In such cases, modifications are required to the update equations that are used for unconstrained matrix factorization. First, one must initialize the entries of U and V to non-negative values in $(0, 1)$. Then, a similar update can be made, as in the section on unconstrained matrix factorization. In fact, the update equations in section 3.6.4.2 can be used directly. The main modification is to ensure that non-negativity is maintained during updates. If any component of U or V violates the non-negativity constraint as a result of the update, then it is set to 0. The updates are performed to convergence as in all stochastic gradient descent methods.

Other solution methodologies are often used to compute optimal solutions to such models. For example, it is possible to adapt an alternating least-squares approach to non-negative matrix factorization. The main difference is that the coefficients of the least-squares regression are constrained to be non-negative. A wide variety of projected gradient descent, coordinate descent, and nonlinear programming methods are also available to handle such optimization models [76, 357].

In the setting where ratings specify both likes and dislikes, non-negative matrix factorization has no special advantages over unconstrained matrix factorization in terms of interpretability. This is because one can no longer interpret the solution from a sum-of-parts perspective. For example, the addition of three dislike ratings cannot be interpreted as leading to a like rating. Furthermore, because of the addition of non-negativity constraints, the quality of the solution is reduced over that of unconstrained matrix factorization when computed over the observed entries. However, this does not always mean that the quality of the solution will be worse when computed over the unobserved entries. In real settings, positive relationships between users and items are more important than negative relationships between users and items. As a result, non-negativity constraints often introduce a bias which is beneficial in avoiding overfitting. As in the case of unconstrained matrix factorization, one can also incorporate user and item biases to further improve the generalization performance.

3.6.7   Understanding the Matrix Factorization Family

It is evident that the various forms of matrix factorization in the previous sections share a lot in common. All of the aforementioned optimization formulations minimize the Frobenius norms of the residual matrix $(R - UV^T)$ subject to various constraints on the factor matrices U and V. Note that the goal of the objective function is to make UV^T approximate the ratings matrix R as closely as possible. The constraints on the factor matrices achieve different interpretability properties. In fact, the broader family of matrix factorization models can use any other objective function or constraint to force a good approximation. This broader family can be written as follows:

$$\text{Optimize } J = [\text{Objective function quantifying matching between } R \text{ and } UV^T]$$
$$\text{subject to:}$$
$$\text{Constraints on } U \text{ and } V$$

The objective function of a matrix factorization method is sometimes referred to as the loss function, when it is in minimization form. Note that the optimization formulation may be either a minimization or a maximization problem, but the goal of the objective function is always to force R to match UV^T as closely as possible. The Frobenius norm is an example of a minimization objective, and some probabilistic matrix factorization methods use a maximization formulation such as the maximum-likelihood objective function. In most cases, regularizers are added to the objective function to prevent overfitting. The various constraints often impose different types of interpretability on the factors. Two examples of such interpretability are orthogonality (which provides geometric interpretability) and non-negativity (which provides sum-of-parts interpretability). Furthermore, even though constraints increase the error on the observed entries, they can sometimes improve the errors on the unobserved entries when they have a meaningful semantic interpretation. This is because constraints reduce the variance[16] on the unobserved entries while increasing bias. As a result, the model has better *generalizability*. For example, fixing the entries in a column in each of U and V to ones almost always results in better performance (cf. section 3.6.4.5). Selecting the right constraints to use is often data-dependent and requires insights into the application-domain at hand.

Other forms of factorization exist in which one can assign probabilistic interpretability to the factors. For example, consider a scenario in which a non-negative unary ratings matrix R is treated as a relative frequency distribution, whose entries sum to 1.

$$\sum_{i=1}^{m} \sum_{j=1}^{n} r_{ij} = 1 \tag{3.31}$$

Note that it is easy to scale R to sum to 1 by dividing it with the sum of its entries. Such a matrix can be factorized in a similar way to SVD:

$$R \approx (Q_k \Sigma_k) P_k^T$$
$$= UV^T$$

As in SVD, the diagonal matrix Σ_k is absorbed in the user factor matrix $U = Q_k \Sigma_k$, and the item factor matrix V is set to P_k. The main difference from SVD is that the columns of Q_k and P_k are not orthogonal, but they are non-negative values summing to 1. Furthermore, the entries of the diagonal matrix Σ_k are non-negative and they also sum to 1.

---

[16]Refer to Chapter 6 for a discussion of the bias-variance trade-off in collaborative filtering.

Table 3.3: The family of matrix factorization methods

Method	Constraints	Objective	Advantages/Disadvantages
Unconstrained	No constraints	Frobenius + regularizer	Highest quality solution Good for most matrices Regularization prevents overfitting Poor interpretability
SVD	Orthogonal Basis	Frobenius + regularizer	Good visual interpretability Out-of-sample recommendations Good for dense matrices Poor semantic interpretability Suboptimal in sparse matrices
Max. Margin	No constraints	Hinge loss + margin regularizer	Highest quality solution Resists overfitting Similar to unconstrained Poor interpretability Good for discrete ratings
NMF	Non-negativity	Frobenius + regularizer	Good quality solution High semantic interpretability Loses interpretability with both like/dislike ratings Less overfitting in some cases Best for implicit feedback
PLSA	Non-negativity	Maximum Likelihood + regularizer	Good quality solution High semantic interpretability Probabilistic interpretation Loses interpretability with both like/dislike ratings Less overfitting in some cases Best for implicit feedback

Such a factorization has a probabilistic interpretation; the matrices Q_k, P_k and Σ_k contain the probabilistic parameters of a generative process that creates the ratings matrix. The objective function learns the parameters of this generative process so that the likelihood of the generative process creating the ratings matrix is as large as possible. Therefore, the objective function is in *maximization form*. Interestingly, this method is referred to as *Probabilistic Latent Semantic Analysis (PLSA)*, and it can be viewed as a probabilistic variant of non-negative matrix factorization. Clearly, the probabilistic nature of this factorization provides it with a different type of interpretability. A detailed discussion of PLSA may be found in [22]. In many of these formulations, optimization techniques such as gradient descent (or ascent) are helpful. Therefore, most of these methods use very similar ideas in terms of formulating the optimization problem and the underlying solution methodology.

Similarly, maximum margin factorization [180, 500, 569, 624] borrows ideas from *support vector machines* to add a maximum margin regularizer to the objective function and some of its variants [500] are particularly effective for discrete ratings. This approach shares a number of conceptual similarities with the regularized matrix factorization method discussed in section 3.6.4. In fact, the maximum margin regularizer is not very different than that used in unconstrained matrix factorization. However, *hinge loss* is used to quantify the errors in the approximation, rather than the Frobenius norm. While it is beyond the scope of this book to discuss these variants in detail, a discussion may be found in [500, 569]. The focus on maximizing the margin often provides higher quality factorization than some of the other models in the presence of overfitting-prone data. In Table 3.3, we have provided a list of various factorization models and their characteristics. In most cases, the addition of constraints such as non-negativity can reduce the quality of the underlying solution on the observed entries, because it reduces the space of feasible solutions. This is the reason that unconstrained and maximum margin factorization are expected to have the highest quality of global optima. Nevertheless, since the global optimum cannot be easily found in most cases by the available (iterative) methods, a constrained method can sometimes perform better than an unconstrained method. Furthermore, the accuracy over observed entries may be different from that over unobserved entries because of the effects of overfitting. In fact, non-negativity constraints can sometimes improve the accuracy over unobserved entries in some domains. Some forms of factorization such as NMF cannot be applied to matrices with negative entries. Clearly, the choice of the model depends on the problem setting, the noise in the data, and the desired level of interpretability. There is no single solution that can achieve all these goals. A careful understanding of the problem domain is important for choosing the correct model.

3.7   Integrating Factorization and Neighborhood Models

Neighborhood-based methods are generally considered inherently different from other optimization models because of their heuristic nature. Nevertheless, it was shown in section 2.6 of Chapter 2 that neighborhood methods can also be understood in the context of optimization models. This is a rather convenient framework because it paves the way for integration of neighborhood models with other optimization models, such as latent factor models. The approach in [309] integrates the item-wise model of section 2.6.2 in Chapter 2 with the SVD++ model of section 3.6.4.6.

Assume that the ratings matrix R is mean centered. In other words, the global mean μ of the ratings matrix is already subtracted from all the entries, and all predictions will be performed on mean-centered values. The global mean μ can be added back to the predicted values in a postprocessing phase. With this assumption on the ratings matrix $R = [r_{ij}]$, we will re-visit the various portions of the model.

3.7.1   Baseline Estimator: A Non-Personalized Bias-Centric Model

The non-personalized bias-centric model predicts the (mean-centered) ratings in R purely as an addition of user and item biases. In other words, ratings are explained completely by user generosity and item popularity, rather than specific and *personalized* interests of users in items. Let b_i^{user} be the bias variable for user i, and b_j^{item} be the bias variable for item j. Then, the prediction of such a model is as follows:

$$\hat{r}_{ij} = b_i^{user} + b_j^{item} \tag{3.32}$$

Let S be the pairs of indices corresponding to the observed entries in the ratings matrix.

$$S = \{(i,j) : r_{ij} \text{ is observed}\} \tag{3.33}$$

Then, b_i^{user} and b_j^{item} can be determined by formulating an objective function over the errors $e_{ij} = r_{ij} - \hat{r}_{ij}$ in the observed entries as follows:

$$\text{Minimize } J = \frac{1}{2} \sum_{(i,j) \in S} \left(r_{ij} - \hat{r}_{ij}\right)^2 + \frac{\lambda}{2} \left(\sum_{u=1}^{m} (b_u^{user})^2 + \sum_{j=1}^{n} (b_j^{item})^2 \right)$$

This optimization problem can be solved via gradient descent using the following update rules over each observed entry (i,j) in S in a stochastic gradient descent method:

$$b_i^{user} \Leftarrow b_i^{user} + \alpha(e_{ij} - \lambda b_i^{user})$$
$$b_j^{item} \Leftarrow b_j^{item} + \alpha(e_{ij} - \lambda b_j^{item})$$

The basic framework of the gradient-descent method is similar to that in Figure 3.9, except for the differences in the choice of optimization variables and corresponding update steps. Interestingly, a pure bias-centric model can often provide reasonable predictions in spite of its non-personalized nature. This is especially the case when the amount of ratings data is limited. After solving for the values of b_i^{user} and b_j^{item}, we set B_{ij} to the predicted value of $\hat{r}_{ij}$ according to Equation 3.32. This value of B_{ij} is then *treated as a constant* throughout this section rather than as a variable. Therefore, the first step of the integrated model solution is to determine the *constant value* B_{ij} by solving the non-personalized model. This non-personalized model can also be viewed as a *baseline* estimator because B_{ij} is a rough baseline estimate to the values of the rating r_{ij}. In general, subtracting the value of B_{ij} from each observed entry r_{ij} results in a new matrix that can often be estimated more robustly by most of the models discussed in earlier sections and chapters. This section provides a specific example of how neighborhood models may be adjusted with the use of the baseline estimator though its applicability is much broader.

3.7.2 Neighborhood Portion of Model

We replicate the neighborhood-based prediction relationship of Equation 2.29 (cf. section 2.6.2 of Chapter 2) as follows:

$$\hat{r}_{ij} = b_i^{user} + b_j^{item} + \frac{\sum_{l \in Q_j(i)} w_{lj}^{item} \cdot (r_{il} - b_i^{user} - b_l^{item})}{\sqrt{|Q_j(i)|}} \qquad (3.34)$$

Although the aforementioned equation is that same as Equation 2.29 of Chapter 2, the subscript notations have been changed to ensure consistency with the latent factor models in this section. Here b_i^{user} is the user bias and b_j^{item} is the item bias. The variable w_{lj}^{item} represents the item-item regression coefficient between item l and item j. The set $Q_j(i)$ represents[17] the subset of the K nearest items to item j, that have been rated by user i. Furthermore, one of the occurrences of $b_i^{user} + b_l^{item}$ in Equation 3.34 is replaced with the *constant* value B_{il} (derived using the approach of the previous section). The resulting prediction is as follows:

$$\hat{r}_{ij} = b_i^{user} + b_j^{item} + \frac{\sum_{l \in Q_j(i)} w_{lj}^{item} \cdot (r_{il} - B_{il})}{\sqrt{|Q_j(i)|}} \qquad (3.35)$$

It is noteworthy that the bias variables b_i^{user} and b_j^{item} are parameters to be optimized, whereas B_{il} is a constant. One can set up an optimization model that sums up the squared errors $e_{ij}^2 = (r_{ij} - \hat{r}_{ij})^2$ in addition to regularization terms. A stochastic gradient-descent approach can be used to determine a solution to the neighborhood portion of the model.

---

[17]Note that we use upper-case variable K to represent the size of the neighborhood that defines $Q_j(i)$. This is a deviation from section 2.6.2 of Chapter 2. We use lower-case variable k to represent the dimensionality of the factor matrices. The values of k and K are generally different.

The resulting gradient-descent steps are as follows:

$$b_i^{user} \Leftarrow b_i^{user} + \alpha(e_{ij} - \lambda b_i^{user})$$

$$b_j^{item} \Leftarrow b_j^{item} + \alpha(e_{ij} - \lambda b_j^{item})$$

$$w_{lj}^{item} \Leftarrow w_{lj}^{item} + \alpha_2 \left(\frac{e_{ij} \cdot (r_{il} - B_{il})}{\sqrt{|Q_j(i)|}} - \lambda_2 \cdot w_{lj}^{item} \right) \quad \forall l \in Q_j(i)$$

This neighborhood model can be enhanced further with implicit feedback by introducing item-item implicit feedback variables c_{lj}. The basic idea is that if item j is rated together with many neighboring items by the same user i, then it should have an impact on the predicted rating $\hat{r}_{ij}$. This impact is irrespective of the actual values of the ratings of these neighbor items of j. This impact is equal to $\frac{\sum_{l \in Q_j(i)} c_{lj}}{\sqrt{|Q_j(i)|}}$. Note that the scaling of the expression with $\sqrt{|Q_j(i)|}$ is done in order to adjust for varying levels of sparsity in different user-item combinations. Then, the neighborhood model with implicit feedback can be written as follows:

$$\hat{r}_{ij} = b_i^{user} + b_j^{item} + \frac{\sum_{l \in Q_j(i)} w_{lj}^{item} \cdot (r_{il} - B_{il})}{\sqrt{|Q_j(i)|}} + \frac{\sum_{l \in Q_j(i)} c_{lj}}{\sqrt{|Q_j(i)|}} \quad (3.36)$$

On creating a least-squares optimization model with respect to the error $e_{ij} = r_{ij} - \hat{r}_{ij}$, one can compute the gradient and derive the stochastic gradient-descent steps. This results in the following modified set of updates:

$$b_i^{user} \Leftarrow b_i^{user} + \alpha(e_{ij} - \lambda b_i^{user})$$

$$b_j^{item} \Leftarrow b_j^{item} + \alpha(e_{ij} - \lambda b_j^{item})$$

$$w_{lj}^{item} \Leftarrow w_{lj}^{item} + \alpha_2 \left(\frac{e_{ij} \cdot (r_{il} - B_{il})}{\sqrt{|Q_j(i)|}} - \lambda_2 \cdot w_{lj}^{item} \right) \quad \forall l \in Q_j(i)$$

$$c_{lj} \Leftarrow c_{lj} + \alpha_2 \left(\frac{e_{ij}}{\sqrt{|Q_j(i)|}} - \lambda_2 \cdot c_{lj} \right) \quad \forall l \in Q_j(i)$$

The work in [309] assumes a more general framework, in which the implicit feedback matrix is not necessarily derived from only the ratings matrix. For example, a retailer might create the implicit ratings matrix based on users who have browsed, rated, or bought an item. This generalization is relatively straightforward to incorporate in our models by changing the final term of Equation 3.36 to $\frac{\sum_{l \in Q_j'(i)} c_{lj}}{\sqrt{|Q_j'(i)|}}$. Here, $Q_j'(i)$ is the set of closest neighbors of user i (based on explicit ratings), who have also provided some form of implicit feedback for item j. This modification can also be applied to the latent factor portion of the model, although we will consistently work with the simplified assumption that the implicit feedback matrix is derived from the ratings matrix.

3.7.3 Latent Factor Portion of Model

The aforementioned prediction is made on the basis of the neighborhood model. A corresponding latent factor model is introduced in section 3.6.4.6, in which implicit feedback is integrated with ratings information to make predictions. We replicate Equation 3.21 from that section here:

$$\hat{r}_{ij} = \sum_{s=1}^{k+2} \left(u_{is} + \sum_{h \in I_i} \frac{y_{hs}}{\sqrt{|I_i|}} \right) \cdot v_{js} \tag{3.37}$$

As in section 3.6.4.6, I_i represents the set of items rated by user i. The $m \times (k+2)$ matrix $Y = [y_{hs}]$ contains the implicit feedback variables, and its construction is described in section 3.6.4.6. Furthermore, the $(k+2)$th column of U contains only 1s, the $(k+1)$th column of V contains only 1s, and the last two columns of Y are 0s. Note that the right-hand side of Equation 3.37 already accounts for the user and item biases. Since the last two columns of the factor matrices contain the bias variables, the component $\sum_{s=1}^{k+2} u_{is} v_{js}$ of Equation 3.37 includes the bias terms.

3.7.4 Integrating the Neighborhood and Latent Factor Portions

One can now integrate the two models in Equations 3.36 and 3.37 to create a single predicted value as follows:

$$\hat{r}_{ij} = \underbrace{\frac{\sum_{l \in Q_j(i)} w_{lj}^{item} \cdot (r_{il} - B_{il})}{\sqrt{|Q_j(i)|}} + \frac{\sum_{l \in Q_j(i)} c_{lj}}{\sqrt{|Q_j(i)|}}}_{\text{Neighborhood Component}} + \underbrace{\sum_{s=1}^{k+2} \left(u_{is} + \sum_{h \in I_i} \frac{y_{hs}}{\sqrt{|I_i|}} \right) \cdot v_{js}}_{\text{Latent Factor Comp.+Bias}} \tag{3.38}$$

Note that the initial bias terms $b_i^{user} + b_j^{item}$ of Equation 3.36 are missing here because they are included in the final term corresponding to the latent factor model. The same user and item biases are now shared by both components of the model.

The corresponding optimization problem, which minimizes the aggregated squared error $e_{ij}^2 = (r_{ij} - \hat{r}_{ij})^2$ over the entries in (observed set) S is as follows:

$$\text{Minimize } J = \frac{1}{2} \sum_{(i,j) \in S} (r_{ij} - \hat{r}_{ij})^2 + \frac{\lambda}{2} \sum_{s=1}^{k+2} \left(\sum_{i=1}^m u_{is}^2 + \sum_{j=1}^n v_{js}^2 + \sum_{j=1}^n y_{js}^2 \right) +$$

$$+ \frac{\lambda_2}{2} \sum_{j=1}^n \sum_{l \in \cup_i Q_j(i)} [(w_{lj}^{item})^2 + c_{lj}^2]$$

subject to:

$(k+2)$th column of U contains only 1s

$(k+1)$th column of V contains only 1s

Last two columns of Y contain only 0s

The value of $\hat{r}_{ij}$ in the aforementioned objective function can be materialized with the help of Equation 3.38. As in all latent factor models, the sum of squares of the optimization variables are included for regularization. Note that the different parameters λ and λ_2 are used for regularizing the sets of variables from the latent factor model and the neighborhood model, respectively, for better flexibility in the optimization process.

3.7.5 Solving the Optimization Model

As in the case of all the other optimization models discussed in this chapter, a gradient descent approach is used to solve the optimization problem. In this case, the optimization model is rather complex because it contains a relatively large number of terms, and a large number of variables. Nevertheless, the approach for solving the optimization model is exactly the same as in the case of the latent factor model of section 3.6.4.6. A partial derivative with respect to each optimization variable is used to derive the update step. We omit the derivation of the gradient descent steps, and simply state them here in terms of the error values $e_{ij} = r_{ij} - \hat{r}_{ij}$. The following rules may be used for each observed entry $(i, j) \in S$ in the ratings matrix:

$$u_{iq} \Leftarrow u_{iq} + \alpha(e_{ij} \cdot v_{jq} - \lambda \cdot u_{iq}) \quad \forall q \in \{1 \ldots k+2\}$$

$$v_{jq} \Leftarrow v_{jq} + \alpha \left(e_{ij} \cdot \left[u_{iq} + \sum_{h \in I_i} \frac{y_{hq}}{\sqrt{|I_i|}} \right] - \lambda \cdot v_{jq} \right) \quad \forall q \in \{1 \ldots k+2\}$$

$$y_{hq} \Leftarrow y_{hq} + \alpha \left(\frac{e_{ij} \cdot v_{jq}}{\sqrt{|I_i|}} - \lambda \cdot y_{hq} \right) \quad \forall q \in \{1 \ldots k+2\}, \forall h \in I_i$$

$$w_{lj}^{item} \Leftarrow w_{lj}^{item} + \alpha_2 \left(\frac{e_{ij} \cdot (r_{il} - B_{il})}{\sqrt{|Q_j(i)|}} - \lambda_2 \cdot w_{lj}^{item} \right) \quad \forall l \in Q_j(i)$$

$$c_{lj} \Leftarrow c_{lj} + \alpha_2 \left(\frac{e_{ij}}{\sqrt{|Q_j(i)|}} - \lambda_2 \cdot c_{lj} \right) \quad \forall l \in Q_j(i)$$

Reset perturbed entries in fixed columns of U, V, and Y

The first three updates can also be written in $(k + 2)$-dimensional vectorized form. Refer to the section on SVD++ for a footnote containing these updates. We repeatedly loop over all the observed ratings in S with a stochastic gradient descent method. The basic algorithmic framework for stochastic gradient descent is described in Figure 3.9. The value of α regulates the step size for variables associated with the latent factor portion of the model, whereas α_2 regulates the step size for variables associated with the neighborhood portion of the model. The fixed columns of U, V, and Y should not be updated by these rules, according to the constraints in the optimization model. This is achieved in practice by always resetting them to their fixed values at the end of an iteration. Furthermore, these columns are always initialized to their fixed values, as required by the constraints of the optimization model. The regularization parameters can be selected by holding out a fraction of the observed entries during training, and tuning the accuracy on the held out entries. A more effective approach is to use the cross-validation method discussed in Chapter 7. It is particularly important to use different step-sizes and regularization parameters for the neighborhood and latent factor portions of the model to avoid poor performance.

3.7.6   Observations about Accuracy

It was shown in [309] that the combined model provided superior results to those of each of the individual models. This is a result of the ability of the combined model to adapt to varying characteristics of different portions of the data set. The basic idea is similar to that used often in hybrid recommender systems (cf. Chapter 6) for combining different types of models. One can try to approximate the results of the integrated model by using a weighted average of the predictions of the two different component models. The relative weights can be learned using the aforementioned hold-out or cross-validation techniques. However, compared to the averaged model, the integrated model of this section is more powerful. One reason is that the bias variables are shared by the two components, which prevents the overfitting of the bias variables to the specific nuances of each model. Furthermore, the use of the prediction function of Equation 3.38 implicitly regulates the importance of each portion of the model by automatically choosing appropriate values for each of the variables in the optimization process. As a result, this type of integration often provides superior accuracy. However, the model provides only slightly superior performance to that given by SVD++, and the results are data-set dependent. One issue to keep in mind is that the neighborhood model has more parameters to be optimized than SVD++. Significant advantages will not be obtained by the neighborhood component unless the data set is sufficiently large. For smaller data sets, increasing the number of parameters often leads to overfitting. In this sense, the proper choice between asymmetric factor models, pure SVD with biases, SVD++, and neighborhood-integrated factorization, often depends on the size of the data set at hand. More complex models require larger data sets to avoid overfitting. For very small data sets, one would do best with asymmetric factor models. For very large data sets, the neighborhood-integrated factorization model is best. SVD++ generally does better than pure SVD (with biases) in most settings.

3.7.7 Integrating Latent Factor Models with Arbitrary Models

The integration of latent factor models with neighborhood-based models provides useful hints about integrating the former with other types of models such as content-based methods. Such an integration naturally leads to the creation of *hybrid recommender systems*. In general, item profiles may be available in the form of descriptions of products. Similarly, users might have explicitly created profiles describing their interests. Assume that the profile for user i is denoted by the keyword vector $\overline{C}_i^{user}$ and the profile for item j is denoted by the keyword vector $\overline{C}_j^{item}$. Furthermore, assume that the observed ratings of user i are denoted by $\overline{R}_i^{user}$, and the observed ratings of item j are denoted by $\overline{R}_j^{item}$. Then, one can write the following general form of the prediction function:

$$\hat{r}_{ij} = \underbrace{[(U + FY)V^T]_{ij}}_{\text{Latent Factor Portion}} + \beta \underbrace{F(\overline{C}_i^{user}, \overline{C}_j^{item}, \overline{R}_i^{user}, \overline{R}_j^{item})}_{\text{Another Prediction Model}} \qquad (3.39)$$

Here, β is a balancing factor that controls the relative importance of the two models. The second term, which is $F(\overline{C}_i^{user}, \overline{C}_j^{item}, \overline{R}_i^{user}, \overline{R}_j^{item})$, is a parameterized function of the user profile, item profile, user ratings, and item ratings. One can optimize the parameters of this function jointly with the latent factors to minimize the error of prediction in Equation 3.39.

The integration of neighborhood and latent factor models can be viewed as a special case of this method in which the function $F()$ is a linear regression function that uses only $\overline{R}_j^{item}$ and ignores all the other arguments. It is, however, possible to design an almost infinite number of variants of this broader approach by varying the choice of function $F()$. It is also possible to broaden the scope of $F()$ by using other sources of data such as social data, location, or time. In fact, virtually any collaborative filtering model, which is posed in the form of a parameterized prediction function, can be integrated with the latent factor model. Many methods have indeed been proposed in the research literature that integrate various types of feature-based regression, topic modeling, or other novel data sources with latent factor models. For example, a social regularization method (cf. section 11.3.8 of Chapter 11) integrates the latent factor model with social trust information to improve predictions. There is significant scope in improving the state of the art in recommender systems by identifying new sources of data, whose predictive power can be integrated with latent factor models using the aforementioned framework.

3.8   Summary

This chapter discusses a number of models for collaborative filtering. The collaborative filtering problem can be viewed as a generalization of the problem of classification. Therefore, many of the models that apply to classification also apply to collaborative filtering with some generalization. A notable exception is that of latent factor models, which are highly tailored to the collaborative filtering problem. Latent factor models use different types of factorization in order to predict ratings. These different types of factorization differ in the nature of their objective functions and the constraints on their basis matrices. Furthermore, they may have different trade-offs in terms of accuracy, overfitting, and interpretability. Latent factor models are the state-of-the-art in collaborative filtering. A wide variety of latent factor models have been proposed, based on the choices of the objective function and optimization constraints. Latent factor models can also be combined with neighborhood methods to create integrated models, which can benefit from the power of both latent factor models and neighborhood methods.

3.9   Bibliographic Notes

The problem of collaborative filtering is closely related to that of classification. Numerous recommender systems have been proposed in the literature; these modify the various classification models to perform recommendations. The relationship between collaborative filtering and classification is discussed in [82]. The earliest association-based methods are described in [524]. Various enhancements of the method, which use support levels specific to the item at hand are discussed in [358, 359, 365]. The first two of these methods leverage user associations rather than item associations [358, 359]. Association rule-based systems have found significant uses in Web-based personalization and recommender systems [441, 552]. Association rule methods can be combined with neighborhood methods in order to extract *localized* associations [25] between items or between users. Localized associations generally provide more refined recommendations than is possible with global rule-based methods. A method for performing collaborative filtering with the use of the Bayes method is discussed in [437]. The use of probabilistic relational models for collaborative filtering is proposed in [219]. Support vector machines for recommender systems are discussed in [638].

Neural networks have also been used recently for collaborative filtering [519, 679]. The *restricted Boltzmann machine (RBM)* is a neural network with one input layer and one hidden layer. This kind of network has been used for collaborative filtering [519], in which the visible units correspond to items, and the training is done over all users in each epoch. The rating of items by users results in the activation of the visible units. Since RBMs can use nonlinearity within the units, they can sometimes achieve superior performance to latent factor models. RBMs use factorized representations of the large parameter space to reduce overfitting, and have been shown to be very accurate in the Netflix Prize contest. The basic idea of factorized parameter representations has also been used other recent methods such as factorization machines [493].

A detailed discussion of various dimensionality reduction methods may be found in [22]. The use of dimensionality reduction methods for neighborhood-based filtering was proposed in [525]. The works in [24, 525], which were proposed independently, also discuss the earliest uses of latent factor models as stand-alone methods for recommendation and missing data imputation. The work in [24] combines an EM-algorithm with latent factor models to impute missing entries. Stand-alone latent factor methods are particularly effective for collaborative filtering and are the state-of-the-art in the literature. Methods for regularizing latent factor methods are discussed by Paterek in [473]. The same work also introduces the notion of user and item biases in latent factor models. An *asymmetric factor model* is discussed in this work, in which users are not explicitly represented by latent factors. In this case, a user factor is represented as a linear combination of the implicit factors of items she has rated. As a result, the number of parameters to be learned is reduced. In fact, Paterek's (relatively under-appreciated) work [473] introduced almost all the basic innovations that were later combined and refined in various ways [309, 311, 313] to create state-of-the-art methods such as SVD++.

The early works [133, 252, 300, 500, 569, 666] showed how different forms of matrix factorization could be used for recommendations. The difference between various forms of matrix factorization is in the nature of the objective (loss) functions and the constraints on the factor matrices. The method in [371] proposes the notion of *kernel collaborative filtering*, which discovers *nonlinear* hyper-planes on which the ratings are distributed. This approach is able to model more complex ratings distributions. These different types of factorization lead to different trade-offs in quality, overfitting, and interpretability. Incremental methods for collaborative filtering for matrix factorization are discussed in [96].

Many variations of the basic objective function and constraints are used in different forms of matrix factorization. The works in [180, 500, 569, 624] explore maximum margin factorization, which is very closely related to unconstrained matrix factorization. The main difference is that a maximum margin regularizer is used with hinge loss in the objective function, rather than using the Frobenius norm of the error matrix to quantify the loss. The works in [252, 666] are non-negative forms of matrix factorization. A detailed discussion of non-negative matrix factorization methods for complete data may be found in [22, 537]. The work in [666] explores the conventional non-negative factorization method with the Frobenius norm, whereas the work in [252, 517] explores probabilistic forms of matrix factorization. Some of the probabilistic versions also minimize the Frobenius norm but also optimize the regularization simultaneously. Methods for combining Bayesian methods with matrix factorization methods (in order to judiciously determine regularization parameters) are discussed in [518]. Gibbs sampling is used to achieve this goal. Initialization techniques for non-negative matrix factorization methods are discussed in [331]. After the popularization of latent factor models by the Netflix Prize contest [73], other factorization-based methods were also proposed for collaborative filtering [309, 312, 313]. One of the earliest latent factor models, which works with implicit feedback data, is presented in [260]. The SVD++ description in this book is borrowed from [309]. A recent work [184] imposes a penalty proportional to the Frobenius norm of UV^T to force unobserved values to have lower ratings. The idea is to penalize higher ratings. This approach imposes stronger biases than [309] because it explicitly assumes that the unobserved ratings have lower values. Furthermore, the ratings in [184] need to be non-negative quantities, so that the Frobenius norm penalizes higher ratings to a greater degree. Some of the latent factor methods [309] show how techniques such as SVD++ can be combined with regression-based neighborhood methods (cf. section 3.7). Therefore, these methods combine linear regression with factorization models. A matrix factorization method that uses singular value decomposition is discussed in [127]. The use of inductive matrix completion methods on collaborative filtering matrices with side information is discussed in [267].

Various regression-based models are discussed in [72, 309, 342, 434, 620, 669]. A general examination of linear classifiers, such as least-squares regression and support vector machines (SVMs), is provided in [669]. This work was one of the earliest evaluations of linear methods, although it was designed only for implicit feedback data sets, such as Web click data or sales data, in which only positive preferences are available. It was observed that collaborative filtering in such cases is similar in form to text categorization. However, because of the noise in the data and the imbalanced nature of the class distribution, a direct use of SVM methods is sometimes not effective. Changes to the loss function are suggested in [669] in order to provide more accurate results. The approach shows that by using a quadratic loss function in SVM optimization, one gets a form that is more similar to the least-squares approach. The modified SVM performs either competitively to, or better than the least-squares approach. The methods in [72, 309] are closely associated with neighborhood-based methods, and they are discussed in section 2.6 of Chapter 2. The work in [620] uses collections of linear models, which are modeled as ordinary least-squares problems. The use of regression-based models, such as *slope-one predictors*, is discussed in [342]. As discussed in section 2.6 of Chapter 2, regression models can be used to show the formal connection between model-based methods and neighborhood-based methods [72, 309]. Other methods for combining regression with latent factor models are discussed in [13]. The works in [321, 455] develop various types of sparse linear models (*SLIM*) that combine the neighborhood approach with regression and matrix factorization. The *SLIM* approach is primarily designed for implicit feedback data sets.

A significant amount of work has been devoted to the choice of methodology for determining the solution to the underlying optimization problems. For example, a discussion of the trade-offs between gradient-descent and stochastic gradient descent is provided in [351], and mini-batches are proposed to bridge the gap between the two. Alternating least-squares methods are discussed in [268, 677]. The original idea of alternating least squares are proposed in the positive matrix factorization of complete matrices [460]. Methods for large-scale and distributed stochastic gradient descent in latent factor models are proposed in [217]. The main trade-off between stochastic descent and alternating least squares is the trade-off between stability and efficiency. The former method is more efficient, whereas the latter is more stable. It has been suggested that coordinate descent methods [650] can be efficient, while retaining stability. It has also been shown [651] that non-parametric methods have several advantages for large-scale collaborative filtering with latent factor models. Methods for addressing cold-start issues in latent factor models are discussed in [676]. The Netflix Prize competition was particularly notable in the history of latent factor models because it resulted in several useful lessons [73] about the proper implementation of such models. Recently, latent factor models have been used to model richer user preferences. For example, the work in [322] shows how one might combine global preferences with interest-specific preferences to make recommendations.

3.10   Exercises

1. Implement a decision tree-based predictor of ratings for an incomplete data sets. Use the dimensionality reduction approach described in the chapter.

2. How would you use a rule-based collaborative filtering system in the case where ratings are real-numbers in $[-1, 1]$.

3. Design an algorithm that combines association rule methods with clustering for recommendations in order to discover localized associations in unary data. What is the advantage of this approach over a vanilla rule-based method?

4. The naive Bayes model discussed in this chapter predicts the ratings of each item using the user's other ratings as a conditional. Design a Bayes model that uses the

item's other ratings as a condition. Discuss the advantages and disadvantages of each model. Identify a case in which each approach would work better. How would you combine the predictions of the two models?

5. Suppose that a merchant had a unary matrix containing the buying behavior of various customers. Each entry in the matrix contains information about whether or not a customer has bought a particular item. Among the users that have not bought an item yet, the merchant wishes to rank all the users in order of their propensity to buy it. Show how to use the Bayes model to achieve this goal.

6. Use the Bayes model on Table 3.1 to determine the probability that John might buy *Bread* in the future. Treat 0s in the table as values that are actually specified for the ratings, rather than as missing values (except for John's ratings for *Bread* and *Beef*). Determine the probability that he might buy *Beef* in the future. Is John more likely to buy *Bread* or *Beef* in the future?

7. Implement the naive Bayes model for collaborative filtering.

8. Perform a straightforward rank-2 SVD of the matrix in Table 3.2 by treating missing values as 0. Based on the use of SVD, what are the predicted ratings for the missing values of user 3? How does this compare with the results shown in the example of section 3.6.5.4, which uses a different initialization? How do the results compare to those obtained using the Bayes model described in the chapter?

9. Suppose you are given a matrix R which can be factorized as $R = UV^T$, where the columns of U are mutually orthogonal and the columns of V are mutually orthogonal. Show how to factorize R into three matrices in the form $Q\Sigma P^T$, where the columns of P and Q are orthonormal and Σ is a non-negative diagonal matrix.

10. Implement the unconstrained matrix factorization method with stochastic gradient descent and batch updates.

11. Discuss the changes required to the alternating least-squares method for unconstrained matrix factorization, when one constrains the last column of the user-factor matrix to contain only 1s, and the second-last column of the item-factor matrix to contain only 1s. This method is useful for incorporating user and item biases in unconstrained matrix factorization.

12. Discuss how you might apply the alternating least-squares method for designing latent factor models with implicit feedback.

13. Let the $m \times k$ matrix F, $n \times k$ matrix V, and and $n \times k$ matrix Y be defined as discussed in the asymmetric factor model portion of section 3.6.4.6. Assume a simplified setting of asymmetric factor models in which we do not need to account for user and item biases.

   (a) Show that the stochastic gradient-descent updates for each observed entry (i, j) in the ratings matrix R are as follows:

   $$v_{jq} \Leftarrow v_{jq} + \alpha \left(e_{ij} \cdot \left[\sum_{h \in I_i} \frac{y_{hq}}{\sqrt{|I_i|}} \right] - \lambda \cdot v_{jq} \right) \quad \forall q \in \{1 \ldots k\}$$

   $$y_{hq} \Leftarrow y_{hq} + \alpha \left(\frac{e_{ij} \cdot v_{jq}}{\sqrt{|I_i|}} - \lambda \cdot y_{hq} \right) \quad \forall q \in \{1 \ldots k\}, \forall h \in I_i$$

Here, $e_{ij} = r_{ij} - \hat{r}_{ij}$ is the error of observed entry (i, j) and I_i is the set of items for which user i has specified ratings.

(b) What changes would need to be made to the definitions of various matrices and to the updates to account for user and item biases?

Chapter 4

Content-Based Recommender Systems

"Form must have a content, and that content must be linked with nature." – Alvar Aalto

4.1 Introduction

The collaborative systems discussed in the previous chapters use the correlations in the ratings patterns across users to make recommendations. On the other hand, these methods do not use item attributes for computing predictions. This would seem rather wasteful; after all, if John likes the futuristic science fiction movie *Terminator*, then there is a very good chance that he might like a movie from a similar genre, such as *Aliens*. In such cases, the ratings of other users may not be required to make meaningful recommendations. Content-based systems are designed to exploit scenarios in which items can be described with descriptive sets of attributes. In such cases, a user's *own* ratings and actions on other movies are sufficient to discover meaningful recommendations. This approach is particularly useful when the item is new, and there are few ratings available for that item.

Content-based recommender systems try to match users to items that are similar to what they have liked in the past. This similarity is not necessarily based on rating correlations across users but on the basis of the *attributes* of the objects liked by the user. Unlike collaborative systems, which explicitly leverage the ratings of *other* users in addition to that of the target user, content-based systems largely focus on the target user's *own* ratings and the attributes of the items liked by the user. Therefore, the other users have little, if any, role to play in content-based systems. In other words, the content-based methodology leverages a different source of data for the recommendation process. As we will see in Chapter 6, many recommender systems leverage the power of both sources. Such recommender systems are referred to as *hybrid* recommender systems.

© Springer International Publishing Switzerland 2016
C.C. Aggarwal, *Recommender Systems: The Textbook*,
DOI 10.1007/978-3-319-29659-3_4

At the most basic level, content-based systems are dependent on two sources of data:

1. The first source of data is a description of various items in terms of content-centric attributes. An example of such a representation could be the text description of an item by the manufacturer.

2. The second source of data is a *user profile*, which is generated from user feedback about various items. The user feedback might be explicit or implicit. Explicit feedback may correspond to ratings, whereas implicit feedback may correspond to user actions. The ratings are collected in a way similar to collaborative systems.

   The user profile relates the attributes of the various items to user interests (ratings). A very basic example of a user profile might simply be a set of labeled training documents of item descriptions, the user ratings as the labels, and a classification or regression model relating the item attributes to the user ratings. The specific user profile is heavily dependent on the methodology at hand. For example, explicit ratings might be used in one setting, and implicit feedback might be used in another. It is also possible for the user to specify her own profile in terms of keywords of interest, and this approach shares some characteristics with *knowledge-based recommender systems*.

It is noteworthy that the ratings of the *other* users usually play no role in a content-based recommendation algorithm. This is both an advantage and a disadvantage, depending on the scenario at hand. On the one hand, in *cold-start scenarios*, where little information about the ratings of other users is available, such an approach can still be used as long as sufficient information about the user's own interests are available. This, at least, partially alleviates the cold-start problem when the number of other users in the recommender system is small. Furthermore, when an item is new, it is not possible to obtain the ratings of other users for that item. Content-based methods enable recommendations in such settings because they can extract the attributes from the new item, and use them to make predictions. On the other hand, the cold-start problem for new *users* cannot be addressed with content-based recommender systems. Furthermore, by not using the ratings of other users, one reduces the diversity and novelty of the recommended items. In many cases, the recommended items may be obvious items for the user, or they may be other items that the user has consumed before. This is because the content attributes will always recommend items with similar attributes to what the user has seen in the past. A recommended item with similar attributes often presents little surprise to the user. These advantages and disadvantages will be discussed in a later section of this chapter.

Content-based systems are largely used in scenarios in which a significant amount of attribute information is available at hand. In many cases, these attributes are keywords, which are extracted from the product descriptions. In fact, the vast majority of content-based systems extract text attributes from the underlying objects. Content-based systems are, therefore, particularly well suited to giving recommendations in text-rich and *unstructured* domains. A classical example of the use of such systems is in the recommendation of Web pages. For example, the previous browsing behavior of a user can be utilized to create a content-based recommender system. However, the use of such systems is not restricted only to the Web domain. Keywords from product descriptions are used to create item and user profiles for the purposes of recommendations in other e-commerce settings. In other settings, relational attributes such as manufacturer, genre, and price, may be used in addition to keywords. Such attributes can be used to create *structured* representations, which can be stored in a relational database. In these cases, it is necessary to combine the structured and unstructured attributes in a single structured representation. The basic principles of content-based systems, however, remain invariant to whether a structured or unstructured representation is used. This is because most learning methods in the structured domain have direct analogs in the unstructured domain, and vice versa. To preserve uniformity in exposition, our discussion in this chapter will be focused on unstructured settings. However, most of these methods can be easily adapted to structured settings.

Content-based systems are closely related to knowledge-based recommender systems. A summary of the relationship between the various types of systems is provided in Table 1.2 of Chapter 1. Like content-based systems, knowledge-based recommender systems use the content attributes of the items to make recommendations. The main difference is that knowledge-based systems support the *explicit specification* of user requirements in conjunction with interactive interfaces between the user and the recommender systems. Knowledge bases are used in conjunction with this interactivity to match user requirements to items. On the other hand, content-based systems generally use a learning-based approach based on *historical ratings*. Therefore, knowledge-based systems provide better control to the user in the recommendation process, whereas content-based systems leverage past behavior more effectively. Nevertheless, these differences are not so significant, and some content-based methods also allow users to explicitly specify their interest profiles. A number of systems leverage both the learning and interactive aspects within a unified framework. Such systems are referred to as *hybrid recommender systems*. Knowledge-based recommender systems are discussed in Chapter 5, whereas hybrid recommender systems are discussed in Chapter 6.

This chapter is organized as follows. The next section provides an overview of the basic components of a content-based recommender system. Feature extraction and selection methods are discussed in section 4.3. The process of learning user profiles and leveraging them for recommendations is discussed in section 4.4. A comparison of the main properties of collaborative and content-based systems is provided in section 4.5. The connections between collaborative filtering and content-based methods are explored in section 4.6. A summary is given in section 4.7.

4.2 Basic Components of Content-Based Systems

Content-based systems have certain basic components, which remain invariant across different instantiations of such systems. Since content-based systems work with a wide variety of item descriptions and knowledge about users, one must convert these different types of unstructured data into standardized descriptions. In most cases, it is preferred to convert the item descriptions into keywords. Therefore, content-based systems largely, but not exclusively, operate in the text domain. Many natural applications of content-based systems are also text-centric. For example, news recommender systems are often content-based systems, and they are also text-centric systems. In general, text classification and regression modeling methods remain the most widely used tools for creating content-based recommender systems.

The main components of content-based systems include the (offline) preprocessing portion, the (offline) learning portion, and the online prediction portion. The offline portions are used to create a summarized model, which is often a classification or regression model. This model is then used for the online generation of recommendations for users. The various components of content-based systems are as follows:

1. *Preprocessing and feature extraction:* Content-based systems are used in a wide variety of domains, such as Web pages, product descriptions, news, music features, and so on. In most cases, features are extracted from these various sources to convert them into

a keyword-based vector-space representation. This is the first step of any content-based recommendation system, and it is highly domain-specific. However, the proper extraction of the most informative features is essential for the effective functioning of any content-based recommender system.

2. *Content-based learning of user profiles:* As discussed earlier, a content-based model is specific to a given user. Therefore, a user-specific *model* is constructed to predict user interests in items, based on their past history of either buying or rating items. In order to achieve this goal, user feedback is leveraged, which may be manifested in the form of previously specified ratings (explicit feedback) or user activity (implicit feedback). Such feedbacks are used in conjunction with the attributes of the items in order to construct the training data. A learning model is constructed on this training data. This stage is often not very different from classification or regression modeling, depending on whether the feedback is categorical (e.g., binary act of selecting an item), or whether the feedback is numerical (e.g., ratings or buying frequency). The resulting model is referred to as the *user profile* because it conceptually relates user interests (ratings) to item attributes.

3. *Filtering and recommendation:* In this step, the learned model from the previous step is used to make recommendations on items for specific users. It is important for this step to be very efficient because the predictions need to be performed in real time.

In the following sections, we will describe each of these phases in detail. The second phase of learning often utilizes off-the-shelf classification models. The field of data classification is a vast area in its own right, and it is not the goal of this book to discuss classification models in detail. Therefore, throughout this chapter, we will assume a working familiarity with classification models. The goal will be to show how a specific classification model can be used as a black-box in the recommender system and the kinds of classification models that are specially suited to content-based recommender systems. A brief description of two of the most commonly used models is included, but this is by no means an exhaustive description. For the reader who is unfamiliar with classification models, pointers to several useful resources are included in the bibliographic notes.

4.3   Preprocessing and Feature Extraction

The first phase in all content-based models is to extract discriminative features for representing the items. Discriminative features are those, which are highly predictive of user interests. This phase is highly dependent on the specific application at hand. For example, a Web page recommendation system will be very different from a product recommendation system.

4.3.1   Feature Extraction

In the feature extraction phase, the descriptions of various items are extracted. Although it is possible to use any kind of representation, such as a multidimensional data representation, the most common approach is to extract keywords from the underlying data. This choice is because unstructured text descriptions are often widely available in a variety of domains, and they remain the most natural representations for describing items. In many cases, the items may have multiple fields describing various aspects of the item. For example, a merchant selling books may have descriptions of the books and keywords describing the content, title, and author. In some cases, these descriptions can be converted into a bag of keywords. In other cases, one might work directly with a multidimensional (structured) representation. The latter is necessary when the attributes contain numerical quantities (e.g., price) or fields that are drawn from a small universe of possibilities (e.g., color).

The various fields need to be weighted appropriately in order to facilitate their use in the classification process. *Feature weighting* is closely related to *feature selection*, in that the former is a soft version of the latter. In the latter case, attributes are either included or not included depending on their relevance, whereas in the former case, features are given differential weight depending on their importance. The issue of feature selection will be discussed in detail in section 4.3.4. Because the feature extraction phase is highly application-specific, we provide the reader with a flavor of the types of features that may need to be extracted in the context of various applications.

4.3.1.1 Example of Product Recommendation

Consider a movie recommendation site[1] such as IMDb [699], that provides personalized recommendations for movies. Each movie is usually associated with a description of the movie such as its synopsis, the director, actors, genre, and so on. A short description of *Shrek* at the IMDb Website is as follows:

> "After his swamp is filled with magical creatures, an ogre agrees to rescue a princess for a villainous lord in order to get his land back."

Many other attributes, such as user tags, are also available, which can be treated as content-centric keywords.

In the case of *Shrek*, one might simply concatenate all the keywords in the various fields to create a text description. The main problem is that the various keywords may not have equal importance in the recommendation process. For example, a particular actor might have greater importance in the recommendation than a word from the synopsis. This can be achieved in two ways:

1. Domain-specific knowledge can be used to decide the relative importance of keywords. For example, the title of the movie and the primary actor may be given more weight than the words in the description. In many cases, this process is done in a heuristic way with trial and error.

2. In many cases, it may be possible to *learn* the relative importance of various features in an automated way. This process is referred to as feature weighting, which is closely related to feature selection. Both feature weighting and feature selection are described in a later section.

4.3.1.2 Example of Web Page Recommendation

Web documents require specialized preprocessing techniques because of some common properties of their structure and the richness of the links inside them. Two major aspects of Web document preprocessing include the removal of specific parts of the documents (e.g., tags) that are not useful and the leveraging of the actual structure of the document.

All fields in a Web document are not equally important. HTML documents have numerous fields in them, such as the title, the meta-data, and the body of the document.

---

[1]The exact recommendation method used by IMDb is proprietary and not known to the author. The description here is intended only for illustrative purposes.

Typically, analytical algorithms treat these fields with different levels of importance, and therefore weight them differently. For example, the title of a document is considered more important than the body and is weighted more heavily. Another example of a specially processed portion of a Web document is anchor text. Anchor text contains a description of the Web page pointed to by a link. Because of its descriptive nature, it is considered important, but it is sometimes not relevant to the topic of the page itself. Therefore, it is often removed from the text of the document. In some cases, where possible, anchor text could even be added to the text of the document *to which it points*. This is because anchor text is often a summary description of the document to which it points. The learning of the importance of these various features can be done through automated methods, as discussed in section 4.3.4.

A Web page may often be organized into content blocks that are not related to the primary subject matter of the page. A typical Web page will have many irrelevant blocks, such as advertisements, disclaimers, or notices, which are not very helpful for mining. It has been shown that the quality of mining results improves when only the text in the main block is used. However, the (automated) determination of main blocks from Web-scale collections is itself a data mining problem of interest. While it is relatively easy to decompose the Web page into blocks, it is sometimes difficult to identify the main block. Most automated methods for determining main blocks rely on the fact that a particular site will typically utilize a similar layout for all its documents. Therefore, the structure of the layout is learned from the documents at the site by extracting *tag trees* from the site. Other main blocks are then extracted through the use of the *tree-matching algorithm* [364, 662]. Machine learning methods can also be used for this task. For example, the problem of labeling the main block in a page can be treated as a classification problem. The bibliographic notes contain pointers to methods for extracting the main block from a Web document.

4.3.1.3 Example of Music Recommendation

Pandora Internet radio [693] is a well-known music recommendation engine, which associates tracks with features extracted from the Music Genome Project [703]. Examples of such features of tracks could be "feature trance roots," "synth riffs," "tonal harmonies," "straight drum beats," and so on. Users can initially specify a single example of a track of their interest to create a "station." Starting with this single training example, similar songs are played for the user. The users can express their likes or dislikes to these songs.

The user feedback is used to build a more refined model for music recommendation. It is noteworthy, that even though the underlying features are quite different in this case, they can still be treated as keywords, and the "document" for a given song corresponds to the bag of keywords associated with it. Alternatively, one can associate specific attributes with these different keywords, which leads to a structural multidimensional representation.

The initial specification of a track of interest is more similar to a knowledge-based recommender system than a content-based recommender system. Such types of knowledge-based recommender systems are referred to as case-based recommender systems. However, when ratings are leveraged to make recommendations, the approach becomes more similar to that of a content-based recommender system. In many cases, Pandora also provides an explanation for the recommendations in terms of the item attributes.

4.3.2 Feature Representation and Cleaning

This process is particularly important when the unstructured format is used for representation. The feature extraction phase is able to determine bags of words from the unstructured descriptions of products or Web pages. However, these representations need to be cleaned and represented in a suitable format for processing. There are several steps in the cleaning process:

1. *Stop-word removal:* Much of the text that is extracted from free-form descriptions of items, will contain many words that are not specific to the item but that are a common part of English vocabulary. Such words are typically high-frequency words. For example, words such as '*a*," "*an*," and "*the*" will not be particularly specific to the item at hand. In the movie recommendation application, it is common to find such words in the synopsis. In general, articles, prepositions, conjunctions, and pronouns are treated as stop-words. In most cases, standardized lists of stop-words are available in various languages.

2. *Stemming:* In stemming, variations of the same word are consolidated. For example, singular and plural forms of a word or different tenses of the same word are consolidated. In some cases, common roots are extracted from various words. For example, words such as "hoping" and "hope" are consolidated into the common root "hop." Of course, stemming can sometimes have a detrimental effect, because a word such as "hop" has a different meaning of its own. Many off-the-shelf tools [710–712] are available for stemming.

3. *Phrase extraction:* The idea is to detect words that occur together in documents on a frequent basis. For example, a phrase such as "hot dog" means something different from its constituent words. Manually defined dictionaries are available for phrase extraction, although automated methods can also be used [144, 364, 400].

After executing these steps, the keywords are converted into a *vector-space representation*. Each word is also referred to as a *term*. In the vector-space representation, documents are represented as bags of words, together with their frequencies. Although it might be tempting to use the raw frequency of word occurrence, this is often not desirable. This is because commonly occurring words are often statistically less discriminative. Therefore, such words are often discounted by down-weighting. This is similar to the principle of stop-words, except that it is done in a soft way by discounting the word, rather than completely removing it.

How are words discounted? This is achieved by using the notion of *inverse document frequency*. The inverse document frequency id_i of the ith term is a decreasing function of the number of documents n_i in which it occurs.

$$id_i = \log(n/n_i) \tag{4.1}$$

Here, the number of documents in the collection is denoted by n.

Furthermore, care needs to be taken that the excessive occurrence of a single word in the collection is not given too much importance. For example, when item descriptions are collected from unreliable sources or open platforms, such as the Web, they are liable to contain a significant amount of spam. To achieve this goal, a damping function $f(\cdot)$, such as the square root or the logarithm, is optionally applied to the frequencies before similarity computation.

$$f(x_i) = \sqrt{x_i}$$
$$f(x_i) = \log(x_i)$$

Frequency damping is optional and is often omitted. Omitting the damping process is equivalent to setting $f(x_i)$ to x_i. The *normalized* frequency $h(x_i)$ for the ith word is defined by combining the inverse document frequency with the damping function:

$$h(x_i) = f(x_i)id_i \qquad (4.2)$$

This model is popularly referred to as the tf-idf model, where tf represents the term frequency and idf represents the inverse document frequency.

4.3.3 Collecting User Likes and Dislikes

Aside from the content about the items, it is also necessary to collect data about the user likes and dislikes for the recommendation process. The data collection is done during the offline phase, whereas recommendations are determined during the online phase when a specific user is interacting with the system. The user for whom the prediction is performed at any given time is referred to as the *active user*. During the online phase, the user's own preferences are combined with the content to create the predictions. The data about user likes and dislikes can take on any of the following forms:

1. *Ratings:* In this case, users specify ratings indicating their preference for the item. Ratings can be binary, interval-based, or ordinal. In rare cases, ratings can even be real valued. The nature of the rating has a significant impact on the model used for learning the user profiles.

2. *Implicit feedback:* Implicit feedback refers to user actions, such as buying or browsing an item. In most cases, only the positive preferences of a user are captured with implicit feedback but not negative preferences.

3. *Text opinions:* In many cases, users may express their opinions in the form of text descriptions. In such cases, implicit ratings can be extracted from these opinions. This form of rating extraction is related to the field of *opinion mining* and *sentiment analysis*. This area is beyond the scope of this book. Interested readers are referred to [364].

4. *Cases:* Users might specify examples (or *cases*) of items that they are interested in. Such cases can be used as implicit feedback with nearest neighbor or Rocchio classifiers. However, when the similarity retrieval is used in conjunction with carefully designed utility functions, these methods are more closely related to case-based recommender systems. Case-based systems are a subclass of knowledge-based recommender systems in which domain knowledge is used to discover matching items, instead of learning algorithms (cf. section 5.3.1 of Chapter 5). It is often difficult to delineate where content-based recommender systems end and knowledge-based recommender systems begin. For example, Pandora Internet Radio often uses an initial case of an interesting music album to set up "radio stations" for users with similar music items. At a later stage, user feedback about likes and dislikes is utilized to refine the recommendations. Therefore, the first part of the approach can be viewed as a knowledge-based system, and the second part of the approach can be viewed as a content-based (or collaborative) system.

In all the aforementioned cases, the likes or dislikes of a user for an item are finally converted into a unary, binary, interval-based, or real rating. This rating can also be viewed as the extraction of a *class label* or *dependent variable*, which is eventually leveraged for learning purposes.

4.3.4 Supervised Feature Selection and Weighting

The goal of feature selection and weighting is to ensure that only the most informative words are retained in the vector-space representation. In fact, many well-known recommender systems [60, 476] explicitly advocate that a size cut-off should be used on the number of keywords. The experimental results in [476], which were performed in a number of domains, suggested that the number of extracted words should be somewhere between 50 and 300. The basic idea is that the noisy words often result in overfitting and should therefore be removed *a priori*. This is particularly important, considering the fact that the number of documents available to learn a particular user profile is often not very large. When the number of documents available for learning is small, the tendency of the model to overfit will be greater. Therefore, it is crucial to reduce the size of the feature space.

There are two distinct aspects to incorporating the feature informativeness in the document representation. One is feature *selection*, which corresponds to the removal of words. The second is feature *weighting*, which involves giving greater importance to words. Note that stop-word removal and the use of inverse-document frequency are examples of feature selection and weighting, respectively. However, these are unsupervised ways of feature selection and weighting, where user feedback is given no importance. In this section, we will study supervised methods for feature selection, which take the user ratings into account for evaluating feature informativeness. Most of these methods evaluate the sensitivity of the dependent variable to a feature in order to evaluate its informativeness.

The measures for computing feature informativeness can be used to either perform a hard selection of features or to heuristically weight the features with a function of the computed quantification of informativeness. The measures used for feature informativeness are also different, depending on whether the user rating is treated as a numeric or categorical value. For example, in the context of binary ratings (or ratings with a small number of discrete values), it makes sense to use categorical rather than numeric representations. We will also describe a few methods that are commonly used for feature weighting. In most of the following descriptions, we will assume an unstructured (textual) representation, although the methods can be generalized easily to structured (multidimensional) representations. This is because the vector-space representation of text can be viewed as a special case of the multidimensional representation. The bibliographic notes contain pointers to more details of feature selection methods.

4.3.4.1 Gini Index

The Gini index is one of the most commonly used measures for feature selection. It is a simple and intuitive measure, which is easy to understand. The Gini index is inherently suited to binary ratings, ordinal ratings, or ratings which are distributed into a small number of intervals. The latter case may sometimes be obtained by discretizing the ratings. The ordering among the ratings is ignored, and each possible value of the rating is treated as an instance of a categorical attribute value. This might seem like a disadvantage because it loses information about the relative ordering of the ratings. However, in practice, the number of possible ratings is usually small and therefore significant accuracy is not lost.

Let t be the total number of possible values of the rating. Among documents containing a particular word w, let $p_1(w) \ldots p_t(w)$ be the fraction of the items rated at each of these t possible values. Then, the Gini index of the word w is defined as follows:

$$\text{Gini}(w) = 1 - \sum_{i=1}^{t} p_i(w)^2 \tag{4.3}$$

The value of Gini(w) always lies in the range $(0, 1-1/t)$, with smaller values being indicative of greater discriminative power. For example, when the presence of the word w always results in the document being rated at the jth possible rating value (i.e., $p_j(w) = 1$), then such a word is very discriminative for rating predictions. Correspondingly, the value of the Gini index in such a case is $1 - 1^2 = 0$. When each value of $p_j(w)$ takes on the same value of $1/t$, the Gini index takes on its maximum value of $1 - \sum_{i=1}^{t}(1/t^2) = 1 - 1/t$.

4.3.4.2 Entropy

Entropy is very similar in principle to the Gini index except that information-theoretic principles are used to design the measure. As in the previous case, let t be the total number of possible values of the rating and $p_1(w) \ldots p_t(w)$ be the fraction of the documents containing a particular word w, which are rated at each of these t possible values. Then, the entropy of the word w is defined as follows:

$$\text{Entropy}(w) = - \sum_{i=1}^{t} p_i(w) \log(p_i(w)) \qquad (4.4)$$

The value of Entropy(w) always lies in the range $(0, 1)$, with smaller values being more indicative of discriminative power. It is easy to see that entropy has similar characteristics with the Gini index. In fact, these two measures often yield very similar results although they have different probabilistic interpretations. The Gini index is easier to understand, whereas entropy measures are more firmly grounded in mathematical principles from information theory.

4.3.4.3 χ^2-Statistic

The χ^2-statistic can be computed by treating the co-occurrence between the word and class as a *contingency table*. For example, consider a scenario where we are trying to determine whether a particular word is relevant to a user's buying interests. Assume that the user has bought about 10% of the items in the collection, and the word w occurs in about 20% of the descriptions. Assume that the total number of items (and corresponding documents) in the collection is 1000. Then, the *expected* number of occurrences of each possible combination of word occurrence and class contingency is as follows:

	Term occurs in description	Term does not occur
User bought item	$1000 * 0.1 * 0.2 = 20$	$1000 * 0.1 * 0.8 = 80$
User did not buy item	$1000 * 0.9 * 0.2 = 180$	$1000 * 0.9 * 0.8 = 720$

The aforementioned expected values are computed under the assumption that the occurrence of the term in the description and the user interest in the corresponding item are independent of one another. If these two quantities are independent, then clearly the term will be irrelevant to the learning process. However, in practice, the item may be highly related to the item at hand. For example, consider a scenario where the contingency table deviates from expected values and the user is very likely to buy the item containing the term. In such a case, the contingency table may appear as follows:

	Term occurs in description	Term does not occur
User bought item	$O_1 = 60$	$O_2 = 40$
User did not buy item	$O_3 = 140$	$O_4 = 760$

The χ^2-statistic measures the normalized deviation between observed and expected values across the various cells of the contingency table. In this case, the contingency table contains $p = 2 \times 2 = 4$ cells. Let O_i be the observed value of the ith cell and E_i be the expected value of the ith cell. Then, the χ^2-statistic is computed as follows:

$$\chi^2 = \sum_{i=1}^{p} \frac{(O_i - E_i)^2}{E_i} \qquad (4.5)$$

Therefore, in the particular example of this table, the χ^2-statistic evaluates to the following:

$$\chi^2 = \frac{(60 - 20)^2}{20} + \frac{(40 - 80)^2}{80} + \frac{(140 - 180)^2}{180} + \frac{(760 - 720)^2}{720}$$
$$= 80 + 20 + 8.89 + 2.22$$
$$= 111.11$$

It is also possible to compute the χ^2-statistic as a function of the observed values in the contingency table without explicitly computing expected values. This is possible because the expected values are functions of the aggregate observed values across rows and columns. A simple arithmetic formula to compute the χ^2-statistic in a 2×2 contingency table is as follows (see Exercise 8):

$$\chi^2 = \frac{(O_1 + O_2 + O_3 + O_4) \cdot (O_1 O_4 - O_2 O_3)^2}{(O_1 + O_2) \cdot (O_3 + O_4) \cdot (O_1 + O_3) \cdot (O_2 + O_4)} \qquad (4.6)$$

Here, $O_1 \ldots O_4$ are the observed frequencies according to the table above. It is easy to verify that this formula yields the same χ^2-statistic of 111.11. Note that the χ^2-test can also be interpreted in terms of the probabilistic level of significance with the use of a χ^2 distribution. However, for practical purposes, it is sufficient to know that larger values of the χ^2-statistic indicate that a particular term and item are related to a greater degree. Note that if the observed values are exactly equal to the expected values, then it implies that the corresponding term is irrelevant to the item at hand. In such a case, the χ^2-statistic will evaluate to its least possible value of 0. Therefore, the top-k features with the largest χ^2-statistic are retained.

4.3.4.4 Normalized Deviation

The problem with most of the aforementioned measures is that they lose information about the relative ordering of ratings. For cases in which the ratings have high granularity, the normalized deviation is an appropriate measure.

Let σ^2 be the variance of the ratings in all the documents. Furthermore, let $\mu^+(w)$ be the average rating of all documents containing the word w, and $\mu^-(w)$ be the average rating of all documents that do not contain the word w. Then, the normalized deviation of the word w is defined as follows:

$$\text{Dev}(w) = \frac{|\mu^+(w) - \mu^-(w)|}{\sigma} \qquad (4.7)$$

Larger values of $\text{Dev}(w)$ are indicative of more discriminatory words.

The aforementioned quantification is based on the relative distribution of the ratings for documents containing a specific word with respect to the ratings distribution of all documents. Such an approach is particularly suitable when ratings are treated as numerical quantities. A related measure is the Fisher's discrimination index, which computes the ratio of the inter-class separation to the intra-class separation in the *feature space* (rather than on the ratings dimension). This measure is described in detail in [22]. Fisher's discriminant index is however, better suited to categorical dependent variables rather than numerical dependent variables, such as ratings.

4.3.4.5 Feature Weighting

Feature Weighting can be viewed as a soft version of feature selection. In the earlier section on feature representation in this chapter, it was already discussed how measures such as the inverse document frequency can be used to weight documents. However, the inverse document frequency is an unsupervised measure that does not depend on user likes or dislikes. A supervised measure can also be used to further weight the vector-space representation in order to yield differential importance to different words. For example, in a movie recommendation application, keywords describing a movie genre or actor name are more important than words selected from the synopsis of the movie. On the other hand, the words in the synopsis are also somewhat indicative of tastes. Therefore, they cannot be excluded either. Feature weighting is a more refined approach for discriminating between various words by using a weight rather than a hard binary decision. The simplest approach to feature weighting is to take any of the feature selection measures and use them to derive the weights. For example, the inverse of the Gini index or entropy could be used. In many cases, a heuristic function can be further applied on the selection measure to control the sensitivity of the weighting process. For example, consider the following weighting function $g(w)$ for word w, where a is a parameter greater than 1.

$$g(w) = a - \text{Gini}(w) \qquad (4.8)$$

The resulting weight $g(w)$ will always lie in the range $(a - 1, a)$. By varying the value of a, the sensitivity of the weighting process can be controlled. Smaller values of a will lead to greater sensitivity. The weight of each word w in the vector-space representation is then multiplied by $g(w)$. Similar weighting functions can be defined with respect to the entropy and the normalized deviation. The process of selecting an appropriate feature weighting is a highly heuristic process that varies significantly corresponding to the application at hand. The value of a can be viewed as a parameter of the weighting function. It is also possible to learn the optimal parameters of such a function using *cross-validation* techniques. Such techniques are discussed in Chapter 7.

4.4   Learning User Profiles and Filtering

The learning of user profiles is closely related to the classification and regression modeling problem. When the ratings are treated as discrete values (e.g., "thumbs up" or "thumbs down"), the problem is similar to that of text classification. On the other hand, when the ratings are treated as a set of numerical entities, the problem is similar to that of regression modeling. Furthermore, the learning problem can be posed in both structured and unstructured domains. For homogeneity in presentation, we will assume that the descriptions of items are in the form of documents. However, the approach can easily be generalized to any type of multidimensional data because text is a special type of multidimensional data.

In each case, we assume that we have a set $\mathcal{D}_L$ of training documents, which are labeled by a specific user. This user is also referred to as the active user when that user obtains a recommendation from the system. The training documents correspond to the descriptions of items, which are extracted in the preprocessing and feature selection phases. Furthermore, the training data contain the ratings assigned by the active user to these documents. These documents are used to construct a training model. Note that the labels assigned by other users (than the active user) are not used in the training process. Therefore, the training models are specific to particular users, and they cannot be used for arbitrarily chosen users. This is different from traditional collaborative filtering, in which methods like matrix factorization build a single model across all users. The training model for a specific user represents the user profile.

The labels on the documents correspond to the numeric, binary, or unary ratings. Assume that the ith document in $\mathcal{D}_L$ has a rating denoted by c_i. We also have a set $\mathcal{D}_U$ of testing documents, which are unlabeled. Note that both $\mathcal{D}_L$ and $\mathcal{D}_U$ are specific to a particular (active) user. The testing documents might correspond to descriptions of items, which might be potentially recommended to the user but which have not yet been bought or rated by the user. In domains such as news recommendation the documents in $\mathcal{D}_U$ might correspond to candidate Web documents for recommendation to the active user. The precise definition of $\mathcal{D}_U$ depends on the domain at hand, but the individual documents in $\mathcal{D}_U$ are extracted in a similar way to those in $\mathcal{D}_L$. The training model on $\mathcal{D}_L$ is used to make recommendations from $\mathcal{D}_U$ to the active user. As in the case of collaborative filtering, the model can be used to provide either a predicted value of the rating or a ranked list of top-k recommendations.

It is immediately evident that this problem is similar to that of classification and regression modeling in the text domain. The reader is referred to a recent survey [21] for a detailed discussion of many of these techniques. In the following, we will discuss some of the common learning methods.

4.4.1 Nearest Neighbor Classification

The nearest neighbor classifier is one of the simplest classification techniques, and it can be implemented in a relatively straightforward way. The first step is to define a similarity function, which is used in the nearest neighbor classifier. The most commonly used similarity function is the cosine function. Let $\overline{X} = (x_1 \ldots x_d)$ and $\overline{Y} = (y_1 \ldots y_d)$ be a pair of documents, in which the normalized frequencies of the ith word are given by x_i and y_i, respectively, in the two documents. Note that these frequencies are normalized or weighted with the use of unsupervised tf-idf weighting or the supervised methods discussed in the previous section. Then, the cosine measure is defined using these normalized frequencies as follows:

$$\text{Cosine}(\overline{X}, \overline{Y}) = \frac{\sum_{i=1}^{d} x_i y_i}{\sqrt{\sum_{i=1}^{d} x_i^2} \sqrt{\sum_{i=1}^{d} y_i^2}} \tag{4.9}$$

The cosine similarity is frequently used in the text domain because of its ability to adjust to the varying lengths of the underlying documents. When this approach is used for other types of structured and multidimensional data, other similarity/distance functions, such as the Euclidean distance and Manhattan distance, are used. For relational data with categorical attributes, various match-based similarity measures are available [22].

This similarity function is useful in making predictions for items (documents) in which the user preference is unknown. For each document in $\mathcal{D}_U$, its k-nearest neighbors in $\mathcal{D}_L$ are determined using the cosine similarity function. The average value of the rating for the k neighbors of each item in $\mathcal{D}_U$ is determined. This average value is the predicted rating for the corresponding item in $\mathcal{D}_U$. An additional heuristic enhancement is that one can weight each rating with the similarity value. In cases where ratings are treated as categorical values, the number of votes for each value of the rating is determined, and the rating value with the largest frequency is predicted. The documents in $\mathcal{D}_U$ are then ranked based on the predicted value of the rating, and the top items are recommended to the user.

The main challenge with the use of this approach is its high computational complexity. Note that the nearest neighbor of each document in $\mathcal{D}_U$ needs to be determined, and the time required for each nearest neighbor determination is linear to the size of $\mathcal{D}_L$. Therefore, the computational complexity is equal to $|\mathcal{D}_L| \times |\mathcal{D}_U|$. One way of making the approach faster is to use clustering to reduce the number of training documents in $\mathcal{D}_L$. For each distinct value of the rating, the corresponding subset of documents in $\mathcal{D}_L$ are clustered into $p \ll |\mathcal{D}_L|$ groups. Therefore, if there are s distinct values of the ratings, then the total number of groups is $p \cdot s$. Typically, a fast centroid-based (i.e., k-means) clustering is used to create each group of p clusters. Note that the number of groups $p \cdot s$ is significantly smaller than the number of training documents. In such cases, each group is converted into a larger document corresponding to the concatenation[2] of the documents in that group. The vector-space representation of this larger document can be extracted by adding up the word frequencies of its constituents. The corresponding rating label associated with the document is equal to the rating of the constituent documents. For each target document T, the closest $k < p$ documents are found from this newly created set of p documents. The average rating of this set of k documents is returned as the label for the target. As in the previous case, the rating is predicted for each item in $\mathcal{D}_U$, and the top-ranked items are returned to the active user. This approach speeds up the classification process, because one must compute the similarity between the target document and a relatively small number of aggregated documents. Even though this approach incurs an additional preprocessing overhead of clustering, this overhead is generally small compared to the savings at recommendation time when the sizes of $\mathcal{D}_L$ and $\mathcal{D}_U$ are large.

A special case of this clustering-based approach is one in which all documents belonging to a particular value of the rating are aggregated into a single group. Thus, the value of p is set to 1. The vector-space representation of the resulting vector of each group is also referred to as the *prototype* vector. For a test document, the rating of the closest document is reported as the relevant one for the target. This approach is closely related to Rocchio classification, which also allows for the notion of *relevance feedback* from the active user. The Rocchio method was originally designed for binary classes, which, in our case, translate to binary ratings. The bibliographic notes contain pointers to the Rocchio method.

4.4.2   Connections with Case-Based Recommender Systems

Nearest neighbor methods are connected to knowledge-based recommender systems in general, and case-based recommender systems in particular. Knowledge-based recommender systems are discussed in detail in Chapter 5. The main difference is that in case-based recommender systems, the user interactively specifies a *single* example of interest, and the nearest neighbors of this example are retrieved as possible items of interest for the user.

---

[2]For structured data, the centroid of the group may be used.

Furthermore, a significant amount of domain knowledge is used in the design of the similarity function, because only a single example is available. This single example can be more appropriately viewed as a user requirement rather than a historical rating, because it is specified interactively. In knowledge-based systems, there is less emphasis on using historical data or ratings. Like the Rocchio method, such methods are also interactive, although the interactivity is far more sophisticated in case-based systems.

4.4.3 Bayes Classifier

The Bayes classifier is discussed in section 3.4 of Chapter 3 in collaborative filtering. However, the discussion in Chapter 3 is a non-standard use of the Bayes model in which the missing entries are predicted from the specified ones. In the context of content-based recommender systems, the problem translates to a more conventional use of the Bayes model for text classification. Therefore, we will revisit the Bayes model in the context of text classification.

In this case, we have a set $\mathcal{D}_L$ containing the training documents, and a set $\mathcal{D}_U$ containing the test documents. For ease in discussion, we will assume that the labels are binary in which users specify either a like or a dislike rating as $+1$ or -1, respectively for each of the training documents in $\mathcal{D}_L$. It is, however, relatively easy to generalize this classifier to the case where the ratings take on more than two values.

As before, assume that the rating of the ith document in $\mathcal{D}_L$ is denoted by $c_i \in \{-1, 1\}$. Therefore, this labeled set represents the user profile. There are two models that are commonly used in text data, which correspond to the Bernoulli and the multinomial models, respectively. In the following, we will discuss only the Bernoulli model. The multinomial model is discussed in detail in [22].

In the Bernoulli model, the frequencies of the words are ignored, and only the presence or absence of the word in the document is considered. Therefore, each document is treated as a binary vector of d words containing only values of 0 and 1. Consider a target document $\overline{X} \in \mathcal{D}_U$, which might correspond to the description of an item. Assume that the d binary features in $\overline{X}$ are denoted by $(x_1 \ldots x_d)$. Informally, we would like to determine $P(\text{Active user likes } \overline{X} | x_1 \ldots x_d)$. Here, each x_i is a 0-1 value, corresponding to whether or not the ith word is present in the document $\overline{X}$. Then, if the class (binary rating) of $\overline{X}$ is denoted by $c(\overline{X})$, this is equivalent to determining the value of $P(c(\overline{X}) = 1 | x_1 \ldots x_d)$. By determining both $P(c(\overline{X}) = 1 | x_1 \ldots x_d)$ and $P(c(\overline{X}) = -1 | x_1 \ldots x_d)$ and selecting the larger of the two, one can determine whether or not the active user likes $\overline{X}$. These expressions can be evaluated by using the Bayes rule and then applying a *naive assumption* as follows:

$$P(c(\overline{X}) = 1 | x_1 \ldots x_d) = \frac{P(c(\overline{X}) = 1) \cdot P(x_1 \ldots x_d | c(\overline{X}) = 1)}{P(x_1 \ldots x_d)}$$

$$\propto P(c(\overline{X}) = 1) \cdot P(x_1 \ldots x_d | c(\overline{X}) = 1)$$

$$= P(c(\overline{X}) = 1) \cdot \prod_{i=1}^{d} P(x_i | c(\overline{X}) = 1) \quad \text{[Naive Assumption]}$$

The naive assumption states that the occurrences of words in documents are conditionally independent events (on a specific class), and therefore one can replace $P(x_1 \ldots x_d | c(\overline{X}) = 1)$ with $\prod_{i=1}^{d} P(x_i | c(\overline{X}) = 1)$. Furthermore, the constant of proportionality is used in the first relationship because the denominator is independent of the class. Therefore, the denominator does not play any role in deciding between the relative order of the classes.

The denominator, however, does play a role in terms of ranking the propensity of *different items (documents)* to be liked by the user. This is relevant to the problem of *ranking* items for a specific user, in order of $P(c(\overline{X}) = 1|x_1 \ldots x_d)$.

In cases where such a ranking of the items is needed, the constant of proportionality is no longer irrelevant. This is particularly common in recommendation applications where it is not sufficient to determine the relative probabilities of items belonging to different rating values, but to actually rank them with respect to one another. In such cases, the constant of proportionality needs to be determined. Assume that the constant of proportionality in the relationship above is denoted by K. The constant of proportionality K can be obtained by using the fact that the sum of the probabilities of all possible instantiations of $c(\overline{X})$ should always be 1. Therefore, we have:

$$K \cdot \left[P(c(\overline{X}) = 1) \cdot \prod_{i=1}^{d} P(x_i|c(\overline{X}) = 1) + P(c(\overline{X}) = 1) \cdot \prod_{i=1}^{d} P(x_i|c(\overline{X}) = 1) \right] = 1$$

Therefore, we can derive the following value for K:

$$K = \frac{1}{P(c(\overline{X}) = 1) \cdot \prod_{i=1}^{d} P(x_i|c(\overline{X}) = 1) + P(c(\overline{X}) = -1) \cdot \prod_{i=1}^{d} P(x_i|c(\overline{X}) = -1)}$$

This approach is used to determine the probability of a user liking each possible item in $\mathcal{D}_U$. The items in $\mathcal{D}_U$ are then ranked according to this probability and presented to the user. These methods are particularly well suited to binary ratings. There are other ways of using the probability to estimate the predicted value of the ratings and rank the items when dealing with ratings that are not necessarily binary. Such methods are discussed in detail in section 3.4 of Chapter 3.

4.4.3.1   Estimating Intermediate Probabilities

The Bayes method requires the computation of intermediate probabilities such as $P(x_i|c(\overline{X}) = 1)$. So far, we have not yet discussed how these probabilities can be estimated in a data-driven manner. The main utility of the aforementioned Bayes rule is that it expresses the prediction probabilities in terms of other probabilities [e.g., $P(x_i|c(\overline{X}) = 1)$] that can be estimated more easily in a data-driven way. We reproduce the Bayes condition above:

$$P(c(\overline{X}) = 1|x_1 \ldots x_d) \propto P(c(\overline{X}) = 1) \cdot \prod_{i=1}^{d} P(x_i|c(\overline{X}) = 1)$$

$$P(c(\overline{X}) = -1|x_1 \ldots x_d) \propto P(c(\overline{X}) = -1) \cdot \prod_{i=1}^{d} P(x_i|c(\overline{X}) = -1)$$

In order to compute the Bayes probabilities, we need to estimate the probabilities on the right-hand side of the equations above. These include the prior class probabilities $P(c(\overline{X}) = 1)$ and $P(c(\overline{X}) = -1)$. Furthermore, the feature-wise conditional probabilities, such as $P(x_i|c(\overline{X}) = 1)$ and $P(x_i|c(\overline{X}) = -1)$, need to be estimated. The probability $P(c(\overline{X}) = 1)$ can be estimated as the fraction of positive training examples $\mathcal{D}_L^+$ in the labeled data $\mathcal{D}_L$. In order to reduce overfitting, Laplacian smoothing is performed by adding values proportional to a small parameter $\alpha > 0$ to the numerator and denominator.

$$P(c(\overline{X}) = 1) = \frac{|\mathcal{D}_L^+| + \alpha}{|\mathcal{D}_L| + 2 \cdot \alpha} \tag{4.10}$$

Table 4.1: Illustration of the Bayes method for a content-based system

Keyword ⇒ Song-Id ⇓	Drums	Guitar	Beat	Classical	Symphony	Orchestra	Like or Dislike
1	1	1	1	0	0	0	Dislike
2	1	1	0	0	0	1	Dislike
3	0	1	1	0	0	0	Dislike
4	0	0	0	1	1	1	Like
5	0	1	0	1	0	1	Like
6	0	0	0	1	1	0	Like
Test-1	0	0	0	1	0	0	?
Test-2	1	0	1	0	0	0	?

The value of $P(c(\overline{X}) = -1)$ is estimated in an exactly similar way. Furthermore, the conditional feature probability $P(x_i|c(\overline{X}) = 1)$ is estimated as the fraction of the instances in the positive class for which the ith feature takes on the value of x_i. Let $q^+(x_i)$ represent the number of instances in the positive class that take on the value of $x_i \in \{0, 1\}$ for the ith feature. Then, we can use a Laplacian smoothing parameter $\beta > 0$ to estimate the probability as follows:

$$P(x_i|c(\overline{X}) = 1) = \frac{q^+(x_i) + \beta}{|\mathcal{D}_L^+| + 2 \cdot \beta} \qquad (4.11)$$

A similar approach can be used to estimate $P(x_i|c(\overline{X}) = -1)$. Note that the Laplacian smoothing is helpful for cases where little training data are available. In the extreme case, where $\mathcal{D}_L^+$ is empty, the probability $P(x_i|c(\overline{X}) = 1)$ would be (appropriately) estimated to be 0.5 as a kind of prior belief. Without smoothing, such an estimation would be indeterminate, because both the numerator and denominator of the ratio would be 0. Laplacian smoothing, like many regularization methods, can be interpreted in terms of the greater importance of prior beliefs when the amount of training data is limited. Although we have presented the aforementioned estimation for the case of binary ratings, it is relatively easy to generalize the estimation when there are k distinct values of the ratings. A similar type of estimation is discussed in the context of collaborative filtering in section 3.4 of Chapter 3.

4.4.3.2 Example of Bayes Model

We provide an example of the use of the Bayes model for a set of 6 training examples and two test examples. In Table 4.1, the columns correspond to features representing properties of various songs. The user like or dislike is illustrated in the final column of the table. Therefore, the final column can be viewed as the rating. The first 6 rows correspond to the training examples, which correspond to the user profile. The final pair of rows correspond to two candidate music tracks that need to be ranked for the specific user at hand. In machine learning parlance, these rows are also referred to as test instances. Note that the final (dependent variable) column is specified only for the training rows because the user like or dislike (ratings) are not known for the test rows. These values need to be predicted.

By examining the features in Table 4.1, it becomes immediately evident that the first three features (columns) might often occur in many popular music genres such as rock music, whereas the final three features typically occur in classical music. The user profile, represented by Table 4.1 clearly seems to suggest a preference for classical music over rock music. Similarly, among the test examples, only the first of the two examples seems to match the user's interests. Let us examine how the Bayes approach is able to derive this fact in a data-driven way. For ease in computation, we will assume that Laplacian smoothing is not used, although it is important to use such smoothing methods in real applications.

By using the Bayes model, we can derive the conditional probabilities for likes and dislikes based on the observed features of the test examples:

$$P(\text{Like}|\,\text{Test-1}) \propto 0.5 \prod_{i=1}^{6} P(\text{Like}|x_i)$$

$$= (0.5) \cdot \frac{3}{4} \cdot \frac{2}{2} \cdot \frac{3}{4} \cdot \frac{3}{3} \cdot \frac{1}{4} \cdot \frac{1}{3}$$

$$= \frac{3}{128}$$

$$P(\text{Dislike}|\,\text{Test-1}) \propto 0.5 \prod_{i=1}^{6} P(\text{Dislike}|x_i)$$

$$= (0.5) \cdot \frac{1}{4} \cdot \frac{0}{2} \cdot \frac{1}{4} \cdot \frac{0}{3} \cdot \frac{3}{4} \cdot \frac{2}{3}$$

$$= 0$$

By normalizing the two probabilities to sum to 1, we obtain the result that $P(\text{Like}|\,\text{Test-1})$ is 1 and $P(\text{Dislike}|\,\text{Test-1})$ is 0. In the case of *Test-2*, exactly the opposite result is obtained where $P(\text{Like}|\,\text{Test-2})$ is 0. Therefore, *Test-1* should be recommended to the active user over *Test-2*. This is the same result that we obtained on visual inspection of this example.

When Laplacian smoothing is used, we will not obtain such binary probability values for the various classes, although one of the classes will obtain a much higher probability than the other. In such cases, all the test examples can be ranked in order of their predicted probability of a "Like" and recommended to the user. Laplacian smoothing is advisable because a single 0-value in the product-wise form of the expression on the right-hand side of the Bayes rule can result in a conditional probability value of 0.

4.4.4 Rule-based Classifiers

Rule-based classifiers can be designed in a variety of ways, including leave-one-out methods, as well as associative methods. A detailed description of the various types of rule-based classifiers is provided in [18, 22]. In the following, we will discuss only associative classifiers because they are based on the simple principles of association rules. A discussion of rule-based methods is provided in section 3.3 of Chapter 3. Refer to that section for the basic definitions of association rules and their measures, such as *support* and *confidence*. The support of a rule defines the fraction of rows satisfying both the antecedent and the consequent of a rule. The confidence of a rule is the fraction of rows satisfying the consequent, from the rows already known to satisfy the antecedent. The concept of a row "satisfying" the antecedent or consequent is described in more detail below.

Rule-based classifiers in content-based systems are similar to rule-based classifiers in collaborative filtering. In the item-item rules of collaborative filtering, both the antecedents and consequents of rules correspond to ratings of items. The main difference is that the antecedents of the rules in collaborative filtering correspond[3] to the ratings of various items,

---

[3]A different approach in collaborative filtering is to leverage user-user rules. For user-user rules, the antecedents and consequents may both contain the ratings of specific users. Refer to section 3.3 of Chapter 3.

whereas the antecedents of the rules in content-based methods correspond to the presence of specific keywords in item descriptions. Therefore, the rules are of the following form:

Item contains keyword set A $\Rightarrow$ Rating= Like

Item contains keyword set B $\Rightarrow$ Rating=Dislike

Therefore, an antecedent of a rule is said to "satisfy" a particular row (keyword representation of item), if all keywords in the antecedent are contained in that row. The consequents correspond to the various ratings, which we have assumed to be binary likes or dislikes for simplicity. A row is said to satisfy the consequent of that rule if the rating value in the consequent matches the dependent variable (rating) of that row.

The first step is to leverage the active user profile (i.e., training documents) to mine all the rules at a desired level of support and confidence. As in all content-based methods, the rules are specific to the active user at hand. For example, in the case of Table 4.1, the active user seems to be interested in classical music. In this case, an example of a relevant rule, which has 33% support and 100% confidence, is as follows:

{Classical, Symphony} $\Rightarrow$ Like

Therefore, the basic idea is to mine all such rules for a given active user. Then, for target items where the user's interests are unknown, it is determined which rules are *fired*. A rule is fired by a target item description if the former's antecedent keywords are included in the latter. Once all such fired rules have been determined for the active user, the average rating in the consequents of these rules is reported as the rating of the target item. Many different heuristics exist for combining the ratings of the consequents. For example, we can choose to weight the rating with the confidence of the rule while computing the average. In the event that no rule is fired, default heuristics need to be used. For example, one can determine the average rating of the active user over all items and also determine the average rating of the target item by all users. The average of these two quantities is reported. Therefore, the overall approach for rule-based classification can be described as follows:

1. **(Training phase:)** Determine all the relevant rules from the user profile at the desired level of minimum support and confidence from the training data set $\mathcal{D}_L$.

2. **(Testing phase)** For each item description in $\mathcal{D}_U$, determine the fired rules and an average rating. Rank the items in $\mathcal{D}_U$ on the basis of this average rating.

One advantage of rule-based systems is the high level of interpretability they provide. For example, for a recommended item, one can use the keywords in the antecedent of the fired rules to give a recommendation to the target user about why she might like a particular item.

4.4.4.1 Example of Rule-based Methods

In order to illustrate the use of rule-based methods, we will provide an example of the rules generated for the active user in Table 4.1. At a support level of 33% and confidence level of

75%, the following rules are generated along with their support-confidence values:

$$\text{Rule 1: } \{\text{Classical}\} \Rightarrow \text{Like} \quad (50\%, 100\%)$$
$$\text{Rule 2: } \{\text{Symphony}\} \Rightarrow \text{Like} \quad (33\%, 100\%)$$
$$\text{Rule 3: } \{\text{Classical, Symphony}\} \Rightarrow \text{Like} \quad (33\%, 100\%)$$
$$\text{Rule 4: } \{\text{Drums, Guitar}\} \Rightarrow \text{Dislike} \quad (33\%, 100\%)$$
$$\text{Rule 5: } \{\text{Drums}\} \Rightarrow \text{Dislike} \quad (33\%, 100\%)$$
$$\text{Rule 6: } \{\text{Beat}\} \Rightarrow \text{Dislike} \quad (33\%, 100\%)$$
$$\text{Rule 7: } \{\text{Guitar}\} \Rightarrow \text{Dislike} \quad (50\%, 75\%)$$

The aforementioned rules are primarily sorted in order of decreasing confidence, with ties broken in order of decreasing support. It is evident that rule 2 is fired by *Test-1*, whereas rules 5 and 6 are fired by *Test-2*. Therefore, *Test-1* should be preferred over *Test-2* as a recommendation to the active user. Note that the rules fired by *Test-1* also provide an understanding of why it should be considered the best recommendation for the active user. Such explanations are often very useful in recommender systems both from the perspective of the customer and the perspective of the merchant.

4.4.5 Regression-Based Models

Regression-based models have the merit that they can be used for various types of ratings such as binary ratings, interval-based ratings, or numerical ratings. Large classes of regression models such as linear models, logistic regression models, and ordered probit models can be used to model various types of ratings. Here, we will describe the simplest model, which is referred to as *linear regression*. The bibliographic notes contain pointers to more sophisticated regression methods.

Let D_L be an $n \times d$ matrix representing the n documents in the labeled training set $\mathcal{D}_L$ on a lexicon of size d. Similarly, let $\overline{y}$ be an n-dimensional column vector containing the ratings of the active user for the n documents in the training set. The basic idea in linear regression is to assume that the ratings can be modeled as a linear function of the word frequencies. Let $\overline{W}$ be a d-dimensional row vector representing the coefficients of each word in the linear function relating word frequencies to the rating. Then, the linear regression model assumes that the word frequencies in the training matrix D_L are related to rating vectors as follows:

$$\overline{y} \approx D_L \overline{W}^T \tag{4.12}$$

Therefore, the vector $(D_L \overline{W}^T - \overline{y})$ is an n-dimensional vector of prediction errors. In order to maximize the quality of the prediction, one must minimize the squared norm of this vector. Furthermore, a regularization term $\lambda ||\overline{W}||^2$ may be added to the objective function in order to reduce overfitting. This form of regularization is also referred to as *Tikhonov regularization*. Here, $\lambda > 0$ is the regularization parameter. Therefore, the objective function O can be expressed as follows:

$$\text{Minimize } O = ||D_L \overline{W}^T - \overline{y}||^2 + \lambda ||\overline{W}||^2 \tag{4.13}$$

The problem can be solved by setting the gradient of this objective function with respect to $\overline{W}$ to 0. This results in the following condition:

$$D_L^T (D_L \overline{W}^T - \overline{y}) + \lambda \overline{W}^T = 0$$
$$(D_L^T D_L + \lambda I) \overline{W}^T = D_L^T \overline{y}$$

Table 4.2: The family of regression models and applicability to various types of ratings

Regression Model	Nature of Rating (Target Variable)
Linear Regression	Real
Polynomial Regression	Real
Kernel Regression	Real
Binary Logistic Regression	Unary, Binary
Multiway Logistic regression	Categorical, Ordinal
Probit	Unary, Binary
Multiway Probit	Categorical, Ordinal
Ordered Probit	Ordinal, Interval-based

The matrix $(D_L^T D_L + \lambda I)$ can be shown to be positive-definite, and therefore invertible (see Exercise 7). Therefore, we can directly solve for the weight vector $\overline{W}$ as follows:

$$\overline{W}^T = (D_L^T D_L + \lambda I)^{-1} D_L^T \overline{y} \tag{4.14}$$

Here, I is a $d \times d$ identity matrix. Therefore, a closed-form solution always exists for $\overline{W}^T$. For any given document vector (item description) $\overline{X}$ from the unlabeled set $\mathcal{D}_U$, its rating can be predicted as the dot product between $\overline{W}$ and $\overline{X}$. Tikhonov regularization uses the L_2-regularization term $\lambda \cdot ||W||^2$. It is also possible to use L_1-regularization, in which this term is replaced with $\lambda \cdot ||W||$. The resulting optimization problem does not have a closed-form solution, and gradient descent methods must be used. This form of regularization, also known as *Lasso* [242], can be used in the dual role of feature selection. This is because such methods have the tendency to select sparse coefficient vectors for $\overline{W}$, in which most components of $\overline{W}$ take on the value of 0. Such features can be discarded. Therefore, L_1-regularization methods provide highly interpretable insights about important subsets of features for the recommendation process. A detailed discussion of these models can be found in [22].

The linear model is one example of a regression model that is suitable for real-valued ratings. In practice, ratings might be unary, binary, interval-based, or categorical (small number of ordinal values). Various linear models have been designed for different types of target class variables. Some examples include logistic regression, probit regression, ordered probit regression, and nonlinear regression. Unary ratings are often treated as binary ratings, in which the unlabeled items are treated as negative instances. However, specialized positive-unlabeled (PU) models exist for such cases [364]. Ordered probit regression is especially useful for interval-based ratings. Furthermore, nonlinear regression models, such as polynomial regression and kernel regression, may be used in cases where the dependency between the features and target variables is nonlinear. When the number of features is large and the number of training samples is small, linear models usually perform quite well and may, in fact, outperform nonlinear models. This is because linear models are less prone to overfitting. Table 4.2 shows the mapping between the various regression models and the nature of the target variable (rating).

4.4.6 Other Learning Models and Comparative Overview

As the problem of content-based filtering is a direct application of classification and regression modeling, many other techniques can be used from the literature. A detailed discussion of various classification models can be found in [18, 86, 242, 436]. The decision-tree model discussed in Chapter 3 can also be applied to content-based methods. However, for very high-dimensional data, such as text, decision trees often do not provide very effective results. Experimental results [477] have shown the poor performance of decision trees compared to other classification methods. Even though rule-based classifiers are closely related to decision trees, they can often provide superior results because they do not assume a strict partitioning of the feature space. Successful results have been obtained with rule-based classifiers for email classification [164, 165]. Among the various models, the Bayes approach has the advantage that it can handle all types of feature variables with the use of an appropriate model. Regression-based models are very robust, and they can handle all forms of target variables. Logistic regression and ordered probit regression are particularly useful for binary and interval-based ratings.

In the case of binary ratings, support vector machines [114] are a popular choice. Support vector machines are very similar to logistic regression; the main difference is that the loss is quantified as a *hinge-loss* rather than with the use of the *logit* function. Support vector machines are highly resistant to overfitting, and numerous off-the-shelf implementations exist. Both linear and kernel-based support vector machines have been used in the literature. For the case of high-dimensional data, such as text, it has been observed that linear support vector machines are sufficient. For such cases, specialized methods with linear performance [283] have been designed. Although neural networks [87] can be used for building arbitrarily complex models, they are not advisable when the amount of available data is small. This is because neural networks are sensitive to the noise in the underlying data, and they can overfit the training data when its size is small.

4.4.7 Explanations in Content-Based Systems

Since content-based systems extract models based on content features, they often provide highly interpretable insights for the recommendation process. For example, in a movie recommendation system, it is often useful to present the user with a reason as to why they might like a specific movie, such as the presence of a particular genre feature, actor feature, or an informative set of keywords. As a result, the active user will be able to make a more informed choice about whether they should watch that movie. Similarly, a descriptive set of keywords in a music recommendation system can provide a better understanding of why a user might like a specific track. As a specific example, Pandora Internet radio [693] provides explanations for recommended tracks, such as the following:

> "We are playing this track because it features trance roots, four-on-the-floor beats, disco influences, a knack for catchy hooks, beats made for dancing, straight drum beats, clear pronunciation, romantic lyrics, storytelling lyrics, subtle buildup/breakdown, a rhythmic intro, use of modal harmonies, the use of chordal patterning, light drum fills, emphasis on instrumental performance, a synth bass riff, synth riffs, subtle use of arpeggiatted synths, heavily effected synths, and synth swoops."

Each of these reported characteristics can be viewed as an important feature, which are responsible for the classification of the test instance as a "like." Note that such detailed explanations are often lacking in collaborative systems, where a recommendation can be explained only in terms of similar items, rather than in terms of detailed *characteristics* of these items. The nature and extent of the insights are, however, highly sensitive to the specific model used. For example, the Bayes model and rule-based systems are very highly interpretable in terms of the specific causality of the classification. Consider the example of Table 4.1 in which the following rule is fired for the example *Test-1*:

$$\{\text{Symphony}\} \Rightarrow \text{Like}$$

It is evident that the user has been recommended the item described by *Test-1* because it is a symphony. Similarly, in the Bayes classification model, it is evident that the contribution of $P(\text{Symphony}|\text{Like})$ is largest in the multiplicative formula for classification. Other models, such as linear and nonlinear regression models, are harder to interpret. Nevertheless, certain instances of these models, such as *Lasso*, provide important insights about the most relevant features for the classification process.

4.5 Content-Based Versus Collaborative Recommendations

It is instructive to compare content-based methods with the collaborative methods discussed in Chapters 2 and 3. Content-based methods have several advantages and disadvantages as compared to collaborative methods. The advantages of content-based methods are as follows:

1. When a new item is added to a ratings matrix, it has no ratings from the various users. None of the memory-based and model-based collaborative filtering methods would recommend such an item, because sufficient ratings are not available for recommendation purposes. On the other hand, in the case of content-based methods, the previous items rated by a given user are leveraged to make recommendations. Therefore, as long as the *user* is not new, meaningful recommendations can always be made in a way that treats the new item in a fair way in comparison to other items. Collaborative systems have cold-start problems *both* for new users and new items, whereas content-based systems have cold-start problems only for new users.

2. As discussed in the previous section, content-based methods provide explanations in terms of the features of items. This is often not possible with collaborative recommendations.

3. Content-based methods can generally be used with off-the-shelf text classifiers. Furthermore, each user-specific classification problem is generally not very large, as in the case of collaborative systems. Therefore, they are particularly easy to use with relatively little engineering effort.

On the other hand, content-based methods also have several disadvantages that are not present in collaborative recommenders.

1. Content-based systems tend to find items that are similar to those the user has seen so far. This problem is referred to as *overspecialization*. It is always desirable to have a certain amount of novelty and serendipity in the recommendations. Novelty refers to the fact that the item is different from one the user has seen in the past. Similarly, serendipity implies that the user would like to discover *surprisingly relevant* items that they might otherwise not have found. This is a problem for content-based systems in which attribute-based classification models tend to recommend very similar items. For example, if a user has never listened to or rated classical music, a content-based

system will typically not recommend such an item to her because classical music will be described by very different attribute values than those that the user has rated so far. On the other hand, a collaborative system might recommend such items by leveraging the interests of her *peer* group. For example, a collaborative system might automatically infer a *surprising* association between certain pop songs and classical songs and recommend the corresponding classical songs to a user who is a pop music lover. Overspecialization and lack of serendipity are the two most significant challenges of content-based recommender systems.

2. Even though content-based systems help in resolving cold-start problems for new *items*, they do not help in resolving these problems for new *users*. In fact, for new users, the problem in content-based systems may be more severe because a text classification model usually requires a sufficient number of training documents to avoid overfitting. It would seem rather wasteful that the training data for all the other users is discarded and only the (small) training data set specific to a single user is leveraged.

In spite of these disadvantages, content-based systems often complement collaborative systems quite well because of their ability to leverage content-based knowledge in the recommendation process. This complementary behavior is often leveraged in *hybrid recommender systems* (cf. Chapter 6), in which the goal is to combine the best of both worlds to create an even more robust recommender system. In general, content-based systems are rarely used in isolation, and they are generally used in combination with other types of recommender systems.

4.6 Using Content-Based Models for Collaborative Filtering

There is an interesting connection between collaborative filtering models and content-based methods. It turns out that content-based methods can be directly used for collaborative filtering. Although the content description of an item refers to its descriptive keywords, it is possible to envision scenarios, where the ratings of users are leveraged to define content-based descriptions. For each item, one can concatenate the user name (or identifier) of a user who has rated the item with the value of this rating to create a new "keyword." Therefore, each item would be described in terms of as many keywords as the number of ratings of that item. For example, consider a scenario where the descriptions of various movies are as follows:

Terminator: John#Like, Alice#Dislike, Tom#Like
Aliens: John#Like, Peter#Dislike, Alice#Dislike, Sayani#Like
Gladiator: Jack#Like, Mary#Like, Alice#Like

The "#" symbol is used to denote the demarcation of the concatenation and ensure a unique keyword for each user-rating combination. This approach is generally more effective, when the number of possible ratings is small (e.g., unary or binary ratings). After such a content-based description has been constructed, it can be used in conjunction with an off-the-shelf content-based algorithm. There is almost a one-to-one mapping between the resulting methods and various collaborative filtering models, depending on the base method used for classification. Although each such technique maps to a collaborative filtering model, the converse is not true because many collaborative filtering methods cannot be captured by this approach. Nevertheless, we provide some examples of the mapping:

1. A nearest neighbor classifier on this representation approximately maps to an item-based neighborhood model for collaborative filtering (cf. section 2.3.2 of Chapter 2).

2. A regression model on the content approximately maps to a user-wise regression model for collaborative filtering (cf. section 2.6.1 of Chapter 2).

3. A rule-based classifier on the content approximately maps to an user-wise rule-based classifier for collaborative filtering (cf. section 3.3.2 of Chapter 3).

4. A Bayes classifier on the content approximately maps to a user-wise Bayes model for collaborative filtering (cf. Exercise 4 of Chapter 3).

Therefore, many methods for collaborative filtering can be captured by defining an appropriate content representation and directly using off-the-shelf content-based methods. These observations are important because they open up numerous opportunities for hybridization. For example, one can combine the ratings-based keywords with actual descriptive keywords to obtain an even more robust model. In fact, this approach is often used in some hybrid recommendation systems. Such an approach no longer wastes the available ratings data from other users, and it combines the power of content-based and collaborative models within a unified framework.

4.6.1 Leveraging User Profiles

Another case in which collaborative filtering-like models can be created with content attributes is when user profiles are available in the form of specified keywords. For example, users may choose to specify their particular interests in the form of keywords. In such cases, instead of creating a local classification model for each user, one can create a global classification model over all users by using the user features. For each user-item *combination*, a content-centric representation can be created by using the Kronecker-product of the attribute vectors of the corresponding user and item [50]. A classification or regression model is constructed on this representation to map user-item *combinations* to ratings. Such an approach is described in detail in section 8.5.3 of Chapter 8.

4.7 Summary

This chapter introduces the methodology of content-based recommender systems in which user-specific training models are created for the recommendation process. The content attributes in item descriptions are combined with user ratings to create user profiles. Classification models are created on the basis of these models. These models are then used to classify item descriptions that have as of yet not been rated by the user. Numerous classification and regression models are used by such systems, such as nearest-neighbor classifiers, rule-based methods, the Bayes method, and linear models. The Bayes method has been used with great success in a variety of scenarios because of its ability to handle various types of content. Content-based systems have the advantage that they can handle cold-start problems with respect to new items, although they cannot handle cold-start problems with respect to new users. The serendipity of content-based systems is relatively low because content-based recommendations are based on the content of the items previously rated by the user.

4.8   Bibliographic Notes

The earliest content-based systems were attributed to the work in [60] and the *Syskill &
Webert* [82, 476–478] systems. *Fab*, however, uses a partial hybridization design in which
the peer group is determined using content-based methods, but the ratings of other users
are leveraged in the recommendation process. The works in [5, 376, 477] provide excellent
overview articles on content-based recommender systems. The latter work was designed for
finding interesting Websites, and therefore numerous text classifiers were tested for their
effectiveness. In particular, the work in [82] provides a number of useful pointers about
the relative performance of various content-based systems. Probabilistic methods for user
modeling are discussed in [83]. The work in [163, 164] is notable for its use of rule-based
systems in e-mail classification. Rocchio's relevance feedback [511] was also used during the
early years, although the work does not have theoretical underpinnings, and it can often
perform poorly in many scenarios. Numerous text classification methods, which can be used
for content-based recommendations, are discussed in [21, 22, 400]. A discussion of the notion
of serendipity in the context of information retrieval is provided in [599]. Some content-
based systems explicitly filter out very similar items in order to improve serendipity [85].
The work in [418] discusses how one can go beyond accuracy metrics to measure the quality
of a recommender system.

Methods for feature extraction, cleaning, and feature selection in text classification are
discussed in [21, 364, 400]. The extraction of the main content block from a Web page
containing multiple blocks is achieved with the help of the tree-matching algorithm can
be found in [364, 662]. The use of visual representations for extracting content structure
from Web pages is described in [126]. A detailed discussion of feature selection measures for
classification may be found in [18]. A recent text classification survey [21] discusses feature
selection algorithms for the specific case of text data.

Numerous real-world systems have been designed with the use of content-based systems.
Some of the earliest are *Fab* [60] and *Syskill & Webert* [477]. An early system, referred to
as *Personal WebWatcher* [438, 439], makes recommendations by learning the interests of
users from the Web pages that they visit. In addition, the Web pages that are linked
to by the visited page are used in the recommendation process. The *Letizia* system [356]
uses a Web-browser extension to track the user's browsing behavior, and uses it to make
recommendations. A system known as *Dynamic-Profiler* uses a pre-defined taxonomy of
categories to make news recommendations to users in real time [636]. In this case, user Web
logs are used to learn the preferences and make personalized recommendations. The *IfWeb*
system [55] represents the user interests in the form of a semantic network. The *WebMate*
system [150] learns user profiles in the form of keyword vectors. This system is designed
for keeping track of positive user interests rather than negative ones. The general principles
in Web recommendations are not very different from those of news filtering. Methods for
performing news recommendations are discussed in [41, 84, 85, 392, 543, 561]. Some of these
methods use enhanced representations, such as *WordNet*, to improve the modeling process.
Web recommender systems are generally more challenging than news recommender systems
because the underlying text is often of lower quality. The *Citeseer* system [91] is able to
discover interesting publications in a bibliographic database by identifying the common
citations among the papers. Thus, it explicitly uses citations as a content mechanism for
determination of similarity.

Content-based systems have also been used in other domains such as books, music, and
movies. Content-based methods for book recommendations are discussed in [448]. The main
challenge in music recommendations is the semantic gap between easily available features

and the likelihood of a user appreciating the music. This is a common characteristic between the music and the image domains. Some progress in bridging the semantic gap has been made in [138, 139]. *Pandora* [693] uses the features extracted in the Music Genome Project to make recommendations. The *ITR* system discusses how one might use text descriptions [178] of items (e.g., book descriptions or movie plots) to make recommendations. Further work [179] shows how one might integrate tags in a content-based recommender. The approach uses linguistic tools such as *WordNet* to extract knowledge for the recommendation process. A movie recommendation system that uses text categorization is the *INTIMATE* system [391]. A method that combines content-based and collaborative recommender systems is discussed in [520]. A broader overview of hybrid recommender systems is provided in [117]. A potential direction of work, mentioned in [376], is to enhance content-based recommender systems with encyclopedic knowledge [174, 210, 211], such as that gained from Wikipedia. A few methods have been designed that use Wikipedia for movie recommendation [341]. Interestingly, this approach does not improve the accuracy of the recommender system. The application of advanced semantic knowledge in content-based recommendations has been mentioned as a direction of future work in [376].

4.9 Exercises

1. Consider a scenario in which a user provides like/dislike ratings of a set of 20 items, in which she rates 9 items as a "like" and the remaining as a "dislike." Suppose that 7 item descriptions contain the word "thriller," and the user dislikes 5 of these items. Compute the Gini index with respect to the original data distribution, and with respect to the subset of items containing the word "thriller." Should feature selection algorithms retain this word in the item descriptions?

2. Implement a rule-based classifier with the use of association pattern mining.

3. Consider a movie recommender system in which movies belong to one or more of the genres illustrated in the table, and a particular user provides the following set of ratings to each of the movies.

Genre ⇒ Movie-Id ⇓	Comedy	Drama	Romance	Thriller	Action	Horror	Like or Dislike
1	1	0	1	0	0	0	Dislike
2	1	1	1	0	1	0	Dislike
3	1	1	0	0	0	0	Dislike
4	0	0	0	1	1	0	Like
5	0	1	0	1	1	1	Like
6	0	0	0	0	1	1	Like
Test-1	0	0	0	1	0	1	?
Test-2	0	1	1	0	0	0	?

Mine all the rules with at least 33% support and 75% confidence. Based on these rules, would you recommend the item *Test-1* or *Test-2* to the user?

4. Implement a Bayes classifier with Laplacian smoothing.

5. Repeat Exercise 3 with the use of a Bayes classifier. Do not use Laplacian smoothing. Explain why Laplacian smoothing is important in this case.

6. Repeat Exercise 3 with the use of a 1-nearest neighbor classifier.

7. For a training data matrix D, regularized least-squares regression requires the inversion of the matrix $(D^T D + \lambda I)$, where $\lambda > 0$. Show that this matrix is always invertible.

8. The χ^2 distribution is defined by the following formula, as discussed in the chapter:

$$\chi^2 = \sum_{i=1}^{p} \frac{(O_i - E_i)^2}{E_i}$$

Show that for a 2×2 contingency table, the aforementioned formula can be rewritten as follows:

$$\chi^2 = \frac{(O_1 + O_2 + O_3 + O_4) \cdot (O_1 O_4 - O_2 O_3)^2}{(O_1 + O_2) \cdot (O_3 + O_4) \cdot (O_1 + O_3) \cdot (O_2 + O_4)}$$

Here, $O_1 \ldots O_4$ are defined in the same way as in the tabular example in the text.

Chapter 5

Knowledge-Based Recommender Systems

"Knowledge is knowing that a tomato is a fruit. Wisdom is knowing not to put it in a fruit salad."–Brian O'Driscoll

5.1 Introduction

Both content-based and collaborative systems require a significant amount of data about past buying and rating experiences. For example, collaborative systems require a reasonably well populated ratings matrix to make future recommendations. In cases where the amount of available data is limited, the recommendations are either poor, or they lack full coverage over the entire spectrum of user-item combinations. This problem is also referred to as the *cold-start problem*. Different systems have varying levels of susceptibility to this problem. For example, collaborative systems are the most susceptible, and they cannot handle new items or new users very well. Content-based recommender systems are somewhat better at handling new items, but they still cannot provide recommendations to new users.

Furthermore, these methods are generally not well suited to domains in which the product is highly *customized*. Examples include items such as real estate, automobiles, tourism requests, financial services, or expensive luxury goods. Such items are bought rarely, and sufficient ratings are often not available. In many cases, the item domain may be complex, and there may be few instances of a specific item with a particular set of properties. For example, one might want to buy a house with a specific number of bedrooms, lawn, locality, and so on. Because of the complexity in describing the item, it may be difficult to obtain a reasonable set of ratings reflecting the past history of a user on a similar item. Similarly, an old rating on a car with a specific set of options may not even be relevant in the present context.

C.C. Aggarwal, *Recommender Systems: The Textbook,*
DOI 10.1007/978-3-319-29659-3_5

How can one handle such customization and paucity of ratings? Knowledge-based recommender systems rely on *explicitly soliciting user requirements* for such items. However, in such complex domains, it is often difficult for users to fully enunciate or even understand how their requirements match the product availability. For example, a user may not even be aware that a car with a certain combination of fuel efficiency and horsepower is available. Therefore, such systems use interactive feedback, which allows the user to explore the inherently complex product space and learn about the trade-offs available between various options. The retrieval and exploration process is facilitated by knowledge bases describing the utilities and/or trade-offs of various features in the product domain. The use of knowledge bases is so important to an effective retrieval and exploration process, that such systems are referred to as knowledge-based recommender systems.

Knowledge-based recommender systems are well suited to the recommendation of items that are not bought on a regular basis. Furthermore, in such item domains, users are generally more active in being explicit about their requirements. A user may often be willing to accept a movie recommendation without much input, but she would be unwilling to accept recommendations about a house or a car without having detailed information about the specific features of the item. Therefore, knowledge-based recommender systems are suited to types of item domains different from those of collaborative and content-based systems. In general, knowledge-based recommender systems are appropriate in the following situations:

1. Customers want to explicitly specify their requirements. Therefore, interactivity is a crucial component of such systems. Note that collaborative and content-based systems do not allow this type of detailed feedback.

2. It is difficult to obtain ratings for a specific type of item because of the greater complexity of the product domain in terms of the types of items and options available.

3. In some domains, such as computers, the ratings may be time-sensitive. The ratings on an old car or computer are not very useful for recommendations because they evolve with changing product availability and corresponding user requirements.

A crucial part of knowledge-based systems is the greater control that the user has in guiding the recommendation process. This greater control is a direct result of the need to be able to specify detailed requirements in an inherently complex problem domain. At a basic level, the conceptual differences in the three categories of recommendations are described in Table 5.1. Note that there are also significant differences in the input data used by various systems. The recommendations of content-based and collaborative systems are primarily based on *historical* data, whereas knowledge-based systems are based on the direct specifications by users of *what they want*. An important distinguishing characteristic of knowledge-based systems is a high level of *customization* to the specific domain. This customization is achieved through the use of a knowledge-base that encodes relevant *domain knowledge* in the form of either constraints or similarity metrics. Some knowledge-based systems might also use user attributes (e.g., demographic attributes) in addition to item attributes, which are specified at query time. In such cases, the domain knowledge might also encode relationships between user attributes and item attributes. The use of such attributes is, however, not universal to knowledge-based systems, in which the greater focus is on user *requirements*.

Knowledge-based recommender systems can be categorized on the basis of user interactive methodology and the corresponding knowledge bases used to facilitate the interaction. There are two primary types of knowledge-based recommender systems:

1. *Constraint-based recommender systems:* In constraint-based systems [196, 197], users typically specify requirements or constraints (e.g., lower or upper limits) on the item

Table 5.1: The conceptual goals of various recommender systems

Approach	Conceptual Goal	Input
Collaborative	Give me recommendations based on a collaborative approach that leverages the ratings and actions of my peers/myself.	User ratings + community ratings
Content-based	Give me recommendations based on the content (attributes) I have favored in my past ratings and actions.	User ratings + item attributes
Knowledge-based	Give me recommendations based on my explicit specification of the kind of content (attributes) I want.	User specification + item attributes + domain knowledge

attributes. Furthermore, domain-specific rules are used to match the user requirements or attributes to item attributes. These rules represent the domain-specific knowledge used by the system. Such rules could take the form of domain-specific constraints on the item attributes (e.g., *"Cars before year 1970 do not have cruise control."*). Furthermore, constraint-based systems often create rules relating user attributes to item attributes (e.g., *"Older investors do not invest in ultrahigh-risk products."*). In such cases, user attributes may also be specified in the search process. Depending on the number and type of returned results, the user might have an opportunity to modify their original requirements. For example, a user might relax some constraints when too few results are returned, or add more constraints when too many results are returned. This search process is interactively repeated until the user arrives at her desired results.

2. *Case-based recommender systems:* In case-based recommender systems [102, 116, 377, 558], specific cases are specified by the user as targets or anchor points. Similarity metrics are defined on the item attributes to retrieve similar items to these targets. The similarity metrics are often carefully defined in a domain-specific way. Therefore, the similarity metrics form the domain knowledge that is used in such systems. The returned results are often used as new target cases with some interactive modifications by the user. For example, when a user sees a returned result that is almost similar to what she wants, she might re-issue a query with that target, but with some of the attributes changed to her liking. Alternatively, a *directional critique* may be specified to prune items with specific attribute values greater (or less) than that of a specific item of interest. This interactive process is used to guide the user towards the final recommendation.

Note that in both cases, the system provides an opportunity for the user to change her specified requirements. However, the way in which this is done is different in the two cases. In case-based systems, examples (or *cases*) are used as anchor points to guide the search in conjunction with *similarity metrics*, whereas in constraint-based systems, specific criteria/rules (or *constraints*) are used to guide the search. In both cases, the presented results are used to modify the criteria for finding further recommendations. Knowledge-based systems derive their name from the fact that they encode various types of *domain knowledge* in the form of constraints, rules, similarity metrics, and utility functions during the search process. For example, the design of a similarity metric or a specific constraint requires domain-specific knowledge, which is crucial to the effective functioning of the recommender system. In general, knowledge-based systems draw on highly heterogeneous, domain-specific sources of knowledge, compared to content-based and collaborative systems, which work with somewhat similar types of input data across various domains. As a result, knowledge-based systems are highly customized, and they are not easily generalizable across various domains. However, the broader principles with which this customization is done are invariant across domains. The goal of this chapter is to discuss these principles.

The interaction between user and recommender may take the form of *conversational systems*, *search-based systems*, or *navigational systems*. Such different forms of guidance may be present either in isolation, or in combination, and they are defined as follows:

1. *Conversational systems:* In this case, the user preferences are determined in the context of a feedback loop. The main reason for this is that the item domain is complex, and the user preferences can be determined only in the context of an iterative conversational system.

2. *Search-based systems:* In search-based systems, user preferences are elicited by using a preset sequence of questions such as the following: "Do you prefer a house in a suburban area or within the city?"

3. *Navigation-based recommendation:* In navigation-based recommendation, the user specifies a number of change requests to the item being currently recommended. Through an iterative set of change requests, it is possible to arrive at a desirable item. An example of a change request specified by the user, when a specific house is being recommended is as follows: "I would like a similar house about 5 miles west of the currently recommended house." Such recommender systems are also referred to as *critiquing recommender systems* [120, 121, 417].

These different forms of guidance are well suited to different types of recommender systems. For example, critiquing systems are naturally designed for case-based recommenders, because one critiques a specific case in order to arrive at the desired outcome. On the other hand, a search-based system can be used to set up user requirements for constraint-based recommenders. Some forms of guidance can be used with both constraint-based and case-based systems. Furthermore, different forms of guidance can also be used in combination in a knowledge-based system. There are no strict rules as to how one might design the interface for a knowledge-based system. The goal is always to guide the user through a complex product space.

Typical examples of the interactive process in constraint-based recommenders and case-based recommenders are illustrated in Figures 5.1(a) and (b), respectively. The overall interactive approach is quite similar. The main difference in the two cases is in terms of how the user specifies the queries and interacts with the system for subsequent refinement. In constraint-based systems, specific requirements (or *constraints*) are specified by the user, whereas in case-based systems, specific targets (or *cases*) are specified. Correspondingly, different types of interactive processes and domain knowledge are used in the two systems. In constraint-based systems, the original query is modified by addition, deletion, modification, or relaxation of the original set of user requirements. In case-based systems, either the target is modified through user interaction, or the search results are pruned through the use of *directional* critiques. In such critiques, the user simply states whether a specific attribute in the search results needs to be increased, decreased, or changed in a certain way. Such an approach represents a more conversational style than simply modifying the target. In both these types of systems, a common motivation is that users are often not in a position to exactly state their requirements up front in a complex product domain. In constraint-based systems, this problem is partially addressed through a knowledge-base of rules, which map user requirements to product attributes. In case-based systems, this problem is addressed

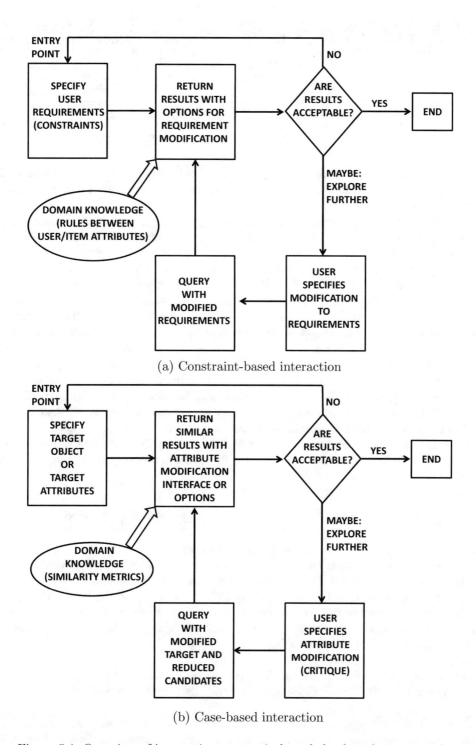

(a) Constraint-based interaction

(b) Case-based interaction

Figure 5.1: Overview of interactive process in knowledge-based recommenders

Table 5.2: Examples of attributes in a recommendation application for buying homes

Item-Id	Beds.	Baths.	Locality	Type	Floor Area	Price
1	3	2	Bronx	Townhouse	1600	220,000
2	5	2.5	Chappaqua	Split-level	3600	973,000
3	4	2	Yorktown	Ranch	2600	630,000
4	2	1.5	Yorktown	Condo	1500	220,000
5	4	2	Ossining	Colonial	2700	430,000

through a conversational style of critiquing. The interactive aspect is common to both systems, and it is crucial in helping the users discover how the items in a complex product domain fit their needs.

It is noteworthy that most forms of knowledge-based recommender systems depend heavily on the descriptions of the items in the form of relational attributes rather than treating them as text keywords like[1] content-based systems. This is a natural consequence of the inherent complexity in knowledge-based recommendations in which domain-specific knowledge can be more easily encoded with relational attributes. For example, the attributes for a set of houses in a real-estate application is illustrated in Table 5.2. In case-based recommenders, the similarity metrics are defined in terms of these attributes in order to provide similar matches to target homes provided by the user. Note that each relational attribute would have a different significance and weight in the matching process, depending on domain-specific criteria. In constraint-based systems, the queries are specified in the form of requirements on these attributes, such as a maximum price on the house, or a specific locality. Therefore, the problem reduces to an instance of the *constraint-satisfaction problem*, where one must identify the relevant set of instances satisfying all the constraints.

This chapter is organized as follows. Constraint-based recommenders are introduced in section 5.2. Case-based recommenders are discussed in section 5.3. The use of persistent personalization in knowledge-based systems is discussed in section 5.4. A summary is given in section 5.5.

5.2   Constraint-Based Recommender Systems

Constraint-based recommender systems allow the users to specify hard requirements or constraints on the item attributes. Furthermore, a set of rules is used in order to match the customer requirements with item attributes. However, the customers may not always specify their queries in terms of the same attributes that describe the items. Therefore, an additional set of rules is required that relates the customer requirements with the product attributes. In relation to the previous home-buying example in Table 5.2, some examples of customer-specified attributes are as follows:

Marital-status (categorical), *Family-Size* (numerical), *suburban-or-city* (binary), *Min-Bedrooms* (numerical), *Max-Bedrooms* (numerical), *Max-Price* (numerical)

These attributes may represent either inherent customer properties (e.g., demographics), or they may specify customer requirements for the product. Such requirements are usually

---

[1]Content-based systems are used both in the information retrieval and the relational settings, whereas knowledge-based systems are used mostly in the relational setting.

specified interactively during the dialog between the customer and the recommender system. Note that many of the requirement attributes are not included in Table 5.2. While the mappings of some of the customer requirement attributes, such as *Max-Price*, to product attributes are obvious, the mappings of others, such as *suburban-or-rural*, are not quite as obvious. Similarly, in a financial application, a customer may specify a product requirement such as "*conservative investments,*" which needs to be mapped to concrete product attributes (e.g., *Asset-type=Treasuries*) directly describing the products. Clearly, one must somehow be able to map these customer attributes/requirements into the product attributes in order to filter products for recommendation. This is achieved through the use of knowledge bases. The knowledge bases contain additional rules that map customer attributes/requirements to the product attributes:

$$Suburban\text{-}or\text{-}rural = Suburban \Rightarrow Locality = \langle List\ of\ relevant\ localities \rangle$$

Such rules are referred to as *filter* conditions because they map user requirements to the item attributes and use this mapping to filter the retrieved results. Note that these types of rules may be either derived from the product domain, or, more rarely, they may be derived by historical mining of such data sets. In this particular case, it is evident that this rule can be derived directly using publicly available geographical information. Another example is the car domain, where certain optional packages may be valid only with certain other attributes. For example, a high-torque engine may be available only in a sports model. Such conditions are also referred to as *compatibility* conditions, because they can be used to quickly discover inconsistencies in the user-specified requirements with the product domain. In many cases, such compatibility constraints can be integrated within the user interface. For example, the car pricing site Edmunds.com prevents users from entering mutually inconsistent requirements within the user interface. In other cases, where inconsistency detection is not possible within the user interface, such inconsistencies can be detected at query processing time by returning empty sets of results.

Some of the other compatibility constraints may relate customer attributes to one another. Such constraints are useful when customers specify personal information (e.g., demographic information) about themselves during the interactive session. For example, demographic attributes may be related to customer product requirements based on either domain-specific constraints, or historical experience. An example of such a constraint is as follows:

$$Marital\text{-}status = single \Rightarrow Min\text{-}Bedrooms \leq 5$$

Presumably, by either domain-specific experience or through data mining of historical data sets, it has been inferred that single individuals do not prefer to buy very large houses. Similarly, a small home might not be suitable for a very large family. This constraint is modeled with the following rule:

$$Family\text{-}Size \geq 5 \Rightarrow Min\text{-}Bedrooms \geq 3$$

Thus, there are three primary types of input to the constraint-based recommender system:

1. The first class of inputs is represented by the attributes describing the inherent properties of the user (e.g., demographics, risk profiles) and specific requirements in the product (e.g., *Min-Bedrooms*). Some of these attributes are easy to relate to product attributes, whereas others can be related to product attributes only through the use of knowledge bases. In most cases, the customer properties and requirements are specified interactively in a session, and they are not persistent across multiple sessions.

Therefore, if another user specifies the same set of requirements in a session, they will obtain the same result. This is different from other types of recommender systems, where the personalization is persistent because it is based on historical data.

2. The second class of inputs is represented by knowledge bases, which map customer attributes/requirements to various product attributes. The mapping can be achieved either directly or indirectly as follows:

   - **Directly:** These rules relate customer requirements to hard requirements on product attributes. An example of such a rule is as follows:

     $$Suburban\text{-}or\text{-}rural{=}Suburban \Rightarrow Locality{=} \langle List\ of\ relevant\ localities\ \rangle$$
     $$Min\text{-}Bedrooms{\geq}3 \Rightarrow Price{\geq}100,000$$

     Such rules are also referred to as filter conditions.

   - **Indirectly:** These rules relate customer attributes/requirements to typically expected product requirements. Therefore, such rules can also be viewed as an indirect way of relating customer attributes to product attributes. Examples of such rules are as follows:

     $$Family\text{-}Size{\geq}5 \Rightarrow Min\text{-}Bedrooms{\geq}3$$
     $$Family\text{-}Size{\geq}5 \Rightarrow Min\text{-}Bathrooms{\geq}2$$

   Note that the conditions on both sides of the rule represent customer attributes, although the ones on the right-hand side are generally customer requirements, which can be mapped to product attributes easily. These constraints represent compatibility constraints. In the event that the compatibility constraints or filter conditions are inconsistent with the customer-specified requirements, the recommended list of items will be empty.

   The aforementioned knowledge bases are derived from publicly available information, domain experts, past experience, or data mining of historical data sets. Therefore, a significant amount of effort is involved in building the knowledge bases.

3. Finally, the product *catalog* contains a list of all the products together with the corresponding item attributes. A snapshot of a product catalog for the home-buying example is illustrated in Table 5.2.

Therefore, the problem boils down to determining all the instances in the available product list that satisfy the customer requirements and the rules in the knowledge base.

5.2.1 Returning Relevant Results

The problem of returning relevant results can be shown to be an instance of the constraint satisfaction problem by viewing each item in the catalog as a constraint on the attributes and expressing the catalog in disjunctive normal form. This expression is then combined with the rules in the knowledge base to determine whether a mutually consistent region of the product space exists.

More simply, the set of rules and requirements can be reduced to a data filtering task on the catalog. All the customer requirements and the active rules relevant to the customer are used to construct a database selection query. The steps for creating such a filtering query are as follows:

1. For each requirement (or personal attribute) specified by the customer in their user interface, it is checked whether it matches the antecedent of a rule in the knowledge base. If such a matching exists, then the consequent of that rule is treated as a valid selection condition. For example, consider the aforementioned real-estate example. If the customer has specified *Family-Size=6* and *ZIP Code=10547* among their personal attributes and preferences in the user interface, then it is detected that *Family-Size=6* triggers the following rules:

$$Family\text{-}Size \geq 5 \Rightarrow Min\text{-}Bedrooms \geq 3$$
$$Family\text{-}Size \geq 5 \Rightarrow Min\text{-}Bathrooms \geq 2$$

Therefore, the consequents of these conditions are added to the user requirements. The rule base is again checked with these expanded requirements, and it is noticed that the newly added constraint $Min\text{-}Bedrooms \geq 3$ triggers the following rules:

$$Min\text{-}Bedrooms \geq 3 \Rightarrow Price \geq 100,000$$
$$Min\text{-}Bedrooms \geq 3 \Rightarrow Bedrooms \geq 3$$
$$Min\text{-}Bathrooms \geq 3 \Rightarrow Bathrooms \geq 2$$

Therefore, the conditions *Price≥100,000*, and the range constraints on the requirement attributes *Min-Bedrooms* and *Min-Bathrooms* are replaced with those on the product attributes *Bedrooms* and *Bathrooms*. In the next iteration, it is found that no further conditions can be added to the user requirements.

2. These expanded requirements are used to construct a database query in conjunctive normal form. This represents a traditional database selection query, which computes the intersection of the following constraints on the product catalog:

$$(Bedrooms \geq 3) \land (Bathrooms \geq 2) \land (Price \geq 100,000) \land (ZIP\ Code=10547)$$

Note that the approach essentially maps all customer attribute constraints and requirement attribute constraints to constraints in the product domain.

3. This selection query is then used to retrieve the instances in the catalog that are relevant to the user requirements.

It is noteworthy that most constraint-based systems enable specification of all user requirements or other attributes (e.g., preferences, demographic information) *during the session itself*. In other words, the specified information is typically not persistent; if a different user specifies the same input, they will get exactly the same result. This characteristic is common to most knowledge-based systems. Section 5.4 will discuss some recent advancements in the *persistent* personalization of knowledge-based systems.

The resulting list of items, which satisfy the constraints, is then presented to the user. The methodology for ranking the items is discussed later in this section. The user may then modify her requirements further to obtain more refined recommendations. The overall process of exploration and refinement often leads the customer to discover recommendations that she might otherwise not have been able to arrive at on her own.

5.2.2 Interaction Approach

The interaction between the user and the recommender system generally proceeds in three phases.

1. An interactive interface is used by the user to specify her initial preferences. A common approach is to use a Web style form in which the desired values of the attributes may be entered. An example of a *hypothetical* interface for home buying, which we will be using as a running example, is provided in Figure 5.2. Alternatively, the user could be asked a series of questions to elicit her initial preferences. For example, the car recommendation site Edmunds.com presents a series of interfaces to the users to specify their preferences about the specific features they might want. The answers to the queries in the first interface may affect the questions in the next interface.

2. The user is presented with a ranked list of matching items. An explanation for why the items are returned is typically provided. In some cases, no items might match the user requirements. In such cases, possible relaxations of the requirements might be suggested. For example, in Figure 5.3, no results are returned by the query, and possible relaxations are suggested. In cases, where too many items are returned, suggestions for possible constraints (user requirements) are included. For example, in Figure 5.4, too many results are returned. Possible constraints are suggested to be added to the query.

3. The user then refines her requirements depending on the returned results. This refinement might take the form of the addition of further requirements, or the removal of some of the requirements. For example, when an empty set is returned, it is evident that some of the requirements need to be relaxed. Constraint satisfaction methods are used to identify possible sets of candidate constraints, which might need to be relaxed. Therefore, the system generally helps the user in making her modifications in a more intelligent and efficient way.

Thus, the overall approach uses an iterative feedback loop to assist the users in making meaningful decisions. It is crucial to design a system that can guide the user towards requirements that increase her awareness regarding the available choices.

There are several aspects of this interaction, in which explicit computation is required in order to help the user. For example, a user will typically not be able specify desired values for all the product attributes. For instance, in our home-buying example, the user may specify constraints only on the number of bedrooms and not specify any constraints on the price. Several solutions are possible under this scenario:

1. The system may leave the other attributes unconstrained and retrieve the results based on only the specified constraints. For example, all possible ranges of prices may be considered in order to provide the first set of responses to the user. Although this may be the most reasonable choice, when the user query has been formulated well, it may not be an effective solution in cases where the number of responses is large.

2. In some cases, default values may be suggested to the user to provide guidance. The default values can be used only to guide the user in selecting values, or they can actually be included in the query if the user does not select any value (including the default) for that attribute. It can be argued that including a default value within the query (without explicit specification) can lead to significant bias within the recommender system, especially when the defaults are not very well researched. In general,

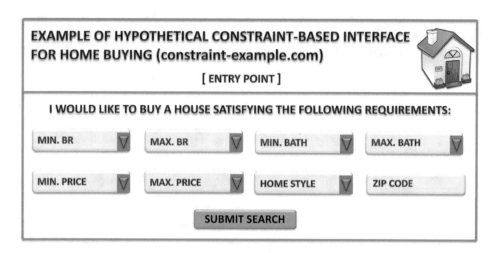

Figure 5.2: A hypothetical example of an initial user interface for a constraint-based recommender (`constraint-example.com`)

default values should be used only as a suggestion for the user. This is because the main goal of defaults should be to *guide* the user towards natural values, rather than to *substitute* for unspecified options.

How are default values determined? In most cases, it is necessary to choose the defaults in a domain-specific way. Furthermore, some values of the defaults may be affected by others. For example, the horsepower of a selected car model might often reflect the desired fuel efficiency. Knowledge bases need to explicitly store the data about such default values. In some cases, where the historical data from user sessions is available, it is possible to learn the default values. For the various users, their specified attribute values in the query sessions may be available, including the missing values. The average values across various sessions may be used as defaults. Consider a query session initiated by Alice for buying cars. Initially, her defaults are computed on the basis of the average values in historical sessions. However, if she specifies the desired horsepower of the car, then the interface automatically adjusts her default value of the fuel efficiency. This new default value is based on the average of fuel efficiency of cars, which were specified in historical sessions for cars with similar horsepower. In some cases, the system might automatically adjust the default values based on feasibility constraints with respect to the knowledge base. As users specify increasingly more values in the interface, the average can be computed only over the sessions within the neighborhood of the current specification.

After the query has been issued, the system provides a ranked list of possible matches from the catalog. Therefore, it is important to be able to meaningfully rank the matches and also provide explanations for the recommended results if needed. In cases, where the returned set of matches is too small or too large, further guidance may be provided to the user on either relaxing or tightening requirements. It is noteworthy that the provision of explanations is also an intelligent way of guiding the user towards more meaningful query refinements. In the following, we will discuss these various aspects of interactive user guidance.

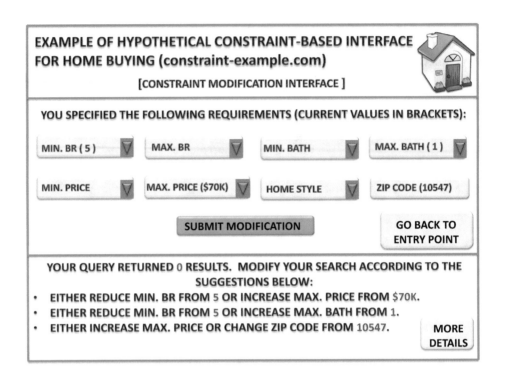

Figure 5.3: A hypothetical example of a user interface for handling empty query results in a constraint-based recommender (`constraint-example.com`)

5.2.3   Ranking the Matched Items

A number of natural methods exist for ranking the items according to user requirements. The simplest approach is to allow the user to specify a single numerical attribute on the basis of which to rank the items. For example, in the home-buying application, the system might provide the user the option to rank the items on the basis of (any one of) the home price, number of bedrooms, or distance from a particular ZIP code. This approach is, in fact, used in many commercial interfaces.

Using a single attribute has the drawback that the importance of other attributes is discounted. A common approach is to use utility functions in order to rank the matched items. Let $\overline{V} = (v_1 \ldots v_d)$ be the vector of values defining the attributes of the matched products. Therefore, the dimensionality of the content space is d. The utility functions may be defined as weighted functions of the utilities of individual attributes. Each attribute has a weight w_j assigned to it, and it has a contribution defined by the function $f_j(v_j)$ depending on the value v_j of the matched attribute. Then, the utility $U(\overline{V})$ of the matched item is given by the following:

$$U(\overline{V}) = \sum_{j=1}^{d} w_j \cdot f_j(v_j) \qquad (5.1)$$

Clearly, one needs to instantiate the values of w_j and $f_j(\cdot)$ in order to learn the utility function. The design of effective utility functions often requires domain-specific knowledge, or learning data from past user interactions. For example, when v_j is numeric, one might assume that the function $f_j(v_j)$ is linear in v_j, and then learn the coefficients of the linear

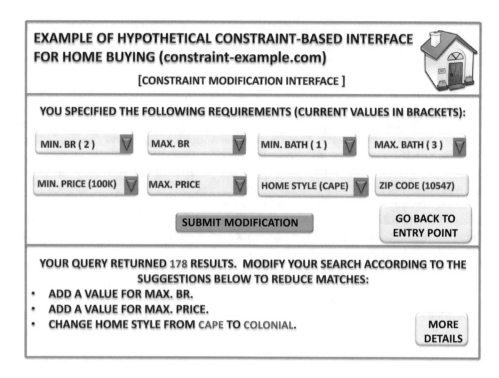

Figure 5.4: A hypothetical example of a user interface for handling too many query results in a constraint-based recommender (`constraint-example.com`)

function as well as w_j by eliciting feedback from various users. Typically, training data is elicited from some users who are given the task of ranking some sample items. These ranks are then used to learn the aforementioned model with the use of regression models. This approach is related to the methodology of *conjoint analysis* [155, 531]. Conjoint analysis defines statistical methods for the formal study of how people value the different attributes that make up an individual product or service. The bibliographic notes contain pointers to some methods that are commonly used for the design of utility functions.

5.2.4 Handling Unacceptable Results or Empty Sets

In many cases, a particular query might return an empty set of results. In other cases, the set of returned results might not be large enough to meet the user requirements. In such cases, a user has two options. If it is deemed that a straightforward way of repairing the constraints does not exist, she may choose to start over from the entry point. Alternatively, she may decide to change or relax the constraints for the next interactive iteration.

How can the user make a meaningful choice on whether to relax the constraints and in what way? In such cases, it is often helpful to provide the user with some guidance on relaxing the current requirements. Such proposals are referred to as *repair proposals*. The idea is to be able to determine minimal sets of inconsistent constraints, and present them to the user. It is easier for the user to assimilate minimal sets of inconsistent constraints, and find ways of relaxing one or more of the constraints in these sets. Consider the home-buying example, in which it may be found that the user has specified many requirements, but the only mutually inconsistent pair of requirements is *Max-Price* $< 100,000$ and *Min-Bedrooms* > 5.

If this pair of constraints is presented to the user, she can understand that she either needs to increase the maximum price she is willing to pay, or she needs to settle for a smaller number of bedrooms. A naive way of finding the minimal set of inconsistent constraints is to perform a bottom-up search of all combinations of user requirements, and determine the smallest sets that are infeasible. In many interactive interfaces, the user might specify only a small number of (say, 5 to 10) requirements, and the number of constraints involving these attributes (in the domain knowledge) might also be small. In such cases, exhaustive exploration of all the possibilities is not an unreasonable approach. By its very nature, interactive requirement specification often results in the specification of a relatively small number of constraints. It is unusual for a user to specify 100 different requirements in an interactive query. In some cases, however, when the number of user-specified requirements is large and the domain knowledge is significant, such an exhaustive bottom-up exploration might not be a feasible option. More sophisticated methods, such as *QUICKXPLAIN* and *MINRELAX*, have also been proposed, which can be used for fast discovery of small conflicting sets and minimal relaxations [198, 273, 274, 289, 419].

Most of these methods use similar principles; small sets of violating constraints are determined, and the most appropriate relaxations are suggested based on some pre-defined criteria. In real applications, however, it is sometimes difficult to suggest concrete criteria for constraint relaxation. Therefore, a simple alternative is to present the user with small sets of inconsistent constraints, which can often provide sufficient intuition to the user in formulating modified constraints.

5.2.5 Adding Constraints

In some cases, the number of returned results may be very large, and the user may need to suggest possible constraints to be added to the query. In such cases, a variety of methods can be used to suggest constraints to the user along with possible default values. The attributes for such constraints are often chosen by mining historical session logs. The historical session logs can either be defined over all users, or over the particular user at hand. The latter provides more personalized results, but may often be unavailable for infrequently bought items (e.g., cars or houses). It is noteworthy that knowledge-based systems are generally designed to not use such persistent and historical information precisely because they are designed to work in cold-start settings; nevertheless, such information can often be very useful in improving the user experience when it is available.

How can historical session data be used? The idea is to select constraints that are popular. For example, if a user has specified the constraints on a set of item attributes, then other sessions containing one or more of these attributes are identified. For example, if a user has specified constraints on the number of bedrooms and the price, previous sessions containing constraints on the bedroom and price are identified. In particular, the top-k nearest neighbor sessions in terms of the number of common attributes are identified. If it is determined that the most popular constraint among these top-k sessions is on the number of bathrooms, then this attribute is suggested by the interface as a candidate for adding additional constraints.

In many cases, the temporal ordering in which users have specified constraints in the past is available. In such cases, it is also possible to use the *order* in which the customer specified the constraints by treating the constraints as an ordered set, rather than as an unordered set [389]. A simple way of achieving this goal is to determine the most frequent attribute that *follows* the current specified set of constrained attributes in previous sessions. Sequential pattern mining can be used to determine such frequent attributes. The works in [389, 390] model the sequential learning problem as a Markov Decision Process (MDP), and use reinforcement learning techniques to measure the impact of various choices. The constraints can be suggested based on their selectivity in the database or based on the average specification of the user in past sessions.

5.3   Case-Based Recommenders

In case-based recommenders, similarity metrics are used to retrieve examples that are similar to the specified targets (or *cases*). For instance, in the real-estate example of Table-5.2, the user might specify a locality, the number of bedrooms, and a desired price to specify a target set of attributes. Unlike constraint-based systems, no *hard* constraints (e.g., minimum or maximum values) are enforced on these attributes. It is also possible to design an initial query interface in which examples of relevant items are used as targets. However, it is more natural to specify desired properties in the initial query interface. A similarity function is used to retrieve the examples that are most similar to the user-specified target. For example, if no homes are found specifying the user requirements exactly, then the similarity function is used to retrieve and rank items that are as similar as possible to the user query. Therefore, unlike constraint-based recommenders, the problem of retrieving empty sets is not an issue in case-based recommenders.

There are also substantial differences between a constraint-based recommender and a case-based recommender in terms of how the results are refined. Constraint-based systems use requirement relaxation, modification, and tightening to refine the results. The earliest case-based systems advocated the repeated modification of user query requirements until a suitable solution could be found. Subsequently, the method of *critiquing* was developed. The general idea of critiquing is that users can select one or more of the retrieved results and specify further queries of the following form:

"Give me more items like X, but they are different in attribute(s) Y according to guidance Z."

A significant variation exists in terms of whether one or more than one attributes is selected for modification and how the guidance for modifying the attributes is specified. The main goal of critiquing is to support interactive browsing of the item space, where the user gradually becomes aware of further options available to them through the retrieved examples. Interactive browsing of the item space has the advantage that it is a learning process for the user during the process of iterative query formulation. It is often possible that through repeated and interactive exploration, the user might be able to arrive at items that could not otherwise have been reached at the very beginning.

For example, consider the home-buying example of Table 5.2. The user might have initially specified a desired price, the number of bedrooms, and a desired locality. Alternatively, the user might specify a target address to provide an example of a possible house she might be interested in. An example of an initial interface in which the user can specify the target in two different ways, is illustrated in Figure 5.5. The top portion of the interface illustrates the specification of target features, whereas the bottom portion of the interface illustrates the specification of a target address. The latter approach is helpful in domains where the users have greater difficulty in specifying technically cryptic features. An example might be the case of digital cameras, where it is harder to specify all the technical features exactly for a non-specialist in photography. Therefore, a user might specify her friend's camera as

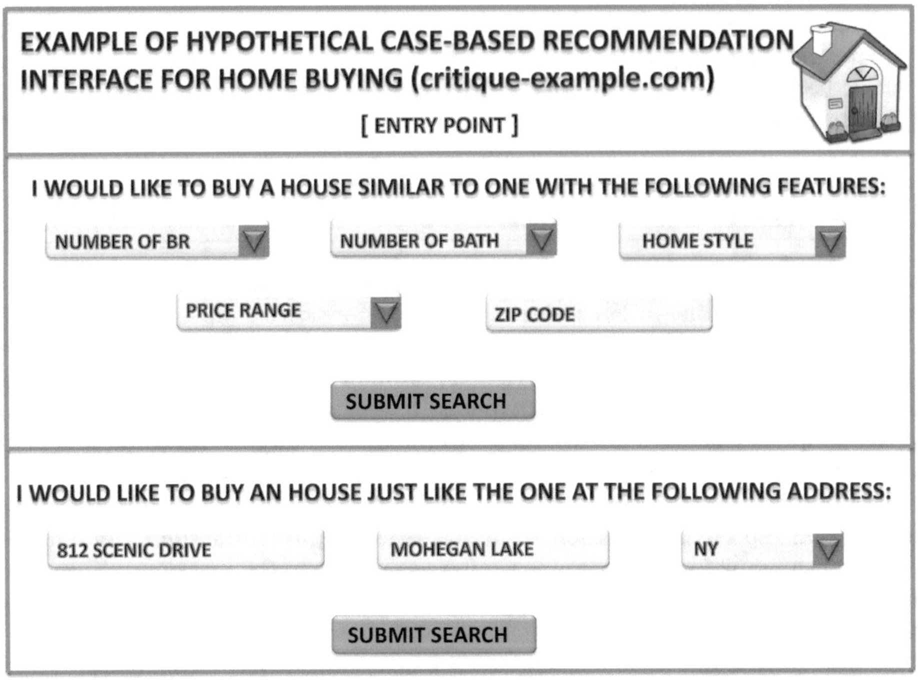

Figure 5.5: A hypothetical example of an initial user interface in a case-based recommender (`critique-example.com`)

the target case, rather than specifying all the technical features. Note that this interface is hypothetically designed for illustrative purposes only, and it is not based on an actual recommender system.

The system uses the target query in conjunction with similarity or utility functions in order to retrieve matching results. Eventually, upon retrieving the results, the user might decide to like a particular house, except that its specifications contain features (e.g., a colonial) that she does not particularly like. At this point, the user might leverage this example as an anchor and specify the particular attributes in it that she wants to be different. Note that the reason that the user is able to make this second set of critiqued query specifications is that she now has a concrete example to work with that she was not aware of earlier. The interfaces for critiquing can be defined in a number of different ways, and they are discussed in detail in section 5.3.2. The system then issues a new query with the modified target, and with a *reduced* set of candidates, which were the results from the previous query. In many cases, the effect is to simply prune the search results of cases that are not considered relevant, rather than provide a re-ranking of the returned results. Therefore, unlike constraint-based systems, the number of returned responses in case-based iterations generally reduces from one cycle to the next. However, it is also possible to design case-based systems in which the candidates are not always reduced from one iteration to the next by expanding the scope of each query to the entire database, rather than the currently retrieved set of candidate results. This type of design choice has its own trade-offs. For example, by expanding the scope of each query, the user will be able to navigate to a final result that is more distant from the current query. On the other hand, it is also possible that the results might become increasingly irrelevant in later iterations. For the purpose of this chapter, we assume that the returned candidates always reduce from one iteration to the next.

Through repeated critiquing, the user may sometimes arrive at a final result that is quite different from the initial query specification. After all, it is often difficult for a user to articulate *all* their desired features at the very beginning. For example, the user might not be aware of an acceptable price point for the desired home features at the beginning of the querying process. This interactive approach bridges the gap between her initial understanding and item availability. It is this power of assisted browsing that makes case-based methods so powerful in increasing user awareness. It is sometimes also possible for the user to arrive at an empty set of candidates through repeated reduction of the candidate set. Such a session may be viewed as a fruitless session, and in this case, the user has to restart from scratch at the entry point. Note that this is different from constraint-based systems, where a user also has the option of relaxing their current set of requirements to enlarge the result set. The reason for this difference is that case-based systems generally reduce the number of candidates from one cycle to the next, whereas constraint-based systems do not.

In order for a case-based recommender system to work effectively, there are two crucial aspects of the system that must be designed effectively:

1. *Similarity metrics:* The effective design of similarity metrics is very important in case-based systems in order to retrieve relevant results. The importance of various attributes must be properly incorporated within the similarity function for the system to work effectively.

2. *Critiquing methods:* The interactive exploration of the item space is supported with the use of critiquing methods. A variety of different critiquing methods are available to support different exploration goals.

In this section, we will discuss both these important aspects of case-based recommender system design.

5.3.1   Similarity Metrics

The proper design of similarity metrics is essential in retrieving meaningful items in response to a particular query. The earliest *FindMe* systems [121] ordered the attributes in decreasing level of importance and first sorted on the most important criterion, then the next most important, and so on. For example, in the *Entree* restaurant recommender system, the first sort might be based on the cuisine type, the second on the price, and so on. While this approach is efficient, its usage may not be effective for every domain. In general, it is desirable to develop a closed-form similarity function whose parameters can either be set by domain experts, or can be tweaked by a learning process.

Consider an application in which the product is described by d attributes. We would like to determine the similarity values between two *partial* attribute vectors defined on a subset S of the universe of d attributes (i.e., $|S| = s \leq d$). Let $\overline{X} = (x_1 \ldots x_d)$ and $\overline{T} = (t_1 \ldots t_d)$ represent two d-dimensional vectors, which might be partially specified. Here, $\overline{T}$ represents the target. It is assumed that at least the attribute subset $S \subseteq \{1 \ldots d\}$ is specified in both vectors. Note that we are using *partial* attribute vectors because such queries are often defined only on a small subset of attributes specified by the user. For example, in the aforementioned real estate example, the user might specify only a small set of query features, such as the number of bedrooms or bathrooms. Then, the similarity function $f(\overline{T}, \overline{X})$ between the two sets of vectors is defined as follows:

$$f(\overline{T}, \overline{X}) = \frac{\sum_{i \in S} w_i \cdot Sim(t_i, x_i)}{\sum_{i \in S} w_i} \tag{5.2}$$

Here, $Sim(t_i, x_i)$ represents the similarity between the values x_i and y_i. The weight w_i represents the weight of the ith attribute, and it regulates the relative importance of that attribute. How can the similarity functions $Sim(t_i, x_i)$ and the attribute importance w_i be learned?

Fist, we will discuss the determination of the similarity function $Sim(t_i, x_i)$. Note that these attributes might be either quantitative or categorical, which further adds to the heterogeneity and complexity of such a system. Furthermore, attributes might be symmetric or asymmetric in terms of higher or lower values [558]. For example, consider the price attribute in the home-buying example of Table 5.2. If a returned product has a lower price than the target value, then it is more easily acceptable than a case in which the returned product has a larger price than the target value. The precise level of asymmetry may be different for different attributes. For example, for an attribute, such as the camera resolution, the user might find larger resolutions more desirable, but the preference might not be quite as strong as in the case of the price. Other attributes might be completely symmetric, in which case the user would want the attribute value exactly at the target value t_i. An example of a symmetric metric is as follows:

$$Sim(t_i, x_i) = 1 - \frac{|t_i - x_i|}{max_i - min_i} \tag{5.3}$$

Here, max_i and min_i represent the maximum or minimum possible values of the attribute i. Alternatively, one might use the standard deviation σ_i (on historical data) to set the similarity function:

$$Sim(t_i, x_i) = \max\left\{0, 1 - \frac{|t_i - x_i|}{3 \cdot \sigma_i}\right\} \tag{5.4}$$

Note that in the case of the symmetric metric, the similarity is entirely defined by the difference between the two attributes. In the case of an asymmetric attribute, one can add an additional *asymmetric* reward, which kicks in depending on whether the target attribute value is smaller or larger. For the case of attributes in which larger values are better, an example of a possible similarity function is as follows:

$$Sim(t_i, x_i) = 1 - \frac{|t_i - x_i|}{max_i - min_i} + \underbrace{\alpha_i \cdot I(x_i > t_i) \cdot \frac{|t_i - x_i|}{max_i - min_i}}_{\text{Asymmetric reward}} \tag{5.5}$$

Here, $\alpha_i \geq 0$ is a user-defined parameter, and $I(x_i > t_i)$ is an indicator function that takes on the value of 1 if $x_i > t_i$, and 0 otherwise. Note that the reward kicks in only when the attribute value x_i (e.g., camera resolution) is greater than the target value t_i. For cases in which smaller values are better (e.g., price), the reward function is similar, except that smaller values are rewarded by the indicator function:

$$Sim(t_i, x_i) = 1 - \frac{|t_i - x_i|}{max_i - min_i} + \underbrace{\alpha_i \cdot I(x_i < t_i) \cdot \frac{|t_i - x_i|}{max_i - min_i}}_{\text{Asymmetric reward}} \tag{5.6}$$

The values of α_i are chosen in a highly domain-specific way. For values of $\alpha_i > 1$, the "similarity" actually increases with greater distance to the target. In such cases, it is helpful to think of $Sim(t_i, x_i)$ as a *utility* function rather than as a similarity function. For example, in the case of price, one would always prefer a lower price to a higher price, although the target price might define an inflection point in the strength with which one prefers a lower price to a higher price. When the value of α_i is exactly 1.0, it implies that one does not care about further change from the target value in one of the directions. An example might be the case of camera resolution, where one might not care about resolutions beyond a certain point. When $\alpha_i \in (0, 1)$, it implies that the user prefers a value at the target over all other values but she may have asymmetric preferences on either side of the target. For example, a user's preference for horsepower might strongly increase up to the target, and she might also have a mild aversion to a horsepower greater than the target because of greater fuel consumption. These examples suggest that there are no simple ways of pre-defining such similarity metrics; a lot of work needs to be done by the domain expert.

Examples of symmetric and asymmetric similarity functions are illustrated in Figure 5.6. The domain range is $[0, 10]$, and a target value of 6 is used. A symmetric similarity function is shown in Figure 5.6(a), where the similarity is linearly dependent on the distance from the target. However, in the horsepower example discussed above, the asymmetric similarity function of Figure 5.6(b) might be more appropriate, where $\alpha_i = 0.5$. For an attribute such as camera resolution, one might decide to allocate no utility beyond the user's target, as a result of which the similarity function might be flat beyond that point. Such a case is illustrated in Figure 5.6(c), where α_i is set to 1. Finally, in the case of price, smaller values are rewarded, although the user's target price might define an inflection point in the utility function. This case is illustrated in Figure 5.6(d), where the value of α_i is set to 1.3, with rewards being awarded for undershooting the target. This particular case is noteworthy, because the "similarity" is actually increasing with greater distance from the target as long as the value is as small as possible. In such cases, the utility interpretation of such functions makes a lot more sense than the similarity interpretation. In this interpretation, the target attribute values represent only key inflection points of the utility function.

For the case of categorical data, the determination of similarity values is often more challenging. Typically, domain hierarchies are constructed in order to determine the similarity values. Two objects that are closer to one another within the context of a domain hierarchy may be considered more similar. This domain hierarchy is sometimes directly available from sources such as the North American Industry Classification System (NAICS), and in other cases it needs to be directly constructed by hand. For example, an attribute such as the movie genre can be classified hierarchically, as shown in Figure 5.7. Note that related genres tend to be closer to one another in the hierarchy. For example, movies for children are considered to be so different from those for general audiences that they bifurcate at the root of the taxonomy. This hierarchy may be used by the domain expert to hand-code similarities. In some cases, learning methods can also be used to facilitate the similarity computation. For example, feedback could be elicited from users about pairs of genres, and learning methods could be used to learn the similarity between pairs of items [18]. The broader learning approach can also be used to determine other parameters of the similarity function, such as the value of α_i in Equations 5.5 and 5.6. It is noteworthy that the specific form of the similarity function may be different from that in Equations 5.5 and 5.6, depending on the data domain. It is here that the domain expert has to invest a significant amount of time in deciding how to model the specific problem setting. This investment is an inherent part of the domain-specific effort that knowledge-based recommender systems demand, and also derive their name from.

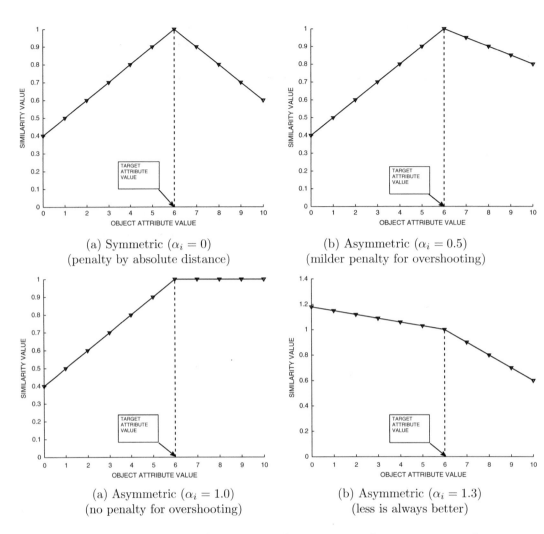

Figure 5.6: Examples of different types of symmetric and asymmetric similarity

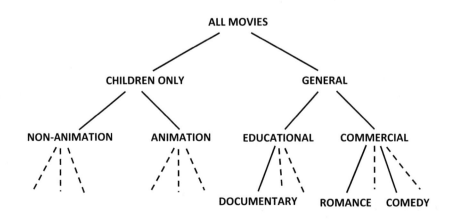

Figure 5.7: An example of hierarchical classification of movie genres

A second issue in the design of similarity functions is the determination of the relative importance of various attributes. The relative importance of the ith attributes is regulated by the parameter w_i in Equation 5.2. One possibility is for a domain expert to hand-code the values of w_i through trial and experience. The other possibility is to learn the values of w_i with user feedback. Pairs of target objects could be presented to users, and users might be asked to rate how similar these target objects are. This feedback can be used in conjunction with a linear regression model to determine the value of w_i. Linear regression models are discussed in detail in section 4.4.5 of Chapter 4, and their usage for similarity function learning is discussed in [18]. A number of other results [97, 163, 563, 627] discuss learning methods with user feedback in the specific context of recommender systems. Many of these methods, such as those in [627], show how feature weighting can be achieved with user feedback. The work in [563] elicits feedback from the user in terms of the relative *ordering* of the returned cases, and uses it to learn the relative feature weights. It is often easier for the user to specify relative orderings rather than to specify explicit similarity values for pairs of objects.

5.3.1.1 Incorporating Diversity in Similarity Computation

As case-based systems use item attributes to retrieve similar products, they face many of the same challenges as content-based systems in returning diverse results. In many cases, the results returned by case-based systems are all very similar. The problem with the lack of diversity is that if a user does not like the top-ranked result, she will often not like the other results, which are all very similar. For example, in the home buying application, it is possible for the recommendation system to return condominium units from the same complex under the same management. Clearly, this scenario reduces the true *choice* available to the user among the top ranked results.

Consider a scenario where it is desired to retrieve the top-k results matching a particular case. One possibility is to retrieve the top $b \cdot k$ results (for $b > 1$) and then randomly select k items from this list. This strategy is also referred to as the *bounded random selection strategy*. However, such a strategy does not seem to work very well in practice.

A more effective approach is the *bounded greedy selection strategy* [560]. In this strategy, we start with the top $b \cdot k$ cases similar to the target, and incrementally build a diverse set of k instances from these $b \cdot k$ cases. Therefore, we start with the empty set R and incrementally build it by adding instances from the base set of $b \cdot k$ cases. The first step is to create a *quality* metric that combines similarity and diversity. Assume without loss of generality that the similarity function $f(\overline{X}, \overline{Y})$ always maps to a value in $(0, 1)$. Then, the diversity $D(\overline{X}, \overline{Y})$ can be viewed as the distance between $\overline{X}$ and $\overline{Y}$:

$$D(\overline{X}, \overline{Y}) = 1 - f(\overline{X}, \overline{Y}) \tag{5.7}$$

Then, the average diversity between the candidate $\overline{X}$, and a set R of currently selected cases is defined as the average diversity between $\overline{X}$ and cases in R:

$$D^{avg}(\overline{X}, R) = \frac{\sum_{\overline{Y} \in R} D(\overline{X}, \overline{Y})}{|R|} \tag{5.8}$$

Then, for target $\overline{T}$, the overall quality $Q(\overline{T}, \overline{X}, R)$ is computed as follows:

$$Q(\overline{T}, \overline{X}, R) = f(\overline{T}, \overline{X}) \cdot D^{avg}(\overline{X}, R) \tag{5.9}$$

The case $\overline{X}$ with the greatest quality is incrementally added to the set R until the cardinality of the set R is k. This set is presented to the user. Refer to the bibliographic notes for other diversity enhancing techniques used in the literature.

5.3.2   Critiquing Methods

Critiques are motivated by the fact that users are often not in a position to state their requirements exactly in the initial query. In some complex domains, they might even find it difficult to translate their needs in a semantically meaningful way to the attribute values in the product domain. It is only after viewing the results of a query that a user might realize that she should have couched her query somewhat differently. Critiques are designed to provide the users this ability after the fact.

After the results have been presented to the users, feedback is typically enabled through the use of *critiques*. In many cases, the interfaces are designed to critique the most similar matching item, although it is technically possible for the user to critique any of the items on the retrieved list of k items. In critiques, the users specify *change* requests on one or more attributes of an item that they may like. For example, in the home-buying application of Figure 5.2, the user might like a particular house, but she may want the house in a different locality or with one more bedroom. Therefore, the user may specify the changes in the features of one of the items she likes. The user may specify a *directional critique* (e.g., "cheaper") or a *replacement critique* (e.g., "different color"). In such cases, examples that do not satisfy the user-specified critiques are eliminated, and examples similar to the user-preferred item (but satisfying the current sequence of critiques) are retrieved. When multiple critiques are specified in sequential recommendation cycles, preference is given to more recent critiques.

At a given moment in time, the user may specify either a single feature or a combination of features for modification. In this context, the critiques are of three different types, corresponding to *simple critiques*, *compound critiques*, and *dynamic critiques*. We will discuss each of these types of critiques in the following sections.

5.3.2.1   Simple Critiques

In a simple critique, the user specifies a single change to one of the features of a recommended item. In Figure 5.8, we have used our earlier case-based scenario (`critique-example.com`) to show an example of a simple critiquing interface. Note that the user can specify a change to only one of the features of the recommended house in this interface. Often, in many systems, such as *FindMe* systems, a more conversational interface is used, where users specify whether to increase or decrease a specific attribute value rather than explicitly modify one of the target attribute values. This is referred to as a *directional critique*. In such cases, the candidate list is simply pruned of those objects for which the critiqued attribute is on the wrong side of the user's stated preference. The advantage of such an approach is that the user is able to state her preference and navigate through the product space without having to specify or change attribute values in a precise way. Such an approach is particularly important in domains where the users might not know the exact value of the attribute to use (e.g., the horsepower of an engine). Another advantage of a directional critique is that it has a simple conversational style, which might be more intuitive and appealing to the user. In cases where the user does not find the current set of retrieved results to be useful at all, she also has the option of going back to the entry point. This represents a fruitless cycle through the critiquing process.

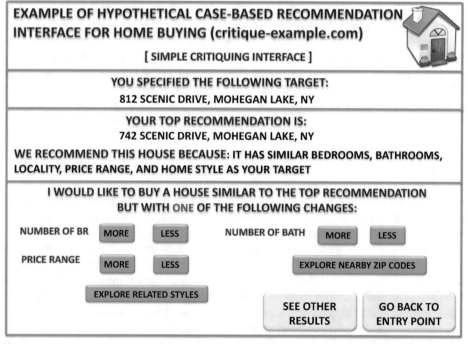

(a) Simple critiquing by directly modifying feature values

(b) The conversational style of directional critiques

Figure 5.8: Hypothetical examples of user interfaces for simple critiquing in a case-based recommender (critique-example.com)

The main problem with the simple critiquing approach is its laborious navigation. If the recommended product contains many features that are required to be changed, then it will lead to a longer chain of subsequent critiques. Furthermore, when one of the features is changed, the recommender system may automatically need to change at least some of the other feature values depending on item availability. In most cases, it is impossible to hold the other feature values at exactly constant values in a given cycle. As a result, when the user has changed a few features to their desired values, they may realize that the other feature values are no longer acceptable. The larger the number of recommendation cycles, the less the control that the user will have on changes in the other feature values that were acceptable in earlier iterations. This problem often results from the user's lack of understanding about the natural trade-offs in the problem domain. For example, a user might not understand the trade-off between horsepower and fuel efficiency and attempt to navigate to a car with high horsepower and also a high fuel efficiency of 50 miles to the gallon [121]. This problem of fruitlessness in long recommendation cycles is discussed in detail in [423]. The main problem in many critiquing interfaces is that the next set of recommended items are based on the *most recent* items being critiqued, and there is no way of navigating back to earlier items. As a result, a long cycle of simple critiques may sometimes drift to a fruitless conclusion.

5.3.2.2   Compound Critiques

Compound critiques were developed to reduce the length of recommendation cycles [414]. In this case, the user is able to specify multiple feature modifications in a single cycle. For example, the *Car Navigator* system [120] allows the user to specify multiple modifications, which are hidden behind informal descriptions that the user can understand (e.g., *classier, roomier, cheaper, sportier*). For example, the domain expert might encode the fact that "*classier*" suggests a certain subset of models with increased price and sophisticated interior structure. Of course, it is also possible for the user to modify the required product features directly, but it increases the burden on her. The point in conversational critiquing is that when a user might wish to have a "*classier*" car, but they might not be easily able to concretely express it in terms of the product features such as the interior structure of the car. On the other hand, a qualification such as "*classier*" is more intuitive, and it can be encoded in terms of the product features by a domain expert. This interactive process is designed to help them learn the complex product space in an intuitive way.

In the home-buying example of Table 5.2, the user might specify a different locality and change in the price in a single cycle. An example of a compound critiquing example for the home-buying example is illustrated in Figure 5.9(a). To make the approach more conversational, an interface like the one in Figure 5.9(b), will automatically encode multiple changes within a single selection. For example, if the user selects "*roomier*," it implies that both the number of bedrooms and the number of bathrooms might need to be increased. For the second type of interface, the domain expert has to expend significant effort in designing the relevant interface and the interpretation of user choices in terms of changes made to multiple product features. This encoding is static, and it is done up front.

The main advantage of compound critiquing is that the user can change multiple features in the target recommendation in order to issue a new query or prune the search results from the previous query. As a result, this approach allows large jumps through the product feature space, and the user often has better control over the critiquing process. This is useful for reducing the number of recommendation cycles and making the exploration process more efficient. It is, however, not clear whether compound critiques always help a user learn the

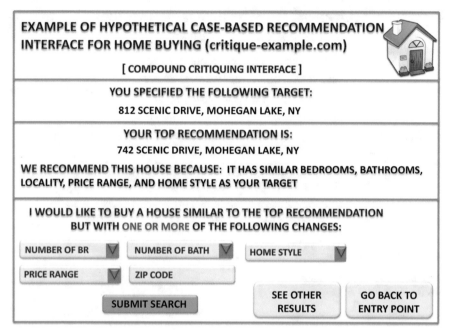

(a) Compound critiquing by modifying multiple feature values

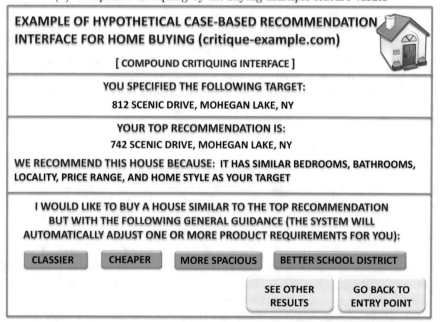

(b) Reducing the user's burden of specifying multiple features with domain knowledge

Figure 5.9: Hypothetical examples of user interfaces for compound critiquing in a case-based recommender (`critique-example.com`)

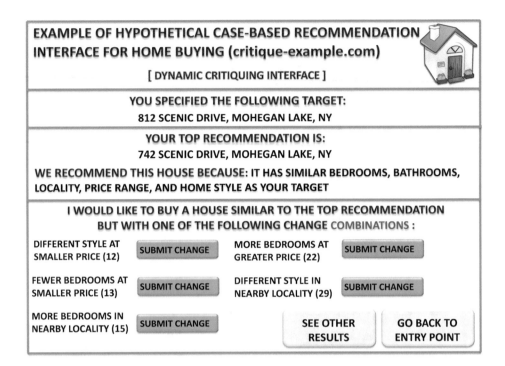

Figure 5.10: A hypothetical example of a user interface for dynamic critiquing in a case-based recommender (`critique-example.com`)

product space better than simple critiques; short critiquing cycles also reduce the likelihood of the user learning different trade-offs and correlations between features in the product space. On the other hand, a user may sometimes learn a lot about the product space by going through the slow and laborious process of simple critiquing.

5.3.2.3 Dynamic Critiques

Although compound critiques allow larger jumps through the navigation space, they do have the drawback that the critiquing options presented to the user are *static* in the sense that they do not depend on the retrieved results. For example, if the user is browsing cars, and she is already browsing the most expensive car with the largest horsepower possible, the option to increase the horsepower and the price will still be shown in the critiquing interface. Clearly, specifying these options will lead to a fruitless search. This is because users are often not fully aware of the inherent trade-offs in the complex product space.

In dynamic critiquing, the goal is to use data mining on the retrieved results to determine the most fruitful avenues of exploration and present them to the user. Thus, dynamic critiques are, by definition, compound critiques because they almost always represent combinations of changes presented to the user. The main difference is that only the subset of the most relevant *possibilities* are presented, based on the currently retrieved results. Therefore, dynamic critiques are designed to provide better guidance to the user during the search process.

An important aspect of dynamic critiquing is the ability to discover frequent combinations of product feature changes. The notion of *support* is adapted from frequent pattern mining [23] in order to determine patterns of frequently co-occurring product features in the retrieved results. The support of a pattern is defined as the fraction of the retrieved results that satisfy that pattern. Refer to Definition 3.3.1 in Chapter 3 for a formal definition of support. Therefore, this approach determines all the patterns of change that specify a pre-defined minimum support value. For example, in the home-buying example of Table 5.2, the system might determine the following dynamic critiques in order of support:

More Bedrooms, Greater Price: Support= 25%
More Bedrooms, More Bathrooms, Greater Price: Support= 20%
Fewer Bedrooms, Smaller Price: Support= 20%
More Bedrooms, Locality=Yonkers: Support= 15%

Note that conflicting options such as "*More Bedrooms, Smaller Price*" have a smaller chance of being included because they might be eliminated based on the minimum support criterion. However, low support patterns are not necessarily uninteresting. In fact, once all the patterns satisfying the minimum support threshold have been determined, many recommender systems order the critiques to the user in ascending order of support. The logic for this approach is that low support critiques are often less obvious patterns that can be used to eliminate a larger number of items from the candidate list. A hypothetical example of a dynamic critiquing interface, based on our earlier home-buying system (`critique-example.com`), is illustrated in Figure 5.10. Note that a numerical quantity is associated with each of the presented options in the interface. This number corresponds to the raw support of the presented options.

A real-world example of a dynamic critiquing approach that uses frequent pattern and association rule mining is the *Qwikshop* system discussed in [491]. An important observation about dynamic critiquing systems is that they increase the cognitive load on the user, when viewed on a *per-cycle basis*, but they reduce the cognitive load over the course of the *entire session* because of their ability to arrive at acceptable recommendations more quickly [416]. This is one of the reasons that the effective design of explanatory processes into the critiquing cycle is more important in dynamic critiquing systems.

5.3.3 Explanation in Critiques

It is always advisable to build explanatory power into the critiquing process, because it helps the user understand the information space better. There are several forms of explanation that are used to improve the quality of critiques. Some examples of such explanations are as follows:

1. In simple critiquing, it is common for a user to navigate in a fruitless way because of a lack of awareness of the inherent trade-offs in the product space. For example, a user might successively increase the horsepower, increase the mileage per gallon, and then try to reduce the desired price. In such cases, the system might not be able to show an acceptable result to the user, and the user will have to start the navigation process afresh. At the end of such a session, it is desirable for the system to automatically determine the nature of the trade-off that resulted in a fruitless session. It is often possible to determine such trade-offs with the use of correlation and co-occurrence statistics. The user can then be provided insights about the conflicts in the critiques entered by them in the previous session. Such an approach is used in some of the *FindMe* systems [121].

2. It has been shown in [492] how explanations can be used in conjunction with dynamic compound critiques during a session. For example, the *Qwikshop* system provides information about the fraction of the instances satisfying each compound critique. This provides the user with a clear idea of the size of the space they are about to explore *before* making a critiquing choice. Providing the user with better explanations *during* the session increases the likelihood that the session will be fruitful.

The main danger in critiquing-based systems is the likelihood of users meandering through the knowledge space in an aimless way without successfully finding what they are looking for. Adding explanations to the interface greatly reduces this likelihood.

5.4 Persistent Personalization in Knowledge-Based Systems

Although knowledge-based systems, such as constraint-based systems, allow the specification of user preferences, characteristics, and/or demographic attributes, the entered information is typically session-specific, and it is not *persistent* across sessions. The only persistent data in most such systems is the domain knowledge in the form of various system-specific databases, such as constraints or similarity metrics. This lack of persistent data is a natural consequence of how knowledge-based systems tend to use historical data only in a limited way compared to content-based and collaborative systems. This is also an advantage of knowledge-based systems, because they tend to suffer less from cold-start issues compared to other systems that are dependent on historical data. In fact, knowledge-based recommender systems are often designed for more expensive and occasionally bought items, which are highly customized. In such cases, historical data should be used with some caution, even when they are available. Nevertheless, a few knowledge-based systems have also been designed to use persistent forms of personalization.

The user's actions over various sessions can be used to build a persistent profile about the user regarding what they have liked or disliked. For example, *CASPER* is an online recruitment system [95] in which the user's actions on retrieved job postings, such as saving the advertisement, e-mailing it to themselves, or applying to the posting, are saved for future reference. Furthermore, users are allowed to negatively rate advertisements when they are irrelevant. Note that this process results in the building of an implicit feedback profile. The recommendation process is a two-step approach. In the first step, the results are retrieved based on the user requirements, as in the case of any knowledge-based recommender. Subsequently, the results are ranked based on similarity to previous profiles that the user has liked. It is also possible to include collaborative information by identifying other users with similar profiles, and using their session information in the learning process.

Many steps in knowledge-based systems can be personalized when user interaction data is available. These steps are as follows:

1. The learning of utility/similarity functions over various attributes can be personalized for both constraint-based recommenders (ranking phase) and in case-based recommenders (retrieval phase). When past feedback from a particular user is available, it is possible to learn the relative importance of various attributes for that user in the utility function.

2. The process of constraint suggestion (cf. section 5.2.5) for a user can be personalized if a significant number of sessions of that user are available.

3. Dynamic critiques for a user can be personalized if sufficient data are available from that user to determine relevant patterns. The only difference from the most common form of dynamic critiquing is that user-specific data are leveraged rather than all the data for determining the frequent patterns. It is also possible to include the sessions of users with similar sessions in the mining process to increase the collaborative power of the recommender.

Although there are many avenues through which personalization can be incorporated within the framework of knowledge-based recommendation, the biggest challenge is usually the unavailability of sufficient session data for a particular user. Knowledge-based systems are inherently designed for highly customized items in a complex domain space. This is the reason that the level of personalization is generally limited in knowledge-based domains.

5.5 Summary

Knowledge-based recommender systems are generally designed for domains in which the items are highly customized, and it is difficult for rating information to directly reflect greater preferences. In such cases, it is desirable to give the user greater control in the recommendation process through requirement specification and interactivity. Knowledge-based recommender systems can be either constraint-based systems, or they can be case-based systems. In constraint-based systems, users specify their requirements, which are combined with domain-specific rules to provide recommendations. Users can add constraints or relax constraints depending on the size of the results. In case-based systems, the users work with targets and candidate lists that are iteratively modified through the process of critiquing. For retrieval, domain-dependent similarity functions are used, which can also be learned. The modifications to the queries are achieved through the use of critiquing. Critiques can be simple, compound, or dynamic. Knowledge-based systems are largely based on user requirements, and they incorporate only a limited amount of historical data. Therefore, they are usually effective at handling cold-start issues. The drawback of this approach is that historical information is not used for "filling in the gaps." In recent years, methods have also been designed for incorporating a greater amount of personalization with the use of historical information from user sessions.

5.6 Bibliographic Notes

Surveys on various knowledge-based recommender systems and preference elicitation methods may be found in [197, 417]. Case-based recommender systems are reviewed in [102, 116, 377, 558]. Surveys of preference elicitation methods and critiquing may be found in [148, 149]. Constraint-based recommender systems are discussed in [196, 197]. Historically, constraint-based recommendation systems were proposed much later than case-based recommenders. In fact, the original paper by Burke [116] on knowledge-based recommender systems mostly describes case-based recommenders. However, some aspects of constraint-based recommenders are also described in this work. Methods for learning utility functions in the context of constraint-based recommender systems are discussed in [155, 531]. Methods for handing empty results in constraint-based systems, such as fast discovery of small conflicting sets, and minimal relaxations are discussed in [198, 199, 273, 274, 289, 419, 574]. These works also discuss how these conflicting sets may be used to provide explanations and repair diagnoses of the user queries. Popularity-based methods for selecting the next constraint attribute are discussed in [196, 389]. The selection of default values for the attribute constraints is discussed in [483]. A well-known constraint-based recommender system is the *VITA* recommender [201], which was built on the basis of the *CWAdvisor* system [200].

The problem of similarity function learning for case-based recommenders is discussed in [18, 97, 163, 563, 627]. The work in [563] is notable in that learns weights of various features for similarity computation. Reinforcement learning methods for learning similarity functions for case-based systems are discussed in [288, 506]. The bounded random selection and bounded greedy selection strategies for increasing the diversity of case-based recommender systems are discussed in [560]. The work in [550] also combines similarity with diversity like the bounded greedy approach, but it applies only diversity on the retrieved set of $b \cdot k$ cases, rather than creating a quality metric combining similarity and diversity. The notions of *similarity layers* and *similarity intervals* for diversity enhancement are discussed in [420]. A compromise-driven approach for diversity enhancement is discussed in [421]. The power of order-based retrieval for similarity diversification is discussed in [101]. Experimental results [94, 560] show the advantages of incorporating diversity into recommender systems. The issue of critiquing in case-based recommender systems is discussed in detail in [417, 422, 423]. Compound critiques were first discussed in [120], although the term was first coined in [414]. A comparative study of various compound critiquing techniques may be found in [664]. The use of explanations in compound critiques is discussed in [492].

The earliest case-based recommenders were proposed in [120, 121] in the context of the *Entree* restaurant recommender. The earliest forms of these systems were also referred to as *FindMe* systems [121], which were shown to be applicable to a wide variety of domains. The Wasabi personal shopper is a case-based recommender system and is discussed in [125]. Case-based systems have been used for travel advisory services [507], online recruitment systems [95], car sales (*Car Navigator*) [120], video sales (*Video Navigator*) [121], movies (*Pick A Flick*) [121], digital camera recommendations (e.g., *Qwikshop*) [279, 491], and rental property accommodation [263].

Most knowledge-based systems leverage user requirements and preferences, as specified in a single session. Therefore, if a different user enters the same input, they will obtain exactly the same result. Although such an approach provides better control to the user, and also does not suffer from cold-start issues, it tends to ignore historical data when they are available. Recent years have also witnessed an increase in long-term and persistent information about the user in knowledge-based recommender systems [95, 454, 558]. An example of such a system is the *CASPER* online recruitment system [95], which builds persistent user profiles for future recommendation. A personalized travel recommendation system with the use of user profiles is discussed in [170]. The sessions of similar users are leveraged for personalized travel recommendations in [507]. Such an approach not only leverages the target user's behavior but also the collaborative information available in a community of users. The work in [641] uses the critiquing information over multiple sessions in a collaborative way to build user profiles. Another relevant work is the *MAUT* approach [665], which is based on multi-attribute utility theory. This approach learns a utility preference function for each user based on their critiques in the previous sessions. Another example of persistent data that can be effectively used in such systems is demographic information. Although demographic recommender systems vary widely in their usage [117, 320], some of the demographic systems can also be considered knowledge-based systems, when profile association rules are used to interactively suggest preferences to users in an online fashion [31, 32]. Such systems allow progressive refinement of the queries in order to derive the most appropriate set of rules for a particular demographic group. Similarly, various types of utility-based recommendation and ranking techniques are used within the context of knowledge-based systems [74].

5.7 Exercises

1. Implement an algorithm to determine whether a set of customer-specified requirements and a set of rules in a knowledge base will retrieve an empty set from a product catalog. Assume that the antecedent and the consequent of each rule both contain a single constraint on the product features. Constraints on numerical attributes are in the form of inequalities (e.g., *Price* $\leq$ 30), whereas constraints on categorical attributes are in the form of unit instantiations (e.g., *Color=Blue*). Furthermore, customer requirements are also expressed as similar constraints in the feature space.

2. Suppose you had data containing information about the utility values of a particular customer for a large set of items in a particular domain (e.g., cars). Assume that the utility value of the jth product is u_j ($j \in \{1 \ldots n\}$). The items are described by a set of d numerical features. Discuss how you will use these data to rank other items in the same product domain for this customer.

Chapter 6

Ensemble-Based and Hybrid Recommender Systems

"What's better, a poetic intuition or an intellectual work? I think they complement each other."– Manuel Puig

6.1 Introduction

In the previous chapters, we discussed three different classes of recommendation methods. Collaborative methods use the ratings of a *community* of users in order to make recommendations, whereas content-based methods use the ratings of a *single* user in conjunction with attribute-centric item descriptions to make recommendations. Knowledge-based methods require the explicit specification of user requirements to make recommendations, and they do not require any historical ratings at all. Therefore, these methods use different sources of data, and they have different strengths and weaknesses. For example, knowledge-based systems can address cold-start issues much better than either content-based or collaborative systems because they do not require ratings. On the other hand, they are weaker than content-based and collaborative systems in terms of using *persistent personalization* from historical data. If a different user enters the same requirements and data in a knowledge-based interactive interface, she might obtain exactly the same result.

All these models seem rather restrictive in isolation, especially when multiple sources of data are available. In general, one would like to make use of all the knowledge available in different data sources and also use the algorithmic power of various recommender systems to make robust inferences. Hybrid recommender systems have been designed to explore these possibilities. There are three primary ways of creating hybrid recommender systems:

C.C. Aggarwal, *Recommender Systems: The Textbook*,
DOI 10.1007/978-3-319-29659-3_6

1. *Ensemble design:* In this design, results from off-the-shelf algorithms are combined into a single and more robust output. For example, one might combine the rating outputs from a content-based and a collaborative recommender into a single output. A significant variation exists in terms of the specific methodologies used for the combination process. The basic principle at work is not very different from the design of ensemble methods in many data mining applications such as clustering, classification, and outlier analysis.

   Ensemble design can be formalized as follows. Let $\hat{R}_k$ be an $m \times n$ matrix containing the *predictions* of the m users for the n items by the kth algorithm, where $k \in \{1 \ldots q\}$. Therefore, a total of q different algorithms are used to arrive at these predictions. The (u, j)th entry of $\hat{R}_k$ contains the predicted rating of user u for item j by the kth algorithm. Note that the observed entries of the original ratings matrix R are replicated in each $\hat{R}_k$, and only the unobserved entries of R vary in different $\hat{R}_k$ because of the different predictions of different algorithms. The final result of the algorithm is obtained by combining the predictions $\hat{R}_1 \ldots \hat{R}_q$ into a single output. This combination can be performed in various ways, such as the computation of the weighted average of the various predictions. Furthermore, in some *sequential* ensemble algorithms, the prediction $\hat{R}_k$ may depend on the results of the previous component $\hat{R}_{k-1}$. In yet other cases, the outputs may not be directly combined. Rather, the output of one system is used as an input to the next as a set of content *features*. The common characteristics of all these systems are that (a) they use *existing* recommenders in *off-the-shelf* fashion, and (b) they produce a unified score or ranking.

2. *Monolithic design:* In this case, an integrated recommendation algorithm is created by using various data types. A clear distinction may sometimes not exist between the various parts (e.g., content and collaborative) of the algorithm. In other cases, existing collaborative or content-based recommendation algorithms may need to be modified to be used within the overall approach, even when there are clear distinctions between the content-based and collaborative stages. Therefore, this approach tends to integrate the various data sources more tightly, and one cannot easily view individual components as off-the-shelf black-boxes.

3. *Mixed systems:* Like ensembles, these systems use multiple recommendation algorithms as black-boxes, but the items recommended by the various systems are presented together side by side. For example, the television program for a whole day is a composite entity containing multiple items. It is meaningless to view the recommendation of a single item in isolation; rather, it is the combination of the items that creates the recommendation.

Therefore, the term "hybrid system" is used in a broader context than the term "ensemble system." All ensemble systems are, by definition, hybrid systems, but the converse is not necessarily true.

Although hybrid recommender systems usually combine the power of different types of recommenders (e.g., content- and knowledge-based), there is no reason why such systems cannot combine models of the same type. Since content-based models are essentially text classifiers, it is well known that a wide variety of ensemble models exist to improve the accuracy of classification. Therefore, any classification-based ensemble system can be used to improve the effectiveness of content-based models. This argument also extends to collaborative recommender models. For example, one can easily combine the predicted results of

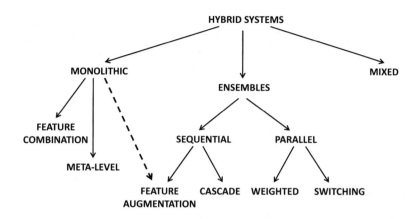

Figure 6.1: The taxonomy of hybrid systems

a latent factor model with those of a neighborhood model to obtain more accurate recommendations [266]. In fact, both[1] the winning entries in the Netflix Prize contest, referred to as "*Bellkor's Pragmatic Chaos*" [311] and "*The Ensemble*" [704], were ensemble systems.

At a broader level, hybrid recommender systems are closely related to the field of ensemble analysis in classification. For example, collaborative models are generalizations of classification models, as discussed in the introduction to Chapter 3. As we will discuss in section 6.2 of this chapter, the theoretical underpinnings of ensemble analysis in classification are similar to those in collaborative filtering. Therefore, this chapter will also focus on how the recommendation approach can be used to improve the effectiveness of collaborative recommender systems in much the same way as one might use ensembles in the field of data classification.

According to Burke [117], hybrid recommender systems can be classified into the following categories:

1. *Weighted:* In this case, the scores of several recommender systems are combined into a single unified score by computing the weighted aggregates of the scores from individual ensemble components. The methodology for weighting the components may be heuristic, or it might use formal statistical models.

2. *Switching:* The algorithm switches between various recommender systems depending on current needs. For example, in earlier phases, one might use a knowledge-based recommender system to avoid cold-start issues. In later phases, when more ratings are available, one might use a content-based or collaborative recommender. Alternatively, the system might adaptively select the specific recommender that provides the most accurate recommendation at a given point in time.

3. *Cascade:* In this case, one recommender system refines the recommendations given by another. In generalized forms of cascades, such as *boosting*, the training process of one recommender system is *biased* by the output of the previous one, and the overall results are combined into a single output.

---

[1]Both entries were tied on the error rate. The award was given to the former because it was submitted 20 minutes earlier.

4. *Feature augmentation:* The output of one recommender system is used to create input features for the next. While the cascade hybrid successively refines the recommendations of the previous system, the feature augmentation approach treats then as features as *input* for the next system. This approach shares a number of intuitive similarities with the notion of *stacking*, which is commonly used in classification. In stacking, the outputs of one classifier are used as features for the next. Because the different recommenders are (generally) used as off-the-shelf black-boxes, the approach is still an ensemble method (in most cases) rather than a monolithic method.

5. *Feature combination:* In this case, the features from different data sources are combined and used in the context of a single recommender system. This approach can be viewed as a monolithic system, and therefore it is not an ensemble method.

6. *Meta-level:* The model used by one recommender system is used as input to another system. The typical combination used is that of a content-based and collaborative system. The collaborative system is modified to use the content features to determine peer groups. Then, the ratings matrix is used in conjunction with this peer group to make predictions. Note that this approach needs to modify the collaborative system to use a content matrix for finding peer groups, although the final predictions are still performed with the ratings matrix. Therefore, the collaborative system needs to be modified, and one cannot use it in an off-the-shelf fashion. This makes the meta-level approach a monolithic system rather than an ensemble system. Some of these methods are also referred to as "collaboration via content" because of the way in which they combine collaborative and content information.

7. *Mixed:* Recommendations from several engines are *presented* to the user at the same time. Strictly speaking, this approach is not an ensemble system, because it does not explicitly combine the scores (of a particular item) from the various components. Furthermore, this approach is often used when the recommendation is a *composite* entity in which multiple items can be recommended as a related set. For example, a composite television program can be constructed from the various recommended items [559]. Therefore, this approach is quite different from all the aforementioned methods. On the one hand, it does use other recommenders as black-boxes (like ensembles), but it does not combine the predicted ratings of the same item from different recommenders. Therefore, mixed recommenders cannot be viewed either as monolithic or ensemble-based methods and are classified into a distinct category of their own. The approach is most relevant in complex item domains, and it is often used in conjunction with knowledge-based recommender systems.

The first four of the aforementioned categories are ensemble systems, the next two are monolithic systems, and the last one is a mixed system. The last category of mixed systems cannot be neatly categorized either as a monolithic or an ensemble system, because it presents multiple recommendations as a composite entity. A hierarchical categorization of these various types of systems is shown in Figure 6.1. Although we have used the higher level categorization of parallel and sequential[2] systems, as introduced by [275], we emphasize that our categorization of Burke's original set of six categories is slightly different from that of [275]. Unlike the taxonomy in [275], which classifies meta-level systems as sequential methods, we view meta-level systems as monolithic because one cannot use off-the-shelf recommendation algorithms, as in the case of a true ensemble. Similarly, the work in [275]

---

[2]This is also referred to as a *pipelined* system [275].

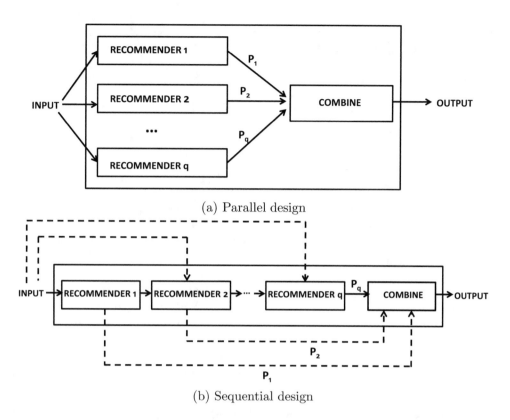

(a) Parallel design

(b) Sequential design

Figure 6.2: Parallel and sequential ensembles

views feature augmentation hybrids as monolithic systems. Although the individual recommenders are combined together in a more complex way in feature augmentation hybrids, the individual recommenders are still used as off-the-shelf black-boxes for the large part. This is the primary distinguishing characteristic of an ensemble system from a monolithic system, and the approach is highly reminiscent of stacking methods in classification. Therefore, we view feature augmentation hybrids as ensemble systems rather than monolithic systems. However, in some cases of feature augmentation hybrids, minor changes are required to the off-the-shelf recommender. In such cases, these systems may be technically considered to have a monolithic design. We have shown this possibility with a dotted line in Figure 6.1.

Aside from the monolithic and mixed methods, which are not truly ensembles, one can view all ensemble methods as having either sequential or parallel designs [275]. In parallel designs, the various recommenders function independently of one another, and the predictions of the individual recommenders are combined at the very end. The weighted and switching methods can be viewed as parallel designs. In sequential designs, the output of one recommender is used as an input to the other. The cascade and meta-level systems can be viewed as examples of sequential methods. A pictorial illustration of the combination process in sequential and parallel systems is shown in Figure 6.2. In this chapter, we will provide a detailed discussion of several recommender systems in each of these categories, although we will use Burke's lower-level taxonomy [117] to organize the discussion.

This chapter is organized as follows. In section 6.2, we discuss the classification perspective of ensemble-based recommender systems. We also explore the level to which the existing theories and methodologies for ensemble methods in the field of classification also

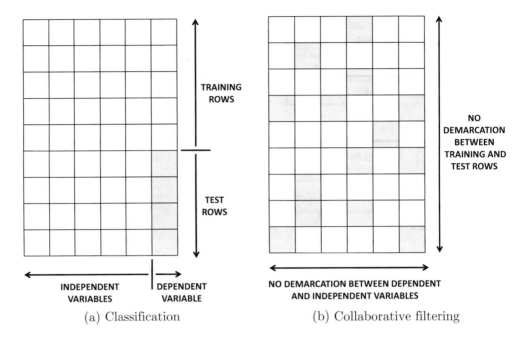

(a) Classification (b) Collaborative filtering

Figure 6.3: Revisiting Figure 1.4 of Chapter 1. Comparing the traditional classification problem with collaborative filtering. Shaded entries are missing and need to be predicted.

apply to recommender systems. In section 6.3, a number of different examples of weighted hybrids are discussed. In section 6.4, a number of switching hybrids are discussed. Cascade hybrids are discussed in section 6.5, whereas feature augmentation hybrids are discussed in section 6.6. Meta-level hybrids are discussed in section 6.7. Feature combination methods are introduced in section 6.8. Mixed systems are discussed in section 6.9. A summary is given in section 6.10.

6.2 Ensemble Methods from the Classification Perspective

Ensemble methods are commonly used in the field of data classification to improve the robustness of learning algorithms. As we will discuss below, much of this theory also applies to various forms of recommender systems. For example, content-based recommender systems are often straightforward applications of text classification algorithms. Therefore, a direct application of existing ensemble methods in classification is usually sufficient to obtain high-quality results.

As discussed in Chapter 1, collaborative filtering is a generalization of the problem of data classification. We have replicated Figure 1.4 of Chapter 1 in Figure 6.3 to illustrate the relationship between the two problems. It is evident from Figure 6.3(a) that the feature variables and class variable are clearly demarcated in classification. The main distinguishing features of collaborative filtering from classification are that the feature variables and the class variable are not clearly demarcated in the former and that the missing entries may occur in any column or row. The fact that missing entries may occur in any row implies that training and test instances are not clearly demarcated, either. A salient question arises as to whether the bias-variance theory developed in the field of classification [242] also applies to recommender systems. Repeated experiments [266, 311] have shown that combining multiple collaborative recommender systems often leads to more accurate results. This is because the bias-variance theory, which is designed for classification, also applies to the collaborative filtering scenario. This means that many traditional ensemble techniques from classification can also be generalized to collaborative filtering. Nevertheless, because of the fact that the missing entries might occur in any row or column of the data, it is sometimes algorithmically challenging to generalize the ensemble algorithms for classification to collaborative filtering.

We first introduce the bias-variance trade-off as it applies to the field of data classification. Consider a simplified classification or regression model, in which a specific field needs to be predicted, as shown in Figure 6.3(a). It can be shown that the error of a classifier in predicting the dependent variable can be decomposed into three components:

1. *Bias:* Every classifier makes its own modeling assumptions about the nature of the decision boundary between classes. For example, a linear SVM classifier assumes that the two classes may be separated by a linear decision boundary. This is, of course, not true in practice. In other words, any given linear support vector machine will have an inherent *bias.* When a classifier has high bias, it will make *consistently incorrect* predictions over particular choices of test instances near the incorrectly modeled decision-boundary, even when different samples of the training data are used for the learning process.

2. *Variance:* Random variations in the choices of the training data will lead to different models. As a result, the dependent variable for a test instance might be inconsistently predicted by different choices of training data sets. Model variance is closely related to overfitting. When a classifier has an overfitting tendency, it will make *inconsistent* predictions for the same test instance over different training data sets.

3. *Noise:* The noise refers to the intrinsic errors in the target class labeling. Because this is an intrinsic aspect of data quality, there is little that one can do to correct it. Therefore, the focus of ensemble analysis is generally on reducing bias and variance.

The expected mean-squared error of a classifier over a set of test instances can be shown to be sum of the bias, variance, and noise. This relationship can be stated as follows:

$$\text{Error} = \text{Bias}^2 + \text{Variance} + \text{Noise} \tag{6.1}$$

It is noteworthy that by reducing either the bias or variance components, one can reduce the overall error of a classifier. For example, classification ensemble methods such as *bagging* [99] reduce the variance, whereas methods such as *boosting* [206] can reduce the bias. It is noteworthy that the only difference between classification and collaborative filtering is that missing entries can occur in any column rather than only in the class variable. Nevertheless, the bias-variance result still holds when applied to the problem of predicting a specific column, whether the other columns are incompletely specified or not. This means that the basic principles of ensemble analysis in classification are also valid for collaborative filtering. Indeed, as we will see later in this chapter, many classical ensemble methods in classification, such as bagging and boosting, have also been adapted to collaborative filtering.

6.3   Weighted Hybrids

Let $R = [r_{uj}]$ be an $m \times n$ ratings matrix. In weighted hybrids, the outputs of various recommender systems are combined using a set of weights. Let $\hat{R}_1 \ldots \hat{R}_q$ be the $m \times n$ *completely specified* ratings matrices, in which the unobserved entries of R are predicted by q different algorithms. Note that the entries r_{uj} that are already observed in the original $m \times n$ ratings matrix R are already fixed to their observed values in each prediction matrix $\hat{R}_k$. Then, for a set of weights $\alpha_1 \ldots \alpha_q$, the weighted hybrid creates a combined prediction matrix $\hat{R} = [\hat{r}_{uj}]$ as follows:

$$\hat{R} = \sum_{i=1}^{q} \alpha_i \hat{R}_i \tag{6.2}$$

In the simplest case, it is possible to choose $\alpha_1 = \alpha_2 = \ldots = \alpha_q = 1/q$. However, it is ideally desired to weight the various systems in a differential way, so as to give greater importance to the more accurate systems. A number of methods exist for such differential weighting. One can also write the aforementioned equation in terms of individual entries of the matrix:

$$\hat{r}_{uj} = \sum_{i=1}^{q} \alpha_i \hat{r}_{uj}^i \tag{6.3}$$

Here $\hat{r}_{uj}^i$ denotes the prediction of the ith ensemble component for user u and item j and $\hat{r}_{uj}$ denotes the final prediction.

In order to determine the optimal weights, it is necessary to be able to evaluate the effectiveness of a particular combination of weights $\alpha_1 \ldots \alpha_q$. While this topic will discussed in more detail in Chapter 7, we will provide a simple evaluation approach here for the purpose of discussion. A simple approach is to hold out a small fraction (e.g., 25%) of the known entries in the $m \times n$ ratings matrix $R = [r_{uj}]$ and create the prediction matrices $\hat{R}_1 \ldots \hat{R}_q$ by applying the q different base algorithms on the remaining 75% of the entries in R. The resulting predictions $\hat{R}_1 \ldots \hat{R}_q$ are then combined to create the ensemble-based prediction $\hat{R}$ according to Equation 6.2. Let the user-item indices (u, j) of these held-out entries be denoted by H. Then, for a given vector $\overline{\alpha} = (\alpha_1 \ldots \alpha_q)$ of weights, the effectiveness of a particular scheme can be evaluated using either the mean-squared error (MSE) or the mean absolute error (MAE) of the predicted matrix $\hat{R} = [\hat{r}_{uj}]_{m \times n}$ over the held-out ratings in H:

$$MSE(\overline{\alpha}) = \frac{\sum_{(u,j) \in H} (\hat{r}_{uj} - r_{uj})^2}{|H|}$$

$$MAE(\overline{\alpha}) = \frac{\sum_{(u,j) \in H} |(\hat{r}_{uj} - r_{uj})|}{|H|}$$

These metrics provide an evaluation of a particular combination of coefficients $\alpha_1 \ldots \alpha_q$. How can we determine the optimal values of $\alpha_1 \ldots \alpha_q$ to minimize these metrics? A simple approach, which works well for the case of MSE, is to use linear regression. It is assumed that the ratings in the held-out set H provide the ground truth values of the dependent variable, and the parameters $\alpha_1 \ldots \alpha_q$ are the independent variables. The idea is to select the independent variables so that the mean-squared error of the linear combination is minimized with respect to the known ratings in the held-out set. The basics of the linear regression model are discussed in section 4.4.5 of Chapter 4, albeit in a different context. The main difference here is in terms of how the dependent and independent variables are defined and in terms of how the linear regression problem is formulated. In this case, the independent variables correspond to the rating predictions of various models for the entry (u, j), and the dependent variable corresponds to the value of each predicted rating $\hat{r}_{uj}$ of the ensemble combination in the held-out set H. Therefore, each observed rating in the held-out set provides a training example for the linear regression model. The regression coefficients correspond to the weights of various component models, and they need to be learned from the (held-out) training examples. After the weights have been learned using linear regression, the individual component models are retrained on the entire training set without any held-out entries. The weights, which were learned using the held-out entries, are used in conjunction with these q models. It is important not to forget this final step in order to ensure that the maximum learning is obtained from all the information available in the ratings. The linear regression approach to model combination is discussed in [266]. A related approach, which can make good use of all the knowledge in the training data, is that of *cross-validation*. Cross-validation methods are discussed in Chapter 7.

Although many systems simply average the results of multiple models, the use of regression is important to ensure that the various models are weighted appropriately. Such regression-based algorithms were included among many of the highly performing entries in the Netflix Prize contest [311, 554], and they are closely related to the concept of stacking in data classification.

The linear regression approach is, however, sensitive to presence of noise and outliers. This is because the squared error function is overly influenced by the largest errors in the data. A variety of *robust regression* methods are available, which are more resistant to the presence of noise and outliers. One such method uses the mean absolute error (MAE) as the objective function as opposed to the mean-squared error. The MAE is well known to be more robust to noise and outliers because it does not overemphasize large errors. A common approach is to use gradient descent method to determine the optimal value of the parameter vector $(\alpha_1 \ldots \alpha_q)$ of Equation 6.3. The algorithm starts by setting $\alpha_1 = \alpha_2 = \ldots = \alpha_q = 1/q$. Subsequently, the gradient is computed over the held-out entries in H as follows:

$$\frac{\partial MAE(\overline{\alpha})}{\partial \alpha_i} = \frac{\sum_{(u,j) \in H} \frac{\partial |(\hat{r}_{uj} - r_{uj})|}{\partial \alpha_i}}{|H|} \tag{6.4}$$

The value of $\hat{r}_{uj}$ can be expanded using Equation 6.3, and the partial derivative may be simplified in terms of the ratings of individual ensemble components as follows:

$$\frac{\partial MAE(\overline{\alpha})}{\partial \alpha_i} = \frac{\sum_{(u,j) \in H} \text{sign}(\hat{r}_{uj} - r_{uj})\hat{r}_{uj}^i}{|H|} \tag{6.5}$$

The gradient can be written in terms of the individual partial derivatives:

$$\overline{\nabla MAE} = \left(\frac{\partial MAE(\overline{\alpha})}{\partial \alpha_1} \ldots \frac{\partial MAE(\overline{\alpha})}{\partial \alpha_q} \right)$$

This gradient is then used to descend through the parameter space $\overline{\alpha}$ with an iterative gradient descent approach as follows:

1. Initialize $\overline{\alpha}^{(0)} = (1/q \ldots 1/q)$ and $t = 0$.

2. **Iterative Step 1:** Update $\overline{\alpha}^{(t+1)} \Leftarrow \overline{\alpha}^{(t)} - \gamma \cdot \overline{\nabla MAE}$. The value of $\gamma > 0$ can be determined using a line search so that the maximum improvement in MAE is achieved.

3. **Iterative Step 2:** Update the iteration index as $t \Leftarrow t + 1$.

4. **Iterative Step 3 (convergence check):** If MAE has improved by at least a minimum amount since the last iteration, then go to iterative step 1.

5. Report $\overline{\alpha}^{(t)}$.

Regularization can be added to prevent overfitting. It is also possible to add other constraints on the various values of α_i such as non-negativity or ensuring that they sum to 1. Such natural constraints improve generalizability to unseen entries. The gradient descent equations can be modified relatively easily to respect these constraints. After the optimal weights have been determined, all ensemble models are retrained on the entire ratings matrix without any held-out entries. The predictions of these models are combined with the use of the weight vector discovered by the iterative approach.

There are other ways of performing parameter searches. A simpler approach is to try several judiciously chosen combinations of parameters on a held-out set of ratings. For example, one might tune the various values of α_i in succession by trying different values and holding the others constant. Such an approach is generally applied to various types of parameter tuning [311], and it can often provide reasonably accurate results. Examples of various search techniques are provided in [162, 659].

These methods can be enhanced further with different types of meta-level content features [65, 66, 554]. These methods are discussed in section 6.8.2. Many of the existing ensemble methods do not use these sophisticated combination schemes. Often, these techniques use a simple average of the predictions of different components. It is particularly important to weight the different components when the predicted utility values are on different scales, or when some of the ensemble components are much more accurate than others. In the following, we will provide specific examples of how different types of models are often combined.

6.3.1   Various Types of Model Combinations

In weighted model combinations, a variety of recommendation engines can be combined. There are typically two forms of model combinations:

1. *Homogeneous data type and model classes:* In this case, different models are applied on the same data. For example, one might apply various collaborative filtering engines such as neighborhood-based methods, *SVD*, and Bayes techniques on a ratings matrix. The results are then aggregated into a single predicted value. Such an approach is robust because it avoids the specific bias of particular algorithms on a given data set even though all the constituent models belong to the same class (e.g., collaborative methods). An example of such a blend is provided in [266]. It was shown in [637], how the combination of an ensemble of three different matrix factorization methods can provide high-quality results. In particular, regularized matrix factorization, non-negative matrix factorization, and maximum margin factorization were used as the ensemble components, and the corresponding results were averaged. An interesting *fusion ensemble*, discussed in [67], uses the same recommendation algorithm for various ensemble components, but with different choices of parameters or algorithmic design choices. For example, different numbers of latent factors may be used in an *SVD* algorithm, different numbers of nearest neighbors may be used in a neighborhood-based algorithm, or the choice of the similarity metric may be varied. A simple average

of the predictions of various systems is used. As shown in [67], this simple approach almost always improved the performance of the base model. An earlier variation of this approach [180] uses ensembles of maximum margin matrix factorization methods but with different parameter settings. The work in [338] combines a user-based and item-based neighborhood algorithm.

2. *Heterogeneous data type and model classes:* In this cases, different *classes of* models are applied to different data sources. For example, one component of the model might be a collaborative recommender that uses a ratings matrix, whereas another component of the model might be a content-based recommender. This approach essentially fuses the power of multiple data sources into the combination process. The idea is to leverage the complementary knowledge in the various data sources in order to provide the most accurate recommendations. For example, the work in [659] combines a collaborative and knowledge-based recommender, whereas the work in [162] combines a content-based and collaborative recommender. When working with different data types, it becomes particularly important to carefully weight the predictions of various ensemble components.

These different forms of models provide excellent flexibility in exploring several types of model combinations.

6.3.2 Adapting Bagging from Classification

As discussed earlier in this chapter, the theoretical results on the bias-variance trade-off also hold for the collaborative filtering problem, because the latter problem is a direct generalization of classification. One of the common weighted combination techniques used in the classification problem is that of bagging. Therefore, this method can be used in collaborative filtering as well. However, the bagging approach needs to be slightly modified in order to adjust for the fact that the collaborative filtering problem is formulated somewhat differently from that of classification. First, we discuss bagging in the context of classification.

The basic idea in bagging is to reduce the variance component of the error in classification. In bagging, q training data sets are created with *bootstrapped sampling*. In bootstrapped sampling, rows of the data matrix are sampled *with replacement* in order to create a new training data set of the same size as the original training data set. This new training data set typically contains many duplicates. Furthermore, it can be shown that the expected fraction of rows from the original data matrix that is not represented in a given bootstrapped sample is given by $1/e$, where e is the base of the natural logarithm. A total of q training models are created with each of the sampled training data sets. For a given test instance, the average prediction from these q models is reported. Bagging generally improves the classification accuracy because it reduces the variance component of the error. A particular variant of bagging, known as *subagging* [111, 112], *subsamples* the rows, rather than sampling with replacement. For example, one can simply use all the distinct rows in a bootstrapped sample for training the models. The bagging and subagging methods can be generalized to collaborative filtering as follows:

1. *Row-wise bootstrapping:* In this case, the rows of the ratings matrix R are sampled with replacement to create a new ratings matrix of the same dimensions. A total of q such ratings matrices $R_1 \ldots R_q$ are thus created. Note that rows may be duplicated in the process of sampling, although they are treated as separate rows. An existing collaborative filtering algorithm (e.g., latent factor model) is then applied to each of

the q training data sets. For each training data set, an item rating can be predicted for a user only if that user is represented at least once in the matrix. In such a case, the predicted rating from that ensemble component is the average rating[3] of that item over the duplicate occurrences of that user. The predicted rating is then averaged over all the ensemble components in which that user is present. Note that for reasonably large values of q, each user will typically be present in at least one ensemble component with a high probability value of $1 - (1/e)^q$. Therefore, all users will be represented with high probability.

2. *Row-wise subsampling:* This approach is similar to row-wise bootstrapping, except that the rows are sampled without replacement. The fraction f of rows sampled is chosen randomly from $(0.1, 0.5)$. The number of ensemble components q should be significantly greater than 10 to ensure that all rows are represented. The main problem with this approach is that it is difficult to predict all the entries in this setting, and therefore one has to average over a smaller number of components. Therefore, the benefits of variance reduction are not fully achieved.

3. *Entry-wise bagging:* In this case, the *entries* of the original ratings matrix are sampled with replacement to create the q different ratings matrices $R_1 \ldots R_q$. Because many entries may be sampled repeatedly, the entries are now associated with weights. Therefore, a base collaborative filtering algorithm is required that can handle entries with weights. Such algorithms are discussed in section 6.5.2.1. As in the case of row-wise bagging, the predicted ratings are averaged over the various ensemble components.

4. *Entry-wise subsampling:* In entry-wise subsampling, a fraction of the entries are retained at random from the ratings matrix R to create a sampled training data set. Typically, a value of f is sampled from $(0.1, 0.5)$, and then a fraction f of the entries in the original ratings matrix are randomly chosen and retained. This approach is repeated to create q training data sets $R_1 \ldots R_q$. Thus, each user and each item is represented in each subsampled matrix, but the number of specified entries in the subsampled matrix is smaller than that in the original training data. A collaborative filtering algorithm (e.g., latent factor model) is applied to each ratings matrix to create a predicted matrix. The final prediction is the simple average of these q different predictions.

In the aforementioned methods, the final step of the ensemble uses a simple average of the predictions rather than a weighted average. The reason for using a simple average is that all model components are created with an identical probabilistic approach and should therefore be weighted equally. In many of these cases, it is important to choose unstable base methods to achieve good performance gains.

Although the aforementioned discussion provides an overview of the various possibilities for variance reduction, only a small subset of these possibilities have actually been explored and evaluated in the research literature. For example, we are not aware of any experimental results on the effectiveness of subsampling methods. Although subsampling methods often provide superior results to bagging in the classification domain [658], their effect on collaborative filtering is difficult to predict in sparse matrices. In sparse matrices, dropping entries could lead to the inability to predict some users or items at all, which can sometimes worsen the overall performance. A discussion of bagging algorithms in the context of collaborative

---

[3]It is possible for the unspecified values in duplicate rows to predicted differently, even though this is relatively unusual for most collaborative filtering algorithms.

filtering can be found in [67]. In this work, a row-wise bootstrapping approach is used, and duplicate rows are treated as weighted rows. Therefore, the approach assumes that the base predictor can handle weighted rows. As discussed in [67], significant improvements in the error were achieved with bagging, although the approach seemed to be somewhat sensitive to the choice of base predictor. In particular, according to the results in [67], the bagging approach improved the accuracy over most of the base predictors, with the exception of the factorized neighborhood model [72]. This might possibly be a result of a high level of correlation between the predictions of the various bagged models, when the factorized neighborhood method is used. In general, it is desirable to use uncorrelated base models with low bias and high variance in order to extract the maximum benefit from bagging. In cases where bagging does not work because of high correlations across base predictors, it may be helpful to explicitly use *randomness injection*.

6.3.3 Randomness Injection

Randomness injection is an approach that shares many principles of random forests in classification [22]. The basic idea is to take a base classifier and explicitly inject randomness into the classifier. Various methods can be used for injecting the randomness. Some examples [67] are as follows:

1. *Injecting randomness into a neighborhood model:* Instead of using the top-k nearest neighbors (users or items) in a user-based or item-based neighborhood model, the top-$\alpha \cdot k$ neighbors are selected for $\alpha \gg 1$. Then, k elements are randomly selected from these $\alpha \cdot k$ neighbors. This approach can, however, be shown to be an indirect variant of row-wise subsampling at factor $1/\alpha$. The average prediction from the various components is returned by the approach.

2. *Injecting randomness into a matrix factorization model:* Matrix factorization methods are inherently randomized methods because they perform gradient descent over the solution space after randomly initializing the factor matrices. Therefore, by choosing different initializations, different solutions are often obtained. The combinations of these different solutions often provide more accurate results.

A simple average of the predictions of the different components is returned by the randomized ensemble. Like random forests, this approach can reduce the variance of the ensemble without affecting the bias significantly. In many cases, this approach works quite well where bagging does not work because of a high level of correlation between various predictors. As shown in [67], the randomness injection approach works quite well when the factorized neighborhood model is used as the base predictor [72]. It is noteworthy that the bagging approach does not work very well in the case of the factorized neighborhood model.

6.4 Switching Hybrids

Switching hybrids are used most commonly in recommender systems in the context of the problem of *model selection*, but they are often not formally recognized as hybrid systems. The original motivation for switching systems [117] was to handle the cold-start problem, where a particular model works better in earlier stages when there is a paucity of available data. However, in later stages, a different model is more effective, and therefore one switches to the more effective model.

It is also possible to view switching models in the more general sense of *model selection*. For example, even the parameter selection step of most recommender models requires the running of the model over multiple parameter values and then selecting the optimal one. This particular form of model selection is adapted from the classification literature, and it is also referred to as the *bucket-of-models*. In the following, we discuss both these types of hybrids.

6.4.1   Switching Mechanisms for Cold-Start Issues

Switching mechanisms are often used to handle the cold-start problem, in which one recommender performs better with less data, whereas the other recommender performs better with more data. One might use a knowledge-based recommender, when few ratings are available because knowledge-based recommender systems can function without any ratings, and they are dependent on user specifications of their needs. However, as more ratings become available, one might switch to a collaborative recommender. One can also combine content-based and collaborative recommenders in this way, because content-based recommenders can work well for new items, whereas collaborative recommenders cannot effectively give recommendations for new items.

The work in [85] proposes the *Daily Learner* system in which various recommenders are used in an ordered strategy. If sufficient recommendations are not found by earlier recommenders, then later recommenders are used. In particular, the work in [85] uses two content-based recommenders and a single collaborative recommender. First, a nearest neighbor content classifier is used, followed by a collaborative system, and finally a naive Bayes content classifier is used to match with the long-term profile. This approach does not fully address the cold-start problem because all the underlying learners need some amount of data. Another work [659] combines hybrid versions of collaborative and knowledge-based systems. The knowledge-based system provides more accurate results during the cold-start phase, whereas the collaborative system provides more accurate results in later stages. Incorporating knowledge-based systems is generally more desirable for handling the cold-start problem.

6.4.2   Bucket-of-Models

In this approach, a fraction (e.g., 25% to 33%) of the specified entries in the ratings matrix are held out, and various models are applied to the resulting matrix. The held-out entries are then used to evaluate the effectiveness of the model in terms of a standard measure, such as the MSE or the MAE. The model that yields the lowest MSE or MAE is used as the relevant one. This approach is also commonly used for parameter tuning. For example, each model may correspond to a different value of the parameter of the algorithm, and the value providing the best result is selected as the relevant one. Once the relevant model has been selected, it is retrained on the *entire* ratings matrix, and the results are reported. Instead of using a hold-out approach, a different technique known as *cross-validation* is also used. You will learn more about hold-out and cross-validation techniques in Chapter 7. The bucket-of-models is the single most useful ensemble approach in recommender systems, although it is rarely recognized as an ensemble system unless the different models are derived from heterogeneous data types. When the bucket-of-models is used in the context of a dynamically changing ratings matrix, it is possible for the system to switch from one component to the other. However, when it is used for static data, the system can also be viewed as a special case of weighted recommenders in which the weight of one component is set to 1, and the weights of the remaining components are set to 0.

6.5 Cascade Hybrids

In Burke's original work [117], cascade hybrids were defined in a somewhat narrow way, in which each recommender actively refines the recommendations made by the previous recommender. Here, we take a broader view of cascade hybrids in which a recommender is allowed to use recommendations of the previous recommender in any way (beyond just direct refinement), and then combine the results to make the final recommendation. This broader definition encompasses larger classes of important hybrids, such as boosting, which would not otherwise be included in any of the categories of hybrids. Correspondingly, we define two different categories of cascade recommenders.

6.5.1 Successive Refinement of Recommendations

In this approach, a recommender system successively refines the output of recommendations from the previous iteration. For example, the first recommender can provide a rough ranking and also eliminate many of the potential items. The second level of recommendation then uses this rough ranking to further refine it and break ties. The resulting ranking is presented to the user. An example of such a recommender system is *EntreeC* [117], which uses knowledge of the user's stated interests to provide a rough ranking. The resulting recommendations are then partitioned into buckets of roughly equal preference. The recommendations within a bucket are therefore considered ties at the end of the first stage. A collaborative technique is used to break the ties and rank the recommendations within each bucket. The first knowledge-based recommender is clearly of higher priority because the second-level recommender cannot change the recommendations made at the first level. The other observation is that the second level recommender is much more efficient because it needs to focus only on the ties within each bucket. Therefore, the item-space of each application of the second recommender is much smaller.

6.5.2 Boosting

Boosting has been used popularly in the context of classification [206] and regression [207]. One of the earliest methods for boosting was the *AdaBoost* algorithm [206]. The regression variant of this algorithm is referred to as *AdaBoost.RT* [207]. The regression variant is more relevant to collaborative filtering because it easier to treat ratings as numeric attributes. In traditional boosting, a sequence of training rounds is used with weighted training examples. The weights in each round are modified depending on the performance of the classifier in the previous round. Specifically, the weights on the training examples with error are increased, whereas the weights on the correctly modeled examples are reduced. As a result, the classifier is biased towards correctly classifying the examples that it was unable to properly classify in the previous round. By using several such rounds, one obtains a sequence of classification models. For a given test instance, all models are applied to it, and the weighted prediction is reported as the relevant one.

Boosting needs to be modified to work for collaborative filtering, in which there is no clear demarcation between the training and test rows, and there is also no clear distinction between the dependent and independent columns. A method for modifying boosting for collaborative filtering is proposed in [67]. Unlike classification and regression modeling, in which weights are associated with rows, the training example weights in collaborative filtering are associated with individual *ratings*. Therefore, if the set S represents the set of observed ratings in the training data, then a total of $|S|$ weights are maintained. Note that

S is a set of positions (u, j) in the $m \times n$ ratings matrix R, such that r_{uj} is observed. It is also assumed that the base collaborative filtering algorithm has the capacity to work with weighted ratings (cf. section 6.3). In each iteration, the weights of each of these ratings are modified depending on how well the collaborative filtering algorithm is able to predict that particular entry.

The overall algorithm is applied for a total of T iterations. In the tth iteration, the weight associated with the (u, j)th entry of the ratings matrix is denoted by $W_t(u, j)$. The algorithm starts by equally weighting each entry and predicts all ratings using a baseline model. The prediction of an entry $(u, j) \in S$ is said to be "incorrect," if the predicted rating $\hat{r}_{uj}$ varies from the actual rating r_{uj} by at least a predefined amount δ. The error rate ϵ_t in the tth iteration is computed as the fraction of specified ratings in S for which the predicted value is incorrect, according to this definition. The weights of correctly predicted examples are reduced by multiplying them with ϵ_t, whereas the weights of the incorrectly predicted examples stay unchanged. In each iteration, the weights are always normalized to sum to 1. Therefore, the *relative* weights of incorrectly classified entries always increase across various iterations. The baseline model is applied again to the re-weighted data. This approach is repeated for T iterations, in order to create T different predictions for the unspecified entries. The weighted average of these T different predictions is used as the final prediction of an entry, where the weight of the tth prediction is $\log\left(\frac{1}{\epsilon_t}\right)$. It is noteworthy that the weight update and model combination rules in [67] are slightly different from those used in classification and regression modeling. However, there are very few studies in this area, beyond the work in [67], on using boosting methods for collaborative filtering. It is conceivable that the simple strategies proposed in [67] can be further improved on with experimentation.

6.5.2.1 Weighted Base Models

The boosting and bagging methods require the use of weighted base models, in which entries are associated with weights. In this section, we show how existing collaborative filtering models can be modified so that they can work with weights.

Let us assume that the weight w_{uk} be associated with a particular entry in the ratings matrix for user u and item k. It is relatively straightforward to modify existing models to work with weights on the entries:

1. *Neighborhood-based algorithms:* The average rating of a user is computed in a weighted way for mean-centering the ratings. Both the Pearson and the cosine measures can be modified to take weights into account. Therefore, Equation 2.2 of Chapter 2 can be modified as follows to compute the Pearson coefficient between users u and v:

$$\text{Pearson}(u, v) = \frac{\sum_{k \in I_u \cap I_v} \max\{w_{uk}, w_{vk}\} \cdot (r_{uk} - \mu_u) \cdot (r_{vk} - \mu_v)}{\sqrt{\sum_{k \in I_u \cap I_v} w_{uk}(r_{uk} - \mu_u)^2} \cdot \sqrt{\sum_{k \in I_u \cap I_v} w_{vk}(r_{vk} - \mu_v)^2}} \quad (6.6)$$

The reader should refer to section 2.3 of Chapter 2 for the details of the notations. A different way[4] of modifying the measure is as follows:

$$\text{Pearson}(u, v) = \frac{\sum_{k \in I_u \cap I_v} w_{uk} \cdot w_{vk} \cdot (r_{uk} - \mu_u) \cdot (r_{vk} - \mu_v)}{\sqrt{\sum_{k \in I_u \cap I_v} w_{uk}^2(r_{uk} - \mu_u)^2} \cdot \sqrt{\sum_{k \in I_u \cap I_v} w_{vk}^2(r_{vk} - \mu_v)^2}} \quad (6.7)$$

---

[4]The work in [67] proposes only the first technique for computing the similarity.

For item-item similarity measures, the adjusted cosine measure can be modified in a similar way. These weighted similarity measures are used both for computing the nearest neighbors and for (weighted) averaging of the ratings in the peer group.

2. *Latent factor models:* Latent factor models are defined as optimization problems in which the sum of the squares of the errors of the specified entries are minimized. In this case, the *weighted* sum of the squares of the optimization problem must be minimized. Therefore, the objective function in section 3.6.4.2 of Chapter 3 can be modified as follows:

$$\text{Minimize } J = \frac{1}{2} \sum_{(i,j) \in S} w_{ij} e_{ij}^2 + \frac{\lambda}{2} \sum_{i=1}^{m} \sum_{s=1}^{k} u_{is}^2 + \frac{\lambda}{2} \sum_{j=1}^{n} \sum_{s=1}^{k} v_{js}^2 \qquad (6.8)$$

Here, $U = [u_{ij}]$ and $V = [v_{ij}]$ are the $m \times k$ and $n \times k$ user-factor and item-factor matrices, respectively. Note the weights associated with the errors on the entries. The corresponding change in the gradient descent method is to weight the relevant updates:

$$u_{iq} \Leftarrow u_{iq} + \alpha(w_{ij} \cdot e_{ij} \cdot v_{jq} - \lambda \cdot u_{iq})$$
$$v_{jq} \Leftarrow v_{jq} + \alpha(w_{ij} \cdot e_{ij} \cdot u_{iq} - \lambda \cdot v_{jq})$$

Many other base collaborative filtering algorithms can be modified to work with weighted entries. These types of weighted base algorithms are useful for many collaborative filtering ensembles, such as boosting and bagging.

6.6 Feature Augmentation Hybrids

The feature augmentation hybrid shares a number of intuitive similarities with the stacking ensemble in classification. In stacking [634], the first level classifier is used to create or augment a set of features for the second level classifier. In many cases, off-the-shelf systems are used like an ensemble. However, in some cases, changes may be required to the component recommender system to work with the modified data, and therefore the hybrid system is not a true ensemble of off-the-shelf systems.

The *Libra* system [448] combines Amazon.com's recommender system with its own Bayes classifier. The approach uses the "related authors" and "related titles" that Amazon generates as features describing the items. Note that Amazon generates these recommendations with the use of a collaborative recommender system. These data are then used in conjunction with a content-based recommender to make the final predictions. Note that any off-the-shelf content-based system can be used in principle, and therefore the approach can be viewed as an ensemble system. The approach in [448] opts for a naive Bayes text classifier. It was found through experiments that the features generated by Amazon's collaborative system were of high quality, and they contributed significantly to better quality recommendations.

Instead of using a collaborative system first, it is also possible to use the content-based system first. The basic idea is to use the content-based system to fill in the missing entries of the ratings matrix so that it is no longer sparse. Thus, the missing entries are estimated by the content-based system to create a denser ratings matrix. These newly added ratings are referred to as *pseudo-ratings*. Then, a collaborative recommender is used on the dense ratings matrix to make rating predictions. Finally, the collaborative prediction is combined with the original content-based prediction in a weighted way to yield the overall prediction of the missing entries in the matrix [431]. The incorporation of missing ratings in the first phase allows for a more robust application of the second phase in terms of similarity computation. However, the similarity computation does need to be modified to give less weight to pseudo-ratings compared to true ratings. This is because pseudo-ratings were inferred, and they might be error-prone.

How can such weights be determined? The weight of a pseudo-rating intuitively represents the algorithm's certainty in the prediction of the first phase, and it is an increasing function of the number of ratings $|I_i|$ of that user. A number of heuristic functions are used to weight various ratings, and the reader is referred to [431] for details. Note that this approach requires modifications to the second phase of collaborative filtering, and off-the-shelf algorithms cannot be used. Such methods can be viewed as monolithic systems.

Feature augmentation has a long history in recommender systems. One of the earliest example of feature augmentation was implemented in the context of the *GroupLens* system [526], in which a knowledge-based system was used to create a database of artificial ratings. The agents, known as filterbots, used specific criteria such as the number of spelling errors or the message size to assign ratings to items, while acting as artificial users. Subsequently, these ratings were used in the context of a collaborative system to make recommendations.

6.7   Meta-Level Hybrids

In a meta-level hybrid, the *model* learned by one recommender is used as input to the next level. An important example of collaboration via content was the early work by Pazzani [475]. A content-based model [363] is constructed that describes the discriminative features predicting restaurants. The discriminative features may be determined using any of the feature selection methods discussed in section 4.3 of Chapter 4. Each user is defined by a vector representation of discriminative words. An example of the possible user-word matrix for a restaurant recommender systems is shown below:

Word ⇒ **User** ⇓	beef	roasted	lamb	fried	eggs
Sayani	0	3	0	2.5	1.7
John	2.3	1.3	0.2	1.4	2.1
Mary	0	2.8	0.9	1.1	2.6
Peter	2.4	1.7	0	3.5	1.9
Jack	1.6	2.2	3.1	1.0	0

The weights in the aforementioned table may be obtained using the descriptions of the items that the user has accessed. Note that the irrelevant words have already been removed because the content-based feature selection in the first phase creates a discriminative vector-space representation for each user. Furthermore, the representation is significantly denser than a typical ratings matrix. Therefore, one can robustly compute the similarities between users with this new representation. The main idea here is that the *content-based* peer group is used to determine the most similar users of the target user. Once the peer group has been determined, then the weighted average of the ratings of the peer group are used to determine the predicted ratings. Note that this approach does require a certain amount of change to the original collaborative recommender, at least in terms of how the similarity is computed. The peer group formation must use the user-word matrix (which was the model created in the first phase), whereas the final recommendation uses the ratings matrix. This is different from a collaborative system in which both stages use the same matrix. Furthermore, the first phase of the approach cannot use off-the-shelf content-based models in their entirety because it is mostly a feature selection (preprocessing) phase. Therefore, in many cases, these systems cannot be considered true ensembles, because they do not use existing methods as off-the-shelf recommenders.

Another example of a meta-level system was *LaboUr* [534] in which an instance-based model is to used to learn the content-based user's profile. The profiles are then compared using a collaborative approach. These models are compared across users to make predictions. Many of these methods fall within the category of "collaboration via content," though that is not the only way in which such hybrids can be constructed.

6.8 Feature Combination Hybrids

In feature combination hybrids, the idea is to combine the input data from various sources (e.g., content and collaborative) into a unified representation before applying a predictive algorithm. In most cases, this predictive algorithm is a content-based algorithm that uses collaborative information as additional features. An example of such an approach was presented in [69], where the *RIPPER* classifier was applied to the augmented data set. It was shown in [69] that the methodology achieved significant improvements over a purely collaborative approach. However, the content features need to be hand picked in order to achieve this result. Therefore, the approach can be sensitive to the choice of data set and feature representation. The approach reduces the sensitivity of the system to the number of users that have rated an item. This is, of course, the property of any content-based system, which is robust to the cold-start problem from the perspective of new items.

Note that it is possible for the combination to be performed in a variety of different ways with different types of background knowledge. For example, consider the case where each item is associated with a higher-level taxonomy representing the genres of the items. The representation profile of the user and items can be augmented in terms of the relevant genres in the hierarchy. The ratings matrix can then be constructed in terms of genres rather than items. In sparse matrices, such an approach can provide more effective results because it reduces the number of columns, and because most entries are likely to be populated in the compressed matrix.

Another approach is to augment a ratings matrix and add columns for keywords in addition to items. Therefore, the ratings matrix becomes an $m \times (n + d)$ matrix, where n is the number of items and d is the number of keywords. The weights of "keyword items" are based on the weighted aggregation of the descriptions of the items accessed, bought, or rated by the user. A traditional neighborhood or matrix factorization approach can be used with this augmented matrix. The relative weights of the two types of columns can be learned through cross-validation (see Chapter 7). This type of combination of two optimization models is common in hybrid settings, where the objective function is set up as follows in terms of a parameter vector $\overline{\theta}$:

$$J = \text{CollaborativeObjective}(\overline{\theta}) + \beta\,\text{ContentObjective}(\overline{\theta}) + \text{Regularization} \qquad (6.9)$$

The objective function is then optimized over the parameter vector $\overline{\theta}$. A specific example, which is discussed below, is the generalization of sparse linear models (cf. section 2.6.5 of Chapter 2) with side information.

6.8.1   Regression and Matrix Factorization

Let R be an $m \times n$ implicit feedback ratings matrix, and C be a $d \times n$ content matrix, in which each item is described by non-negative frequencies of d words. Examples include descriptions of items or short reviews of items. Since R is an implicit feedback matrix, missing entries are assumed to be 0s. As in section 2.6.5, let W be an $n \times n$ item-item coefficient matrix in which the ratings are predicted as $\hat{R} = RW$. However, in this case we can also predict the ratings as $\hat{R} = CW$. Therefore, instead of optimizing only $||R - RW||^2$, we add an additional content-based term $||R - CW||^2$. Together with elastic-net regularization, and non-negativity/diagonal constraints, the enhanced optimization model is stated as follows [456]:

$$\text{Minimize } J = ||R - RW|||^2 + \beta \cdot ||R - CW||^2 + \lambda ||W||^2 + \lambda_1 \cdot ||W||_1$$
$$\text{subject to:}$$
$$W \geq 0$$
$$\text{Diagonal}(W) = 0$$

The weight parameter β can be determined in a tuning phase. Although the ratings can be predicted either as $\hat{R} = RW$ or as $\hat{R} = CW$, only the former prediction function is used. Therefore, the term $||R - CW||^2$ is used only to refine the objective function as an additional regularizer. In other words, the goal of the additional term is to improve the *generalization power* of the model for predicting future (and as yet unknown) actions of the user. Some variations of this basic objective function are discussed in [456].

This type of approach can be used for combining any other type of collaborative filtering (optimization) model with content-based methods. For example, in the case of matrix factorization, one can use an $m \times k$ user factor matrix U, an $n \times k$ *shared* item factor matrix V, and a $d \times k$ content factor matrix Z to set up the optimization model as follows [557]:

$$\text{Minimize } J = ||R - UV^T|||^2 + \beta \cdot ||C - ZV^T||^2 + \lambda(||U||^2 + ||V||^2 + ||Z||^2)$$

Note that the item factor matrix V is shared between the factorizations of the ratings matrix and content matrix. Such shared matrix factorization models are also used for incorporating other types of side information such as social trust data (cf. section 11.3.8 of Chapter 11). An overview of combining matrix factorization methods with arbitrary models is provided in section 3.7.7 of Chapter 3.

6.8.2   Meta-level Features

It is not necessary to use feature combination in the context of multiple types of recommenders (e.g., content and collaborative). New meta-features can be extracted from features of a particular type of recommender and then combined within the ensemble model. For example, one can extract *meta-level features* from a ratings matrix corresponding to the number of ratings given by various users and items. When a user rates many movies, or when a movie is rated by many users, it affects the recommendation accuracy of the various algorithms in different ways. Different recommender systems will be more or less sensitive to these characteristics, and will therefore do better or worse for various users and items. The basic idea of meta-level features is to account for these *entry-specific* differences in the model combination process with the use of meta-features. The resulting meta-features can be paired with other ensemble algorithms to create an ensemble design, which incorporates characteristics from various types of hybrids, but it does not neatly fall into any of Burke's seven original categories [117]. However, it is most closely related to feature combination hybrids in the sense that it combines meta-features with ratings.

Id.	Description
1	Constant value of 1 (using only this feature amounts to using the global linear regression model of section 6.3)
2	A binary variable indicating whether the user rated more than 3 movies on this particular date
3	The log of the number of times a movie has been rated
4	The log of the number of distinct dates on which a user has rated movies
5	A Bayesian estimate of the mean rating of the movie after having subtracted out the user's Bayesian estimated mean
6	The log of the number of user ratings
16	The standard deviation of the user ratings
17	The standard deviation of the movie ratings
18	The log of (Rating Date − First User Rating Date +1)
19	The log of the number of user ratings on the date +1

Table 6.1: A subset of the meta-features used in [554] for ensemble combination on the Netflix Prize data set

The meta-feature approach has proven to be a potentially powerful method for robust ensemble design. In fact, both winning entries in the Netflix Prize content, corresponding to *Bellkor's Pragmatic Chaos* [311] and *The Ensemble* [704], used such an approach. We will describe the use of such meta-level features in collaborative filtering algorithms. In particular, we will discuss the methodology of *feature weighted linear stacking* [554], which combines such meta-level features with the stacking methods discussed earlier in section 6.3. This approach is based on the blending technique used in *The Ensemble* [704]. A subset of the meta-features used in [554] for the stacking process on the Netflix Prize data set is provided in Table 6.1 for illustrative purposes. The identifier in the left column corresponds to the identifier used in the original paper [554]. These features are particularly instructive because one can usually extract analogous features for other ratings data sets. Note that each feature in Table 6.1 is specific to an entry in the ratings matrix.

Let us assume that a total of l (numeric) meta-features have been extracted, and their values are $z_1^{ut} \ldots z_l^{ut}$ for user-item pair (u, t). Therefore, the meta-features are specific to each entry (u, t) in the ratings matrix, although some features may take on the same values for varying values of u or varying values of t. For example, feature 3 in Table 6.1 will not vary with the user u, but it will vary with the item t.

Let us assume that there are a total of q base recommendation methods, and the weights associated with the q recommendation methods are denoted by $w_1 \ldots w_q$. Then, for a given entry (u, t) in the ratings matrix, if the predictions of the q components are $\hat{r}_{ut}^1 \ldots \hat{r}_{ut}^q$, then the prediction $\hat{r}_{ut}$ of the overall ensemble is given by the following:

$$\hat{r}_{ut} = \sum_{i=1}^{q} w_i \hat{r}_{ut}^i \tag{6.10}$$

We would like the estimated prediction $\hat{r}_{ut}$ of the ensemble to match the observed ratings r_{ut} as closely as possible. Note that the approach in section 6.3 uses a linear regression model to learn the weights $w_1 \ldots w_q$ by holding out a pre-defined fraction of the entries during the process of training the q models, and then using the held-out entries as the observed values in the linear regression model. Such an approach is pure stacking, and it can be considered a weighted hybrid. However, it can be enhanced further using meta-features. The main idea is that *the linear regression weights $w_1 \ldots w_q$ are specific to each entry in the ratings matrix and they are themselves linear functions of the meta-features.* In other words, the weights now need to be super-scripted with (u, t) to account for the fact that they are specific to each entry (u, t) in the ratings matrix:

$$\hat{r}_{ut} = \sum_{i=1}^{q} w_i^{ut} \hat{r}_{ut}^i \qquad (6.11)$$

This is a more refined model because the nature of the combination is local to each entry in the ratings matrix, and it is not blindly global to the entire matrix. The problem is that the number, $m \times n \times q$, of different parameters w_i^{ut} becomes too large to be learned in a robust way. In fact, the number of parameters (weights) is larger than the number of observed ratings, as a result of which overfitting will occur. Therefore, the weights are assumed to be linear combinations of the meta-features under the assumption that these meta-features regulate the relative importance of the various models for the individual user-item combinations. Therefore, we introduce the parameters v_{ij} that regulate the importance of the jth meta-feature to the ith model. The weights for entry (u, t) can now be expressed as linear combination of the meta-feature values of entry (u, t) as follows:

$$w_i^{ut} = \sum_{j=1}^{l} v_{ij} z_j^{ut} \qquad (6.12)$$

We can now express the regression modeling problem in terms of a fewer number, $q \times l$, of parameters v_{ij}, where v_{ij} regulates the impact of the jth meta-feature on the relative importance of the ith ensemble model. Substituting the value of w_i^{ut} from Equation 6.12 in Equation 6.11, we obtain the relationship between the ensemble rating and the component ratings as follows:

$$\hat{r}_{ut} = \sum_{i=1}^{q} \sum_{j=1}^{l} v_{ij} z_j^{ut} \hat{r}_{ut}^i \qquad (6.13)$$

Note that this is still a linear regression problem in $q \times l$ coefficients corresponding to the variables v_{ij}. A standard least-squares regression model can be used to learn the values of v_{ij} on the held-out[5] ratings. The independent variables of this regression are given by the quantities $z_j^{ut} \hat{r}_{ut}^i$. Regularization can be used to reduce overfitting. After the weights have been learned using linear regression, the individual component models are retrained on the entire training set without any held-out entries. The weights, which were learned using the held-out entries, are used in conjunction with these q models.

6.9   Mixed Hybrids

The main characteristic of mixed recommender systems is that they combine the scores from different components in terms of *presentation*, rather than in terms of combining the predicted scores. In many cases, the recommended items are presented next to one another [121, 623]. Therefore, the main distinguishing characteristic of such systems is the combination of presentation rather than the combination of predicted scores.

---

[5]In the context of the Netflix Prize contest, this was achieved on a special part of the data set, referred to as the *probe set*. The probe set was not used for building the component ensemble models.

Most of the other hybrid systems focus on creating a unified rating, which is extracted from the various systems. A classical example is illustrated in [559], in which a personalized television listing is created using a mixed system. Typically, a *composite* program is presented to the user. This composite program is created by combining items recommended by different systems. Such composite programs are typical in the use of mixed systems, although the applicability of mixed systems goes beyond such scenarios. In many of these cases, the basic idea is that the recommendation is designed for a relatively complex item containing many components, and it is not meaningful to recommend the individual items. The new item startup problem is often alleviated with a mixed recommender system. Because a television program has many slots, either the content-based or collaborative recommender might be successful in filling the different slots. In some cases, a sufficient number of recommendations for the slots may be achieved only with multiple recommenders of different types, especially at the very beginning when there is a paucity in the available data. However, conflict resolution may be required in some cases where more choices are available than the available slots.

Another example of a mixed hybrid has been proposed in the tourism domain [660, 661]. In this case, bundles of recommendations are created, where each bundle contains multiple categories of items. For example, in a tourism recommender system, the different categories may correspond to accommodations, leisure activities, air-tickets, and so on. Tourists will typically buy bundles of these items from various categories in order to create their trips. For each category, a different recommender system is employed. The basic idea here is that the recommender system that is most appropriate for obtaining the best accommodations, may not be the one that is most appropriate for recommending tourism activities. Therefore, each of these different aspects is treated as a different category for which a different recommender system is employed. Furthermore, it is important to recommend bundles in which the items from multiple categories are not mutually inconsistent. For example, if a tourist is recommended a leisure activity that is very far away from her place of accommodation, then the overall recommendation bundle will not be very convenient for the tourist. Therefore, a knowledge base containing a set of domain constraints is incorporated for the bundling process. The constraints are deigned to resolve inconsistencies in the product domain. A constraint satisfaction problem is employed to determine a mutually consistent bundle. More details of the approach are discussed in [660, 661].

It is noteworthy that many of the mixed hybrids are often used in conjunction with knowledge-based recommender systems as one of the components [121, 660]. This is not a coincidence. Mixed hybrids are generally designed for complex product domains with multiple components like knowledge-based recommender systems.

6.10 Summary

Hybrid recommender systems are used either to leverage the power of multiple data sources or to improve the performance of existing recommender systems within a particular data modality. An important motivation for the construction of hybrid recommender systems is that different types of recommender systems, such as collaborative, content-based, and knowledge-based methods, have different strengths and weaknesses. Some recommender systems work more effectively at cold start, whereas other work more effectively when sufficient data are available. Hybrid recommender systems attempt to leverage the complementary strengths of these systems to create a system with greater overall robustness.

Ensemble methods are also used to improve the accuracy of collaborative filtering methods in which the different components use the same ratings matrix. In these cases, the individual models use the same base data rather than different sources of data. Such methods are much closer to the existing ideas on ensemble analysis in the classification domain. The basic idea is to use the various models to incorporate diversity and reduce model bias. Many of the existing theoretical results on the bias-variance trade-off in classification are also applicable to collaborative filtering applications. Therefore, many techniques, such as bagging and boosting, can be adapted with relatively minor modifications.

Hybrid systems are designed as monolithic systems, ensemble systems, or mixed systems. Ensemble systems are typically designed by using either sequential or parallel arrangement of recommenders. In monolithic design, either existing recommenders are modified, or entirely new recommenders are created by combining the features from multiple data modalities. In mixed systems, recommendations from multiple engines are presented simultaneously. In many cases, meta-features can also be extracted from a particular data modality in order to combine the predictions of various recommenders in an entry-specific way. The great strength of hybrid and ensemble systems arises from their ability to leverage complementary strengths in various systems. The top entries in the Netflix Prize contest were all ensemble systems.

6.11   Bibliographic Notes

Although hybrid systems have a long and rich history in the development of recommender systems, a formal categorization of these methods was not performed until the survey by Burke [117]. A discussion of hybrid recommender systems in the specific context of the Web is provided in [118]. Burke originally categorized recommender systems into seven different categories. Subsequently, Jannach *et al.* [275] created a higher-level categorization of these lower-level categories into pipelined and parallel systems. The hierarchical taxonomy in this book roughly follows the work of [275] and [117], although it makes a number of modifications to include several important methods, such as boosting, into one of these categories. It is important to note that this taxonomy is not exhaustive because many ensemble systems, such as those winning the Netflix Prize, use ideas from many types of hybrids. Nevertheless, Burke's original categorization is very instructive, because it covers most of the important building blocks. Recently, ensemble methods have received a lot of attention, especially after the winning entries in the Netflix Prize contest were both ensemble systems [311, 704].

Ensemble methods have been used extensively in the classification literature. A detailed discussion of the bias-variance trade-off in the context of the classification problem is provided in [22]. Bagging and subsampling methods for classification are discussed in [111–113]. A recent work [67] shows how one might leverage ensemble methods from the classification literature to recommender systems by adapting methods such as bagging and *AdaBoost.RT*. While some ensemble systems are developed with this motivation, other systems combine the power of different data types. Weighted models are among the most popular classes of models. Some of the models combine models built on homogeneous data types. Methods for constructing homogeneous weighted ensembles are discussed in [67, 266]. The winners [311, 704] of the Netflix Prize contest also used a weighted ensemble system, although the combination uses additional meta-features, which imbues it with some of the properties of a feature combination approach. The work in [180] uses ensembles of maximum margin matrix factorization methods with different parameter settings. User-based and item-based neighborhood algorithms are combined in [338]. Other recent work on weighted models shows how to combine systems built on top of different data types. The work in [659] combines a collaborative and knowledge-based recommender, whereas the work in [162] combines a content-based and collaborative recommender.

A performance-based switching hybrid is discussed in [601]. An interesting machine-learning approach to switching mechanisms is discussed in [610]. Other switching mechanisms for handling cold-start issues are discussed in [85]. Another combination of a knowledge-based and collaborative system to create a switching hybrid is discussed in [659].

Cascade systems use sequential processing of the ratings to make recommendations. Such systems can either use refinements or they can use boosting methods. The *EntreeC* recommender [117] is the most well-known example of a cascade system that uses refinements. A cascade system that uses boosting is discussed in [67]. The latter methods uses a weighted version of the *AdaBoost.RT* algorithm in order to create the hybrid recommender.

Feature augmentation hybrids use the recommenders of one type to augment the features of another. The *Libra* system [448] combines Amazon.com's recommender system with its own Bayes classifier. The output of the Amazon system is used to create a content-based recommender. The method in [431] uses a content-based system to estimate the missing entries of the ratings matrix and uses the estimated values in the context of a collaborative system. In the *GroupLens* system [526], a knowledge-based system was used to create a database of artificial ratings. These ratings were used in the context of a collaborative system to make recommendations. The work in [600] shows how to use a feature augmentation hybrid to recommend research papers.

Many techniques have been used recently to create fused feature spaces or unified representations from ratings matrices and content matrices. This unified representation or feature space forms the basis on which machine learning tools can be applied. One of the earliest works along this line constructs joint feature maps [68] from rating and content information and then uses machine learning models in order to perform the prediction. A tensor-based approach is used to achieve this goal. An analogous approach is also used in [557], which jointly factorizes the user-item purchase profile matrix and the item-feature content matrix into a common latent space. This latent representation is then used for learning. The work in [411] uses a latent factor model in which the review text is combined with ratings. A regression-based latent factor model is proposed in [14] for rating prediction, which uses content features for factor estimation. The user and item latent factors are estimated through independent regression on user and item features. Then, a multiplicative function is used on the user and item factors for prediction. Sparse regression models have also been used for fused prediction in [456]. Finally, graph-based models have been used to create unified representations. The work in [238] leans the interaction weights between user actions and various features such as user-item profile information and side information. Unified Boltzmann machines are used to perform the prediction. A unified graph-based representation has been proposed in [129]. A Bayesian network is created containing item nodes, user nodes, and item feature nodes. This Bayesian network is used to perform combined content-based and collaborative recommendations.

In a meta-level hybrid, the *model* learned by one recommender is used as input to the next level. In the early work by Pazzani [475], a content-based model [363] is constructed that describes the discriminative features predicting restaurants. Each user is defined by a vector representation of discriminative words. The content-based model is used to determine the peer group, which is then used for the purpose of recommendation. Meta-level combinations of content-based and collaborative systems are discussed in [475, 534]. A two-stage Bayesian meta-level hybrid is discussed in [166]. A different type of hierarchical Bayes model that combines collaborative and content-based systems is presented in [652]. Methods for stacking recommender systems with meta-features are discussed in [65, 66, 311, 554]. The *STREAM* system [65, 66] was one of the earliest systems to leverage meta-level features.

A number of mixed recommender systems have been proposed in [121, 559, 623, 660, 661]. A mixed recommender system for creating television programs is discussed in [559], whereas a system for providing tourism bundles is discussed in [660]. It is noteworthy that many mixed recommender systems are used in complex product domains like knowledge-based recommender systems [121, 660].

6.12   Exercises

1. How does the rank of the latent factor model affect the bias-variance trade-off in a recommender system? If you had to use a latent factor model as the base model for a bagging ensemble, would you choose a model with high rank or low rank?

2. Does your answer to Exercise 1 change if you had to use boosting in conjunction with a latent factor model?

3. Implement an entry-wise bagging model by using a weighted latent factor model as the base model.

4. Suppose that you created a collaborative system in which the user-item matrix contained word frequencies as additional rows of the matrix. Each additional row is a word, and the value of the word-item combination is a frequency. An item-based neighborhood model is used with this augmented representation. What kind of hybrid would this be considered? Discuss the possible impact of using such a model on the accuracy and diversity of the recommender system.

5. Discuss how you would control the relative strength of collaborative and content-based portions in Exercise 4 with a single weight parameter. How would you determine the optimal value of the weight parameter in a data-driven way?

Chapter 7

Evaluating Recommender Systems

"True genius resides in the capacity for evaluation of uncertain, hazardous, and conflicting information."– Winston Churchill

7.1 Introduction

The evaluation of collaborative filtering shares a number of similarities with that of classification. This similarity is due to the fact that collaborative filtering can be viewed as a generalization of the classification and regression modeling problem (cf. section 1.3.1.3 of Chapter 1). Nevertheless, there are many aspects to the evaluation process that are unique to collaborative filtering applications. The evaluation of content-based methods is even more similar to that of classification and regression modeling, because content-based methods often use text classification methods under the covers. This chapter will introduce various mechanisms for evaluating various recommendation algorithms and also relate these techniques to the analogous methods used in classification and regression modeling.

A proper design of the evaluation system is crucial in order to obtain an understanding of the effectiveness of various recommendation algorithms. As we will see later in this chapter, the evaluation of recommender systems is often multifaceted, and a single criterion cannot capture many of the goals of the designer. An incorrect design of the experimental evaluation can lead to either gross underestimation or overestimation of the true accuracy of a particular algorithm or model.

Recommender systems can be evaluated using either *online* methods or *offline* methods. In an online system, the user reactions are measured with respect to the presented recommendations. Therefore, user participation is essential in online systems. For example, in an online evaluation of a news recommender system, one might measure the *conversion rate* of users clicking on articles that were recommended. Such testing methods are referred to as *A/B testing*, and they measure the direct impact of the recommender system

© Springer International Publishing Switzerland 2016
C.C. Aggarwal, *Recommender Systems: The Textbook*,
DOI 10.1007/978-3-319-29659-3_7

on the end user. At the end of the day, increasing the conversion rate on profitable items is the most important goal of a recommender system, and it can provide a true measure of the effectiveness of the system. However, since online evaluations require active user participation, it is often not feasible to use them in benchmarking and research. There are usually significant challenges in gaining access to user conversion data from systems with large-scale user participation. Even if such access is gained, it is usually specific to a single large-scale system. On the other hand, one often desires to use data sets of different types, and from multiple domains. Testing over multiple data sets is particularly important for assuring greater generalization power of the recommender system so that one can be assured that the algorithm works under a variety of settings. In such cases, offline evaluations with historical data sets are used. Offline methods are, by far, the most common methods for evaluating recommender systems from a research and practice perspective. Therefore, most of this chapter will focus on offline methods, although some discussion of online methods is also included for completeness.

When working with offline methods, accuracy measures can often provide an incomplete picture of the true conversion rate of a recommender system. Several other secondary measures also play a role. Therefore, it is important to design the evaluation system carefully so that the measured metrics truly reflect the effectiveness of the system from the user perspective. In particular, the following issues are important from the perspective of designing evaluation methods for recommender systems:

1. *Evaluation goals:* While it is tempting to use accuracy metrics for evaluating recommender systems, such an approach can often provide an incomplete picture of the user experience. Although accuracy metrics are arguably the most important components of the evaluation, many secondary goals such as novelty, trust, coverage, and serendipity are important to the user experience. This is because these metrics have important short- and long-term impacts on the conversion rates. Nevertheless, the actual quantification of some of these factors is often quite subjective, and there are often no hard measures to provide a numerical metric.

2. *Experimental design issues:* Even when accuracy is used as the metric, it is crucial to design the experiments so that the accuracy is not overestimated or underestimated. For example, if the same set of specified ratings is used both for model construction and for accuracy evaluation, then the accuracy will be grossly overestimated. In this context, careful experimental design is important.

3. *Accuracy metrics:* In spite of the importance of other secondary measures, accuracy metrics continue to be the single most important component in the evaluation. Recommender systems can be evaluated either in terms of the prediction accuracy of a rating or the accuracy of ranking the items. Therefore, a number of common metrics such as the *mean absolute error* and *mean squared error* are used frequently. The evaluation of rankings can be performed with the use of various methods, such as utility-based computations, rank-correlation coefficients, and the *receiver operating characteristic* curve.

In this chapter, we will first begin by discussing the general goals of evaluating recommender systems beyond the most basic criterion of accuracy. Examples of such goals include diversity and novelty. The main challenge with *quantifying* such goals is that they are often subjective goals based on user experience. From a quantification perspective, accuracy is a concrete goal that is relatively easy to measure and is therefore used more frequently for bench-marking and testing. A few quantification methods do exist for evaluating the secondary goals such as diversity and novelty. Although the majority of this chapter will focus on accuracy metrics, various quantification measures for the secondary goals will also be discussed.

This chapter is organized as follows. An overview of the different types of evaluation systems is provided in section 7.2. Section 7.3 studies the general goals of evaluating recommender systems. The appropriate design of accuracy testing methods is discussed in section 7.4. Accuracy metrics for recommender systems are discussed in section 7.5. The limitations of evaluation measures are discussed in 7.6. A summary is given in section 7.7.

7.2 Evaluation Paradigms

There are three primary types of evaluation of recommender systems, corresponding to user studies, online evaluations, and offline evaluations with historical data sets. The first two types involve users, although they are conducted in slightly different ways. The main differences between the first two settings lie in how the users are recruited for the studies. Although online evaluations provide useful insights about the true effects of a recommendation algorithm, there are often significant practical impediments in their deployment. In the following, an overview of these different types of evaluation is provided.

7.2.1 User Studies

In user studies, test subjects are actively recruited, and they are asked to interact with the recommender system to perform specific tasks. Feedback can be collected from the user before and after the interaction, and the system also collects information about their interaction with the recommender system. These data are then used to make inferences about the likes or dislikes of the user. For example, users could be asked to interact with the recommendations at a product site and give their feedback about the quality of the recommendations. Such an approach could then be used to judge the effectiveness of the underlying algorithms. Alternatively, users could be asked to listen to several songs, and then provide their feedback on these songs in the form of ratings.

An important advantage of user studies is that they allow for the collection of information about the user interaction with the system. Various scenarios can be tested about the effect of changing the recommender system on the user interaction, such as the effect of changing a particular algorithm or user-interface. On the other hand, the active awareness of the user about the testing of the recommender system can often bias her choices and actions. It is also difficult and expensive to recruit large cohorts of users for evaluation purposes. In many cases, the recruited users are not representative of the general population because the recruitment process is itself a bias-centric filter, which cannot be fully controlled. Not all users would be willing to participate in such a study, and those who do agree might have unrepresentative interests with respect to the remaining population. For example, in the case of the example of rating songs, the (voluntary) participants are likely to be music enthusiasts. Furthermore, the fact that users are actively aware of their recruitment for a particular study is likely to affect their responses. Therefore, the results from user evaluations cannot be fully trusted.

7.2.2 Online Evaluation

Online evaluations also leverage user studies except that the users are often real users in a fully deployed or commercial system. This approach is sometimes less susceptible to bias from the recruitment process, because the users are often directly using the system in the natural course of affairs. Such systems can often be used to evaluate the comparative performance of various algorithms [305]. Typically, users can be sampled randomly, and the various algorithms can be tested with each sample of users. A typical example of a metric, which is used to measure the effectiveness of the recommender system on the users, is the *conversion rate*. The conversion rate measures the frequency with which a user selects a recommended item. For example, in a news recommender system, one might compute the fraction of times that a user selects a recommended article. If desired, expected costs or profits can be added to the items to make the measurement sensitive to the importance of the item. These methods are also referred to as *A/B testing*, and they measure the direct impact of the recommender system on the end user. The basic idea in these methods is to compare two algorithms as follows:

1. Segment the users into two groups A and B.

2. Use one algorithm for group A and another algorithm for group B for a period of time, while keeping all other conditions (e.g., selection process of users) across the two groups as similar as possible.

3. At the end of the process, compare the conversion rate (or other payoff metric) of the two groups.

This approach is very similar to what is used for clinical trials in medicine. Such an approach is the most accurate one for testing the long-term performance of the system directly in terms of goals such as profit. These methods can also be leveraged for the user studies discussed in the previous section.

One observation is that it is not necessary to strictly segment the users into groups in cases where the payoff of each interaction between the user and the recommender can be measured separately. In such cases, the same user can be shown one of the algorithms at random, and the payoff from that specific interaction can be measured. Such methods of evaluating recommender systems have also been generalized to the development of more effective recommendation algorithms. The resulting algorithms are referred to as *multi-arm bandit algorithms*. The basic idea is similar to that of a gambler (recommender system) who is faced with a choice of selecting one of a set of slot machines (recommendation algorithms) at the casino. The gambler suspects that one of these machines has a better payoff (conversion rate) than others. Therefore, the gambler tries a slot machine at random 10% of the time in order to *explore* the relative payoffs of the machines. The gambler greedily selects the best paying slot machine the remaining 90% of the time in order to *exploit* the knowledge learned in the exploratory trials. The process of exploration and exploitation is fully interleaved in a random way. Furthermore, the gambler may choose to give greater weight to recent results as compared to older results for evaluation. This general approach is related to the notion of *reinforcement learning*, which can often be paired with online systems. Although reinforcement learning has been studied extensively in the classification and regression modeling literature [579], the corresponding work in the recommendation domain is rather limited [389, 390, 585]. A significant research opportunity exists for the further development of such algorithms.

The main disadvantage is that such systems cannot be realistically deployed unless a large number of users are already enrolled. Therefore, it is hard to use this method during the start up phase. Furthermore, such systems are usually not openly accessible, and they are only accessible to the owner of the specific commercial system at hand. Therefore, such tests can be performed only by the commercial entity, and for the limited number of scenarios handled by their system. This means that the tests are often not generalizable to system-independent benchmarking by scientists and practitioners. In many cases, it is desirable to test the robustness of a recommendation algorithm by stress-testing it under a variety of settings and data domains. By using multiple settings, one can obtain an idea of the generalizability of the system. Unfortunately, online methods are not designed for addressing such needs. A part of the problem is that one cannot fully control the actions of the test users in the evaluation process.

7.2.3   Offline Evaluation with Historical Data Sets

In offline testing, historical data, such as ratings, are used. In some cases, temporal information may also be associated with the ratings, such as the time-stamp at which each user has rated the item. A well known example of a historical data set is the Netflix Prize data set [311]. This data set was originally released in the context of an online contest, and has since been used as a standardized benchmark for testing many algorithms. The main advantage of the use of historical data sets is that they do not require access to a large user base. Once a data set has been collected, it can be used as a standardized benchmark to compare various algorithms across a variety of settings. Furthermore, multiple data sets from various domains (e.g., music, movies, news) can be used to test the generalizability of the recommender system.

Offline methods are among the most popular techniques for testing recommendation algorithms, because standardized frameworks and evaluation measures have been developed for such cases. Therefore, much of this chapter will be devoted to the study of offline evaluation. The main disadvantage of offline evaluations is that they do not measure the actual propensity of the user to react to the recommender system in the future. For example, the data might evolve over time, and the current predictions may not reflect the most appropriate predictions for the future. Furthermore, measures such as accuracy do not capture important characteristics of recommendations, such as *serendipity* and *novelty*. Such recommendations have important long-term effects on the conversion rate of the recommendations. Nevertheless, in spite of these disadvantages, offline methods continue to be the most widely accepted techniques for recommender system evaluation. This is because of the statistically robust and easily understandable quantifications available through such testing methods.

7.3   General Goals of Evaluation Design

In this section, we will study some of the general goals in evaluating recommender systems. Aside from the well known goal of accuracy, other general goals include factors such as diversity, serendipity, novelty, robustness, and scalability. Some of these goals can be concretely quantified, whereas others are subjective goals based on user experience. In such cases, the only way of measuring such goals is through user surveys. In this section, we will study these different goals.

7.3.1   Accuracy

Accuracy is one of the most fundamental measures through which recommender systems are evaluated. In this section, we provide a brief introduction to this measure. A detailed discussion is provided in section 7.5 of this chapter. In the most general case, ratings are numeric quantities that need to be estimated. Therefore, the accuracy metrics are often similar to those used in regression modeling. Let R be the ratings matrix in which r_{uj} is the known rating of user u for item j. Consider the case where a recommendation algorithm estimates this rating as $\hat{r}_{uj}$. Then, the *entry-specific* error of the estimation is given by the quantity $e_{uj} = \hat{r}_{uj} - r_{uj}$. The overall error is computed by averaging the entry-specific errors either in terms of absolute values or in terms of squared values. Furthermore, many systems do not predict ratings; rather they only output rankings of top-k recommended items. This is particularly common in implicit feedback data sets. Different methods are used to evaluate the accuracy of ratings predictions and the accuracy of rankings.

As the various methods for computing accuracy are discussed in detail in section 7.5, they are not discussed in detail here. The goal of this short section is to briefly introduce a few measures to ensure continuity in further discussion. The main components of accuracy evaluation are as follows:

1. *Designing the accuracy evaluation:* All the observed entries of a ratings matrix cannot be used both for training the model and for accuracy evaluation. Doing so would grossly overestimate the accuracy because of overfitting. It is important to use only a different set of entries for evaluation than was used for training. If S is the observed entries in the ratings matrix, then a small subset $E \subset S$ is used for evaluation, and the set $S - E$ is used for training. This issue is identical to that encountered in the evaluation of classification algorithms. After all, as discussed in earlier chapters, collaborative filtering is a direct generalization of the classification and regression modeling problem. Therefore, the standard methods that are used in classification and regression modeling, such as hold-out and cross-validation, are also used in the evaluation of recommendation algorithms. These issues will be discussed in greater detail in section 7.4.

2. *Accuracy metrics:* Accuracy metrics are used to evaluate either the prediction accuracy of estimating the ratings of specific user-item combinations or the accuracy of the top-k ranking predicted by a recommender system. Typically, the ratings of a set E of entries in the ratings matrix are hidden, and the accuracy is evaluated over these hidden entries. Different classes of methods are used for the two cases:

   - *Accuracy of estimating ratings:* As discussed above, the entry-specific error is given by $e_{uj} = \hat{r}_{uj} - r_{uj}$ for user u and item j. This error can be leveraged in various ways to compute the overall error over the set E of entries in the ratings matrix on which the evaluation is performed. An example is the *mean squared error*, which is denoted by *MSE*:

$$MSE = \frac{\sum_{(u,j)\in E} e_{uj}^2}{|E|} \tag{7.1}$$

   The square-root of the aforementioned quantity is referred to as the *root mean squared error*, or *RMSE*.

$$RMSE = \sqrt{\frac{\sum_{(u,j)\in E} e_{uj}^2}{|E|}} \tag{7.2}$$

   Most of these measures are borrowed from the literature on regression modeling. Other important ways of measuring the error, such as the mean absolute error, are discussed in section 7.5.

- *Accuracy of estimating rankings:* Many recommender systems do not directly estimate ratings; instead, they provide estimates of the underlying ranks. Depending on the nature of the ground-truth, one can use rank-correlation measures, utility-based measures, or the receiver operating characteristic. The latter two methods are designed for unary (implicit feedback) data sets. These methods are discussed in detail in section 7.5.

Some measures of accuracy are also designed to maximize the profit for the merchant because all items are not equally important from the perspective of the recommendation process. These metrics incorporate item-specific costs into the computation. The main problem with accuracy metrics is that they often do not measure the true effectiveness of a recommender system in real settings. For example, an obvious recommendation might be accurate, but a user might have eventually bought that item anyway. Therefore, such a recommendation might have little usefulness in terms of improving the conversion rate of the system. A discussion of the challenges associated with the use of accuracy metrics may be found in [418].

7.3.2 Coverage

Even when a recommender system is highly accurate, it may often not be able to ever recommend a certain proportion of the items, or it may not be able to ever recommend to a certain proportion of the users. This measure is referred to as *coverage*. This limitation of recommender systems is an artifact of the fact that ratings matrices are sparse. For example, in a rating matrix contains a single entry for each row and each column, then no meaningful recommendations can be made by almost *any* algorithm. Nevertheless, different recommender systems have different levels of propensity in providing coverage. In practical settings, the systems often have 100% coverage because of the use of defaults for ratings that are not possible to predict. An example of such a default would be to report the average of all the ratings of a user for an item when the rating for a specific user-item combination cannot be predicted. Therefore, the trade-off between accuracy and coverage always needs to be incorporated into the evaluation process. There are two types of coverage, which are referred to as *user-space coverage* and *item-space coverage*, respectively.

User-space coverage measures the fraction of users for which at least k ratings may be predicted. The value of k should be set to the expected size of the recommendation list. When fewer than k ratings can be predicted for a user, it is no longer possible to present a meaningful recommendation list of size k to the user. Such a situation could occur when a user has specified very few ratings in common with other users. Consider a user-based neighborhood algorithm. It is difficult to robustly compute the peers of that user, because of very few mutually specified ratings with other users. Therefore, it is often difficult to make sufficient recommendations for that user. For very high levels of sparsity, it is possible that no algorithm may be able to predict even one rating for that user. However, different algorithms may have different levels of coverage, and the coverage of a user can be estimated by running each algorithm and determining the number of items for which a prediction is made. A tricky aspect of user-space coverage is that any algorithm can provide full coverage by simply predicting random ratings for user-item combinations, whose ratings it cannot reliably predict. Therefore, user-space coverage should always be evaluated in terms of the trade-off between accuracy and coverage. For example, in a neighborhood-based recommender increasing the size of the neighborhood provides a curve showing the trade-off between coverage and accuracy.

An alternative definition of user-space coverage is in terms of the minimum amount of profile that must be built for a user before it is possible to make recommendations for that user. For a particular algorithm, it is possible to estimate through experiments the minimum number of observed ratings of any user for which a recommendation could be made. However, it is often difficult to evaluate this quantity because the metric is sensitive to the identity of the items for which the user specifies ratings.

The notion of *item-space coverage* is analogous to that of user-space coverage. Item-space coverage measures the fraction of items for which the ratings of at least k users can be predicted. In practice, however, this notion is rarely used, because recommender systems generally provide recommendation lists for users, and they are only rarely used for generating recommended users for items.

A different form of item-space coverage evaluation is defined by the notion of *catalog coverage*, which is specifically suited to recommendation *lists*. Note that the aforementioned definition was tailored to the prediction of the values of ratings. Imagine a scenario where every entry in the ratings matrix can be predicted by an algorithm, but the same set of top-k items is always recommended to every user. Therefore, even though the aforementioned definition of item-space coverage would suggest good performance, the actual coverage across all users is very limited. In other words, the recommendations are not diverse across users, and the catalog of items is not fully covered. Let T_u represent the list of top-k items recommended to user $u \in \{1 \ldots m\}$. The catalog coverage CC is defined as the fraction of items that are recommended to at least one user.

$$CC = \frac{|\cup_{u=1}^{m} T_u|}{n} \tag{7.3}$$

Here, the notation n represents the number of items. It is easy to estimate this fraction through the use of experiments.

7.3.3 Confidence and Trust

The estimation of ratings is an inexact process that can vary significantly with the specific training data at hand. Furthermore, the algorithmic methodology might also have a signifi-cant impact on the predicted ratings. This always leads to uncertainty in the user about the accuracy of the predictions. Many recommender systems may report ratings together with confidence estimates. For example, a confidence interval on the range of predicted ratings may be provided. In general, recommender systems that can accurately recommend smaller confidence intervals are more desirable because they bolster the user's trust in the system. For two algorithms that use the same method for reporting confidence, it is possible to mea-sure how well the predicted error matches these confidence intervals. For example, if two recommender systems provide 95% confidence intervals for each rating, one can measure the absolute width of the intervals reported by the two algorithms. The algorithm with the smaller confidence interval width will win as long as both algorithms are correct (i.e., within the specified intervals) at least 95% of the time on the hidden ratings. If one of the algo-rithms falls below the required 95% accuracy, then it automatically loses. Unfortunately, if one system uses 95% confidence intervals and another uses 99% confidence intervals, it is not possible to meaningfully compare them. Therefore, it is possible to use such systems only by setting the same level of confidence in both cases.

While confidence measures the system's faith in the recommendation, trust measures the user's faith in the evaluation. The notion of social trust is discussed in more detail in Chapter 11. Broadly speaking, trust measures the level of faith that the *user* has in the reported ratings. Even if the predicted ratings are accurate, they are often not useful if the user fails to trust the provided ratings. Trust is closely related to, but not quite the same as, accuracy. For example, when explanations are provided by the recommender system, the user is more likely to trust the system, especially if the explanations are logical.

Trust often does not serve the same goals as the usefulness (utility) of a recommendation. For example, if a recommender system suggests a few items already liked and known by the user, it can be argued that there is little utility provided to the user from such a recommendation. On the other hand, such items can increase the trust of the user in the system. This goal is directly in contradiction to other goals such as novelty in which recommendations already known by the user are undesirable. It is common for the various goals in recommender systems to trade-off against one another. The simplest way to measure trust is to conduct user surveys during the experiments in which the users are explicitly queried about their trust in the results. Such experiments are also referred to as *online experiments*. Numerous online methods for trust evaluation are discussed in [171, 175, 248, 486]. Generally, it is hard to measure trust through offline experiments.

7.3.4 Novelty

The novelty of a recommender system evaluates the likelihood of a recommender system to give recommendations to the user that they are not aware of, or that they have not seen before. A discussion of the notion of novelty is provided in [308]. Unseen recommendations often increase the ability of the user to discover important insights into their likes and dislikes that they did not know previously. This is more important than discovering items that they were already aware of but they have not rated. In many types of recommender systems, such as content-based methods, the recommendations tend to be somewhat obvious because of the propensity of the system to recommend expected items. While a small number of such recommendations can improve the trust of the end user in the underlying system, they are not always useful in terms of improving conversion rates. The most natural way of measuring novelty is through online experimentation in which users are explicitly asked whether they were aware of an item previously.

As discussed in the introduction, online experimentation is not always feasible because of the lack of access to a system supporting a large base of online users. Fortunately, it is possible to approximately estimate novelty with offline methods, as long as time stamps are available with the ratings. The basic idea is that novel systems are better at recommending items that are more likely to be selected by the user in the *future*, rather than at the present time. Therefore, all ratings that were created after a certain point in time t_0 are removed from the training data. Furthermore, some of the ratings occurring before t_0 are also removed. The system is then trained with these ratings removed. These removed items are then used for scoring purposes. For each item rated before time t_0 and correctly recommended, the novelty evaluation score is penalized. On the other hand, for each item rated after time t_0 and correctly recommended, the novelty evaluation score is rewarded. Therefore, this evaluation measures a type of *differential* accuracy between future and past predictions. In some measures of novelty, it is assumed that popular items are less likely to be novel, and less credit is given for recommending popular items.

7.3.5 Serendipity

The word "serendipity" literally means "lucky discovery." Therefore, serendipity is a measure of the level of surprise in successful recommendations. In other words, recommendations need to be *unexpected.* In contrast, novelty only requires that the user was not *aware* of the recommendation earlier. Serendipity is a stronger condition than novelty. All serendipitious recommendations are novel, but the converse is not always true. Consider the case where a particular user frequently eats at Indian restaurants. The recommendation of a new Pakistani restaurant to that user might be novel if that user has not eaten at that restaurant earlier. However, such a recommendation is not serendipitious, because it is well known that Indian and Pakistani food are almost identical. On the other hand, if the recommender system suggests a new Ethiopian restaurant to the user, then such a recommendation is serendipitious because it is less obvious. Therefore, one way of viewing serendipity is as a departure from "obviousness."

There are several ways of measuring serendipity in recommender systems. This notion also appears in the context of information retrieval applications [670]. The work in [214] proposed both online and offline methods for evaluating serendipity:

1. *Online methods:* The recommender system collects user feedback both on the usefulness of a recommendation and its obviousness. The fraction of recommendations that are both useful and non-obvious, is used as a measure of the serendipity.

2. *Offline methods:* One can also use a primitive recommender to generate the information about the obviousness of a recommendation in an automated way. The primitive recommender is typically selected as a content-based recommender, which has a high propensity for recommending obvious items. Then, the fraction of the recommended items in the top-k lists that are correct (i.e., high values of hidden ratings), and are also not recommended by the primitive recommender are determined. This fraction provides a measure of the serendipity.

It is noteworthy that it is not sufficient to measure the fraction of non-obvious items, because a system might recommend unrelated items. Therefore, the usefulness of the items is always incorporated in the measurement of serendipity. Serendipity has important long-term effects on improving the conversion rate of a recommender system, even when it is opposed to the immediate goal of maximizing accuracy. A number of metrics for serendipity evaluation are discussed in [214, 450].

7.3.6 Diversity

The notion of diversity implies that the set of proposed recommendations *within a single recommended list* should be as diverse as possible. For example, consider the case where three movies are recommended to a user in the list of top-3 items. If all three movies are of a particular genre and contain similar actors, then there is little diversity in the recommendations. If the user dislikes the top choice, then there is a good chance that she might dislike all of them. Presenting different types of movies can often increase the chance that the user might select one of them. Note that the diversity is always measured over a *set* of recommendations, and it is closely related to novelty and serendipity. Ensuring greater diversity can often increase the novelty and serendipity of the recommendations. Furthermore, greater diversity of recommendations can also increase the sales diversity and catalog coverage of the system.

Diversity can be measured in terms of the content-centric similarity between pairs of items. The vector-space representation of each item description is used for the similarity computation. For example, if a set of k items are recommended to the user, then the pairwise similarity is computed between every pair of items in the list. The average similarity between all pairs can be reported as the diversity. Lower values of the average similarity indicate greater diversity. Diversity can often provide very different results from those of accuracy metrics. A discussion of the connection of diversity and similarity is provided in [560].

7.3.7 Robustness and Stability

A recommender system is stable and robust when the recommendations are not significantly affected in the presence of attacks such as fake ratings or when the patterns in the data evolve significantly over time. In general, significant profit-driven motivations exist for some users to enter fake ratings [158, 329, 393, 444]. For example, the author or publisher of a book might enter fake positive ratings about a book at Amazon.com, or they might enter fake negative ratings about the books of a rival. Attack models for recommender systems are discussed in Chapter 12. The evaluation of such models is also studied in the same chapter. The corresponding measures can be used to estimate the robustness and stability of such systems against attacks.

7.3.8 Scalability

In recent years, it has become increasingly easy to collect large numbers of ratings and implicit feedback information from various users. In such cases, the sizes of the data sets continue to increase over time. As a result, it has become increasingly essential to design recommender systems that can perform effectively and efficiently in the presence of large amounts of data [527, 528, 587]. A variety of measures are used for determining the scalability of a system:

1. *Training time:* Most recommender systems require a training phase, which is separate from the testing phase. For example, a neighborhood-based collaborative filtering algorithm might require pre-computation of the peer group of a user, and a matrix factorization system requires the determination of the latent factors. The overall time required to train a model is used as one of the measures. In most cases, the training is done offline. Therefore, as long as the training time is of the order of a few hours, it is quite acceptable in most real settings.

2. *Prediction time:* Once a model has been trained, it is used to determine the top recommendations for a particular customer. It is crucial for the prediction time to be low, because it determines the latency with which the user receives the responses.

3. *Memory requirements:* When the ratings matrices are large, it is sometimes a challenge to hold the entire matrix in the main memory. In such cases, it is essential to design the algorithm to minimize memory requirements. When the memory requirements become very high, it is difficult to use the systems in large-scale and practical settings.

The importance of scalability has become particularly great in recent years because of the increasing importance of the "big-data" paradigm.

7.4 Design Issues in Offline Recommender Evaluation

In this section, we will discuss the issue of recommender evaluation design. The discussions in this section and the next pertain to *accuracy* evaluation of offline and historical data sets. It is crucial to design recommender systems in such a way that the accuracy is not grossly overestimated or underestimated. For example, one cannot use the same set of specified ratings for both training and evaluation. Doing so would grossly overestimate the accuracy of the underlying algorithm. Therefore, only a part of the data is used for training, and the remainder is often used for testing. The ratings matrix is typically sampled in an *entry-wise fashion*. In other words, a subset of the entries are used for training, and the remaining entries are used for accuracy evaluation. Note that this approach is similar to that used for testing classification and regression modeling algorithms. The main difference is that classification and regression modeling methods sample *rows* of the labeled data, rather than sampling the *entries*. This difference is because the unspecified entries are always restricted to the class variable in classification, whereas any entry of the ratings matrix can be unspecified. The design of recommender evaluation systems is very similar to that of classifier evaluation systems because of the similarity between the recommendation and classification problems.

A common mistake made by analysts in the benchmarking of recommender systems is to use the same data for parameter tuning and for testing. Such an approach grossly overestimates the accuracy because parameter tuning is a part of training, and the use of test data in the training process leads to overfitting. To guard against this possibility, the data are often divided into three parts:

1. *Training data:* This part of the data is used to build the training model. For example, in a latent factor model, this part of the data is used to create the latent factors from the ratings matrix. One might even use these data to create multiple models in order to eventually select the model that works best for the data set at hand.

2. *Validation data:* This part of the data is used for model selection and parameter tuning. For example, the regularization parameters in a latent factor model may be determined by testing the accuracy over the validation data. In the event that multiple models have been built from the training data, the validation data are used to determine the accuracy of each model and select the best one.

3. *Testing data:* This part of the data is used to test the accuracy of the final (tuned) model. It is important that the testing data are not even looked at during the process of parameter tuning and model selection to prevent overfitting. The testing data are *used only once at the very end of the process*. Furthermore, if the analyst uses the results on the test data to adjust the model in some way, then the results will be contaminated with knowledge from the testing data.

An example of a division of the ratings matrix into training, validation, and testing data is illustrated in Figure 7.1(a). Note that the validation data may also be considered a part of the training data because they are used to create the final tuned model. The division of the ratings matrix into the ratios 2:1:1 is particularly common. In other words, half the specified ratings are used for model-building, and a quarter may be used for each of model-selection and testing, respectively. However, when the sizes of the ratings matrices are large, it is possible to use much smaller proportions for validation and testing. This was the case for the Netflix Prize data set.

7.4.1 Case Study of the Netflix Prize Data Set

A particularly instructive example of a well-known data set used in collaborative filtering is the Netflix Prize data set, because it demonstrates the extraordinary lengths to which Netflix went to prevent overfitting on the test set from the contest participants. In the Netflix data

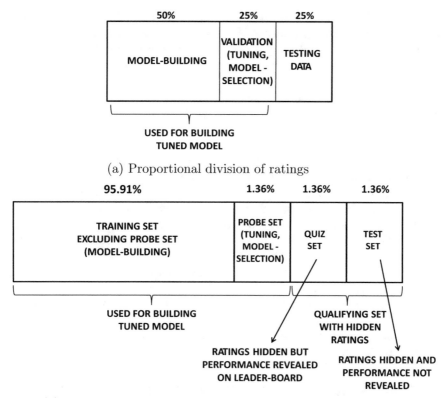

(a) Proportional division of ratings

(b) Division in Netflix Prize data set (not drawn to scale)

Figure 7.1: Partitioning a ratings matrix for evaluation design

set, the largest portion of the data set contained 95.91% of the ratings. This portion of the data set was typically used by the contest participants for model-building. Another 1.36% of the data set was revealed to the participants as a *probe set*. Therefore, the model-building portion of the data and the probe data together contained $95.91 + 1.36 = 97.27\%$ of the data. The probe set was typically used by contests for various forms of parameter tuning and model selection, and therefore it served a very similar purpose as a validation set. However, different contestants used the probe set in various ways, especially since the ratings in the probe set were more recent, and the statistical distribution of the ratings in the training and probe sets were slightly different. For the case of ensemble methods [554], the probe set was often used to learn the weights of various ensemble components. The combined data set with revealed ratings (including the probe set) corresponds to the full training data, because it was used to build the final tuned model. An important peculiarity of the training data was that the distributions of the probe set and the model-building portion of the training set were not exactly identical, although the probe set reflected the statistical characteristics of the *qualifying set* with hidden ratings. The reason for this difference was that most of the ratings data were often quite old and they did not reflect the true distribution of the more recent or future ratings. The probe and qualifying sets were based on more recent ratings, compared to the 95.91% of the ratings in the first part of the training data.

The ratings of the remaining 2.7% of the data were hidden, and only triplets of the form $\langle User, Movie, GradeDate \rangle$ were supplied without actual ratings. The main difference from a test set was that participants could submit their performance on the qualifying set to Netflix, and the performance on half the qualifying data, known as the *quiz set*, was revealed to the participants on a *leader-board*. Although revealing the performance on the quiz set to the participants was important in order to give them an idea of the quality of their results, the problem with doing so was that participants could use the knowledge of the performance of their algorithm on the leader-board to over-train their algorithm on the quiz set with repeated submissions. Clearly, doing so results in contamination of the results from knowledge of the performance on the quiz set, even when ratings are hidden. Therefore, the part of the qualifying set that was *not* in the quiz set was used as the test set, and the results on *only* this part of the qualifying set were used to determine the final performance for the purpose of prize determination. The performance on the quiz set had no bearing on the final contest, except to give the participants a continuous idea of their performance during the contest period. Furthermore, the participants were not informed about which part of the qualifying set was the quiz set. This arrangement ensured that a truly out-of-sample data set was used to determine the final winners of the contest.

The overall division of the Netflix data set is shown in Figure 7.1(b). The only difference from the division in Figure 7.1(a) is the presence of an additional quiz set. It is, in fact, possible to remove the quiz set entirely without affecting the Netflix contest in any significant way, except that participants would no longer be able to obtain an idea of the quality of their submissions. Indeed, the Netflix Prize evaluation design is an excellent example of the importance of not using any knowledge of the performance on the test set at any stage of the training process *until the very end*. Benchmarking in research and practice often fails to meet these standards in one form or the other.

7.4.2   Segmenting the Ratings for Training and Testing

In practice, real data sets are not pre-partitioned into training, validation, and test data sets. Therefore, it is important to be able to divide the entries of a ratings matrix into these portions automatically. Most of the available division methods, such as *hold-out* and *cross-validation*, are used to divide[1] the data set into *two* portions instead of *three*. However, it is possible to obtain three portions as follows. By first dividing the rating entries into training and test portions, and then further segmenting the validation portion from the training data, it is possible to obtain the required three segments. Therefore, in the following, we will discuss the segmentation of the ratings matrix into training and testing portions of the entries using methods such as hold-out and cross-validation. However, these methods are also used for dividing the training data into the model-building and validation portions. This hierarchical division is illustrated in Figure 7.2. In the following, we will consistently use the terminology of the first level of division in Figure 7.2 into "training" and "testing" data, even though the same approach can also be used for the second level division into model building and validation portions. This consistency in terminology is followed to avoid confusion.

7.4.2.1   Hold-Out

In the hold-out method, a fraction of the entries in the ratings matrix are hidden, and the remaining entries are used to build the training model. The accuracy of predicting the hidden entries is then reported as the overall accuracy. Such an approach ensures that the reported accuracy is not a result of overfitting to the specific data set, because the entries

---

[1] The actual design in methods such as cross-validation is slightly more complex because the data are segmented in multiple ways, even though they are always divided into two parts during a particular execution phase of training.

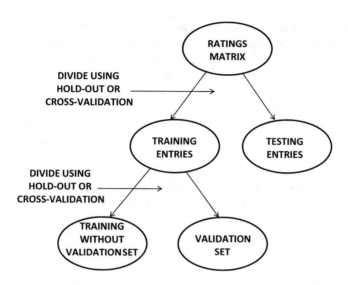

Figure 7.2: Hierarchical division of rated entries into training, validation, and testing portions

used for evaluation are hidden during training. Such an approach, however, underestimates the true accuracy. First, all entries are not used in training, and therefore the full power of the data is not used. Second, consider the case where the held-out entries have a higher average rating than the full ratings matrix. This means that the held-in entries have a lower average rating than the ratings matrix, and also the held-out entries. This will lead to a pessimistic bias in the evaluation.

7.4.2.2 Cross-Validation

In the cross-validation method, the ratings entries are divided into q equal sets. Therefore, if S is the set of specified entries in the ratings matrix R, then the size of each set, in terms of the number of entries, is $|S|/q$. One of the q segments is used for testing, and the remaining $(q-1)$ segments are used for training. In other words, a total of $|S|/q$ entries are hidden during *each* such training process, and the accuracy is then evaluated over these entries. This process is repeated q times by using each of the q segments as the test set. The average accuracy over the q different test sets is reported. Note that this approach can closely estimate the true accuracy when the value of q is large. A special case is one where q is chosen to be equal to the number of specified entries in the ratings matrix. Therefore, $|S| - 1$ rating entries are used for training, and the one entry is used for testing. This approach is referred to as *leave-one-out cross-validation*. Although such an approach can closely approximate the accuracy, it is usually too expensive to train the model $|S|$ times. In practice, the value of q is fixed to a number such as 10. Nevertheless, leave-one-out cross-validation is not very difficult to implement for the specific case of neighborhood-based collaborative filtering algorithms.

7.4.3 Comparison with Classification Design

The evaluation design in collaborative filtering is very similar to that in classification. This is not a coincidence. Collaborative filtering is a generalization of the classification problem, in which any missing entry can be predicted rather than simply a particular variable, which is designated as the dependent variable. The main difference from classification is that the data are segmented on a row-wise basis (between training and test *rows*) in classification, whereas the data are segmented on an entry-wise basis (between training and test *entries*) in collaborative filtering. This difference closely mirrors the nature of the relationship between the classification and the collaborative filtering problems. Discussions of evaluation designs in the context of the classification problem can be found in [18, 22].

One difference from classification design is that the performance on hidden entries often does not reflect the true performance of the system in real settings. This is because the hidden ratings are not chosen at random from the matrix. Rather, the hidden ratings are typically items that the user has chosen to consume. Therefore, such entries are likely to have higher values of the ratings as compared to truly missing values. This is a problem of *sample selection bias*. Although this problem could also arise in classification, it is far more pervasive in collaborative filtering applications. A brief discussion of this issue is provided in section 7.6.

7.5 Accuracy Metrics in Offline Evaluation

Offline evaluation can be performed by measuring the accuracy of predicting rating values (e.g., with *RMSE*) or by measuring the accuracy of ranking the recommended items. The logic for the latter set of measures is that recommender systems often provide ranked lists of items without explicitly predicting ratings. Ranking-based measures often focus on the accuracy of only the ranks of the top-k items rather than all the items. This is particularly true in the case of implicit feedback data sets. Even in the case of explicit ratings, the ranking-based evaluations provide a more realistic perspective of the true usefulness of the recommender system because the user only views the top-k items rather than all the items. However, for bench-marking, the accuracy of ratings predictions is generally preferred because of its simplicity. In the Netflix Prize competition, the *RMSE* measure was used for final evaluation. In the following, both forms of accuracy evaluation will be discussed.

7.5.1 Measuring the Accuracy of Ratings Prediction

Once the evaluation design for an offline experiment has been finalized, the accuracy needs to be measured over the test set. As discussed earlier, let S be the set of specified (observed) entries, and $E \subset S$ be the set of entries in the test set used for evaluation. Each entry in E is a user-item index pair of the form (u, j) corresponding to a position in the ratings matrix. Note that the set E may correspond to the held out entries in the hold-out method, or it may correspond to one of the partitions of size $|S|/q$ during cross-validation.

Let r_{uj} be the value of the (hidden) rating of entry $(u, j) \in E$, which is used in the test set. Furthermore, let $\hat{r}_{uj}$ be the predicted rating of the entry (u, j) by the specific training algorithm being used. The entry-specific error is given by $e_{uj} = \hat{r}_{uj} - r_{uj}$. This error can be leveraged in various ways to compute the overall error over the set E of entries on which the evaluation is performed. An example is the *mean squared error,* denoted by *MSE*:

$$MSE = \frac{\sum_{(u,j) \in E} e_{uj}^2}{|E|} \tag{7.4}$$

Clearly, smaller values of the *MSE* are indicative of superior performance. The square-root of this value is referred to as the *root mean squared error (RMSE),* and it is often used instead of the *MSE*:

$$RMSE = \sqrt{\frac{\sum_{(u,j)\in E} e_{uj}^2}{|E|}} \tag{7.5}$$

The *RMSE* is in units of ratings, rather than in units of squared ratings like the *MSE*. The *RMSE* was used as the standard metric for the Netflix Prize contest. One characteristic of the *RMSE* is that it tends to disproportionately penalize large errors because of the squared term within the summation. One measure, known as the *mean-absolute-error (MAE)*, does not disproportionately penalize larger errors:

$$MAE = \frac{\sum_{(u,j)\in E} |e_{uj}|}{|E|} \tag{7.6}$$

Other related measures such as the normalized *RMSE* (*NRMSE*) and normalized *MAE* (*NMAE*) are defined in a similar way, except that each of them is divided by the range $r_{max} - r_{min}$ of the ratings:

$$NRMSE = \frac{RMSE}{r_{max} - r_{min}}$$

$$NMAE = \frac{MAE}{r_{max} - r_{min}}$$

The normalized values of the *RMSE* and *MAE* always lie in the range $(0, 1)$, and therefore they are more interpretable from an intuitive point of view. It is also possible to use these values to compare the performance of a particular algorithm over different data sets with varying scales of ratings.

7.5.1.1 RMSE versus MAE

Is *RMSE* or *MAE* better as an evaluation measure? There is no clear answer to this question, as this depends on the application at hand. As the *RMSE* sums up the squared errors, it is more significantly affected by large error values or outliers. A few badly predicted ratings can significantly ruin the *RMSE* measure. In applications where robustness of prediction across various ratings is very important, the *RMSE* may be a more appropriate measure. On the other hand, the *MAE* is a better reflection of the accuracy when the importance of outliers in the evaluation is limited. The main problem with *RMSE* is that it is not a true reflection of the average error, and it can sometimes lead to misleading results [632]. Clearly, the specific choice should depend on the application at hand. A discussion of the relative benefits of the two kinds of measures can be found in [141].

7.5.1.2 Impact of the Long Tail

One problem with these metrics is that they are heavily influenced by the ratings on the popular items. The items that receive very few ratings are ignored. As discussed in Chapter 2, ratings matrices exhibit a long-tail property, in which the vast majority of items are bought (or rated) rarely. We have replicated Figure 2.1 of Chapter 2 in Figure 7.3. The X-axis represents the indices of the items in decreasing order of popularity, and the Y-axis indicates the rating frequency. It is evident that only a few items receive a large number of ratings, whereas most of the remaining items receive few ratings. The latter constitute the long tail. Unfortunately, items in the long tail often contribute to the vast majority of profit for merchants [49]. As a result, the most important items often get weighted the least in the evaluation process. Furthermore, it is often much harder to predict the values of

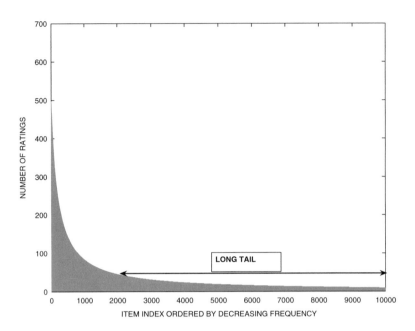

Figure 7.3: The long tail of rating frequencies (Revisiting Figure 2.1 of Chapter 2)

the ratings in the long tail because of greater local sparsity [173]. Therefore, the prediction accuracies on sparse items will typically be different from those on popular items. One way of handling this problem is to compute the *RMSE* or *MAE* separately for all the hidden ratings associated with each item, and then average over the different items in a weighted way. In other words, the accuracy computations of Equations 7.5 and 7.6 can be weighted with an item-specific weight, depending on the relative importance, profit, or utility to the merchant. It is also possible to perform these computations with user-specific weighting (rather than item-specific weighting), although the practical applicability of user-specific weighting is limited.

7.5.2   Evaluating Ranking via Correlation

The aforementioned measures are designed to evaluate the prediction accuracy of the actual rating value of a user-item combination. In practice, the recommender system creates a ranking of items for a user, and the top-k items are recommended. The value of k may vary with the system, item, and user at hand. In general, it is desirable for highly rated items to be ranked above items which are not highly rated. Consider a user u, for which the ratings of the set I_u of items have been hidden by a hold-out or cross-validation strategy. For example, if the ratings of the first, third, and fifth items (columns) of user (row) u are hidden for evaluation purposes, then we have $I_u = \{1, 3, 5\}$.

We would like to measure how well the ground-truth orderings of the ratings in I_u are related to the ordering predicted by the recommender system for the set I_u. An important issue to keep in mind is that ratings are typically chosen from a discrete scale, and many ties exist in the ground truth. Therefore, it is important for the ranking measures to not penalize the system for ranking one item above another when they are tied in the ground truth. The most common class of methods is to use *rank correlation coefficients*. The two most commonly used rank correlation coefficients are as follows:

1. *Spearman rank correlation coefficient:* The first step is to rank all items from 1 to $|I_u|$, both for the recommender system prediction and for the ground-truth. The Spearman correlation coefficient is simply equal to the Pearson correlation coefficient applied on these ranks. The computed value always ranges in $(-1, +1)$, and large positive values are more desirable.

   The Spearman correlation coefficient is specific to user u, and it can then be averaged over all users to obtain a global value. Alternatively, the Spearman rank correlation can be computed over all the hidden ratings over all users in one shot, rather than computing user-specific values and averaging them.

   One problem with this approach is that the ground truth will contain many ties, and therefore random tie-breaking might lead to some noise in the evaluation. For this purpose, an approach referred to as *tie-corrected Spearman* is used. One way of performing the correction is to use the average rank of all the ties, rather than using random tie-breaking. For example, if the ground-truth rating of the top-2 ratings is identical in a list of four items, then instead of using the ranks $\{1, 2, 3, 4\}$, one might use the ranks $\{1.5, 1.5, 3, 4\}$.

2. *Kendall rank correlation coefficient:* For each pair of items $j, k \in I_i$, the following credit $C(j, k)$ is computed by comparing the predicted ranking with the ground-truth ranking of these items:

$$
C(j, k) = \begin{cases}
+1 & \text{if items } j \text{ and } k \text{ are in the same relative order in} \\
& \text{ground-truth ranking and predicted ranking (concordant)} \\
-1 & \text{if items } j \text{ and } k \text{ are in a different relative order in} \\
& \text{ground-truth ranking and predicted ranking (discordant)} \\
0 & \text{if items } j \text{ and } k \text{ are tied in either the} \\
& \text{ground-truth ranking or predicted ranking}
\end{cases}
\tag{7.7}
$$

Then, the Kendall rank correlation coefficient τ_u, which is specific to user u, is computed as the average value of $C(j, k)$ over all the $|I_u|(|I_u| - 1)/2$ pairs of test items for user i:

$$
\tau_u = \frac{\sum_{j < k} C(j, k)}{|I_u| \cdot (|I_u| - 1)/2}
\tag{7.8}
$$

A different way of understanding the Kendall rank correlation coefficient is as follows:

$$
\tau_u = \frac{\text{Number of concordant pairs} - \text{Number of discordant pairs}}{\text{Number of pairs in } I_u}
\tag{7.9}
$$

Note that this value is a *customer-specific* value of the Kendall coefficient. The value of τ_u may be averaged over all users u to obtain a heuristic global measure. Alternatively, one can perform the Kendall coefficient computation of Equation 7.8 over all hidden user-item pairs, rather than only the ones for customer u, in order to obtain a global value τ.

A number of other measures, such as the *normalized distance-based performance measure (NDPM)*, have been proposed in the literature. Refer to the bibliographic notes.

7.5.3   Evaluating Ranking via Utility

In the previous discussion, the ground-truth ranking is compared to the recommender system's ranking. Utility-based methods use the ground-truth *rating* in combination with the recommender system's ranking. For the case of implicit feedback data sets, the rating is substituted with a 0-1 value, depending on whether or not the customer has consumed the item. The overall goal of utility-based methods is to create a crisp quantification of how *useful* the customer might find the recommender system's ranking. An important principle underlying such methods is that recommendation lists are short compared to the total number of items. Therefore, most of the utility of a particular ranking should be based on the relevance of items, which are high in the recommended list. In this sense, the *RMSE* measure has a weakness because it equally weights the errors on the low-ranked items as compared to those on the highly-ranked items. It has been suggested [713] that small changes in *RMSE* such as 1%, can lead to large changes of more than 15% in the *identities* of the top-rated items. These are the only items that the end-user of the recommender system will actually see. Correspondingly, utility-based measures quantify the utility of a recommendation list by giving greater importance to the top-ranked items.

As in the previous sections, it is assumed that the ground-truth rating of each item in I_u is hidden from the recommender system before evaluation. Here, I_u represents the set of items rated by user u, which are hidden from the recommender system before evaluation. We will develop both user-specific and global utility quantifications.

In utility-based ranking, the basic idea is that each item in I_u has a utility to the user, which depends both on its position in the recommended list and its ground-truth rating. An item that has a higher ground-truth rating obviously has greater utility to the user. Furthermore, items ranked higher in the recommended list have greater utility to the user i because they are more likely to be noticed (by virtue of their position) and eventually selected. Ideally, one would like items with higher ground-truth rating to be placed as high on the recommendation list as possible.

How are these rating-based and ranking-based components defined? For any item $j \in I_u$, its rating-based utility to the user i is assumed to be $\max\{r_{uj} - C_u, 0\}$, where C_u is a break-even (neutral) rating value for user u. For example, C_u might be set to the mean rating of user u. On the other hand, the ranking-based utility of the item is $2^{-(v_j-1)/\alpha}$, where v_j is the rank of item j in the list of recommended items and α is a half-life parameter. In other words, the ranking-based utility exponentially decays with its rank, and moving down the ranks by α reduces the utility by a factor of 2. The logic of the decay-based ranking component is to ensure that the final utility of a particular ranking is regulated primarily by the top few items. After all, the user rarely browses the items that are very low in the list. The utility $F(u, j)$ of item $j \in I_u$ to user u is defined as the product of the rating-based and ranking-based utility values:

$$F(u, j) = \frac{\max\{r_{uj} - C_u, 0\}}{2^{(v_j-1)/\alpha}} \qquad (7.10)$$

The R-score, which is specific to user u, is the sum of $F(u, j)$ over all the hidden ratings in I_u:

$$\text{R-score}(u) = \sum_{j \in I_u} F(u, j) \qquad (7.11)$$

Note that the value of v_j can take on any value from 1 to n, where n is the total number of items. However, in practice, one often restricts the size of the recommended list to a maximum value of L. One can therefore compute the R-score over a recommended list of specific size L instead of using all the items, as follows:

$$\text{R-score}(u) = \sum_{j \in I_u, v_j \leq L} F(u, j) \tag{7.12}$$

The idea here is that ranks below L have no utility to the user because the recommended list is of size L. This variation is based on the principle that recommended lists are often very short compared to the total number of items. The overall R-score may be computed by summing this value over all the users.

$$\text{R-score} = \sum_{u=1}^{m} \text{R-score}(u) \tag{7.13}$$

The exponential decay in the utility implies that users are only interested in top-ranked items, and they do not pay much attention to lower-ranked items. This may not be true in all applications, especially in news recommender systems, where users typically browse multiple items lower down the list of recommended items. In such cases, the discount rate should be set in a milder way. An example of such a measure is the discounted cumulative gain (DCG). In this case, the discount factor of item j is set to $\log_2(v_j + 1)$, where v_j is the rank of item j in the test set I_u. Then, the discounted cumulative gain is defined as follows:

$$DCG = \frac{1}{m} \sum_{u=1}^{m} \sum_{j \in I_u} \frac{g_{uj}}{\log_2(v_j + 1)} \tag{7.14}$$

Here, g_{uj} represents the utility (or gain) of the user u in consuming item j. Typically, the value of g_{uj} is set to an exponential function of the relevance (e.g., non-negative ratings or user hit rates):

$$g_{uj} = 2^{rel_{uj}} - 1 \tag{7.15}$$

Here, rel_{uj} is the ground-truth relevance of item j for user u, which is computed as a heuristic function of the ratings or hits. In many settings, the raw ratings are used. It is common to compute the discounted cumulative gain over a recommendation list of specific size L, rather than using all the items:

$$DCG = \frac{1}{m} \sum_{u=1}^{m} \sum_{j \in I_u, v_j \leq L} \frac{g_{uj}}{\log_2(v_j + 1)} \tag{7.16}$$

The basic idea is that recommended lists have size no larger than L.

Then, the normalized discounted cumulative gain (NDCG) is defined as ratio of the discounted cumulative gain to its ideal value, which is also referred to as ideal discounted cumulative gain (IDCG).

$$NDCG = \frac{DCG}{IDCG} \tag{7.17}$$

The ideal discounted cumulative gain is computed by repeating the computation for DCG, except that the ground-truth rankings are used in the computation.

Another measure that is commonly used, is the average reciprocal hit rate (ARHR) [181]. This measure is designed for implicit feedback data sets, in which each value of $r_{uj} \in \{0, 1\}$. Therefore, a value of $r_{uj} = 1$ represents a "hit" where a customer has bought or clicked on an item. A value of $r_{uj} = 0$ corresponds to a situation where a customer has not bought or clicked on an item. In this implicit feedback setting, missing values in the ratings matrix are assumed to be 0.

In this case, the rank-based discount rate is $1/v_j$, where v_j is the rank of item j in the recommended list, and the item utility is simply the hidden "rating" value $r_{uj} \in \{0, 1\}$. Note that the discount rate is not as rapid as the R-score metric, but it is faster than DCG. Therefore, the combined utility of an item is given by r_{uj}/v_j. This expression represents the contribution of item $j \in I_u$ to the utility. Then, the ARHR metric for the user i is defined by summing up these values over all the hidden items in I_u:

$$ARHR(u) = \sum_{j \in I_u} \frac{r_{uj}}{v_j} \qquad (7.18)$$

It is also possible to define the average reciprocal hit-rate for a recommended list of size L by adding only those utility values for which $v_j \leq L$.

$$ARHR(u) = \sum_{j \in I_u, v_j \leq L} \frac{r_{uj}}{v_j} \qquad (7.19)$$

One quirk of the average reciprocal hit-rate is that it is typically used when the value of $|I_u|$ is exactly 1, and when the value r_{uj} of the corresponding (hidden) item $j \in I_u$ is always 1. Therefore, there is exactly one hidden item for each user, and the user has always bought or clicked on this item. In other words, the average reciprocal hit-rate rewards the utility (in a rank-reciprocal way) for recommending the single correct answer at a high position on the recommended list. This was the setting in which this measure was introduced [181], although one can generalize it to arbitrary settings in terms of the number of hidden items and explicit-feedback settings. The aforementioned expression provides this generalized definition because one can use a set I_u of arbitrary size in an explicit feedback setting. The global ARHR value is computed by averaging this value over the m users:

$$ARHR = \frac{\sum_{u=1}^{m} ARHR(u)}{m} \qquad (7.20)$$

The average reciprocal hit-rate is also referred to as the *mean reciprocal rank (MRR)*. In cases where the value of $|I_u|$ is 1, the average reciprocal hit-rate always lies in the range $(0, 1)$. In such cases, the hidden entry is usually an item for which $r_{uj} = 1$ and the length of the recommendation list is restricted to L. Note that only "hits" contribute to the utility in these cases. A simplification of this measure is the *hit-rate*, in which the rank-reciprocal weighting is not used, and the value of $|I_u|$ is exactly 1. Therefore, the hit-rate (HR) is simply the fraction of users for which the correct answer is included in the recommendation list of length L. The disadvantage of the hit-rate is that it gives equal importance to a hit, irrespective of its position in the recommended list.

The ARHR and HR are almost always used in implicit feedback data sets, in which missing values are treated as 0. Nevertheless, the definition of Equation 7.19 is stated in a more general way. Such a definition can also be used in the context of explicit feedback data sets, in which the values of r_{uj} need not be drawn from $\{0, 1\}$. In such cases, the ratings of any number of items of each user are hidden, and the values of the hidden ratings can be arbitrary. Furthermore, the missing values need not be treated as 0s, and I_u is always selected from the observed items.

A related measure is the *mean average precision (MAP)*, which computes the fraction of relevant items in a recommended list of length L for a given user. Various equally spaced

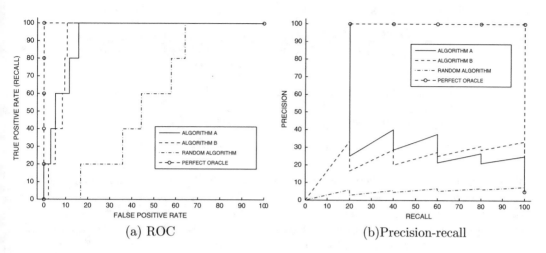

(a) ROC   (b)Precision-recall

Figure 7.4: ROC curve and precision-recall curves

values of L are used, and the precision is averaged over these recommendation lists of varying lengths. The resulting precision is then averaged over all the users.

Numerous other measures have been proposed in the literature to evaluate the effectiveness of rankings. For example, the *lift index* [361] divides the ranked items into deciles to compute a utility score. Refer to the bibliographic notes.

7.5.4 Evaluating Ranking via Receiver Operating Characteristic

Ranking methods are used frequently in the evaluation of the actual *consumption* of items. For example, Netflix might recommend a set of ranked items for a user, and the user might eventually consume only a subset of these items. Therefore, these methods are well suited to implicit feedback data sets, such as sales, click-throughs, or movie views. Such actions can be represented in the form of unary ratings matrices, in which missing values are considered to be equivalent to 0. Therefore, the ground-truth is of a binary nature.

The items that are eventually consumed are also referred to as the *ground-truth positives* or *true positives*. The recommendation algorithm can provide a ranked list of *any* number of items. What percentage of these items is relevant? A key issue here is that the answer to this question depends on the size of the recommended list. Changing the number of recommended items in the ranked list has a direct effect on the trade-off between the fraction of recommended items that are actually consumed and the fraction of consumed items that are captured by the recommender system. This trade-off can be measured in two different ways with the use of a precision-recall or a *receiver operating characteristic (ROC)* curve. Such trade-off plots are commonly used in rare class detection, outlier analysis evaluation, and information retrieval. In fact, such trade-off plots can be used in any application where a binary ground truth is compared to a ranked list discovered by an algorithm.

The basic assumption is that it is possible to rank all the items using a numerical score, which is the output of the algorithm at hand. Only the top items are recommended. By varying the size of the recommended list, one can then examine the fraction of relevant (ground-truth positive) items in the list, and the fraction of relevant items that are missed by the list. If the recommended list is too small, then the algorithm will miss relevant items (false-negatives). On the other hand, if a very large list is recommended, this will lead to too many spurious recommendations that are never used by the user (false-positives).

Table 7.1: Rank of ground-truth positive instances

Algorithm	Rank of items that are truly used (ground-truth positives)
Algorithm A	1, 5, 8, 15, 20
Algorithm B	3, 7, 11, 13, 15
Random Algorithm	17, 36, 45, 59, 66
Perfect Oracle	1, 2, 3, 4, 5

This leads to a trade-off between the false-positives and false-negatives. The problem is that the correct size of the recommendation list is never known exactly in a real scenario. However, the entire trade-off curve can be quantified using a variety of measures, and two algorithms can be compared over the entire trade-off curve. Two examples of such curves are the *precision-recall* curve and the *receiver operating characteristic (ROC)* curve.

Assume that one selects the top-t set of ranked items to recommend to the user. For any given value t of the size of the recommended list, the set of recommended items is denoted by $\mathcal{S}(t)$. Note that $|\mathcal{S}(t)| = t$. Therefore, as t changes, the size of $\mathcal{S}(t)$ changes as well. Let $\mathcal{G}$ represent the true set of relevant items (ground-truth positives) that are consumed by the user. Then, for any given size t of the recommended list, the *precision* is defined as the percentage of recommended items that truly turn out to be relevant (i.e., consumed by the user).

$$Precision(t) = 100 \cdot \frac{|\mathcal{S}(t) \cap \mathcal{G}|}{|\mathcal{S}(t)|}$$

The value of $Precision(t)$ is *not* necessarily monotonic in t because both the numerator and denominator may change with t differently. The *recall* is correspondingly defined as the percentage of *ground-truth* positives that have been recommended as positive for a list of size t.

$$Recall(t) = 100 \cdot \frac{|\mathcal{S}(t) \cap \mathcal{G}|}{|\mathcal{G}|}$$

While a natural trade-off exists between precision and recall, this trade-off is not necessarily monotonic. In other words, an increase in recall does not always lead to a reduction in precision. One way of creating a single measure that summarizes both precision and recall is the F_1-measure, which is the harmonic mean between the precision and the recall.

$$F_1(t) = \frac{2 \cdot Precision(t) \cdot Recall(t)}{Precision(t) + Recall(t)} \tag{7.21}$$

While the $F_1(t)$ measure provides a better quantification than either precision or recall, it is still dependent on the size t of the recommended list and is therefore still not a complete representation of the trade-off between precision and recall. It is possible to visually examine the entire trade-off between precision and recall by varying the value of t and plotting the precision versus the recall. As shown later with an example, the lack of monotonicity of the precision makes the results harder to intuitively interpret.

A second way of generating the trade-off in a more intuitive way is through the use of the ROC curve. The *true-positive rate*, which is the same as the recall, is defined as the percentage of ground-truth positives that have been included in the recommendation list of size t.

$$TPR(t) = Recall(t) = 100 \cdot \frac{|\mathcal{S}(t) \cap \mathcal{G}|}{|\mathcal{G}|}$$

The false-positive rate $FPR(t)$ is the percentage of the falsely reported positives in the recommended list out of the ground-truth negatives (i.e., irrelevant items not consumed by the user). Therefore, if $\mathcal{U}$ represents the universe of all items, the ground-truth negative set is given by $(\mathcal{U} - \mathcal{G})$, and the falsely reported part in the recommendation list is $(\mathcal{S}(t) - \mathcal{G})$. Therefore, the false-positive rate is defined as follows:

$$FPR(t) = 100 \cdot \frac{|\mathcal{S}(t) - \mathcal{G}|}{|\mathcal{U} - \mathcal{G}|} \tag{7.22}$$

The false-positive rate can be viewed as a kind of "bad" recall, in which the fraction of the ground-truth negatives (i.e., items not consumed), which are incorrectly captured in the recommended list $\mathcal{S}(t)$, is reported. The ROC curve is defined by plotting the $FPR(t)$ on the X-axis and $TPR(t)$ on the Y-axis for varying values of t. In other words, the ROC curve plots the "good" recall against the "bad" recall. Note that both forms of recall will be at 100% when $\mathcal{S}(t)$ is set to the entire universe of items. Therefore, the end points of the ROC curve are always at $(0, 0)$ and $(100, 100)$, and a random method is expected to exhibit performance along the diagonal line connecting these points. The *lift* obtained above this diagonal line provides an idea of the accuracy of the approach. The area under the ROC curve provides a concrete quantitative evaluation of the effectiveness of a particular method. Although one can directly use the area shown in Figure 7.4(a), the staircase-like ROC curve is often modified to use local linear segments which are not parallel to either the X-axis or the Y-axis. The trapezoidal rule [195] is then used to compute the area slightly more accurately. From a practical point of view, this change often makes very little difference to the final computation.

To illustrate the insights gained from these different graphical representations, consider an example of a scenario with 100 items, in which 5 items are truly relevant. Two algorithms A and B are applied to this data set that rank all items from 1 to 100, with lower ranks being selected first in the recommended list. Thus, the true-positive rate and false-positive rate values can be generated from the ranks of the 5 relevant items. In Table 7.1, some hypothetical ranks for the 5 truly relevant items have been illustrated for the different algorithms. In addition, the ranks of the ground-truth positive items for a random algorithm have been indicated. This algorithm ranks all the items randomly. Similarly, the ranks for a "perfect oracle" algorithm, which ranks the correct top 5 items in the recommended list, have also been illustrated in the table. The resulting ROC curves are illustrated in Figure 7.4(a). The corresponding precision-recall curves are illustrated in Figure 7.4(b). Note that the ROC curves are always increasing monotonically, whereas the precision-recall curves are not monotonic. While the precision-recall curves are not quite as nicely interpretable as the ROC curves, it is easy to see that the *relative trends* between different algorithms are the same in both cases. In general, ROC curves are used more frequently because of greater ease in interpretability.

What do these curves really tell us? For cases in which one curve strictly dominates another, it is clear that the algorithm for the former curve is superior. For example, it is immediately evident that the oracle algorithm is superior to all algorithms and that the random algorithm is inferior to all the other algorithms. On the other hand, algorithms A and B show domination at different parts of the ROC curve. In such cases, it is hard to say that one algorithm is strictly superior. From Table 7.1, it is clear that Algorithm A ranks three relevant items very highly, but the remaining two items are ranked poorly. In the case of Algorithm B, the highest ranked items are not as well ranked as Algorithm A, though all 5 relevant items are determined much earlier in terms of rank threshold. Correspondingly, Algorithm A dominates on the earlier part of the ROC curve, whereas

Algorithm B dominates on the later part. It is possible to use the area under the ROC curve as a proxy for the overall effectiveness of the algorithm. However, not all parts of the ROC curve are equally important because there are usually practical limits on the size of the recommended list.

The aforementioned description illustrates the generation of *customer-specific ROC curves*, because the ROC curves are specific to each user. It is also possible to generate *global ROC curves* by ranking user-item pairs and then using the same approach as discussed above. In order to rank user-item pairs, it is assumed that the algorithm has a mechanism to rank them by using predicted affinity values. For example, the predicted ratings for user-item pairs can be used to rank them.

7.5.5 Which Ranking Measure is Best?

Although ROC curves are often used for evaluating recommender systems, they do not always reflect the performance from the end-user perspective. In many settings, the end user sees only a small subset of top-ranked items. Measures such as ROC and Kendall coefficient, which treat higher and lower ranked items equally, are unable to capture the greater importance of higher ranked items. For example, the relative ranking between two items ranked first and second on the recommendation list is far more important than the relative ranking of two items, which are ranked 100th and 101st on the list. In this context, utility-based measures such as NDCG do a much better job than rank-correlation coefficients or ROC measures at distinguishing between higher-ranked and lower-ranked items.

7.6 Limitations of Evaluation Measures

Accuracy-based evaluation measures have a number of weaknesses that arise out of selection bias in recommender systems. In particular, the missing entries in a ratings matrix are not random because users have the tendency of rating more popular items. As shown in Figure 7.3, a few items are rated by many users, whereas the vast majority of items may be found in the *long tail*. The distributions of the ratings on popular items are often different from those on items in the long tail. When an item is very popular, it is most likely because of the notable content in it. This factor will affect[2] the rating of that item as well. As a result, the accuracy of most recommendation algorithms is different on the more popular items versus the items in the long tail [564]. More generally, the fact that a particular user has *chosen* not to rate a particular item thus far has a significant impact on what her rating would be if the user were forced to rate all items. This issue is stated in [184] in a somewhat different context as follows:

> "Intuitively, a simple process could explain the results: users chose to rate songs they listen to, and listen to music they expect to like, while avoiding genres they dislike. Therefore, most of the songs that would get a bad rating are not voluntarily rated by the users. Since people rarely listen to random songs, or rarely watch random movies, we should expect to observe in many areas a difference between the distribution of ratings for random items and the corresponding distribution for the items selected by the users."

---

[2]A related effect is that observed ratings are likely to be specified by users who are frequent raters. Frequent raters may show different patterns of rating values compared to infrequent raters.

These factors cause problems of bias in the evaluation process. After all, in order to perform the evaluation on a given data set, one cannot use truly missing ratings; rather, one must simulate missing items with the use of hold-out or cross-validation mechanisms on ratings that are already specified. Therefore, the *simulated* missing items may not show similar accuracy to that one would theoretically obtain on the *truly* consumed items in the future. The items that are consumed in the future will not be randomly selected from the missing entries for the reasons discussed above. This property of rating distributions is also known as *Missing Not At Random (MNAR)*, or *selection bias* [402, 565]. This property can lead to an incorrect *relative* evaluation of algorithms. For example, a popularity-based model in which items with the highest mean rating are recommended might do better in terms of gaining more revenue for the merchant than its evaluation on the basis of randomly missing ratings might suggest. This problem is aggravated by the fact that items in the long tail are especially important to the recommender system, because a disproportionate portion of the profits in such systems are realized through such items.

There are several solutions to this issue. The simplest solution is to not select the missing ratings at random but to use a model for selecting the test ratings based on their likelihood of being rated in the future. Another solution is to not divide the ratings at random between training and test, but to divide them *temporally* by using more recent ratings as a part of the test data; indeed, the Netflix Prize contest used more recent ratings in the qualifying set, although some of the recent ratings were also provided as a part of the probe set. An approach that has been used in recent years, is to correct for this bias by modeling the bias in the missing rating distribution within the evaluation measures [565, 566]. Although such an approach has some merits, it does have the drawback that the evaluation process itself now *assumes* a model of how the ratings behave. Such an approach might inadvertently favor algorithms that use a model similar to that used for the prediction of ratings as for the evaluation process. It is noteworthy that many recent algorithms [309] use implicit feedback within the *prediction* process. This raises the possibility that a future *prediction* algorithm might be designed to be tailored to the model used for adjusting for the effect of user selection bias within the *evaluation*. Although the assumptions in [565], which relate the missing ratings to their relevance, are quite reasonable, the addition of more assumptions (or complexity) to evaluation mechanisms increases the possibility of "gaming" during benchmarking. At the end of the day, it is important to realize that these limitations in collaborative filtering evaluation are inherent; the quality of any evaluation system is fundamentally limited by the quality of the available ground truth. In general, it has been shown through experiments on Netflix data [309] that the use of straightforward RMSE measures on the observed ratings often correlate quite well with the precision on all items.

Another source of evaluation bias is the fact that user interests may evolve with time. As a result, the performance on a hold-out set might not reflect future performance. Although it is not a perfect solution, the use of *temporal* divisions between training and test ratings seems like a reasonable choice. Even though temporal division results in training and testing tests with somewhat different distributions, it also reflects the real-world setting more closely. In this sense, the Netflix Prize contest again provides an excellent model of realistic evaluation design. Several other variations of temporal methods in the evaluation process are discussed in [335].

7.6.1   Avoiding Evaluation Gaming

The fact that missing ratings are not random can sometimes lead to unintended (or intended) gaming of the evaluations in settings where the user-item pairs of the test entries are specified. For example, in the Netflix Prize contest, the *coordinates* of the user-item pairs in the qualifying set were specified, although the *values* of the ratings were not specified. By incorporating the coordinates of the user-item pairs within the qualifying set as implicit feedback (i.e., matrix F in section 3.6.4.6), one can improve the quality of recommendations. It can be argued that such an algorithm would have an unfair advantage over one that did not include any information about the identities of rated items in the qualifying set. The reason is that in real-life settings, one would never have any information about future coordinates of rated items, as are easily available in the qualifying set of the Netflix Prize data. Therefore, the additional advantage of incorporating such implicit feedback would disappear in real-life settings. One solution would be to not specify the coordinates of test entries and thereby evaluate over all entries. However, if the ratings matrix has very large dimensions (e.g., $10^7 \times 10^5$), it may be impractical to perform the prediction over all entries. Furthermore, it would be difficult to store and upload such a large number of predictions in an online contest like the Netflix Prize. In such cases, an alternative would be to include (spurious) unrated entries within the test set. Such entries are not used for evaluation but they have the effect of preventing the use of coordinates of the test entries as implicit feedback.

7.7   Summary

The evaluation of recommender systems is crucial in order to obtain a clear idea about the quality of different algorithms. The most direct method of measuring the effectiveness of a recommender system is to compute the conversion rate at which recommended items are converted to actual usages. This can be done through either user studies or online studies. Such studies are often difficult for researchers and practitioners because of the difficulty in obtaining access to the relevant infrastructure with large groups of users. Offline methods have the advantage that they can be used with multiple historical data sets. In such cases, it is dangerous to use accuracy as the only criterion, because maximizing accuracy does not always lead to long-term maximization of conversion rates. A variety of criteria, such as coverage, novelty, serendipity, stability, and scalability, are used to evaluate the effectiveness of recommender systems.

The proper design of recommender evaluation systems is necessary to ensure that there are no biases in the evaluation process. For example, in a collaborative filtering application, it is important to ensure that all the ratings are evaluated with an out-of-sample approach. A variety of methods such as hold-out and cross-validation are used in order to ensure out-of-sample evaluation. The error is computed with measures, such as the MAE, MSE, and RMSE. In some measures, items are weighted differently to account for their differential importance. In order to evaluate the effectiveness of ranking methods, rank correlation, utility-based measures or usage-based measures may be used. For usage-based measures, precision and recall are used to characterize the trade-off inherent in varying the size of the recommended list. The F1-measure is also used, which is the harmonic mean between the precision and the recall.

7.8 Bibliographic Notes

Excellent discussions on evaluating recommender systems may be found in [246, 275, 538]. Evaluation can be performed either with user studies or with historical data sets. The earliest work on evaluation with user studies may be found in [339, 385, 433]. An early study of evaluation of recommendation algorithms with historical data sets may be found in [98]. Metrics for evaluating recommender systems in the presence of cold-start are discussed in [533]. Controlled experiments for online evaluation in Web applications are discussed in [305]. A general study of online evaluation design is provided in [93]. The evaluation of multi-armed bandit systems is discussed in [349]. A comparison of online recommender systems with respect to human decisions is provided in [317].

The work in [246] presents several variants of accuracy metrics for evaluation. This article is perhaps one of the foremost authorities on the evaluation of recommender systems. The pitfalls of using the *RMSE* as an evaluation measure are presented in [632]. A brief technical note on the relative merits of using *MAE* and *RMSE* as accuracy measures may be found in [141]. The challenges and pitfalls in the use of accuracy metrics are discussed in [418]. Alternative methods for evaluating recommender systems are provided in [459]. A discussion of the importance of novelty is provided in [308]. Online methods for measuring the novelty of a recommender system are provided in [140, 286]. The use of popularity in the measurement of novelty is discussed in [140, 539, 680]. The work in [670] showed that serendipity can be achieved in a recommender system with the help of user labeling. Metrics for serendipity evaluation are discussed in [214, 450]. The work in [214] also studies the use of coverage metrics. Diversity metrics are discussed in [560]. The impact of recommender systems on sales diversity is discussed in [203]. Robustness and stability metrics for recommender systems are discussed in [158, 329, 393, 444]. A study of the evaluation of classification systems may be found in [18, 22]. The discussions in these books provide an understanding of the standard techniques used, such as hold-out and cross-validation.

Rank correlation methods are discussed in [298, 299]. The normalized distance preference measure is discussed in [505]. The R-score for the utility-based evaluation of rankings is discussed in [98]. The NDCG measure is discussed in [59]. The lift index is discussed in [361], whereas the average reciprocal hit rate (ARHR) is proposed in [181]. A discussion of ROC curves in the context of classification may be found in [195], although the same ideas are also applicable to the case of recommender systems. The use of customer-specific and global ROC curves is discussed in [533].

One limitation of recommender systems is that the values on the ratings are related to their relative frequency and that missing items are often in the long tail. Therefore, the use of cross-validation or hold-out mechanisms leads to a selection bias against less frequent items. A number of recent methods for correcting for missing item bias are discussed in [402, 564–566]. The approach in [565] proposes the use of different assumptions for the relevant and non-relevant items, in terms of deciding which ratings are missing. A training algorithm is also designed in [565] based on these assumptions. A temporal framework for realistic evaluation is discussed in [335]. Recommender systems also need to be evaluated somewhat differently in various settings, such as in the presence of specific contexts. These contexts could include time, location, or social information. An evaluation framework for recommender systems in the context of temporal data is provided in [130]. A recent workshop that was devoted exclusively to recommender systems' evaluation may be found in [4].

7.9   Exercises

1. Suppose that a merchant knows the amount of profit q_i made on the sale of the ith item. Design an error metric for a collaborative filtering system that weights the importance of each item with its profit.

2. Suppose that you designed an algorithm for collaborative filtering and found that it was performing poorly on ratings with value 5, but it was performing well on the other ratings. You used this insight to modify your algorithm and tested the algorithm again. Discuss the pitfalls with the second evaluation. Relate your answer to why Netflix chose to separate the quiz set and the test set in the Netflix Prize data set.

3. Implement an algorithm for constructing the ROC and the precision-recall curves.

4. Suppose you have an implicit feedback data set in which the ratings are unary. Would an ROC curve provide more meaningful results or would the *RMSE* metric?

5. Consider a user John, for whom you have hidden his ratings for *Aliens* (5), *Terminator* (5), *Nero* (1), and *Gladiator* (6). The values in brackets represent his hidden ratings, and higher values are better. Now consider a scenario where the recommender system ranks these movies in the order *Terminator, Aliens, Gladiator, Nero*.

    (a) Compute the Spearman rank correlation coefficient as a measure of recommendation ranking quality.

    (b) Compute the Kendall rank correlation coefficient as a measure of ranking quality.

6. For the problem in Exercise 5, John's utility for a movie j is given by $\max\{r_{ij} - 3, 0\}$, where r_{ij} is his rating.

    (a) Under this utility assumption, compute the R-score specific to John. Assume a half-life value of $\alpha = 1$.

    (b) For the same utility assumption, compute the component of the discounted cumulative gain (DCG) specific to John, if there are a total of 10 users in the system.

7. For the problem in Exercise 5, assume that the only hidden ratings belong to John, and the predicted ratings from the recommender system are *Aliens* (4.3), *Terminator* (5.4), *Nero* (1.3), and *Gladiator* (5). The values in brackets represent the predicted ratings.

    (a) Compute the MSE of the predicted ratings.

    (b) Compute the MAE of the predicted ratings.

    (c) Compute the RMSE of the predicted ratings.

    (d) Compute the normalized MAE and RMSE, assuming that all ratings lie in the range $\{1 \ldots 6\}$.

Chapter 8

Context-Sensitive Recommender Systems

"For me context is the key – from that comes the understanding of everything." – Kenneth Noland

8.1 Introduction

Context-sensitive recommender systems tailor their recommendations to additional information that defines the specific situation under which recommendations are made. This additional information is referred to as the *context*. Some examples of context are as follows:

1. *Time:* Recommendations can be affected by many aspects of time, such as weekdays, weekends, holidays, and so on. A recommendation that is relevant to the morning context, may not be relevant in the evening and vice versa. Clothing recommendations during summer and winter may be different. A number of time-sensitive recommendation methods are discussed in Chapter 9. In fact, some of the methods discussed in this chapter, such as pre-filtering and post-filtering, are re-examined in Chapter 9 in the temporal context.

2. *Location:* With the increasing popularity of GPS-enabled mobile phones, location-sensitive recommendations have gained increasing importance in recent years. For example, a traveling user might wish to determine a recommendation for a restaurant in her locality. Context-sensitive systems can provide more relevant recommendations by using the location as a context. The next chapter will provide several examples of location-aware systems.

© Springer International Publishing Switzerland 2016
C.C. Aggarwal, *Recommender Systems: The Textbook,*
DOI 10.1007/978-3-319-29659-3_8

3. *Social information:* The social context is often important from the perspective of recommender systems. For example, the choice of a user's friends, tags, and social circles can affect the recommendation process. Similarly, a person might choose to watch a different movie depending on whether she is watching it with her parents or with her boyfriend [5]. Social recommender systems are discussed in Chapters 10 and 11. Some of these systems can be considered contextual systems as well.

The context of a user can be detected in a variety of ways. In some cases, it can be learned with little effort because the data is already available. For example, the GPS receiver on a mobile phone would indicate a customer's location, and the time-stamp of a customer transaction indicates the time. Such methods are referred to as implicit collection methods [466]. In other cases, the context is not quite as readily available. For example, it can be learned explicitly by gathering information through surveys or other means. Finally, in some cases, data mining and inference tools can be used to gather contextual information.

In traditional recommender systems with user set U and item set I, the set of possibilities in $U \times I$ is mapped to a rating. This mapping results in (an incompletely specified) ratings matrix of size $|U| \times |I|$. In a context-aware system, an additional set of contextual possibilities is present in the set C. For example, the set C might be {*morning, afternoon, night*}, with the context corresponding to the time of day. In this case, it is no longer possible to map the possibilities in $U \times I$ to the ratings, because the same user might have different preferences for an item depending on whether the time is in the morning, afternoon, or night. The context must be included in the mapping in order to provide a more refined and accurate recommendation. Therefore, in context-sensitive recommender systems, the possibilities in $U \times I \times C$ are mapped to the ratings. Formally, the function h_R, which maps the user, items, and context to the rating, can be written as follows:

$$h_R : U \times I \times C \rightarrow rating$$

The function h_R is subscripted with R to denote the data set to which it is applied. In this case, the ratings data R is a 3-dimensional ratings data *cube* corresponding to the user, item, and context. It is possible to use multiple types of context within a single recommendation application. For example, aside from the time, one might use location, weather, or the social context. Therefore, one might have multiple contextual *dimensions*. This would result in an multidimensional cube for representing the ratings. As we will see later in this chapter, the notion of multidimensional representation can be used to seamlessly represent a variety of different contexts. In this chapter, we will work with such an multidimensional contextual model.

This chapter is organized as follows. Section 8.2 discusses the multidimensional model for contextual recommendations. A contextual pre-filtering and reduction approach is described in section 8.3. Post-filtering methods are described in section 8.4. The process of incorporating context directly into the recommendation process is discussed in section 8.5. A summary is given in section 8.6.

8.2   The Multidimensional Approach

The traditional problem of recommendations can be viewed as that of learning a mapping function from the user-item combinations to the ratings. The corresponding function f_R may be defined as follows:

$$f_R : U \times I \rightarrow rating \tag{8.1}$$

The 2-dimensional ratings matrix is used by this function to create the mapping. Therefore, this function maps a data point in the 2-dimensional space of users and items to ratings. Of course, the dimensions could, in principle, correspond not just to users or items, but to any type of context. This general principle motivates the *multidimensional approach* [6] to recommendations, in which the rating problem is seen as that of mapping a set of w different dimensional values to a rating.

$$g_R : D_1 \times D_2 \ldots \times D_w \Rightarrow rating$$

In this case, the ratings data R contain w different dimensions that are mapped to ratings, just as the 2-dimensional user-item combinations are mapped to ratings in the traditional setting. This results in a w-dimensional cube rather than a 2-dimensional matrix. The w different dimensions are denoted by $D_1 \ldots D_w$. Note that two of these dimensions will always be users and items, as in the classical case of multidimensional recommendations, but the other values of D_i might correspond to other contexts. For example, these contexts could correspond to time, location, and so on. Therefore, the traditional recommendation problem can be viewed as a special case of the multidimensional approach in which the only two dimensions are users and items. A nice way of viewing this generalization is as an online analytical processing (OLAP) data cube [145], which is traditionally used in data warehousing. An example of such an OLAP cube with three dimensions is shown in Figure 8.1 corresponding to user, item (movie), and time. Each cell in this cube contains a rating for a particular user, item, and time combination. Although the context in this case is an ordered variable (time), it is usually treated as a discrete value during the analytical process. Furthermore, some representations of time, such as weekday, weekend, or season, are not ordered. Similarly, the contextual dimension could very well have been location, which is not an ordered variable. Treating the contextual dimensions in a discrete setting is essential to the data cube paradigm.

The rating function g_R is defined as a partial function, in which the number of arguments is equal to the number of dimensions w. In the example of Figure 8.1, the rating function $g_R(David, Terminator, 9\ PM)$ refers to the rating of user David when he watches the movie *Terminator* at 9 PM. This cell is shaded in Figure 8.1. The mapping function g_R is referred to as *partial* because it is defined only for the subset of cells corresponding to observed rating values. The remaining values need to be learned in a data-driven manner for making contextual recommendations. Note that the context can be a property of the user, a property of the item, a property of the user-item combination, or a completely independent property. For example, when David watches *Terminator* at 9 PM, the context of 9 PM relates to both, because the user watches the movie at that specific time and the time does not exclusively relate to either the user or the item. However, it is also possible for the context to relate to only one of the two. For example, consider a movie recommendation application in which the movies are recommended to a user based on the ratings matrix and also her demographic characteristics. In such a case, the context is clearly related to the user. In general, however, it is not important who the context is related to, because it is treated as a completely independent entity from the user or the item. Therefore, a separate dimension is assigned to each context, just as there are individual dimensions assigned to the user and item, respectively. This abstraction helps in addressing the most general cases of context-sensitive recommendations.

At a more general level, this idea can be related to that of querying for top-ranked combinations of values with the use of two disjoint subsets from $D_1 \ldots D_w$. The selected subsets of dimensions in $D_1 \ldots D_w$ are either "what" dimensions, or they are "for whom"

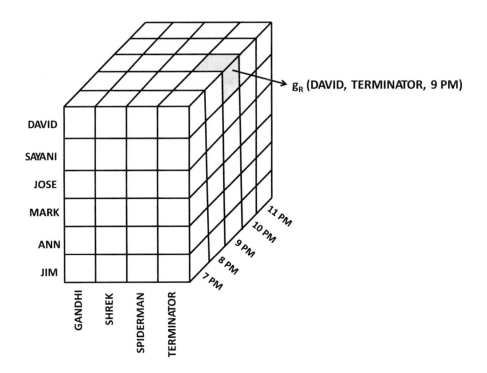

Figure 8.1: The multidimensional rating cube

dimensions. Each of the dimensions belongs to one of the two categories but it cannot belong to both categories. A typical query is of the following form:

Determine the top-k possibilities in the "what" dimensions for a particular set of specified values in the "for whom" dimensions.

In traditional recommender systems, the item dimension always belongs to the former category, whereas the user dimension always belongs to the latter category. However, in multidimensional recommender systems, this constraint does not apply. Formally, the problem of multidimensional recommendations can be defined as follows [6]:

Definition 8.2.1 (Multidimensional Recommendations) *Given the recommendation space $D_1 \times D_2 \times \ldots D_w$ and the rating function $g_R : D_1 \times D_2 \ldots \times D_w \rightarrow$ rating, the recommendation problem is defined by selecting certain "what" dimensions $D_{i_1} \ldots D_{i_p}$ and certain "for whom" dimensions $D_{j_1} \ldots D_{j_q}$ that do not overlap, and recommending for a query tuple $(d_{j_1} \ldots d_{j_q}) \in D_{j_1} \times \ldots \times D_{j_q}$ the top-k tuples $(d_{i_1} \ldots d_{i_p}) \in D_{i_1} \times \ldots \times D_{i_p}$ with the maximum predicted value of the rating $g_R(d_1, d_2, \ldots, d_w)$.*

In other words, a ranked list of the "what" dimension combinations are recommended in response to "for whom" queries. The traditional 2-dimensional model of recommendations is a special case of this scenario in which items are recommended to users. Therefore, items always belong to the "what" category and users always belong to the "for whom" category. In a multidimensional recommender system, a more general framework is used, where the segmentation between the "what" and the "for whom" items might be arbitrary. For example, one might recommend best item-time combinations for each user, or one might recommend best user-time combinations for each item. Alternatively, one might recommend the best time(s) for each user-item combination. Note that both users and items belong to the "for whom" category in the last case. In a social application, one might wish to recommend the best companions to watch a movie with for a particular user-movie combination. Note that the union of the "what" and "for whom" dimensions might be a proper subset of the full set of w dimensions. For example, consider the case where $w = 4$ and we have the time and location context in addition to the user and item dimensions. It is possible for the query to completely ignore the time and only use the location context to make recommendations.

As is reflected in the above descriptions, the multidimensional model is particularly rich, and it allows broad leeway in deciding the formulation of the recommendations. In fact, a query language [9], referred to as *Recommendation Query Language (RQL)*, has been developed for formulating different types of recommendation requests in a multidimensional recommender system. Such query languages are particularly useful for selecting different subsets of "what" and "for whom" dimensions in the querying process, and in developing a systematic query response methodology.

8.2.1   The Importance of Hierarchies

In the traditional OLAP model, hierarchies are often defined over various dimensions. For example, in a sales application, the various cells of the data cube correspond to sales values, and the location dimensions may have various hierarchical levels, such as city, state, region, and so on. One can aggregate the sales at the level of state, region, or the country. Furthermore, one can combine the location dimension with the time dimension by aggregating the sales in a particular region over a particular period of time. Such aggregation can also be performed in multidimensional recommender systems. Hierarchies are also useful in the context-sensitive recommender systems because they provide various levels of abstraction in which one might perform aggregated analysis.

In order to perform aggregated analysis, it is assumed that some or all of the dimensions have hierarchies associated with them. These hierarchies are part of the input to the recommender system. The nature of the hierarchy is highly domain-specific, and it depends on the application at hand. Some examples are as follows:

1. The location dimension can have a hierarchy corresponding to city, state, region, country, and so on.

2. If demographic information is associated with users, then one can also arrange the person dimension in a hierarchy of demographic attributes, such as age or occupation. A dimension such as age can be discretized into various hierarchical levels of granularity.

3. The item dimension can use a standard industry hierarchy, such as the North American Industry Classification System (NAICS). Alternatively, one can use a variety of genres or subgenres to represent the items in a number of product domains (with the movie domain being an example).

4. Dimensions such as time can be represented in various granular levels of hierarchy, such as hours, days, weeks, months, and so on.

Clearly, the user needs to make careful choices up front about the hierarchy to use, so that the most relevant analysis may be performed in a given application. It is also important to select the most relevant contextual dimensions $D_1 \ldots D_w$ for the application at hand. This problem is closely related to that of feature selection [18, 22] in the traditional classification and machine learning literature. Alternatively, these dimensions can be selected by domain experts.

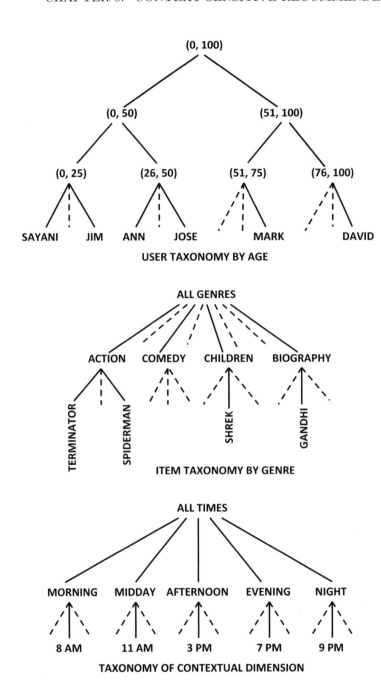

Figure 8.2: Taxonomies of users, items, and context

Examples of possible hierarchies for users, items (movies), and time are shown in Figure 8.2. The users are categorized by age, the movies are categorized by genre, and the time is categorized by the time-of-day. Now consider the case where these hierarchies are used in the example of Figure 8.1. With these hierarchies, one can now make more general (aggregated) queries, such as $g_R(David, Terminator, Evening)$, instead of

$g_R(David, Terminator, 7\ PM)$. The former provides an *average* prediction of how much David likes the movie *Terminator*, if he watches it at any time in the evening, whereas the latter provides a prediction of how much he would like it, if he saw the movie in the 7 PM show. At the extreme end, a query such as $g_R(David, Action, Any\ Time)$ corresponds to completely ignoring the time context and focusing on a specific genre of movies. This query estimates the average rating of David for action movies that are watched at any time. Therefore, the hierarchies are useful not only from the contextual perspective, but also from the perspective of hierarchical analysis on the user and item dimensions.

It is possible to combine the hierarchical analysis on the user and item dimensions. For example, one can aggregate further by querying how much users in the age range $[20, 30]$ like action movies rather than focusing on a specific user such as David. This is achieved with the function $g_R(Users \in Age[20, 30], Action, Any\ Time)$. Note that the hierarchy in Figure 8.2 groups the users by age. Such aggregated queries can be viewed as a sort of aggregation in multidimensional recommender systems. For example, one can view the aggregated rating $g_R(David, Action, Any\ Time)$ in terms of an aggregation function as follows:

$$g_R(David, Action, Any\ Time) = \text{AGGR}_{(x \in Action, \text{All } y)} g_R(David, x, y) \qquad (8.2)$$

In traditional OLAP applications, one can obtain the relevant aggregation by summing the relevant values in the cells. This is also referred to as the "roll-up" operator in traditional OLAP systems. However, in recommender systems, it is more meaningful to talk of *averages* rather than additions. One might determine either the average rating of David for action movies or the average of his top-k ratings for action movies. The main challenge here is that the ratings are not completely observed in the original data cube, which is defined as a partial function. In most cases, the ratings are specified in a very sparse way at the bottom level of the hierarchy. In some cases, it is also possible for the observed ratings to be specified at the higher levels. For example, in some systems, David might be able to directly specify his interests for action or comedy movies, rather than providing ratings for individual movies. Multidimensional recommender systems are designed to address these scenarios as well. Therefore, a crucial step is to able to estimate the missing ratings at all levels of the hierarchy. These estimated ratings, together with the originally specified ratings, can be used to provide responses to various queries. Therefore, the *multi-level multidimensional rating estimation problem* is stated as follows:

Definition 8.2.2 (Multi - level Multidimensional Rating Estimation Problem)
Given an initial set of user-assigned ratings specified at different levels of the multidimensional cube of ratings, the task is to estimate all other ratings in the cube at all the levels of the OLAP hierarchies.

Although full use of the hierarchical information is not always possible with various strategies, most techniques are able to predict ratings at the lowest level from other ratings. The techniques for performing contextual recommendation fall into one of three categories:

1. *Contextual pre-filtering:* In these methods, a segment of the ratings is pre-filtered corresponding to the relevant context. This relevant segment of ratings is then used to make targeted recommendations.

2. *Contextual post-filtering:* In these methods, the recommendations are first performed on the entire global set of ratings. Subsequently, the ranked recommendation lists are filtered or adjusted as a post-processing step with the use of temporal context.

3. *Contextual modeling:* In this case, the contextual information is incorporated directly into the prediction function, rather than as a pre-filtering or post-filtering step. This

is fundamentally different from the previous case, in which traditional 2-dimensional recommender systems are used under the covers. Contextual modeling is the most general approach in which one *directly* works with the w-dimensional representation of the ratings matrix in the modeling process. This approach provides the most integrated results, but it is sometimes computationally intensive or otherwise difficult to execute in a high-dimensional setting.

In the following sections, we will discuss these different classes of techniques for making recommendations. It is noteworthy that some of these techniques, such as post-filtering, use additional side-information about the various dimensions. These pieces of side information are referred[1] to as *attributes*. For example, a user might have demographic information associated with them, such as their name, address, age, gender, or profession. An item, such as a movie, might have side information associated with it, such as the title, actors, directors, and so on. Attributes are associated not just with the user and item dimensions, but also with the contextual dimensions. For example, consider a case where a user wishes to watch a movie with a specific companion. The companion dimension might contain a name, companion type (e.g., friend or parent), and age. As we will see later in this chapter, these types of side information are often important for some types of contextual recommendation applications. The set of attributes associated with a dimension is referred to as its *profile*. Note that item profiles and user profiles are used frequently for learning content-based recommendation models (cf. Chapter 4). Such attributes are also useful in many algorithms for the contextual setting.

8.3 Contextual Pre-filtering: A Reduction-Based Approach

Contextual pre-filtering is also referred to as *reduction* [6]. In the reduction-based approach, the idea is to reduce the w-dimensional estimation problem to a set of 2-dimensional estimations. The 2-dimensional estimation problem is equivalent to that in traditional collaborative filtering systems.

In order to understand this point, we will use an example of a 3-dimensional recommender system. Consider the case where the three attributes are users (U), movie items (I), and time (T). In such a case, the rating function g_R is defined as follows:

$$g_R : U \times I \times T \to rating$$

Note that the data set R is a 3-dimensional cube in this case. Consider a traditional 2-dimensional recommender system in which the mapping $f_{R'}$ is as follows:

$$f_{R'} : U \times I \to rating$$

In this case, the data cube R' is a 2-dimensional cube, in which only the two dimensions U and I are present. Clearly, ignoring the contextual dimension is equivalent to using a 2-dimensional recommender system. The 3-dimensional prediction function can be expressed in terms of the 2-dimensional prediction function by using a reduced derivative of the 3-dimensional ratings matrix. At any queried time t, this is achieved by deriving a

---

[1] In the traditional database context, the notions of dimension and attribute mean the same thing. In this case, however, they do not mean the same thing. A set of attributes is associated with a dimension.

2-dimensional ratings matrix $R'(t)$ from R with a pair of standard database operations:

$$R'(t) = \text{Project}_{U,I}(\text{Select}_{T=t}(R))$$
$$= \pi_{U,I}(\sigma_{T=t}(R))$$

Note that projection and selection are standard database operators. The matrix $R'(t)$ is obtained by first selecting the ratings in which the time is fixed to t, and then projecting down to the user and item dimensions. In other words, the 2-dimensional *slice* of the data cube in which the time is fixed to t corresponds to $R'(t)$. This is shown in Figure 8.3, where the entire user-item slice at 9 PM is shaded. Note that this 2-dimensional slice creates a user-item matrix, which can be used with traditional collaborative filtering algorithms. Such an approach can be used to perform the ratings prediction with the context fixed at 9 PM. In general, the 3-dimensional ratings estimation can be systematically reduced to 2-dimensional ratings estimation on this slice with the following relationship between the 3-dimensional function g_R and the traditional 2-dimensional collaborative filtering function $f_{R'(t)}$:

$$\forall (u, i, t) \in U \times I \times T, \quad g_R(u, i, t) = f_{R'(t)}(u, i)$$

This approach can easily be generalized to the case where there are $w > 3$ dimensions $D_1 \ldots D_w$ by fixing the remaining $w - 2$ dimensions. The two dimensions, which are not fixed, are referred to as the *main* dimensions, whereas the other dimensions are the *contextual* dimensions. In typical applications, the users and items are the main dimensions. By fixing the values of the contextual dimensions, we can extract specific *slices* or *segments* of the data that are defined on only the two main dimensions. Traditional collaborative filtering algorithms can be used on such segments.

Because only a small subset of the ratings are used in a given slice, one may sometimes not have sufficient ratings to perform an accurate recommendation. In such cases, one may aggregate the rating at t with other adjacent time slices to create more accurate recommendations. For example, instead of using $t = 9$ PM, one might use all values of t in the evening, from 7 PM to 11 PM, and then average the ratings in these slices to create the resulting matrix. The 2-dimensional recommender is then applied to this averaged slice.

The main advantage of the approach is that it performs the collaborative filtering only over the *relevant* ratings in which the ratings have been selected with the use of the context. This can lead to improved accuracy in many cases, although the trade-off is that fewer ratings are now being used for prediction. Averaging over adjacent slices allows retention of a limited amount of local relevance, while reducing sparsity. The problem of sparsity can, nevertheless, be significant in many cases, and it may not always be possible to use such an averaging in order to increase the number of ratings. When fewer ratings are available, overfitting becomes more likely, and it is easy to envision scenarios in which the accuracy of the approach is not very high.

There are many natural solutions to reducing the sparsity problem at the expense of losing refinement. At one extreme, one might ignore the context altogether, and the ratings matrix is averaged over all the possible (combinations of) values of the contextual dimension(s). Such an approach will contain less relevant ratings than the *local* model that pre-selects the ratings based on context. This is clearly an extreme generalization of the approach discussed in the previous section where one averages only over adjacent values of the contextual variable (e.g., averaging the ratings in slices from 7 PM to 11 PM instead of using only the slice at 9 PM). We refer to this extreme approach as the *global* approach. Although the global approach uses less relevant ratings than the approach of using contextually localized slices, it will be able to use more ratings from the averaged slice. The comparative accuracy between the two alternatives depends on the nature of the

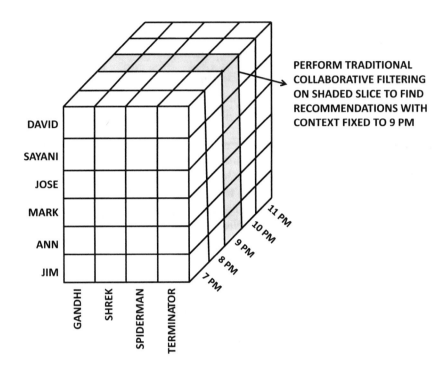

Figure 8.3: Extracting a 2-dimensional slice by fixing the context in reduction methods

trade-off between relevance and data sparsity. In many practical applications, it has been observed that either of the two alternatives might be better, depending on the part of the ratings matrix that one is looking at.

8.3.1   Ensemble-Based Improvements

Because of the unpredictability in the relative quality obtained with the global and local method, a question arises, as to how one might combine the two methods to obtain a technique that provides high accuracy in most scenarios. Although the local method provides more relevant results, it could also cause overfitting when there are too few relevant ratings for that context. We will discuss an ensemble-based method to improve the quality of the prediction. The goal of the ensemble-based method is to use the best of both worlds in the prediction process. In other words, either the local or the global matrix may be used, depending on the part of the ratings matrix that one is looking at. This approach results in the best trade-off between sparsity and local relevance. In this context, the *bucket-of-models* hybrid (cf. section 6.4.2 of Chapter 6) is very useful because it can help decide between alternative models of varying quality. The approach, however, needs to be tailored so that instead of selecting the best model, it selects the best data segment to train the model.

In the following discussion, a define a data segment of the ratings cube with the use of a combination of values of the contextual variables. For example, if the main variables are users and items, and the contextual variables are location and time, then each possible value of a location-time pair defines a data segment. When a recommendation problem is posed with a particular location-time context, it is important for the recommendation algorithm to determine whether or not using that context is indeed helpful.

In the training step, the approach first identifies the cross-validated accuracy on each data segment. For example, when the context is location and time, one determines the cross-validated accuracy of using all location-time pairs. In cases, where a hierarchical tree-like structure of the context is available, the higher-level nodes of the tree can be included as possibilities for the location-time pairs. A table is created for each location-time possibility, which contains its best generalization to use to obtain the highest accuracy. For example, if the contextual variables are location and time, examples of generalizations of (*9 PM, Boston*) might be (*night, Boston*), (*9 PM, Massachusetts*), (*night, Massachusetts*), (*9 PM*, ∗), (∗, *Boston*), (*night*, ∗), (∗, *Massachusetts*), and (∗, ∗). For each contextual possibility such as (*9 PM, Boston*), the table will contain the correct level of generalization to use in order to obtain the best accuracy. This level is determined using a cross-validation procedure on the training data. Segments that contain too few ratings are ignored. In the testing step, the appropriate data segment is identified for the test instance using this table. Only the specific data segment is used, which provides the best quality results.

How is the cross-validation procedure performed? For example, for the context (*9 PM, Boston*), the relevant ratings in the training cube are identified, which correspond to the context of 9 PM and Boston. These ratings are segmented into folds using the cross-validation approach described in Chapter 7. The same folds are used to test the various alternative generalizations of (*9 PM, Boston*), which are enumerated above. The fold providing the best accuracy is retained. In practice, a more refined way of choosing the best segment is used, while keeping in mind that overfitting is more likely for more specialized segments. A local segment is selected over its generalization only if it *significantly* outperforms the latter according to standardized statistical tests.

One problem with this approach is that it can be very expensive when the number of contextual possibilities is large. For example, in the aforementioned example, one must test the accuracy of all generalizations of all possible local-time combinations. Clearly, this is possible only in the cases in which the number of possible contextual alternatives to the tested is small. Otherwise, the training phase can become too expensive. In some cases, simpler heuristics are used instead of explicitly learning the accuracy of various generalizations. Instead of using the *accuracy*, one might simply determine the number of training entries (i.e., ratings) for each possible generalization of a particular context. The lowest-level (i.e., most specific) generalization that contains the required minimum number of ratings is used. The basic idea is to ensure that overfitting is avoided because of limited training data.

8.3.2 Multi-level Estimation

So far, we have only discussed how to estimate the ratings at the lowest level of the hierarchy from other ratings at the lowest level. However, in some cases, a user might have expressed ratings at upper levels of the hierarchy. For example, a user might have specified ratings for genres of movies, as opposed to individual movies. A question arises as to how one might use these higher-level ratings to improve the estimation process. The basic idea here is to assign ratings at the lowest level, so that the computed average of the observed and predicted ratings at the lower levels (descendent nodes) is as close as possible to the observed rating at the higher level (ancestor node). For example, David might have specified ratings for action movies such as "*Terminator*" and also for the genre of action movies. How can one integrate the ratings at various levels to provide a holistic prediction?

Let the rating of David for action movies be r_a. In such cases, David's predicted ratings of action movies at the lowest level should be such that the average value of his observed and predicted ratings is as close as possible to his rating r_a of action movies. In the extreme case, we can impose exact equality. In other words, we impose the constraint that the sum of the predicted and observed ratings of David on action movies is equal to $n_a \cdot r_a$, where n_a is the total number of action movies. Note that this is a linear constraint on the variables of the collaborative filtering problem. There will be many such constraints for various users, and the ratings specified at various levels. Therefore, in addition to the standard techniques used for collaborative filtering, genre-specific constraints are introduced for predicting David's rating. This problem can be formulated as an optimization problem with added linear constraints. The work in [6] does not provide further details of how one might use this approach in an actual collaborative filtering algorithm, and leaves the solution of this problem to further research. This type of optimization modeling could provide a promising direction for future research; the main caveat is that sufficient ratings must be available to prevent overfitting.

8.4 Post-Filtering Methods

In pre-filtering methods, relevant segments (slices) of the data are extracted and the collaborative filtering algorithm is applied on these extracted slices. Therefore, the filtering is performed on the *input* data, *before* applying the collaborative filtering algorithm. The qualifier "pre-" in pre-filtering derives its name from this fact. In post-filtering, the filtering steps are applied to the *output* obtained *after* applying a global collaborative filtering algorithm that ignores the contextual information in the data set.

In post-filtering methods, the contextual information is ignored, and a global 2-dimensional ratings matrix is created by aggregating the ratings over all the possible contextual values. For example, the rating for each user-item combination may be derived by averaging the available ratings over all contextual alternatives of that combination. Subsequently the ratings are adjusted with the use of context. Therefore, the approach comprises two steps:

1. Recommendations are generated on all the data by applying a conventional collaborative filtering model on an *aggregated* user-item matrix. Thus, context is ignored in the first step.

2. Context is then used to adjust or filter the recommended list.

How is the multidimensional ratings cube aggregated into a 2-dimensional ratings matrix? In the case of explicit ratings, the aggregation process refers to the averaging of (observed) ratings, whereas in the case of implicit feedback matrices (e.g., units sold), the process of aggregation refers to the sum of the values. Note that the use of sums or averages will not yield the same result in general because of the varying number of observed values across different user-item combinations. In implicit feedback matrices, it is more appropriate to use the sum rather than averages because the number of nonzero values is highly indicative of the user interest in the item.

Consider a scenario where a user has provided a separate rating for the same item over three different contexts (e.g., morning, afternoon, evening). In this case, the ratings are averaged over these contexts in order to create a global 2-dimensional user-item ratings matrix. For implicit feedback matrices, the number of 1s over the different contexts need to be added up. The resulting matrix no longer contains context-specific information because the contextual dimension has been aggregated. For the case of w-dimensional cubes, the rating values need to be aggregated over all the $(w-2)$-dimensional combinations. For example, if there are two contexts corresponding to location and time, then the user ratings of the same item over various location-time combinations need to be aggregated. If the user has never rated the item in any context, then the corresponding entry is also missing in the resulting aggregated matrix. The final result is always a 2-dimensional matrix, which is similar to that of traditional collaborative filtering.

Traditional collaborative filtering algorithms can be applied to this aggregated matrix in order to create predicted ratings $\hat{r}_{uj}$ and a corresponding ranked list of items for each user u. However, this ranked list is not sensitive to the contextual information, because the contextual dimension was ignored in the recommendation process. The post-filtering strategy adjusts the results *after* the estimations have been made. The adjustments can be made in one of two different ways. The first method corresponds to *filtering* out irrelevant items, and the second method corresponds to *adjusting* the ranking of the recommendations in the list based on the underlying context. The latter approach can be viewed as a soft version of the former. Both forms of post-filtering adjust the predicted rating $\hat{r}_{uj}$ for a given user-item combination.

One approach is to use heuristic methods to adjust or filter the recommended list based on the attributes associated with users and items. The notion of attributes associated with dimensions is discussed at the end of section 8.2. For example, if the context corresponds to {*summer, winter*}, then a clothing merchant might want to filter out sweaters and heavy jackets in the summer context, even if they are high in the list of recommended items. Such items can be detected with the use of attribute information.

For example, the attribute "wool" for a clothing item may be relevant to the context of the season attribute. One heuristic approach is to find the common item attributes that are relevant to a given context. Those items that do not have a significant number of these attributes are then filtered. In a more refined version of the approach, one can actually build a *predictive model* that uses the attributes to estimate the probability of relevance of the item to the context at hand. This approach is desirable because one can now use many of the traditional machine-learning techniques to build these predictive models. Those items that have very low probability of relevance are then filtered out. Such an approach is akin to determining a context-based probability $P(u, j, C)$ of the user u liking item j in context C, with the use of content-based models. The value of $P(u, j, C)$ need not be estimated with the use of content-based models. For example, one can even use a collaborative approach in conjunction with pre-filtering to estimate $P(u, j, C)$. This is identical to the pre-filtering approach discussed in the previous section. However, instead of using the pre-filtered prediction directly as the final result, it is normalized to the range $(0, 1)$, and then multiplied with the predicted rating $\hat{r}_{uj}$, which was estimated with the *global* data. The value of $P(u, j, C) \cdot \hat{r}_{uj}$ now defines an adjusted value of the prediction after post-filtering, and it may be used to adjust the ranking. Alternatively, one might simply remove the items j from the ranked list, for which the value of $P(u, j, C) \cdot \hat{r}_{uj}$ is very small. Post-filtering can be more robust than pre-filtering in a larger number of situations because one combines the local information $P(u, j, C)$ with the rating $\hat{r}_{uj}$ determined using all the data.

In cases where the amount of data available for context C is very limited, the value of $P(u, j, C)$ can be determined independent of the user u. In other words, the training data is used over all users to relate the item j to context C with a content-based model. For each item k, its attributes are treated as the feature variables, and the fraction of time that the item k is consumed in context C is treated as a numeric dependent variable. A linear regression model is constructed to relate the attributes to context C. Then, for each item j, this linear regression model can be used to estimate $P(*, j, C)$. Note that we use a "*" (don't care) as the user argument, because this model is independent of the user at hand. The final predicted value of the rating of user i for item j after the post-filtering step is given by $P(*, j, C) \cdot \hat{r}_{uj}$.

8.5   Contextual Modeling

In both pre-filtering and post-filtering, the collaborative filtering problem is reduced to the 2-dimensional setting, and the context is used during pre-processing or post-processing. The main disadvantage of this approach is that context is not integrated very tightly into the recommendation algorithm. Such an approach prevents the full use of the relationships between various user-item combinations and contextual values. Contextual modeling methods have been designed to explore this possibility.

It is possible to incorporate context directly into the recommendation process by modifying existing models (such as neighborhood-based methods) to the w-dimensional setting. Such an approach provides the most flexible and generalized view of context-sensitive recommendations, which is independent of the shackles of 2-dimensional algorithms. The following sections will review some of these methods.

8.5.1   Neighborhood-Based Methods

It is possible to adapt ideas from existing neighborhood-based methods to perform context-sensitive recommendations. An example of such an approach was presented in [7, 8]. However, the approach can sometimes be subtly different from traditional user-user or item-item methods because of the use of contextual dimensions in the similarity computation process. For the purpose of discussion, let us consider the case where the specific context used is time. Therefore, we have three dimensions corresponding to users, items, and time. The first step is to compute the distances separately on the users, items, and time. Consider two points in the 3-dimensional cube, corresponding to $A = (u, i, t)$ and $B = (u', i', t')$ respectively. Then, the distance between A and B can be defined as the sum of the weighted distances between the individual dimensions. In other words, we have:

$$Dist(A, B) = w_1 \cdot Dist(u, u') + w_2 \cdot Dist(i, i') + w_3 \cdot Dist(t, t') \tag{8.3}$$

Here, w_1, w_2, and w_3, respectively, reflect the relative importance of the user, item, and context (time) dimensions. Note that one can add as many contextual dimensions as are of interest to the aforementioned summation, rather than only time. Alternatively, one can also use the weighted Euclidean metric:

$$Dist(A, B) = \sqrt{w_1 \cdot Dist(u, u')^2 + w_2 \cdot Dist(i, i')^2 + w_3 \cdot Dist(t, t')^2} \tag{8.4}$$

Then, for a given cell of the 3-dimensional matrix, the closest r (observed) ratings are determined by using this metric. The weighted average of these ratings is reported as the predicted rating. The weighting used is the similarity between A and B, which is also defined as $1/Dist(A, B)$. In order to perform the recommendation for a given user u and context t, one would need to apply this process to each item, and then report the top-k items as the recommendations.

A question arises as to how $Dist(u, u')$, $Dist(i, i')$, and $Dist(t, t')$ may be determined. There are several different ways of doing this:

1. *Collaborative:* In this case, one can use the Pearson method or the adjusted cosine to calculate $Dist(u, u')$, $Dist(i, i')$, and $Dist(t, t')$. For example, in order to compute the distance between users u and u', one can extract the 2-dimensional slices corresponding to the users u and u'. We can generalize the neighborhood-based similarity measure (cf. Chapter 2)) to compute the Pearson coefficients between all the ratings with the users fixed at u and u', respectively. Therefore, the individually observed ratings of users u and u' over the entire *Item × Context* grid are used in the Pearson computation. The inverse of the similarity is used to determine the distance values. A similar approach can be used to compute the item-wise and context-wise distances $Dist(i, i')$ and $Dist(t, t')$.

2. *Content-based:* In this case, the attributes associated with the dimensions (i.e., user profiles and item profiles) are used to compute the profile. A variety of text-based measures, such as the cosine, may be used. A similar approach may be used to compute $Dist(t, t')$ by associating each context with its frequently co-occurring content attributes. Alternatively, the attributes associated with a specific context, such as season, weekday, and so on, may be used. This approach may be viewed as a monolithic hybrid method because the representation is content-centric, but the overall approach uses the framework of collaborative methods.

3. *Combined:* It is possible to combine the collaborative and content-based measures to obtain a more robust measure of similarity. The relative weighting may be inferred with the use of cross-validation methods, so as to maximize the prediction accuracy.

Significant variations exist in terms of how the distance function may be designed for the specific application at hand. The aforementioned approach describes the broader idea although the specific implementation may vary with the application at hand. It is noteworthy that this approach can be viewed as a contextual generalization of the user-item approach discussed in section 2.3.6 of Chapter 2.

8.5.2 Latent Factor Models

Tensor factorization can be considered a generalization of matrix factorization, in which an n-dimensional data cube is factorized, rather than a 2-dimensional matrix. The traditional context-sensitive representation is indeed an w-dimensional cube, and therefore it is particularly well suited to tensor factorization. In this sense, tensor factorization methods can be considered contextual generalizations of conventional matrix factorization methods in recommender systems. While a detailed discussion of tensors is beyond the scope of this book, the reader is referred to [212, 294, 332, 495, 496] for details of such methods. A particularly notable example of a higher-order tensor factorization method is the *Multiverse Recommendation* model [294]. The Multiverse Recommendation model uses the higher-order Tucker decomposition [605], whose complexity increases *exponentially* with the order of the decomposition.

Such an application of tensor factorization is computationally very intensive, especially when the underlying data cubes are large. In most cases, the use of *higher-order* tensor factorization is an overkill in such settings [496]. There are, however, other simplified ways of applying the principles of latent factor models in multidimensional settings. Some simplified forms of these factorization methods use only *pairwise* interactions between the different dimensions [496, 498].

Here, we describe one such pairwise interaction approach. A closely related ranking method, referred to as *Pairwise Interaction Tensor Factorization (PITF)* [496], has also been used for tag recommendation. This description can be viewed as a very special case of the notion of *factorization machines* discussed in [496]. Let $R = [r_{ijc}]$ be a 3-dimensional ratings cube of size $m \times n \times d$ with m users, n items, and d different values of the contextual dimension. For example, in Figure 8.1, we have $m = 6$, $n = 4$, and $d = 5$. Let $U = [u_{is}]$, $V = [v_{js}]$, and $W = [w_{cs}]$ be $m \times k$, $n \times k$, and $d \times k$, matrices. Here, U denotes the user-factor matrix, V denotes the item-factor matrix, and W denotes the context-factor matrix. The notation k denotes the rank of the latent factor model. Then, the basic principle of the simplified prediction function of the (i, j, k)th element of the data cube is based on pairwise interactions between users, items, and contexts. This implies the following prediction function:

$$\hat{r}_{ijc} = (UV^T)_{ij} + (VW^T)_{jc} + (UW^T)_{ic} \tag{8.5}$$

$$= \sum_{s=1}^{k} (u_{is}v_{js} + v_{js}w_{cs} + u_{is}w_{cs}) \tag{8.6}$$

It is easy to see that this prediction function is a straightforward generalization of latent factor models. One can now use this prediction function to set up the optimization problem as in all latent factor models. Let S be the set of all observed entries in R.

$$S = \{(i, j, c) : r_{ijc} \text{ is observed}\} \tag{8.7}$$

In cases where R is an implicit feedback matrix, a sample of unobserved entries also need to be included within S, under the assumption that these entries are observed as 0s. The detailed reasons for doing so are described in section 3.6.6.2 of Chapter 3.

Then, the errors over all the observed entries needs to be minimized as follows:

$$\text{Minimize } J = \frac{1}{2} \sum_{(i,j,c) \in S} (r_{ijc} - \hat{r}_{ijc})^2 + \frac{\lambda}{2} \sum_{s=1}^{k} \left(\sum_{i=1}^{m} u_{is}^2 + \sum_{j=1}^{n} v_{js}^2 + \sum_{c=1}^{d} w_{cs}^2 \right)$$

$$= \frac{1}{2} \sum_{(i,j,c) \in S} \left(r_{ijc} - \sum_{s=1}^{k} [u_{is}v_{js} + v_{js}w_{cs} + u_{is}w_{cs}] \right)^2 +$$

$$\frac{\lambda}{2} \sum_{s=1}^{k} \left(\sum_{i=1}^{m} u_{is}^2 + \sum_{j=1}^{n} v_{js}^2 + \sum_{c=1}^{d} w_{cs}^2 \right)$$

The last term is the regularization term, where $\lambda > 0$ is the regularization parameter.

We need to solve for the optimum values of parameters in U, V, and W. One can determine the partial derivative of J with respect to the individual elements in U, V, and W, in order to derive the update directions for a gradient-descent method. Therefore, all elements of U, V, and W are updated *simultaneously* as follows:

$$u_{iq} \Leftarrow u_{iq} - \alpha \frac{\partial J}{\partial u_{iq}} \quad \forall i \; \forall q \in \{1 \ldots k\}$$

$$v_{jq} \Leftarrow v_{jq} - \alpha \frac{\partial J}{\partial v_{jq}} \quad \forall j \; \forall q \in \{1 \ldots k\}$$

$$w_{cq} \Leftarrow w_{cq} - \alpha \frac{\partial J}{\partial w_{cq}} \quad \forall c \; \forall q \in \{1 \ldots k\}$$

Here, $\alpha > 0$ is the step size. As in traditional latent factor models, the descent direction depends on the error $e_{ijc} = r_{ijc} - \hat{r}_{ijc}$ over the observed entries in S. The corresponding updates are as follows:

$$u_{iq} \Leftarrow u_{iq} + \alpha \left(\sum_{j,c:(i,j,c) \in S} e_{ijc} \cdot (v_{jq} + w_{cq}) - \lambda \cdot u_{iq} \right) \quad \forall i \; \forall q \in \{1 \ldots k\}$$

$$v_{jq} \Leftarrow v_{jq} + \alpha \left(\sum_{i,c:(i,j,c) \in S} e_{ijc} \cdot (u_{iq} + w_{cq}) - \lambda \cdot v_{jq} \right) \quad \forall j \; \forall q \in \{1 \ldots k\}$$

$$w_{cq} \Leftarrow w_{cq} + \alpha \left(\sum_{i,j:(i,j,c) \in S} e_{ijc} \cdot (u_{iq} + v_{jq}) - \lambda \cdot w_{cq} \right) \quad \forall c \; \forall q \in \{1 \ldots k\}$$

A faster alternative is to use stochastic gradient descent. In stochastic-gradient descent, instead of descending over the gradient with respect to all the errors in S simultaneously, one descends with respect to the error in a single observed entry $(i, j, c) \in S$, which is randomly chosen:

$$u_{iq} \Leftarrow u_{iq} - \alpha \left[\frac{\partial J}{\partial u_{iq}} \right]_{\text{Contributed by } (i,j,c)} \quad \forall q \in \{1 \ldots k\}$$

$$v_{jq} \Leftarrow v_{jq} - \alpha \left[\frac{\partial J}{\partial v_{jq}} \right]_{\text{Contributed by } (i,j,c)} \quad \forall q \in \{1 \ldots k\}$$

$$w_{cq} \Leftarrow w_{cq} - \alpha \left[\frac{\partial J}{\partial w_{cq}} \right]_{\text{Contributed by } (i,j,c)} \quad \forall q \in \{1 \ldots k\}$$

On computing these contributions, the following steps may be executed for each specified entry $(i, j, c) \in S$ and the qth latent component ($1 \leq q \leq k$):

$$u_{iq} \Leftarrow u_{iq} + \alpha \left(e_{ijc} \cdot (v_{jq} + w_{cq}) - \frac{\lambda \cdot u_{iq}}{n_i^{user}} \right) \quad \forall q \in \{1 \ldots k\}$$

$$v_{jq} \Leftarrow v_{jq} + \alpha \left(e_{ijc} \cdot (u_{iq} + w_{cq}) - \frac{\lambda \cdot v_{jq}}{n_j^{item}} \right) \quad \forall q \in \{1 \ldots k\}$$

$$w_{cq} \Leftarrow w_{cq} + \alpha \left(e_{ijc} \cdot (u_{iq} + v_{jq}) - \frac{\lambda \cdot w_{cq}}{n_c^{context}} \right) \quad \forall q \in \{1 \ldots k\}$$

Here, n_i^{user}, n_j^{item}, and $n_c^{context}$ represent the number of observed entries in the data cube for user i, item j, and context c, respectively. Using these terms to normalize the regularization terms results in better convergence, although they can be (heuristically) omitted and a smaller value of λ may be used instead. It is necessary to repeatedly cycle through the specified entries in S with each of the aforementioned updates. These gradient descent steps may be executed to convergence to obtain the matrices U, V, and W. The resulting updates are similar to the case of traditional matrix factorization discussed in section 3.6.4 of Chapter 3. Refer to Figure 3.9 for the algorithmic framework of stochastic gradient descent. The main change to that pseudocode is the use of an additional set of context factors, and the corresponding changes to the update equations. One now needs to cycle through each observed *triplet* (i, j, c) within the algorithmic framework of Figure 3.9 in order to execute the updates. Better convergence may be achieved by selecting different regularization parameters for each of the matrices U, V, and W. The values of these regularization parameters may be learned using cross-validation. One can also incorporate bias into the model by using a straightforward 3-dimensional generalization of the baseline predictors in section 3.7.1 of Chapter 3. The resulting baseline predictions B_{ijc} for user i, item j, and context c may be subtracted from the corresponding (observed) entries in the data cube before applying the factorization process. These values may be added back to the predictions in a post-processing phase.

This approach is less complex than higher-order tensor factorization models, and it can work particularly well in sparse matrices. It uses the 2-dimensional interactions in an additive way without going through the higher-order interactions, which can unnecessarily hamper the model both in terms of computational time and overfitting. In real settings, the ratings cube is generally too sparse to take full advantage of higher-order models. These issues are stated as the primary criticisms of *Multiverse Recommendation* methods [496].

This general principle can also be extended to w-dimensional cubes for $w > 3$. Consider an w-dimensional data cube R, in which a rating entry is denoted by $r_{i_1 \ldots i_w}$, with corresponding matrix dimensions $n_1 \ldots n_w$. Then, one can express the predicted rating value in terms of w different latent factor matrices U_{i_a} of respective sizes $n_a \times k$ ($a \in \{1 \ldots w\}$) as follows:

$$\hat{r}_{i_1 \ldots i_w} = \sum_{a < b \leq w} [U_a (U_b)^T]_{i_a i_b} \tag{8.8}$$

A least-squares optimization problem can be set up, as in the case of the 3-dimensional cube. A standard gradient descent approach can be used to solve this problem. The derivation of the update equations in this case is provided in Exercise 6.

8.5.2.1   Factorization Machines

The latent factor approach in the previous section can be viewed as a special case of *factorization machines*. Large classes of models (such as SVD and SVD++) are special cases of factorization machines. In factorization machines, the basic idea is to model each rating as a linear combination of interactions between input variables. The input variables are derived from the original ratings matrix. For example, consider the case in which we have a 3-dimensional cube containing m users, n items, and d values of the contextual dimension, and each rating is associated with a unique triplet. One can then "flatten" this 3-dimensional cube into a set of $(m + n + d)$-dimensional rows, such that each row corresponds to the user, item, and contextual value of an observed rating. Therefore, there are as many rows as the number of observed ratings. In this *specific example*, each row is a vector of binary indicator variables, in which exactly three of the values are 1s, depending on the specific user-item-context triplet relevant to that observed rating. All the remaining values are 0s. We represent the variables of the row by $x_1 \ldots x_{m+n+d}$, all of which are either 0s or 1s. Furthermore, the target variable for that row corresponds to the rating represented by

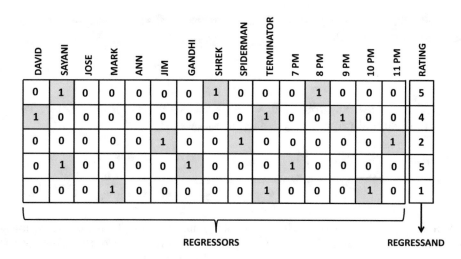

Figure 8.4: The flattened representation of five observed ratings in the data cube of Figure 8.1. Most recommendation problems can be transformed to sparse classification and regression problems.

that row. In Figure 8.4, we have shown the flattened representation of five observed ratings of the data cube in Figure 8.1. At first sight, it would seem that we could use a classification or regression predictor on this flattened representation in a straightforward way; however, it would not work very well because of the extraordinary data sparsity, in which there are only three nonzero entries in each row. It is here that factorization machines rescue us from the perils of sparsity.

The basic idea is to associate a k-dimensional latent factor with each of the $p = (m+n+d)$ decision variables $x_1 \ldots x_p$. Assume that the factor vector associated with the ith variable is denoted by $\overline{v_i} = (v_{i1} \ldots v_{ik})$. Similarly, the ith column has a bias b_i associated with it, and we also have a global bias variable g. The rating prediction $\hat{y}(\overline{x})$ of *second-order* factorization machines uses pairwise interactions between the factors as follows:

$$\hat{y}(\overline{x}) = g + \sum_{i=1}^{p} b_i x_i + \sum_{i=1}^{p} \sum_{j=i+1}^{p} (\overline{v_i} \cdot \overline{v_j}) x_i x_j \tag{8.9}$$

The variables to be learned are g, the different values of b_i, and each of the vectors $\overline{v_i}$. Although the number of interaction terms might seem alarmingly large, most of them will evaluate to zero in sparse settings. In the example shown in Figure 8.4, only three of the interaction terms will be nonzero, which seems suspiciously similar[2] to the three factorization terms in Equation 8.6. In fact, it can be easily shown that Equation 8.9 is a generalization of Equation 8.6 with added bias variables, and we can set up a similar least-squares model. As in matrix factorization, the stochastic-gradient descent method cycles through the observed ratings to estimate the aforementioned parameters. The update step with respect

---

[2]This similarity might not be obvious at first because the two equations do not use the same notation. Each k-dimensional factor vector $\overline{v_i}$ of the factorization machine is equivalent to one of the k-dimensional rows of either the user, item, or context factor matrix in Equation 8.6.

to any particular model parameter θ depends on the error $e(\overline{x}) = y(\overline{x}) - \hat{y}(\overline{x})$ between the predicted and observed values:

$$\theta \Leftarrow \theta(1 - \alpha \cdot \lambda) + \alpha \cdot e(\overline{x})\frac{\partial \hat{y}(\overline{x})}{\partial \theta} \tag{8.10}$$

Here, $\alpha > 0$ is the learning rate, and $\lambda > 0$ is the regularization parameter. The partial derivative in the update equation is defined as follows:

$$\frac{\partial \hat{y}(\overline{x})}{\partial \theta} = \begin{cases} 1 & \text{if } \theta \text{ is } g \\ x_i & \text{if } \theta \text{ is } b_i \\ x_i \sum_{j=1}^{p} v_{js} \cdot x_j - v_{is} \cdot x_i^2 & \text{if } \theta \text{ is } v_{is} \end{cases} \tag{8.11}$$

The term $L_s = \sum_{j=1}^{p} v_{js} \cdot x_j$ in the third case is noteworthy. To avoid redundant effort, this term can be pre-stored while evaluating $\hat{y}(\overline{x})$ for computation of the error term $e(\overline{x}) = y(\overline{x}) - \hat{y}(\overline{x})$. This is because Equation 8.9 can be algebraically rearranged as follows:

$$\hat{y}(\overline{x}) = g + \sum_{i=1}^{p} b_i x_i + \frac{1}{2} \sum_{s=1}^{k} \left([\sum_{j=1}^{p} v_{js} \cdot x_j]^2 - \sum_{j=1}^{p} v_{js}^2 \cdot x_j^2 \right)$$

$$= g + \sum_{i=1}^{p} b_i x_i + \frac{1}{2} \sum_{s=1}^{k} \left(L_s^2 - \sum_{j=1}^{p} v_{js}^2 \cdot x_j^2 \right)$$

Furthermore, the parameters $\overline{v}_i$ and b_i do not need to be updated when $x_i = 0$. This allows for an efficient update process in sparse settings, which is linear in both the number of nonzero entries and the value of k.

In this specific example, we have assumed that the vector $\overline{x}$ contains indicator variables with exactly three 1s. Factorization machines, however, allow arbitrary values of $\overline{x}$ to increase expressiveness. For example, it is possible for the values of $\overline{x}$ to be real, or to contain multiple nonzero values from the same dimension (e.g., context). This flexibility also allows the interaction between latent factors of pairs of users or pairs of contexts. There could be settings in which a context could correspond to a *set* of keywords or a set of entities. In the traditional data-cube model, there is no mechanism to represent such *set-wise* attributes. For example, consider the case where the context represents the companions with which a user watches a movie, and therefore each rating is associated with a set of companions (as context). In this case, the context variables x_i correspond to individual companions. If John watches a movie with Alice, Bob, and Jack, then the value of x_i for each of the three companions will be 1/3. This scenario is not quite so simple to represent with the straightforward latent factor approach, and it provides an example of the greater expressiveness of factorization machines. It is also relatively easy to see that this approach can be generalized to a setting in which each context is a document with associated word frequencies. For any given ratings matrix, all we have to do is to spend some time in a *feature engineering* effort. The observed ratings (target variables) are associated with a set of carefully designed attributes, some of which might be given (e.g., user, item, and context), and others might be extracted (e.g., implicit feedback). The versatility of factorization machines is striking. For example, by removing the contextual columns in Figure 8.4, one obtains traditional matrix factorization. By replacing the contextual columns in Figure 8.4 with implicit feedback variables, one roughly obtains SVD++ (with a few additional terms).

Factorization machines can be used for any (massively sparse) classification or regression task; ratings prediction in recommender systems is only one example of a natural application. Although the model is inherently designed for regression, binary classification can be handled by applying the logistic function on the numerical predictions to derive the probability whether $\hat{y}(\overline{x})$ is $+1$ or -1. Applications to classification and pairwise ranking are discussed in section 13.2.1 of Chapter 13. In fact, factorization machines can be shown to be sparsity-resistant generalizations of polynomial regression [493]. Note that Equation 8.9 is not very different from the prediction function of second-order polynomial regression. The most important difference is that the regression coefficients w_{ij} of pairwise interactions $x_i x_j$ are assumed to satisfy the low-rank assumption, and can therefore be expressed as $w_{ij} = \overline{v_i} \cdot \overline{v_j}$. For example, we could have tried to directly learn w_{ij} without making this low-rank assumption; this would be almost equivalent to using kernel regression with a second-order polynomial kernel. Since there are $O(p^2) = O([m + n + d]^2)$ such values of w_{ij}, overfitting is extremely likely. Factorization machines assume that the $p \times p$ regression coefficient matrix $W = [w_{ij}]$ is of low rank and can be expressed as VV^T. This reduces the $O(p^2)$ coefficients in $W = [w_{ij}]$ to the $O(p \cdot k)$ coefficients in $V = [v_{js}]$ and therefore helps in reducing overfitting. Under the covers, *factorization machines are polynomial regression models with an added low-rank assumption on the coefficients*. The basic idea here is that it would be difficult to accurately estimate the interaction coefficient w_{ij} between Jim and *Terminator* (with off-the-shelf polynomial regression) if Jim has never rated *Terminator*. However, the low-rank assumption enables accurate estimation of such regression coefficients by forcing inter-coefficient relationships in the parameter space. This is particularly useful in sparse settings.

The description in this section is based on second-order factorization machines that are popularly used in practice. In third-order polynomial regression, we would have $O(p^3)$ additional regression coefficients of the form w_{ijk}, which correspond to interaction terms of the form $x_i x_j x_k$. These coefficients would define a massive third-order *tensor*, which can be compressed with tensor factorization. Although higher-order factorization machines have also been developed, they are often impractical because of greater computational complexity and overfitting. A software library, referred to as *libFM* [494], provides an excellent set of factorization machine implementations. The main task in using *libFM* is an initial feature engineering effort, and the effectiveness of the model mainly depends on the skill of the analyst in extracting the correct set of features.

8.5.2.2 A Generalized View of Second-Order Factorization Machines

Although second-order factorization machines assume that all pairs of variables x_i and x_j interact with one another, this may not always be desirable. For example, when the contextual variables correspond to word frequencies of documents, it may not always be desirable for the word frequencies to interact with one another. In some cases, such as SVD++, implicit feedback variables might interact with item factors, but not with the user factors. Similarly, implicit feedback variables do not interact with one another in SVD++. In order to handle this setting, we define an interaction indicator δ_{ij}, which indicates pairs of variables that are allowed to interact with one another:

$$\delta_{ij} = \begin{cases} 1 & \text{if } x_i \text{ and } x_j \text{ are allowed to interact} \\ 0 & \text{otherwise} \end{cases} \tag{8.12}$$

The interaction indicators are typically defined on the basis of the *block structure* of the variables, and therefore all p^2 values do not need to be explicitly stored. For example, user variables might not be allowed to interact with context variables, contexts might not be allowed to interact with other contexts, and so on. This provides the analyst flexibility in specifying domain knowledge about important pairs of interacting blocks of variables. We can use this indicator to generalize Equation 8.9 as follows:

$$\hat{y}(\overline{x}) = g + \sum_{i=1}^{p} b_i x_i + \sum_{i=1}^{p} \sum_{j=i}^{p} (\overline{v_i} \cdot \overline{v_j}) \delta_{ij} x_i x_j \tag{8.13}$$

Unlike Equation 8.9, this equation allows x_i to interact with itself when δ_{ii} is nonzero. This can be useful in some versions of polynomial regression, when x_i is real. This model can be used to *exactly* simulate SVD++ by defining m user-indicator variables, n item-indicator variables, and an additional set of n implicit feedback variables associated with items. Therefore, there are two sets of item variables corresponding to explicit and implicit feedback, respectively. For implicit feedback variables, the feature values are nonzero only for items in the set I_u rated by the relevant user u. These nonzero values are all set to $1/\sqrt{I_u}$. The value of δ_{ij} is set to 1 only for interactions between users and explicit feedback (item) variables, and between implicit and explicit feedback (item) variables. With this definition, one can easily show that Equation 8.13 is *exactly* SVD++.

The solution methodology is almost the same as that of factorization machines. The update step of Equation 8.10 can be used for stochastic gradient-descent. The only difference is that the partial derivative of the predicted variable with respect to each model parameter θ needs to be modified:

$$\frac{\partial \hat{y}(\overline{x})}{\partial \theta} = \begin{cases} 1 & \text{if } \theta \text{ is } g \\ x_i & \text{if } \theta \text{ is } b_i \\ x_i \sum_{j=1}^{p} \delta_{ij} \cdot v_{js} \cdot x_j & \text{if } \theta \text{ is } v_{is} \end{cases} \tag{8.14}$$

A recent method, referred to as *SVDFeature* [151], can also be shown to be a special case of this setting by defining δ_{ij} appropriately. *SVDFeature* and factorization machines were the top two finishers in a network recommendation task at the KDD Cup contest (2012) [715].

8.5.2.3   Other Applications of Latent Parametrization

Factorization machines impose a low-rank structure on a large parameter space in order to reduce overfitting. A rarely noticed fact is that this general principle had been used earlier in the completely different context of *conditionally factored* Restricted Boltzmann Machines (RBMs) for collaborative filtering [519]. The basic idea is that the number of weights between two successive layers of a neural network can be represented as a matrix $W = [w_{ij}]$ (see Figure 8.5). The size of the matrix can be rather large in the collaborative filtering setting because the size of the input layer scales with the number of items and the size of the hidden layer can be of the order of several hundred units. The size of W is defined by the product of these two values. A large parameter space will inevitably lead to overfitting. Therefore, the work in [519] assumes that the matrix $W = UV^T$ is the product of two low-rank matrices U and V^T. Instead of learning W, the approach then learns the parameters of U and V. It has been shown in [519] that this type of low-rank reduction of the parameter space has significant advantages both in terms of accuracy and running time. These results show that a natural approach for effectively handling large and matrix-structured parameter spaces is by imposing a low-rank structure on them. Although the approach in [519] has been designed for conventional collaborative filtering, it can be

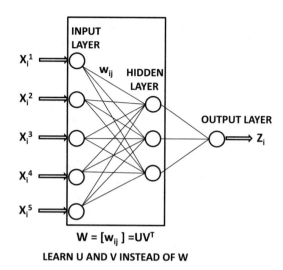

Figure 8.5: Low-rank parametrization of neural networks to avoid overfitting

easily extended to context-sensitive scenarios by adding appropriate input nodes for context features. An untapped potential of this approach is to leverage the low-rank parametrization in deep-learning methods for collaborative filtering. Deep neural networks with many layers can benefit [516] from the low-rank decomposition of the matrices corresponding to the weights between successive layers of neurons. This can be particularly useful because of the pervasive problem of overfitting in deep-learning methods.

8.5.3 Content-Based Models

A variety of machine learning models, such as support vector machines and linear regression, are used in conjunction with context-sensitive recommender systems. These methods can be viewed as generalizations of content-based models because they use the attributes associated with users, items, and context. Recall that content-based models use only the attributes associated with the items. However, in this case, we make the more general assumption that attributes are associated with any of the dimensions.

In most of these cases, the users or items are represented as vectors in feature space, and the ratings correspond to the dependent or class variable. One of the earliest methods, which used support vector machines, was proposed in [458]. In this case, a restaurant recommender system is proposed, in which additional contextual dimensions, such as weather, companion, and time, are used. In this case, each item-context combination is represented as a feature vector that is either liked or disliked. A support vector machine is constructed that separates the liked items from the disliked items. A previously unseen item-context combination will be recommended as liked if it falls on the liked side of the support-vector separation. This model can be viewed as a direct generalization of content-centric models because a separate model is built for each user, and the prediction for the model is specific to each user. Furthermore, the attributes of the user are not used in this model. However, it is possible, in principle, to use a global model which is constructed for prediction over all users.

An example of such a model is discussed in [50] where a single linear regression model is constructed to predict the ratings of any user-item combination. In this case, the attributes of the user are also used. For ease in discussion, we assume that the features are treated as the frequencies of discrete keywords, although one can also use numeric attributes after discretization. Note that the features of users and items are contained in the user or item profile.

First, we describe a simple linear regression model that does not use any contextual information. Later, we will show how to extend this model with the use of context. Consider the following linear regression model, which estimates the rating r_{ij} as a linear function of the user features, item features, and Kronecker cross-product features:

$$\hat{r}_{ij} = \overline{W_1} \cdot \overline{y_i} + \overline{W_2} \cdot \overline{z_j} + \overline{W_3} \cdot (\overline{y_i} \otimes \overline{z_j}) \tag{8.15}$$

Here, $\overline{W_1}$, $\overline{W_2}$, and $\overline{W_3}$ are linear regression coefficient *vectors* of the appropriate length. For example, the length of the coefficient vector $\overline{W_1}$ is the same as that of the feature space representing all the different $\overline{y_i}$. Furthermore, $\overline{y_i}$ corresponds to the feature variable vector of user i (e.g., gender or race), $\overline{z_j}$ corresponds to the feature variable vector of item j (e.g., movie genre and production studio), and $(\overline{y_i} \otimes \overline{z_j})$ corresponds to the entries in the Kronecker product between the feature vectors of user i and item j. The Kronecker product is defined by all possible cross-product combinations between the feature values of user i and item j. In the aforementioned example, combinations correspond to various possibilities for gender-genre, race-genre, gender-studio, and race-studio. For a particular user-item instance, the relevant combinations might be male-comedy, Caucasian-comedy, male-Sony, and Caucasian-Sony. The values of these combinations are set to 1, and those of all other possible combinations (e.g., female-comedy) are set to 0. In this case, all feature values are binary, although it is also possible to work with arbitrary frequencies by multiplying the frequencies of the corresponding value-pairs. For example, if the keyword "golf" has a frequency of 2 in the user profile and the keyword "cart" has a frequency of 3 in the item profile, the corresponding frequency of the keyword pair is $2 \times 3 = 6$. The basic idea here is to represent a rating of a user-item combination (i, j) in terms of the features of user i, the features of item j, and the interaction effects among them.

The observed ratings are used as the training data in order to create the model and learn the coefficient vectors $\overline{W_1}, \overline{W_2}$ and $\overline{W_3}$. The interaction coefficients tell us how various combinations of user-item features affect the model. The coefficients of the user features and the item features tell us about the specific biases of the user and the item at hand. The work in [50] uses a Markov Chain Monte Carlo (MCMC) method in order to estimate these coefficients and learn the model from the observed ratings. This approach is a generalization of linear regression models for content-based methods, which create *user-specific* models with *item* features (cf. section 4.4.5 of Chapter 4). Here, the model is constructed over all users and items, and the features are also extracted from the user-item combination, which is being rated. Therefore, this approach is richer than an off-the-shelf content-based model.

The approach can be generalized easily to the contextual scenario by introducing additional feature variables for the contextual dimensions [7, 607]. As a specific example, consider the case where time is used as a contextual variable, and the feature variables associated with the kth possible value of the time dimension are denoted by the vector $\overline{v_k}$. The features associated with time might correspond to various descriptions, such as the time of the day, whether it is a weekday, the season, and so on. Since we have three dimensions, the ratings r_{ijk} are now subscripted by three different values. Here i denotes the index of the user, j denotes the index of the item, and k denotes the index of the time dimension. Then, the predicted value $\hat{r}_{ijk}$ of the rating can be computed as a linear function of the feature variables and interaction variables as follows:

$$\hat{r}_{ijk} = \overline{W_1} \cdot \overline{y_i} + \overline{W_2} \cdot \overline{z_j} + \overline{W_3} \cdot \overline{v_k} + \overline{W_4} \cdot (\overline{y_i} \otimes \overline{z_j}) + \overline{W_5} \cdot (\overline{z_j} \otimes \overline{v_k}) + \overline{W_6} \cdot (\overline{y_i} \otimes \overline{v_k}) + \overline{W_7} \cdot (\overline{y_i} \otimes \overline{z_j} \otimes \overline{v_k}) \tag{8.16}$$

To reduce the number of coefficients, it is possible to set the third-order coefficients $\overline{W_7}$ to 0. Such a model is in a similar form as second-order factorization machines, although interactions only occur between attributes from different dimensions (e.g., users and items). This is similar to the model of Equation 8.13. A similar gradient-descent approach can be used.

This generic approach can, in fact, be used in conjunction with *any* off-the-shelf machine learning model, and not just regression-based models. The overall approach is as follows:

1. Generate a multi-dimensional data record $\overline{X}_{ijk}$ for each observed rating r_{ijk}, where the class label of the record is the value r_{ijk}.

2. Generate the frequencies of the discrete features corresponding to $\overline{y_i}$, $\overline{z_j}$, $\overline{v_k}$, $\overline{y_i} \otimes \overline{z_j}$, $\overline{z_j} \otimes \overline{v_k}$, and $\overline{y_i} \otimes \overline{v_k}$. Let these frequencies represent the feature vector for $\overline{X}_{ijk}$.

3. Use the pairs $(\overline{X}_{ijk}, r_{ijk})$ in conjunction with any off-the-shelf supervised learning algorithm to build a model $\mathcal{M}$.

4. For any entry (i_1, j_1, k_1) in the ratings cube for which the value is unknown, extract the feature representation $\overline{X}_{i_1 j_1 k_1}$ using the aforementioned approach, and apply the machine learning model $\mathcal{M}$ to predict the value of the rating.

As the dimensionality of the context-sensitive system increases, the model becomes more likely to overfit. Furthermore, the scalability of the system is adversely affected. This can sometimes be a significant challenge in such systems and is, of course, a general drawback of contextual modeling methods, which directly try to work with the w-dimensional ratings matrix rather than reducing it to a 2-dimensional problem with pre- or post-filtering methods. Nevertheless, if sufficient ratings data are available, then direct contextual modeling is likely to provide the most robust results. Such methods are likely to become increasingly relevant in the "big-data" era.

8.6 Summary

Various types of context, such as location, time, and social information, have a significant influence on the recommendation process. The multidimensional model is used frequently to create a general framework for various types of context-aware recommendations. There are three primary ways in which context-aware recommendations are performed. In pre-filtering, the problem is reduced to a 2-dimensional collaborative filtering problem by filtering the w-dimensional cube to a 2-dimensional ratings matrix *before* applying the collaborative filtering algorithm. In post-filtering, the context is ignored during the first phase of collaborative filtering. Subsequently, the results are adjusted with the use of a predictive model that regulates the relative importance of context. Finally, a recent approach is to incorporate the context directly into the model by treating it as a w-dimensional prediction problem. Generalizations of matrix factorization and linear regression models have been proposed in this setting. This approach is computationally intensive, but it is a generic approach with the best potential when a large amount of data is available.

8.7   Bibliographic Notes

Some of the earliest work on context-aware recommendation systems was performed in the context of mobile applications [2, 3], such as creating a mobile context-aware tour guide. An early survey of context-aware computing research for mobile systems may be found in [147]. A recent survey on context-based recommender systems may be found in [7]. Context-sensitive systems have been used in a wide variety of domains, such as news recommendation [134], Web search [336], tourist recommendations [2, 3], and database querying [39]. A survey of context-aware recommender systems, which use technology-enhanced learning technologies, may be found in [612].

The notion of multidimensional recommender systems was proposed in [6]. An interesting discussion is also found in [466]. A query language, referred to as *Recommendation Query Language (RQL)* [9], was proposed for context-based systems. Another recent query language in the context of personalizing recommendations is *REQUEST* [10].

The use of pre-filtering methods has a rich history in recommender systems. The reduction-based approach of [6] is one of the seminal techniques for pre-filtering. Many subsequent methods based on this broad methodology have been developed. The work in [62] uses the notion of item splitting, in which a single item is split into several fictitious items corresponding to various contexts. The work in [61] develops the notion of micro-profiles, each of which is relevant to a specific context. Specifically, a different model is constructed for the user in different contexts. This approach was used for time-sensitive recommender systems, and it is discussed in section 9.2.2.1 of Chapter 9. The basic idea in [61] is similar to the reduction-based approach described in this chapter. A mobile advertisement recommender system, which uses pre-filtering, is discussed in [40]. The application of the approach in an online retailing application is discussed in [374]. A comparison of pre- and post-filtering methods in context-sensitive systems is provided in [471]. Results on the accuracy and diversity of context-sensitive systems are provided in [470]. The use of neighborhood-based methods for context-sensitive recommendations is discussed in [7, 8]. Many matrix and tensor factorization methods have been proposed for contextual recommendations in which time is treated as a discrete contextual value [212, 294, 332, 495, 496]. The notion of *factorization machines* [493] has found significant popularity in these settings. Factorization machines can be viewed as generalizations of large classes of latent factor models, and they have found increasing popularity for context-sensitive recommendation applications. Alternating least-squares methods for factorization machines are discussed in [496]. A related model, referred to as *SVDFeature*, is proposed in [151].

The support vector machine method for model construction was proposed in [458]. The work in [63] proposes a family of matrix-factorization methods for context-aware recommendation, although the approach discussed in this book is more general than the methods discussed in this family. Scalable algorithms for building context-sensitive recommender systems are discussed in [607].

A major issue is the selection of appropriate attributes for contextual methods. A discussion of how appropriate attributes may be selected for contextual methods is provided in [188]. The use of latent contextual information as a possible representation is discussed in [47].

8.8 Exercises

1. Discuss how you might decide whether pre-filtering, post-filtering, or contextual modeling is most appropriate for a particular data set.

2. Discuss how you might use hybrid recommender systems to combine the power of pre-filtering, post-filtering, and contextual modeling. Propose as many schemes as you can imagine. How would you decide which of these to use?

3. Implement the pre-filtering algorithm with a single contextual attribute. Use an item-based (neighborhood) collaborative filtering as the base method.

4. Suppose that you have three contextual attributes (say, location, time, and companion), each of which has its own taxonomy. Your system is designed to recommend items for a given user in the context of location, time, and companion. For a given context at the lowest level of the hierarchy, you might have the sparsity problem because only a modest number (say, 500) of the observed ratings might be available in which the three contexts are fixed to the queried values. This can cause overfitting in a pre-filtering method if only 500 ratings are used for the training process. You decide that you will use a more general level of the taxonomy from each of the three contexts, in order to extract the relevant segments and increase the amount of training data. Describe how to determine *the specific level of the* taxonomy to use for each contextual attribute. Once you have extracted the taxonomy level for each context, describe the collaborative filtering algorithm.

5. Consider the w-dimensional matrix factorization discussed in section 8.5.2, in which the prediction function is of the following form:

$$\hat{r}_{i_1 \ldots i_w} = \sum_{a < b \leq w} [U_a (U_b)^T]_{i_a i_b}$$

(a) Let S be the set of all the w-dimensional coordinates of the specified entries in the w-dimensional data cube. Show that the objective function for optimization, with regularization, is of the following form:

$$J = \sum_{(i_1 \ldots i_w) \in S} \left(r_{i_1 \ldots i_w} - \sum_{a < b \leq w} [U_a (U_b)^T]_{i_a i_b} \right)^2 + \lambda \sum_{a=1}^{w} ||U_a||^2$$

(b) How would you use this objective function to derive the gradient descent method?

(c) Let $e_{i_1 \ldots i_w} = r_{i_1 \ldots i_w} - \hat{r}_{i_1 \ldots i_w}$ represent the prediction error of an entry $(i_1 \ldots i_w)$ at an intermediate stage of the gradient descent updates. Show that the gradient-descent update equations to each U_a $(1 \leq a \leq w)$ are of the following form for each specified entry $(i_1 \ldots i_w) \in S$:

$$[U_a]_{i_a q} \Leftarrow [U_a]_{i_a q} + \alpha \left(e_{i_1 \ldots i_w} \cdot \sum_{b \neq a} [U_b]_{i_b q} \right) - \lambda \cdot [U_a]_{i_a q}) \quad \forall q \in \{1 \ldots k\}$$

Chapter 9

Time- and Location-Sensitive Recommender Systems

"Time is the wisest counselor of all."–Pericles

9.1 Introduction

In many real scenarios, the buying and rating behaviors of customers are associated with temporal information. For example, the ratings in the Netflix Prize data set are associated with a *"GradeDate"* variable, and it was eventually shown [310] how the temporal component could be used to improve the rating predictions. Similarly, many forms of user activity, such as buying behavior and Web clickstreams, are inherently temporal. In general, recommender systems use the temporal aspects of user activity in two different ways:

1. *Explicit ratings:* In this case, dates are associated with explicit ratings. These dates can be used to improve the accuracy of the prediction process either through the use of forecasting methods, or through periodic and seasonal information (e.g., day of week).

2. *Implicit feedback:* These scenarios correspond to customer *actions*, such as buying items or clicking on Web pages. The historical sequence of user actions is used to make predictions about future actions. The underlying methods often share many similarities with forecasting based on sequential patterns. Such techniques are used frequently in many scenarios such as Web clickstreams or Web log analysis. The techniques can also be used to make predictions about future customer buying behavior.

Generally, it is much harder to use the temporal information in ratings to make recommendations. As we will see later in this chapter, existing temporal models [310] use the temporal

© Springer International Publishing Switzerland 2016
C.C. Aggarwal, *Recommender Systems: The Textbook*,
DOI 10.1007/978-3-319-29659-3_9

information in ratings in a limited and carefully calibrated way. On the other hand, the literature on implicit feedback and discrete models is quite rich because it has been explored extensively in the context of Web clickstreams and logs. The latter problem is closely related to that of forecasting of sequential data with categorical attributes. In this case, discrete data mining methods, such as Markovian models and sequential pattern-mining, are very useful. In this chapter, we will study both types of recommenders.

Time can be viewed from a recency and forecasting perspective, or from a contextual (e.g., seasonal) perspective. From a recency perspective, the basic idea is that recent ratings are more important than older ratings. Therefore, various aging and filtering strategies are used to assign greater importance to more recent data. In the contextual perspective, various periodic aspects, such as season or month, may be used.

The latter scenario is closely related to context-aware recommender systems. In context-aware recommender systems [7], an additional variable, such as location or time, is used to refine the recommendation. In standard collaborative filtering with user set U and item set I, the user-item possibilities in $U \times I$ are mapped to ratings. This mapping is learned from the available data. However, the presence of a context C requires us to learn the mapping from the possibilities in $U \times I \times C$ to the ratings. Note that the context C may itself contain multiple attributes such as location, time, weather, season, and so on. These properties could either be dependent or independent of one another. In this particular chapter, we will study the specific case in which the contextual property is a single attribute corresponding to time. When time is viewed as a continuous variable, the recommendations are often created as functions of time. The temporal context can be viewed from a periodic, recency, or modeling point of view. In *periodic* contexts, a specific periodic aspect of the time, such as weekday, time of day, or month is used in order to sharpen the recommendation. For example, it makes more sense for a North American clothing retailer to recommend winter clothing in December rather than in July. Context-aware recommenders are discussed in a generic sense in Chapter 8. However, we have allocated a separate chapter to the time dimension because of the large amount of literature associated with it. Furthermore, many of the temporal methods, such as forecasting-based ratings prediction and discrete sequence-based methods, cannot be easily generalized to other context-sensitive methods and scenarios. Therefore, the temporal aspect of recommender systems needs to be treated separately from context-aware systems, although the connections to context-based methods are highlighted throughout the relevant parts of the chapter.

Time can be treated as a modeling variable by explicitly expressing the predicted ratings as a function of time. The parameters of this function can be learned in a data-driven manner by minimizing the squared error of the predicted ratings with respect to the observed ratings. An example of such a model is *time-SVD++*, which expresses the predicted ratings as a function of temporally parameterized biases and factor matrices. This approach is considered one of the state-of-the-art techniques for temporal prediction. The main advantage of this approach is that it can capture *future* trends, which are not easily captured by recency, decay-based, or periodic models.

Many data domains such as Web clickstreams do not contain explicit ratings, but they contain discrete action sequences. Such data can be viewed as the temporal version of implicit feedback data sets. The methods used in such domains are often quite distinct from those used in the case of ratings. In particular, Markovian models and sequential pattern-mining methods are commonly used. Such methods have been studied extensively in the Web-mining domain because Web logs are widely available for mining purposes. In this chapter, we will also review discrete-sequence mining methods for recommendations in applications such as Web clickstreams.

Like time, location is another commonly used context in recommender systems. With the increasing popularity of GPS-enabled mobile phones, the use of the location context is useful in a variety of scenarios, such as finding movie theaters, restaurants, or other entertainment locations. In some cases, the location context can be combined with time. This chapter will use the location-based scenario as an important example of context-based systems.

This chapter is organized as follows. In section 9.2, we will introduce methodologies for temporal collaborative filtering with ordered ratings. In particular, we will introduce three different types of models. These correspond to recency-based models, periodic models, and more complex parameterized models. An example of the latter is the *time-SVD++* model, which is considered the state-of-the-art for temporal recommendation. The connections of various models with the context-based models of Chapter 8 are also discussed. Section 9.3 discusses how discrete models can be extended to the temporal scenario in cases where the user actions represent discrete selections such as clicks. Markovian models and sequential pattern-mining methods are discussed in this section. Location-aware recommender systems are discussed in section 9.4. A summary is given in section 9.5.

9.2 Temporal Collaborative Filtering

In this section, we will study the use of temporal recommendations with ratings. Temporal information can be used in one of two ways in order to improve the effectiveness of prediction:

1. *Recency-based models:* Some models consider recent ratings more important than older ratings. In these cases, window-based and decay-based models are used for more accurate prediction. The basic idea in all these models is that the recent ratings are given more importance within the collaborative filtering model.

2. *Periodic context-based models:* In periodic context-based models, the specific property of a period, such as the time at the level of specificity of the hour, day, week, month, or season, is used to perform the recommendation. For example, a clothing retailer would make very different recommendations depending upon whether it was summer or winter [567]. Similarly, the movie recommendations during the Christmas week might be very different from those in the week leading to the Oscars [100]. In these methods, time becomes a contextual variable that is exploited in order to make recommendations. These models are closely related to contextual recommender systems, as introduced in Chapter 8.

3. *Models that explicitly use time as an independent variable:* A recent approach, referred to as *time-SVD++*, uses time as an independent variable within the modeling process. Such an approach uses more refined user-specific and item-specific trends to handle local temporal variations and it can also account for intermittent temporal noise in the ratings. Generally, such models are more sophisticated than recency-based models because they include an element of forecasting.

Window-based and decay-based models have the merit that they are simple and easy to implement in a wide variety of settings. On the other hand, they cannot capture the refined temporal characteristics captured by the *time-SVD++* model. Therefore, the latter method is considered the state-of-the-art in temporal collaborative filtering. Nevertheless, recency-based models have the advantage of being easy to implement. Furthermore, a wider variety of models can be generalized to these cases. On the other hand, only a small number of models have been proposed in the second category.

9.2.1 Recency-Based Models

In recency-based models, recent ratings are given greater importance than older ones. The greater importance of recency can be addressed with either decay-based methods or with window-based methods. In decay-based methods, older ratings are given less importance with the use of a decay function. Window-based methods can be viewed as special cases of decay-based methods, in which binary decay functions are used to completely disregard data points that are older than a specific amount of time. In other words, the binary decay function ensures that older ratings are given a weight of 0, whereas recent ratings are given a weight of 1.

9.2.1.1 Decay-Based Methods

In decay-based methods, a time-stamp t_{uj} is associated with each observed rating of user u and item j in the $m \times n$ ratings matrix R. Therefore, the number of observed values of t_{uj} is exactly equal to the number of observed ratings in R. It is assumed that all recommendations should be made at a future time t_f. This future time is also referred to as the *target time*. Then, the weight $w_{uj}(t_f)$ of the rating r_{uj} at target time t_f is defined with the use of a decay function, that penalizes larger distances between t_{uj} and t_f. A decay function, which is commonly used [185], is the exponential function:

$$w_{uj}(t_f) = \exp[-\lambda(t_f - t_{uj})] \tag{9.1}$$

The decay-rate λ is a user-defined parameter that regulates the importance of time. Larger values of λ de-emphasize older ratings to a greater degree. These weights can be used in neighborhood-based methods to regulate the importance of a rating during the prediction phase.

A method proposed in [185] modifies user-based neighborhood methods by changing the final prediction function. The simple approach used in [185] first determines the k-nearest neighbors of each user. The determination of the nearest neighbors (users) is identical to off-the-shelf user-based neighborhood methods. Subsequently, the only difference from other neighborhood-based methods is that the ratings of other users are weighted with $w_{uj}(t_f)$ during the aggregation process. Specifically, Equation 2.4 of Chapter 2 can now be modified as follows for predicting the rating of item j for user u at time t_f as follows:

$$\hat{r}_{uj}(t_f) = \mu_u + \frac{\sum_{v \in P_u(j)} w_{vj}(t_f) \cdot \text{Sim}(u, v) \cdot (r_{vj} - \mu_v)}{\sum_{v \in P_u(j)} w_{vj}(t_f) \cdot |\text{Sim}(u, v)|} \tag{9.2}$$

Here, $P_u(j)$ represents the k closest users to user u that have specified ratings for item j. Note that the main difference of the aforementioned equation from traditional collaborative filtering is the presence of weights in the prediction function. These weights bias the solutions in favor of more recent trends by discounting stale ratings.

The approach can be easily applied to both user-based and item-based models with a small modification in the final step. In both cases, the final prediction step needs to be augmented with recency-based weights. The optimal value of λ can be learned using cross-validation methods, although such an approach is not discussed in [185].

The work in [186] provides a slightly more refined model in which an item-based neighborhood method is used for collaborative filtering. Aside from weighting each item with item-to-item similarity in the prediction process, a temporal discount factor is also multiplied to the rating of each item within the prediction function. This is, of course, similar to the method used in [185] (and also discussed above). Unlike the work in [185], however, this discount factor is not a simple exponential decay function. Each item is assigned a discount factor by estimating its expected future error and then assigning a weight that is inversely proportional to this error.

Consider a scenario in which the set of peer items of the target item j for which the user u has specified ratings is $Q_j(u)$. The process of determining $Q_j(u)$ is identical to that of item-based neighborhood methods. Then, the discount factor D_{ui} for each item $i \in Q_j(u)$ needs to be determined in order to modify the final prediction function. Note that the discount factor in local to the user u at hand and therefore contains u in the subscript. The prediction of the rating of item j by user u is computed[1] with a discounted version of the item-based prediction function:

$$\hat{r}_{uj} = \frac{\sum_{i \in Q_j(u)} \text{Sim}(i,j) \cdot D_{ui} \cdot r_{ui}}{\sum_{i \in Q_j(u)} |\text{Sim}(i,j)| \cdot D_{ui}} \qquad (9.3)$$

How is each discount factor D_{ui} computed? This is achieved by computing the normalized difference in ratings between the user rating r_{ui} on each item $i \in Q_j(u)$, and the average rating O_{ui} of user u on similar items to item i. The similar items to item i are identified using item-to-item similarity computation. The discount factor (weight) $D_{ui} \in (0,1)$ for each user u and item $i \in Q_j(u)$ is computed as a function of these two quantities.

$$D_{ui} = \left(1 - \frac{|O_{ui} - r_{ui}|}{r_{max} - r_{min}}\right)^{\alpha} \qquad (9.4)$$

Here, r_{max} and r_{min} are the maximum and minimum values on the scale of ratings. α is a tuning parameter, which can be chosen using cross-validation. The basic idea here is that the difference in the user's rating between that of item i and the average rating of the same user on similar items is a manifestation of the error caused by temporal evolution. Furthermore, different users may have different rates of evolution; therefore, the discount factor is local to the specific user at hand.

The methods in [185, 186] do not incorporate the decay weights and discount factors in the similarity computation, and these weights are used only in the prediction phase. It is, however, also possible to compute the similarity in a weighted way, as discussed in section 6.5.2.1 of Chapter 6. In fact, any of these weighted models may be used, once $w_{ij}(t)$ has been defined. While these weighted models were developed in [67] in the context of ensemble methods such as bagging and boosting, they can easily be adapted to the temporal scenario. Note that the way in which matrix factorization models can be generalized to the weighted case is also shown in section 6.5.2.1. Given this, matrix factorization methods can also be easily generalized to recency-based techniques.

---

[1]The original work [186] does not use a modulus in the denominator. We have added it in Equation 9.3 because omitting it does not make much sense in the case of negative similarity. Nevertheless, negative similarities in the peer item-group are rare in practical settings because the peers are defined as the most similar items.

9.2.1.2 Window-Based Methods

In window-based methods, ratings that are older than a particular time are pruned from consideration. This approach can be viewed as a special case of *pre-filtering* or *post-filtering* methods in context-based models. Such methods are discussed in a generic sense in Chapter 8. Furthermore, these methods can also be viewed as (discrete) special cases of decay-based methods. There are several ways in which windows can be modeled:

1. If the difference between the target time t_f and the rating time t_{ij} is larger than a particular threshold, then the rating is dropped. The collaborative filtering model is otherwise identical to any of the methods discussed in Chapters 2 and 3. This approach can be viewed as an extreme case of the decay-based model, in which the decay function is binary. It is often suggested [131] that all of the ratings should be used for similarity computation in neighborhood-based methods. Window-based pruning is used only within the prediction function after the similarities have been computed with all the data. Such an approach can sometimes provide better robustness because of the sparsity of the ratings, in which any type of pruning can make similarity computations unstable. Pruning at the time of similarity computation can lead to overfitting.

2. In some cases, it is possible to obtain some insight into the active periods for various items depending on the underlying domain. In such cases, the windows can be set in a domain- and item-specific way. For example, the method in [131] uses not only the most recent ratings, but also the ratings from the same month in the previous years. Therefore, this approach combines window-based models with some periodic information. This method is referred to as *time-periodic biased k-NN* approach.

So far, all the temporal models are based on the time at which an item was *rated*. A somewhat different approach is to associate weights with a different temporal attribute than the rating time. For example, the work in [595] discusses how one might use the production time of a movie to drop it from consideration. A movie, which is too old, might not be relevant for a user looking for more recent movies. Note that such an approach prunes *all* ratings for the item, because the production time is associated with the item and not the user-item combination. Pruning all ratings for an item is equivalent to dropping the item from the ratings matrix. Therefore, such an approach reduces the dimensionality of the data set by effectively removing them from consideration. However, such an approach should be used with caution, and only for items which are inherently time-sensitive on certain well-known traits.

9.2.2 Handling Periodic Context

Periodic context is designed to handle cases in which the time dimension may refer to a specific period in time, such as hour of the day, day of the week, season, or the time intervals in the vicinity of specific periodic events (e.g., Christmas). Such cases are best handled with the use of multidimensional contextual models, as proposed in [6]. These methods are discussed in detail in Chapter 8.

In this case, the target recommendation time defines the context in which the recommendation is made. This context can sometimes play an all-important role in the recommendation process. For example, for a supermarket, the recommendations targeted for the weekend before the Thanksgiving holiday would be very different from those targeted during other times. Several natural ways of handling periodic context are discussed in the following sections.

9.2.2.1 Pre-Filtering and Post-Filtering

There are two types of filtering methods used in the context-based methods that are referred to as *pre-filtering* and *post-filtering*, respectively. These methods are discussed in detail in sections 8.3 and 8.4 of Chapter 8. Here, we provide a brief overview in the context of temporal recommender systems.

In pre-filtering, a significant part of the ratings data are removed that are not relevant to the specific target time (i.e., context) within which the recommendation is being performed or executed. For example, one might use only the ratings from the fortnight before Thanksgiving of each year, in order to construct the models for making recommendations on the weekend before Thanksgiving. A particularly interesting approach along this direction was the use of *contextual microfiles* [61], which segmented the ratings by context. This kind of segmentation effectively filters out the irrelevant ratings from each segment. Some examples of possible segmentations include {Morning, Evening}, {WeekDay, Weekend}, and so on. Within each context, a separate model is constructed for prediction. After filtering, any non-contextual method may be used to make predictions on the pruned data within each segment. The main challenge associated with pre-filtering methods is that the pruned data set is even more sparse than the original data, and therefore the accuracy of the recommendation process is impacted negatively. This is a direct result of overfitting. The success of pre-filtering often depends on the sparsity of the pruned data set. Therefore, the approach cannot be used easily for contexts that are too fine-grained (e.g., day of year). In many cases, hierarchies are used on the periodic context to improve the accuracy of recommendation. For example, consider a scenario where the context is set to 7 AM. Instead of using ratings received between 7 AM and 8 AM, one might use all the ratings received between 6 AM to 9 AM. This will result in the use of a larger number of ratings, and the approach will therefore help to prevent overfitting.

In post-filtering, the recommendations are adjusted based on the context, after a non-contextual method has been used to generate the recommendation on all the data. Therefore, the basic approach of post-filtering uses the following two steps:

1. Generate the recommendations using a conventional collaborative filtering approach on all data, while ignoring the temporal context.

2. Adjust the generated recommendation list with the use of temporal context as a post-processing step. Either the ranks of the recommended list may be adjusted, or the list may be pruned of contextually irrelevant items.

After the recommendation lists have been generated, the ranking is either re-adjusted by weighting with a *contextual relevance weight*, or the items with very low contextual relevance weights are removed. Let $\hat{r}_{uj}$ represent the predicted rating of user u for item j using all the data, before contextual post-filtering has been applied. The resulting ratings (and rankings) are then adjusted with the use of contextual relevance weights $P(u, j, C)$, where C is the context. Therefore, the adjusted rating is given by $\hat{r}_{uj} \cdot P(u, j, C)$.

How is the contextual relevance weight determined? In contrast to pre-filtering methods, which work only with ratings, post-filtering methods often use content properties of items in order to determine the contextual relevance weights. Under the covers, post-filtering methods sometimes incorporate pre-filtering techniques to a minor degree in the process of determining contextual relevance weights. For a given user u for whom a prediction needs to be made, her ratings for the specific period of interest are pre-filtered, and an off-the-shelf recommendation model is constructed on the pre-filtered ratings, to predict her ratings for that specific periodic context C. For example, if movie recommendations are to be made in the context of weekends, the relevance of the user for each movie in the weekend is determined with the use of either a collaborative model or a content-based model on the pre-filtered data. The work in [471] uses a very simple scheme in which the fraction of neighbors of a user who have watched a particular movie in the pre-filtered data is used in order to compute the contextual relevance weight. This relevance weight $P(u, j, C)$ is assumed (or scaled) to be a probability value in $(0, 1)$, where larger values imply greater interest. Then, the predicted rating $\hat{r}_{uj}$ is multiplied with the relevance weight $P(u, j, C)$, or the item is simply removed from the recommended list when $P(u, j, C)$ is very small. These two methods are referred to as *weighting* or *filtering* methods in contextual post-filtering [471]. Post-filtering methods hedge their bets between the robustness of (larger) global data sets and the refined accuracy of the pruned data by making use of both in the recommendation process.

In many cases, the estimation of $P(u, j, C)$ is executed independent of the user u, by using only the content information in the item j. For example, if comedy movies and Steven Spielberg movies are often watched over weekends by all users, the genre/actors/director of the movie can be used as the content, and the label is either *weekends* or *weekdays*. The training data can contain the data over all users and not just user u. A machine learning model then estimates the value of $P(*, j, C)$ by using this training data, where "*" denotes a "don't care." Such an approach is less personalized in the computation of $P(u, j, C)$, but it can handle sparsity more effectively. Note that the final predicted value $\hat{r}_{uj} \cdot P(*, j, C)$ is still personalized to u because of how $\hat{r}_{uj}$ is determined. The specific choice of model used for estimating $P(u, j, C)$ depends on the data set at hand and its sparsity. Readers are advised to refer to Chapter 8 for more details about both methods. In particular, post-filtering methods are discussed in section 8.4.

9.2.2.2 Direct Incorporation of Temporal Context

In pre-filtering and post-filtering methods, the incorporation of context is done either strictly before or strictly after the recommendation process. In both cases, the approach reduces the problem to a 2-dimensional model. However, it is also possible to directly modify existing models such as neighborhood methods in order to incorporate temporal context. In such cases, one works directly with the 3-dimensional representation corresponding to user, item, and context. For example, in the user-based neighborhood scheme, one might modify the distance computation between two users with the use of contextual attributes. If two users give the same rating to an item during weekends, they should be considered more similar than a pair of users that have specified these ratings in different temporal contexts. By using the modified distance computation, the context is automatically incorporated into the recommendation process. One can also modify regression and latent-factor models to incorporate the temporal context directly. These methods apply generally to any context-based scenario (e.g., location), and not just the temporal context. Therefore, this topic is discussed in detail in Chapter 8 on context-based methods. Refer to section 8.5 of that chapter.

9.2.3 Modeling Ratings as a Function of Time

In these methods, the ratings are modeled as a function of time and the parameters of the model are learned in a data-dependent way. A few methods that use time-series models to make predictions are discussed in the bibliographic notes. In this section, we will study the use of *temporal factor models*, which are considered state-of-the-art in this domain.

These methods can intelligently separate long-term trends from transient and noisy trends. Furthermore, the models have a natural element of forecasting built into them. These distinctions are important for making such temporal models robust. Such robustness cannot be achieved with a mere decay-based or filtering approach to the temporal model. In this section, we will study the *time-SVD++* model, on which a lot of subsequent work in the field is based.

9.2.3.1 The Time-SVD++ Model

The *time-SVD++* model can be viewed as a temporal enhancement of the *SVD++* model. Readers are advised to revisit section 3.6.4.6 of Chapter 3, as the discussion in this section relies on that previous passage. We will also briefly discuss the *SVD++* model here; this will provide us with the opportunity to introduce slightly different notations from those in Chapter 3. These notations are relevant to the temporal version of the model.

As in the case of the *SVD++* model, we can assume without loss of generality that we are working with a ratings matrix in which the mean of all ratings in the training data is 0. Note that when the mean (denoted by μ) of all ratings is nonzero, it can be subtracted from all the entries, and the analysis can be performed with this centered matrix in order to predicted correspondingly centered ratings. Later, the mean can be added back to the predicted value of the rating.

Recall that the factor model of section 3.6.4.5, which incorporates bias, expresses the ratings matrix $R = [r_{ij}]_{m \times n}$ in terms of the user biases, the item biases, and the factor matrices. The predicted rating $\hat{r}_{ij}$ is expressed in terms of these variables as follows:

$$\hat{r}_{ij} = o_i + p_j + \sum_{s=1}^{k} u_{is} \cdot v_{js} \tag{9.5}$$

Here, o_i is the bias variable for user i, p_j is the bias variable for item j, $U = [u_{is}]_{m \times k}$, and $V = [v_{js}]_{n \times k}$ are factor matrices of rank k. The part $(o_i + p_j)$ does not use any personalized rating data, but it simply relies on global properties of the ratings. Intuitively, the variable o_i indicates the propensity of user i to rate all items highly, whereas the variable p_j denotes the propensity of item j to be rated highly. For example, a generous and optimistic user is likely to have large positive values of o_i, and a box office hit is likely to have large positive values of p_j. This basic bias-based model is further enhanced in section 3.6.4.6 with the notion of implicit feedback variables $Y = [y_{ij}]_{n \times k}$ for each user-item pair. These variables encode the propensity of each factor-item combination to contribute to implicit feedback. For example, if $|y_{ij}|$ is large, then it means that simply the act of rating item i contains significant information about the affinity of that user for the jth latent component (no matter what the actual value of the rating might be). In other words, the jth latent component of any user who has rated item i should be adjusted based on the value of y_{ij}.

Let I_i be the set of items rated by user i. Then, the predicted value of the rating, which includes implicit feedback, can be expressed as follows:

$$\hat{r}_{ij} = o_i + p_j + \sum_{s=1}^{k} \left(u_{is} + \sum_{h \in I_i} \frac{y_{hs}}{\sqrt{|I_i|}} \right) \cdot v_{js} \tag{9.6}$$

Note that the term $\sum_{h \in I_i} \frac{y_{hs}}{\sqrt{|I_i|}}$ on the right-hand side of the aforementioned equation adjusts the sth latent factor u_{is} of user i, based on the implicit feedback. Refer to section 3.6.4.6 of Chapter 3 for a more detailed explanation of this adjustment. Equation 9.6

is identical to Equation 3.21 of Chapter 3, except that we use slightly different[2] notations here in explicitly separating out the bias variables.

The main difference between the *SVD++* model and the *time-SVD++* model is that *some* of the model parameters are assumed to be functions of time in the latter. Specifically, the *time-SVD++* model assumes that the user biases o_i, item biases p_j, and the user factors u_{is} are functions of time. Therefore, these terms will be expressed as $o_i(t)$, $p_j(t)$, and $u_{is}(t)$ to denote the fact that they are functions of time. By using these temporal variables, one now obtains the time-varying predicted value $\hat{r}_{ij}(t)$ of the (i, j)th entry of the ratings matrix at time t as follows:

$$\hat{r}_{ij}(t) = o_i(t) + p_j(t) + \sum_{s=1}^{k} \left(u_{is}(t) + \sum_{h \in I_i} \frac{y_{hs}}{\sqrt{|I_i|}} \right) \cdot v_{js} \qquad (9.7)$$

It is noteworthy that the item variables v_{js} and the implicit feedback variables y_{hs} have not been temporally parameterized, and are assumed to stay constant with time. It is possible in principle[3] to temporally parameterize these variables as well, although the *time-SVD++* model chooses a simplified approach in which each temporal parametrization can be justified with intuitive arguments. These intuitions are discussed below, together with the specific form of the temporal parametrization of each of the variables $o_i(t)$, $p_j(t)$, and $u_{is}(t)$, respectively:

1. The intuition for choosing the temporal form of the item bias $p_j(t)$ is that the popularity of an item can vary significantly with time, but it shows a high level of continuity and stability over shorter periods. For example, a box-office hit will have an approximately stable distribution of ratings in the short period after release, but it might be rated very differently after a couple of years have elapsed. Therefore, the time horizon can be split into bins of equal size, and the ratings belonging to a particular bin have the same bias. Smaller bin sizes lead to better granularity but it may also result in overfitting because enough ratings may not be present in each bin. In the original work on Netflix movie ratings [310], a total of 30 bins were used, and each bin represented about 10 consecutive weeks of ratings. The item bias $p_j(t)$ can now be split into a constant part and an offset parameter, which is bin-specific depending on the time t at which item j is rated:

$$p_j(t) = C_j + \text{Offset}_{j, Bin(t)} \qquad (9.8)$$

   Note that both the constant part C_j and offsets are parameters that need to be learned in a data-driven manner. The optimization problem for this learning process will be discussed later. Note that the value of $p_j(t)$ will be different for different ratings, depending on *when* they were rated. Unlike users, items can be more successfully binned in this way because most items usually have sufficient ratings.

2. A different approach is used to parameterize the user bias $o_i(t)$. The binning approach will not work for users because many users may not have sufficient ratings. Therefore, a functional form may be used to parameterize the user bias, which captures the

---

[2]In the discussion of section 3.6.4.6, the bias variables are absorbed within the factor matrices U and V by increasing the number of columns in each of the two factor matrices from k to $(k + 2)$. However, in this exposition, we do not absorb the bias variables in the columns of the factor matrices. This is because of the more complex and special way in which bias variables are treated in temporal models. For example, Equation 3.21 of Chapter 3 and Equation 9.6 are identical, but they use somewhat different notations. It is important to keep these notational distinctions in mind to avoid confusion.

[3]The work in [293] uses time-varying item factors.

concept drift of the user over time. Let the mean date of all ratings of user i be denoted by ν_i. Then, the concept drift $\text{dev}_i(t)$ of user i at time t can be computed as a function of t as follows:

$$\text{dev}_i(t) = \text{sign}(t - \nu_i) \cdot |t - \nu_i|^\beta \qquad (9.9)$$

The parameter β is selected using cross-validation. A typical value of β is around 0.4. In addition, the transient noise at each time t is captured with the parameters ϵ_{it}. Then, the user bias $o_i(t)$ is split into a constant part, a time-dependent part, and transient noise, as follows:

$$o_i(t) = K_i + \alpha_i \cdot \text{dev}_i(t) + \epsilon_{it} \qquad (9.10)$$

In practice, the time is often discretized on a day-specific basis. Therefore, ϵ_{it} corresponds to the transient day-specific variability. As in the case of item bias parameters, the parameters K_i, α_i, and ϵ_{it} must be learned in a data-driven manner. The idea here is that the average rating of the user can vary significantly from that at the mean date of rating. The user might be rating most items positively (or negatively) now, but her mean rating might decrease (or increase) in a couple of years. This portion of variability is captured by $\alpha_i \cdot \text{dev}_i(t)$. However, transient mood changes from day to day might lead to sudden and unpredictable spikes or dips in ratings. A user might rate all items poorly when she is having a bad day. Such variations are captured by ϵ_{it}.

3. The user factors $u_{is}(t)$ correspond to the affinity of users towards various concepts. For example, a young user who appreciates action movies today might be interested in documentaries after a few years. As in the case of user biases, the amount of elapsed time is a crucial factor in deciding the amount of drift. Therefore, a similar approach to user biases is used for modeling the temporal change in the user factors:

$$u_{is}(t) = K'_{is} + \alpha'_{is} \cdot \text{dev}_i(t) + \epsilon'_{ist} \qquad (9.11)$$

As in the case of user biases, the constant effects, long-term effects, and transient effects are modeled by the three terms. Although we have used similar looking symbols as user biases to emphasize the similarity between the two modeling cases, we have added an apostrophe superscript to each variable to emphasize the fact that the variables in Equations 9.10 and 9.11 are distinct. Note that the same user-specific deviation function $\text{dev}_i(t)$ is used in the two cases although it is possible to use different forms of this function for the two cases.

How does one use the aforementioned model to set up the optimization problem? We assume that the observation time of all the ratings is known. Therefore, for an entry (i, j) whose rating has been observed at time t_{ij}, one must compare the observed value r_{ij} with the predicted value $\hat{r}_{ij}(t_{ij})$ in order to compute the error in prediction. In this case, one needs to minimize the squared error function $[r_{ij} - \hat{r}_{ij}(t_{ij})]^2$ over all the observed ratings in the data. The value of $\hat{r}_{ij}(t_{ij})$ is derived with the help of Equation 9.7. In addition, the squared regularization terms for the various parameters need to be added to the objective function. In other words, if S contains the set of user-item pairs for which ratings are specified in the matrix $R = [r_{ij}]_{m \times n}$, then one must solve the following optimization problem:

$$\text{Minimize } J = \frac{1}{2} \sum_{(i,j) \in S} [r_{ij} - \hat{r}_{ij}(t_{ij})]^2 + \lambda \cdot (\text{Regularization Term})$$

The regularization term contains the sum of the squares of all the variables in the model. As in the case of all the factorization models discussed in Chapter 3, a gradient descent approach can be used in order to optimize the objective function J and learn the relevant parameters. The partial derivative of J is computed with respect to each parameter to determine the relevant gradient directions. These learned parameters are then used for prediction. The details of these learning steps are omitted. Readers are referred to the original work [310] for more details. Here, we will discuss how the model can be used once the parameters have been learned.

Using the Model for Prediction

After the parameters of the model have been learned, how can they be used for predictions? For a given user i and item j, one can use Equation 9.7 to determine the predicted rating $\hat{r}_{ij}(t)$ at future time t by substituting the learned values of the parameters. The main problem with doing so is the presence of day-specific parameters, such as ϵ_{it} and ϵ_{ist}. These parameters can only be learned for past days from the historical data, but they cannot be learned for future days. However, these parameters only correspond to transient noise, which cannot, by definition, be learned in a data-driven manner. Therefore, these values are set to 0 for future days under the assumption that a noise-free prediction is being made; accordingly, the learned values of these parameters are not used for prediction. Although these parameters are not used in the final prediction, they are still quite important for the modeling process because they absorb the transient noise and spikes in the ratings. For example, if a user provides very low ratings to all items because she is having a bad day, the presence of these parameters will dampen the effects of this transient noise in the historical data. Therefore, the parameters ϵ_{it} and ϵ_{ist} help in learning the *other* parameters in a more robust way by removing transient spikes and noise. In other words, the day-specific parameters ϵ_{it} and ϵ_{ist} play the role of cleaning the training data in the modeling process.

Practical Issues

An immediate observation is that the aforementioned model has a very large number of parameters as compared to those in Chapter 3. Therefore, it is crucial to have sufficient data, so that overfitting may not remain an issue. This can be a problem for smaller data sets. However, the approach seems to perform quite well [310] for the Netflix data set, which is quite large. Interestingly, the survey in [312] shows that one can obtain reasonably good results on the Netflix Prize data set by dropping the factorization completely and using only the bias terms. The use of only bias terms yielded results that were almost comparable to Netflix's *Cinematch* recommender system. This is because the non-personalized aspect of the ratings (i.e., user-specific and item-specific bias) can explain a very large part of the ratings. These results suggest the importance of incorporating the bias terms in latent factor models, as discussed in section 3.6.4.5 of Chapter 3.

Furthermore, the time-dependent terms in $o_i(t)$ and $u_{is}(t)$ can be modeled using other functional forms such as splines or with the use of periodic trends. These different functional forms can capture different data-specific temporal scenarios. We have restricted our discussion to the simplest possible choices for ease in discussion. A detailed discussion of these alternatives is provided in [312].

Observations

It is noteworthy that user factors vary in the temporal sense but item factors do not. This choice can be intuitively justified. Recall from the discussion in Chapter 3 that the user factors correspond to user affinities towards various concepts, whereas the item factors correspond to item affinities towards various concepts. The basic idea here is that user moods and preferences can change over time, which will be reflected in the changes of their affinities towards the various concepts. On the other hand, the affinity of an item to a concept is inherent to that item, and it can be assumed to be stable over time. Therefore, it is not necessary to increase the complexity of the model by temporally parameterizing the item factors. Unnecessary temporal parametrization increases the complexity of the model and leads to overfitting. Nevertheless, the work in [293] shows how one might use time-varying item factors as well. It is an open question as to whether the use of temporal parametrization of item biases will lead to overall improvement in accuracy over most data sets.

9.3 Discrete Temporal Models

Discrete temporal models are relevant to the case where the underlying data is received as discrete sequences. Such data can be encountered in a variety of application scenarios, most of which are associated with implicit user feedback rather than explicit ratings. Some examples of such application scenarios are as follows:

1. *Web logs and clickstreams:* The user accesses to Web logs can typically be represented as sequential patterns. User patterns can often show predictable access patterns. For example, users will frequently access particular sequences of Web pages. The frequent sequence information can be used to make recommendations [182, 208, 440, 442, 443, 562].

2. *Supermarket transactions:* The customer buying behavior in supermarkets is a form of sequential data. In fact, the problem of sequential pattern-mining was defined [37] to handle this scenario. In fact, because the activity time-stamps are usually available in supermarket data sets, they can be converted into user-specific sequential patterns of buying activity. The *temporal order* is often quite important. For example, it makes sense to recommend a printer cartridge after a user has bought a printer, but not vice versa.

3. *Query recommendations:* Many Web sites record the user queries at their site. The sequence of queries can be used to make recommendations of other more useful queries.

In this section, two types of models will be discussed. The first of these methods is based on Markovian models, whereas the second is based on sequential pattern-mining.

9.3.1 Markovian Models

An interesting Markovian model to predict Web page accesses was proposed in [182]. Although the approach is discussed in the context of Web page accesses, it can be generalized to recommending any type of action, as long as the temporal ordering of the user actions is available. The discussion in this section is based on this work [182].

In Markovian models, the sequential information is encoded in the form of *states*, which are used for predictive purposes. A kth order Markov model defines a state based on the last k actions performed by the user. An *action* is defined in an application-specific way. It might correspond to the user visiting a particular Web page, or it might correspond to the user buying a specific item. The actions are represented by a set of symbols Σ. As the actions are application-specific, the symbol set Σ is application-specific as well. For example, the symbol set Σ could correspond to the indices of the universe of items in an e-commerce application, or it could correspond to the URLs of Web pages in a Web log mining application. We assume that the symbol set Σ contains the symbols $\Sigma = \{\sigma_1 \ldots \sigma_{|\Sigma|}\}$. Therefore, a state $Q = a_1 \ldots a_k$ is defined by sequence of k actions, such that each a_i is drawn from Σ. A state with k actions is drawn from an order-k Markovian model. For example, consider the case in which the symbols in Σ correspond to the act of watching various movies. Furthermore, consider the following state Q:

$$Q = \textit{Julius Caesar, Nero, Gladiator}$$

This state has three different actions corresponding to the user watching these movies in a particular sequence. Therefore, this state is drawn from an order-3 Markov model. Furthermore, the default assumption in such Markov models is that the movies have been watched *consecutively*. There are a total of $|\Sigma|^k$ possible states in an order-k Markovian model, although many of these states might not occur frequently in a particular data set.

In general, a sequence defines the transitions[4] in a *Markov chain*. In an order-k model, the current state is defined by the last k actions in the Markov chain. Consider a sequence of actions (e.g., Web page accesses), in which t actions $a_1 a_2 \ldots a_t$ have occurred so far in sequence, where $a_i \in \Sigma$. Then, the current state of the order-k Markov model at time t is $a_{t-k+1} a_{t-k+2} \ldots a_t$. The last action in this sequence is a_t, which resulted in a transition from the state $a_{t-k} a_{t-k+1} \ldots a_{t-1}$ to the state $a_{t-k+1} a_{t-k+2} \ldots a_t$. Therefore, the states in a Markov chain are connected by edges, corresponding to transitions. Each edge is annotated with an action drawn from Σ, and a probability of transition. In this particular example, the transition from the state $a_{t-k} a_{t-k+1} \ldots a_{t-1}$ to the state $a_{t-k+1} a_{t-k+2} \ldots a_t$ is associated with the action a_t. As there are $|\Sigma|$ possible transitions out of each of the $|\Sigma|^k$ states, the total number of edges in a complete Markov model of order k is equal to $|\Sigma|^{k+1}$. Any incoming edge of a state $a_{t-k+1} a_{t-k+2} \ldots a_t$ in an order-k Markov chain is always annotated with the last action a_t. The sum of the probabilities of the transitions out of a state is always 1. The probabilities of transitions are learned from the training data (e.g., a sequence of previous Web page accesses). We have shown an order-1 Markov chain drawn on the alphabet $\{A, B, C, D\}$ in Figure 9.1. Note that this Markov chain has 4 states and $4 \times 4 = 16$ edges. The sequence of actions $AABCBCA$ corresponds to the following path of states in the Markov chain:

$$A \Rightarrow A \Rightarrow B \Rightarrow C \Rightarrow B \Rightarrow C \Rightarrow A$$

Note that a Markov model of order-2 would contain $4^2 = 16$ states and $4^3 = 64$ edges. This is already too large to neatly show in a diagram like Figure 9.1. The corresponding sequence of transitions for the action sequence $AABCBCA$ is given by the following:

$$AA \Rightarrow AB \Rightarrow BC \Rightarrow CB \Rightarrow BC \Rightarrow CA$$

Consider a situation, where we have trained a Markov model of order k, and we need to make a prediction for the next action after the sequence $a_1 \ldots a_t$. Then, for each action $\sigma_i \in \Sigma$, we need to estimate the value of the action σ_i, given the current state of the last k actions. In

---

[4]Refer to the bibliographic notes for background on Markov chains.

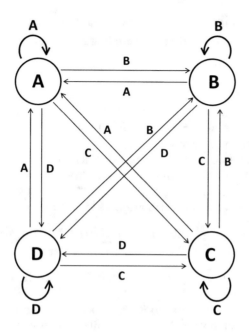

Figure 9.1: An order-1 Markovian model

other words, we need to estimate the probability $P(a_{t+1} = \sigma_i | a_{t-k+1} a_{t-k+2} \dots a_t)$ for each $\sigma_i \in \Sigma$. The top-r actions with the largest probabilities can be returned as the predictions. Note that the probabilities $P(a_{t+1} = \sigma_i | a_{t-k+1} a_{t-k+2} \dots a_t)$ need to be estimated from the training data. This suggests the following simple approach for training and prediction on a Markov model of order k.

1. **(Training phase)** Let $\mathcal{S}$ be the set of $|\Sigma|^k$ possible sequences of length k. For each possible sequence (state) $S \in \mathcal{S}$, use the training data to learn the $|\Sigma|$ probabilities $P(\sigma_i | S)$ corresponding to each candidate action $\sigma_i \in \Sigma$. Note that a total of $|\Sigma|^{k+1}$ probabilities need to be learned, which is also equal to the number of edges in an order-k Markov model. Each of the learned probabilities corresponds to the transition probability of an edge in the Markov model.

2. **(Prediction phase)** For a current sequence of user actions, determine the relevant state S_t in the Markov chain using the last k actions of the user. Report the top-r actions in Σ with the largest values of $P(\sigma_i | S_t)$ as the recommendations.

The Markovian approach relies on the *short memory assumption* of typical user action sequences. The idea is that the user actions depend only on the set of k immediately preceding actions. While this assumption may not be completely true in practice, it often approximates many real-world scenarios.

It remains to be explained how the probabilities are estimated from a given training data set. This can be achieved by extracting all of the k-sequences from the training database and determining the fraction of times that each action in σ_i occurs after this sequence. This estimate is determined as the relevant probability. Consider a sequence S, which is one of the $|\Sigma|^k$ possible sequences. If this sequence occurs $F(S)$ times in the training data, and the sequence S is followed by the action σ_i for a total of $f(S, \sigma_i) \leq F(S)$ times in the data, then the probability $P(\sigma_i|S)$ is estimated as follows:

$$P(\sigma_i|S) = \frac{f(S, \sigma_i)}{F(S)} \qquad (9.12)$$

Note that the training data may contain either one long sequence or multiple sequences. In either case, the frequencies $f(S, \sigma_i)$ and $F(S)$ count the repeated occurrences within a single sequence as multiple occurrences.

This estimation may sometimes be difficult when the value of $F(S)$ is small. In fact, when the value of $F(S)$ is 0, the estimated probability becomes indeterminate. In order to address this problem, we use Laplacian smoothing. A Laplacian smoothing parameter α is used to modify the aforementioned estimation as follows:

$$P(\sigma_i|S) = \frac{f(S, \sigma_i) + \alpha}{F(S) + |\Sigma| \cdot \alpha} \qquad (9.13)$$

Typically, the value of α is set to a small quantity. Note that when the value of $F(S)$ is 0, each action is estimated to the probability value of $1/|\Sigma|$. This is quite reasonable when we do not have sufficient data about the specific action after a particular sequence. The notion of Laplacian smoothing serves a similar function to that of regularization by avoiding overfitting from limited training data. In practice, states with zero frequency are not represented at all in the Markov model. This means that some of the states are missing, and it may not be possible to find a matching state for a particular test sequence. Such test instances are said to be *uncovered* by the Markov model. How are such states handled?

The work in [182] builds all Markov models up to maximum order l, and then uses the highest-order model that covers the test instance. In other words, if all models up to order 3 are constructed, the approach first tries to find a matching state in the order-3 Markov model. If such a state is found, then it is used for prediction. Otherwise, the Markov model of order 2 is tested, and then the model of order 1 is tested. For most training data sets of reasonable sizes, all possible $|\Sigma|$ states are present in the Markov model of order 1, and therefore it serves as a default model for difficult cases in which matching models of higher order are not found. If needed, a default catch-all prediction corresponding to the most frequent action can be returned if no matching state is found.

9.3.1.1 Selective Markov Models

One of the problems with the approach outlined in the previous section is that the number of possible states may be too large, and most of them may not even be present in a particular training data set. The large number of states also makes it expensive to train the model, necessitating the estimation of as many as $|\Sigma|^{k+1}$ possible probabilities for an order-k Markov model. For larger values of k, it may be impractical to train such a model. Furthermore, many of the states with little presence in the training data may be unreliable for training purposes.

The main idea in [182] is to propose the notion of *selective* Markov models, in which many of the irrelevant states are pruned up front during model construction. This pruning may be accomplished in several ways:

1. *Support-pruned Markov model:* The *support* of a state (or k-sequence) is the frequency of its presence in the training data. The basic assumption is that states with low support are unreliable in terms of their predictive power on unseen test data.

In particular, the estimated probabilities of states with low support might be unreliable because of overfitting. Support pruning can drastically reduce the number of states in the models of higher orders. The support threshold is defined as an absolute frequency (rather than as a fraction), and is defined as the same value across models of all orders. Higher-order models have lower support values, and are therefore more likely to have their states pruned. This approach greatly reduces the state-space complexity of the model because the number of possible states increases exponentially with the order of the model.

2. *Confidence-pruned Markov model:* The confidence-pruned Markov model tends to favor states in which the highest probability of an outgoing edge from a state is as large as possible. Note that if all the transition probabilities on the edges exiting a state are similar, one cannot confidently claim that any of the actions in Σ is significantly more likely than the others. In the other extreme case, if one of the edges exiting a state has a probability of almost 1 and the others are almost 0, then one can confidently predict the next action at this state. Such states are more useful. How can one determine the appropriate confidence thresholds for pruning?

This approach computes the $100 \cdot (1 - \alpha)$ confidence interval around the most probable action, and then determines if the second-highest probability lies in this interval. Consider a candidate state for pruning, which has raw frequency n in the training data. Let p_1 and p_2 be the transition probabilities of the first- and second-highest probable edges exiting that state. We are already assured that $p_2 \leq p_1$ because p_1 is probability of the most probable edge. Let $z_{\alpha/2}$ be the absolute value of the Z-number matching the upper $(\alpha/2)$-percentage point of the standard normal distribution. Then, in order for the state to be pruned, the following condition must hold:

$$p_2 \geq p_1 - z_{\alpha/2} \sqrt{\frac{p_1(1 - p_1)}{n}} \tag{9.14}$$

Note that $\sqrt{\frac{p_1(1-p_1)}{n}}$ represents the standard deviation of the average of n i.i.d. Bernoulli variables, each of which have probability p_1 of success. The level of pruning is controlled by the confidence threshold α.

3. *Error-pruned Markov model:* In the error-pruned Markov model, a validation set is held out from the training data and is not used for building the Markov model. This validation set is used to test the accuracy of the model. The accuracy specific to each state is computed with the validation set. For each higher-order state, its immediate lower-order prediction alternatives are determined. For example, for an order-4 state $a_1 a_2 a_3 a_4$, the same action sequence can be predicted with the lower-order states $a_2 a_3 a_4$, $a_3 a_4$, and a_4. If the error rate of a higher-order state is greater than that of any of its lower-order alternatives, then it is pruned. This process is recursively applied to states of all orders, starting from the higher to the lower, until no more states can be pruned. States of order 1 are always retained in order to maximize coverage.

Although the aforementioned approach compares the errors of higher- and lower-order states, it does not use the same validation examples to compare the accuracy of a pair of states. A second approach for error pruning that uses the same set of validation examples for comparing the error of two states is as follows. First, all the validation examples that can be predicted with a higher-order state are determined. Then the same validation examples are tested with respect to the lower-order states. If the error with the use of the higher-order state is greater than that of any of the lower-order states *on the same validation examples*, then the higher-order state is pruned. This approach is applied recursively to all states of lower order, except for states of order 1.

These alternatives were experimentally tested in [182]. It was shown that all forms of pruning provided some advantage, although the greatest advantage was obtained with the use of error-pruned models. There was little statistical difference between the support-pruned and confidence-pruned Markov models.

9.3.1.2 Other Markovian Alternatives

In the Markovian models of this section, a *contiguous* sequence of actions is used to predict the next action. Furthermore, the states are fully visible and can be directly explained in terms of the last k user actions. A more sophisticated alternative is the use of *Hidden* Markov Models (HMMs) in which the states are hidden. In such cases, non-contiguous subsequences can be used to make predictions. The HMM approach is beyond the scope of this book; refer to the bibliographic notes.

9.3.2 Sequential Pattern Mining

Sequential pattern-mining was originally proposed as a method for mining patterns from sequences of supermarket data. Sequential patterns can be used to create rule-based predictive models for temporal sequences. Such methods can be considered the temporal analog of rule-based methods discussed in section 3.3 of Chapter 3. First, we define the notions of *subsequences* and *frequent subsequences*.

Definition 9.3.1 (Subsequence) *A sequence of symbols $a_1 a_2 \ldots a_k$ is said to be a subsequence of the sequence $b_1 b_2 \ldots b_n$, if we can find k elements $b_{i_1} \ldots b_{i_k}$, such that $i_1 < i_2 < \ldots < i_k$, and $a_r = b_{i_r}$.*

In the original definition of sequential pattern mining [37], the elements are themselves allowed to be sets, and the condition $a_r = b_{i_r}$ is replaced with the condition $a_r \subseteq b_{i_r}$. However, in most recommender applications, this complex definition is not necessary, and we can work with sequences of individual symbols. Therefore, we will use this simplified definition in this chapter. It is noteworthy that the definition of a subsequence allows gaps in the matching. The allowance of such gaps is useful in accounting for the noise in sequences.

In sequential pattern-mining methods, the goal is to determine the frequently occurring subsequences in the data at a support level s. The frequency is defined with respect to a database $\mathcal{D}$ of multiple sequences.

Definition 9.3.2 (Frequent Subsequences) *A subsequence $a_1 \ldots a_k$ is said to be a frequent subsequence with respect to a database $\mathcal{D}$ of sequences at minimum support s, if it is a subsequence of at least a fraction s of the sequences in the data.*

Note that the support s is always a fraction by definition. One can also define the confidence of a rule in sequential pattern mining. Traditionally, the notion of confidence is defined only for non-temporal association rules, but one can also extend the definition to sequential pattern mining in various ways.

Definition 9.3.3 (Confidence) *The confidence of a rule $a_1 \ldots a_k \Rightarrow a_{k+1}$ is equal to the conditional probability that $a_1 \ldots a_{k+1}$ is a sequence in the database, given that $a_1 \ldots a_k$ is a sequence. In other words, if $f(S)$ denotes the support of sequence S, then the confidence of the rule $a_1 \ldots a_k \Rightarrow a_{k+1}$ is defined as follows:*

$$Confidence(a_1 \ldots a_k \Rightarrow a_{k+1}) = \frac{f(a_1 \ldots a_{k+1})}{f(a_1 \ldots a_k)}$$

Note that the definition of confidence in sequential rule mining is adapted from association rule mining. The notion of confidence can be defined in other ways depending on the application at hand. For example, one can impose the constraint that a_{k+1} immediately follows a_k without a gap in some applications.

The definitions of support and confidence can be used to define sequential pattern-based rules.

Definition 9.3.4 (Sequential Pattern-Based Rule) *A rule $a_1 \ldots a_k \Rightarrow a_{k+1}$ is said to be valid at minimum support s and minimum confidence c, if both the following conditions are satisfied:*

1. *The support of $a_1 \ldots a_{k+1}$ is at least s.*

2. *The confidence of $a_1 \ldots a_k \Rightarrow a_{k+1}$ is at least c.*

Algorithms for determining frequent sequential patterns are discussed in [23]. After the sequential patterns have been determined, one can also determine the rules at the desired level of minimum support and confidence. The training phase in sequential pattern-mining methods finds all the rules at the specified level of minimum support and confidence. After the rules have been determined, the following approach is used for the prediction of the relevant ranked list of items (e.g., clicks in a Web clickstream) for a current test sequence T:

1. Identify all the matching rules for the test sequence T.

2. Rank the items in the consequents of the matching rules in decreasing order of confidence. Heuristic methods can be used to aggregate the predictions when multiple rules contain the same item in the consequent.

In some cases, it may be desirable to restrict the gaps between successive elements. For example, when the sequences are very long, it generally becomes more desirable to impose gap constraints on the sequences during the training and prediction process. Depending on the specific application at hand, many variations of this basic approach may be used. These variations are as follows:

1. Maximum gap constraints may be imposed during the process of finding the frequent sequences. In other words, the matching process may allow the maximum gap between a pair of adjacent sequences to be at most δ. Alternatively, one can impose a maximum constraint on the time difference between the first and last elements of the sequence. Such constraints can be handled by *constrained sequential pattern-mining* methods, and they are particularly important when the individual sequences in the database are very long. A discussion of constrained sequential pattern-mining methods may be found in [22].

2. The entire test sequence T may not be necessary for prediction. Rather, only the most recent window from the test sequence of a pre-defined size may be used. The windowing approach is necessary when the lengths of the individual sequences are long.

The most appropriate variation of this methodology depends on the specific application at hand. The bibliographic notes contain pointers to various recommender systems that use sequential pattern mining. Many of these systems have been developed in the context of Web clickstreams. Sequential-pattern-mining methods have the advantage that one can use numerous off-the-shelf tools for efficiently finding the patterns in large databases.

9.4 Location-Aware Recommender Systems

Location-aware recommender systems can be viewed as special cases of context-aware recommender systems, in which the context is defined by location. Location can influence the recommendation process in a wide variety of ways, of which the following two ways are particularly common:

1. The global geographical location of a user can have a significant influence on her preferences in terms of taste, culture, clothing, eating habits, and so on. For example, an analysis [343] of the MovieLens data set shows that the top genre preference of users from Wisconsin is *War*, whereas the top genre preference of users from Florida is *Fantasy*. Similar results are also shown on the Foursquare data set. This property is referred to as *preference locality*. In this case, the locality is inherently associated with the user, but not with the item. Therefore, in this case, the users are spatial, whereas the items are not.

2. Mobile users often want to discover restaurants or leisure places in the vicinity of their current location. In this case, the recommended items are inherently spatial. This property is referred to as *travel locality*. For example, an analysis [343] of the Foursquare data set shows that 45% of the users travel 10 miles or less, and 75% of the users travel 50 miles or less to visit a restaurant in their locality. In these applications, the location is inherently associated with the item (e.g., restaurant). Although users might specify their *current* location, this transient property is specified only during querying, and it is not inherently associated with the ratings specified by the user. Therefore, in this case, items are spatial, whereas users are not.

3. It is possible to imagine scenarios in which both users and items are spatial. For example, a traveling user might set up a profile, which indicates their permanent address. At the same time, they may record their ratings for spatial items such as restaurants. For example, consider two users from New Orleans and Boston, respectively, who are spending a vacation in Hawaii. These tourists might specify their ratings for restaurants in Hawaii. In this case, both users and items are spatial because their choices of their restaurants will be affected by their place of origin. At the same time, travel locality preferences will also play a role in their choice of restaurant, when the users query from specific locations in Hawaii during their vacation.

Location-aware recommender systems can be treated as special cases of context-sensitive methods. One can use the multidimensional techniques discussed in the previous sections in order to handle context within the recommender systems. This is especially true for the notion of preference locality, in which the multidimensional model of [6] may be used by treating location as a context, associating a hierarchical taxonomy of grid regions with the spatial location, and then reducing the problem to a traditional collaborative filtering application within one of the hierarchical regions of the grid. In fact, the *Location Aware Recommender System (LARS)* [343] does use a similar reduction-based approach for handling preference locality. However, the approach in [343] is much more sophisticated than a straightforward application of the multidimensional methodology of [6]. To represent the hierarchical taxonomy of grid regions, it uses a multidimensional indexing structure. This indexing structure can support incremental addition of ratings, and can therefore work well in settings that require scalability. Furthermore, the same work also proposes methods for handling travel locality and for combining travel and preference locality.

9.4.1 Preference Locality

As discussed earlier, the notion of preference locality shares a number of characteristics of the reduction-based multidimensional model of recommender systems [6]. For example, consider the example of the MovieLens data set, where the user locations are available in addition to ratings information. For a user in California, we might use only the ratings entered by other users of California in order to provide the recommendations for that user. This approach is equivalent to extracting a slice of the *User × Item × Location* data cube, by fixing the location to California. Then a 2-dimensional recommender system can be used on this slice. This is a direct application of a reduction-based system [6].

Of course, such an approach is rather crude because the locality information may be available to a greater degree of granularity. For example, one might have the address of each user at hand. Users in southern California might show different preferences from those in northern California. On the other hand, for a small state or locality, enough rating data might not be available to make a robust recommendation. Therefore, one might need to combine the data from multiple adjacent regions. How can one meaningfully address such trade-offs?

The *LARS* approach [343] divides the entire spatial region in hierarchical fashion using a pyramid-tree or quad-tree [53, 202]. Note that this approach partitions the data *space*, rather than the data *points*, in order to ensure that every point in the space is included in one of the partitions. This ensures that new test locations can be effectively handled in the querying process, even if they are not represented in the data. The pyramid-tree decomposes the space into H levels. For any level $h \in \{0 \ldots H - 1\}$, the space is partitioned into 4^h grid cells. The top-level at $h = 0$ contains only one cell, and it contains the entire data space. For example, consider the case where the top level of the model contains the region corresponding to the entire United States. Then, the next level divides the United States into four regions, with a separate model for each. The next level divides each of these regions into four more regions, and so on. Each grid cell contains a collaborative filtering model *only* for the region of the data space bounded by the corresponding rectangle. Therefore, the top-level grid cell contains a traditional (non-localized) collaborative filtering model containing all of the ratings. An example of the hierarchical partitioning of the pyramid-tree is illustrated in Figure 9.2. In the figure, the cell-identifier is denoted by CID, and the table entry to the left of it contains the pointer to the relevant collaborative filtering model for that cell. This data structure is maintained dynamically, so that ratings can be inserted or deleted from the system. One challenge in the dynamic update process is that it is sometimes not possible to maintain the models for a subset of the cells because of dynamic merges or splits of the cells during the updates. Note that the models for these new cells need to be recreated from scratch if cells merge or split during updates. This can sometimes be computationally expensive. However, if the tree is built up front without dynamic updates, then it is possible to maintain the models for all the entries. Therefore, the approach is straightforward when only static data is considered. The approach can also be extended to dynamic updates with some modifications. Readers are referred to [343] for details of the dynamic update process.

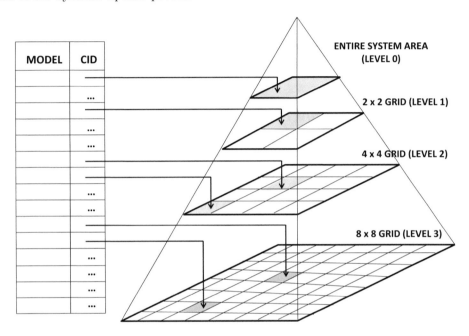

Figure 9.2: The pyramid-tree for location-aware query processing [343]

The query processing approach uses this pyramid data structure. In order to recommend items for a given user, the *LARS* approach determines the lowest level cell, which is maintained in the pyramid structure. The localized collaborative filtering model at this level is used to predict the rating. An item-based (neighborhood) collaborative filtering technique is used to perform the recommendations. Note that any conventional collaborative filtering model can be used in principle. The model does need to be incrementally updated as new ratings come in. Therefore, it is important to choose a base model that is amenable to incremental updates.

The approach is also able to support *continuous* queries where the location of a user changes with time. Note that the rate of change of the location of the user is highly application-specific. For cases in which the user-location corresponds to their address, the rate of change is very slow. However, it is possible to envision other definitions of location in which the change occurs faster over time. However, preference locality usually does not change very rapidly as a rule. In continuous queries, an initial recommendation is made as discussed above. Then, the system waits for the user location to change sufficiently, so that it crosses a cell boundary. When a cell-boundary is crossed, the lowest level cell is again used to updated the recommendation. Therefore, the last reported answer can be incrementally updated over time.

Finally, it is also possible for the users to optionally specify the geographic level of granularity at which their recommendation process should be executed. Instead of using the lowest maintained grid-cell, one might work with a user-specified level in the pyramid tree. For example, by specifying a level of 0, only the root node can be used. This results in a traditional collaborative filtering model, which does not use locality at all. This approach allows the user to specifying varying levels of geographic resolution in her queries.

9.4.2 Travel Locality

In this case, the locations are associated with items and not the users. For example, in a restaurant recommender system, the locations are associated with the restaurants. However, the users might specify their *current* location in a particular query. Clearly, it is desirable to provide responses that are close to the specified location in the query. This is achieved in *LARS* with the notion of *travel penalty*. The distance $\Delta(i, j)$ between the query location of user i and the location of item j is computed. The rating $\hat{r}_{ij}$ of user i for item j is first predicted with a traditional collaborative filtering model on the entire data. Then, the predicted rating is penalized with a function $F(\cdot)$ of $\Delta(i, j)$. The adjusted rating $\hat{r}_{ij}^{\Delta}$ is computed as follows:

$$\hat{r}_{ij}^{\Delta} = \hat{r}_{ij} - F(\Delta(i, j)) \tag{9.15}$$

Here, $F(\cdot)$ is a non-decreasing function of the distance $\Delta(i, j)$, so that the penalty is normalized to the rating scale. The exact nature of the penalty function $F(\cdot)$ is heuristic in nature. The approach in [343] uses straightforward normalization of the travel distance to the rating scale in order to define the function. If desired, it can even be assumed to be a specific function (e.g., linear function) of the distance, and the coefficients of this function can be optimally chosen with cross-validation. The choice of optimal function is an interesting research problem that can be explored in future work, because it directly affects the accuracy of the system. It is likely that the optimal choice of function is specific to the data set at hand.

9.4.3 Combined Preference and Travel Locality

It is possible to have situations in which the locations are associated with both users and items. For example, a tourist with a primary address in New Orleans might have a different restaurant preference as compared to one with a primary address in Boston, when the two of them visit Hawaii for a vacation. At the same time, the recommender system should also take into account their *transient* query location within Hawaii during the recommendation process. In this case, the methods associated with preference locality and travel locality can be combined. First, the pyramid-tree structure is used, based on the primary user location, in order to predict the ratings. Then, the transient query location is used in conjunction with the aforementioned travel penalty. The top-ranked items are then returned to the user.

9.5 Summary

Many types of time- and location-aware systems fall under the category of context-aware recommender systems. The notion of time can greatly enhance the effectiveness of recommender systems. Temporally sensitive recommender systems can use recency-based methods, context-based methods, or they may use time as a modeling variable. One of the most well-known methods of the last type is the *time-SVD++* model, which proposes a latent factor model for recommendations. Recommendation methods have also been proposed for data that are expressed as discrete sequences. For example, Web clickstreams or supermarket data contain discrete sequences of activity. These scenarios often arise in the context of implicit feedback data sets. A variety of discrete sequential methods are used to perform recommendations in such cases. Discrete Markovian models and sequential pattern-mining methods are used to perform recommendations in such cases.

Location-aware recommender systems are special cases of context-aware systems, in which the spatial location provides the context in which the recommendations are made. In location-based systems, the location could be associated with the user, the item, or both. These different forms of context lead to distinctly different methods for performing the recommendations.

9.6   Bibliographic Notes

Temporal recommendations belong to the class of context-aware recommendations, which are discussed in a generic sense in Chapter 8. A recent survey on time-aware recommender systems may be found in [130]. Some of the earliest time-weighted and decay-based collaborative filtering models are discussed in [185, 186]. A variety of decay functions are tested in [635]. The work in [249] also incorporates the time-similarity between ratings into the computation. Methods based on time-windows are proposed in [230] in which ratings from inactive intervals are essentially pruned. The work in [595] performs movie recommendations by pruning according to the year of *production*. Such an approach reduces the dimensionality of the data set because it drops a subset of the items as opposed to a method that only prunes older ratings.

Methods for extending neighborhood models with evolutionary models are discussed in [366]. Another technique that uses adaptive neighborhoods for temporal collaborative filtering is discussed in [333]. This work also showed that many of the existing recommendation algorithms did not seem to work very well on the Netflix Prize data set when using past ratings in order to predict future ratings. Time-sensitive methods for location-based recommendation are discussed in [655]. Methods for performing temporal recommendations with the use of random walk methods are discussed in [639]. An interesting class of algorithms, related to temporal collaborative filtering, is the multi-arm bandit class of algorithms in which the recommender trades off exploration vs exploitation in the recommendation space [92, 348]. This methodology is also closely related to active learning, which is discussed in Chapter 13.

A generic method for performing time-aware recommendation is to treat the time information as a discrete contextual value to create a multidimensional representation [6, 7]. Subsequent work [626] specifically addressed the temporal context within this framework. Various forms of context were tested in [61] for performing music recommendations. Although some forms of context, such as "morning" and "evening," were shown to improve the recommendation, the greatest improvement was shown using a meaningless split such as "odd hours" and "even hours." This might be the result of data-specific characteristics, and therefore further research is needed to understand these effects.

Realistic methods for evaluating temporal recommender systems are discussed in [335]. A recent survey [130] points to the significant importance of evaluation methods in disambiguating the contradictory findings of the recent results and also proposes a number of evaluation metrics for temporal recommender systems. The combination of multiple variables such as *season*, *timeOfDay*, and *dayOfWeek* was discussed in [337]. Other ways of combining temporal dimensions in a more complex way are discussed in [231, 471]. The work in [100] studies the use of periodic context in movie recommendations. For example, the movie recommendations during Christmas week are very different from those in the week leading to the Oscars. The use of contextual methods for improved recommendation of seasonal products is discussed in [567]. A temporal regression approach was used in this work. The *Context-Aware Movie Recommendation Challenge (CAMRA)* [515] was the platform on which the work in [100] was tested. This challenge studied contexts of various types and not just the temporal context. A contextual method [131] evaluates the effect of various time dimensions, including the hour of the day, day of the week, and the date of the rating. The work in [458] used support vector machines to model various types of context, such as time, weather, and company.

The use of time-series models been used in the context of ratings has been investigated in multiple studies [136, 435]. In these methods, a time series of user ratings is used to predict the current user interests. Time-series methods have also been used in cases of implicit feedback, where explicit ratings are not available. For example, the work in [684] encodes the Web logs as time series, and time-series techniques are used for the purpose of forecasting. The work in [266] builds several time-unaware models for different time buckets and then uses a blending approach to combine the predictions of these models. The use of factorization models in temporal recommendations was first proposed in [310]. Similar models have also been applied to the music recommendation scenario [304]. The work in [310] does not differentiate the item factors based on time. A more refined model is proposed in [293] in which differentiated item factors are learned based on the rating time-stamps. Subsequently, many matrix and tensor factorization methods were also proposed for contextual recommendation in which time is treated as a discrete contextual value [212, 294, 332, 495, 496]. These methods can be viewed as a generic implementation of the multidimensional contextual model of [7].

Discrete methods are common in the context of the Web domain, where the personalization needs to be performed with Web clickstreams [109]. A primer on finite Markov chains is provided in [296]. The sequential pattern-mining problem was defined in the context of supermarket data [37]. An overview of common algorithms for sequential pattern mining can be found in [22, 23]. In order to use these methods on Web logs, a significant amount of data preparation is required [169]. Discrete Markovian methods for predicting Web page accesses are discussed in [182]. The required background in Markov chains may be found in [265]. Sequential pattern-mining methods for predicting accesses in Web logs are discussed in [208, 440, 442, 443, 562]. The use of long repeating subsequences to predict Web page accesses is discussed in [479]. The use of path profiles for predicting Web page requests is discussed in [532]. An evaluation of various pattern-mining approaches for next-request prediction is found in [218]. A detailed discussion on Hidden Markov Models may be found in [319], and a simplified discussion on applications to data mining may be found in [22].

A significant amount of recent work has focussed on location-aware recommender systems [64, 108, 343, 447, 464, 645, 649]. Much of this work has been motivated by hardware enhancements in mobile phone technology and GPS-enabled phones. As a result, the field of mobile recommender systems [504] has gained increasing prominence. One of the earliest works [54] proposes methods to use the data from GPS-enabled mobile phones to predict user movement across various locations. Context-aware media recommendations for smart phones are discussed in [654]. A mobile advertisement recommender system with the use of collaborative filtering is proposed in [40]. Numerous tourist guide applications, such as INTRIGUE [52], GUIDE [156], MyMap [177], SPETA [213], MobiDENK [318], COMPASS [611], Archeoguide [618], and LISTEN [685], have been proposed in the literature. Some of the location-based recommender systems [633, 649] use hybrid systems in order to perform context-aware recommendations. The work by Bohnert *et al.* [89] uses the sequence of patterns in user visits to various locations to predict the next location.

It was also explored how a hybrid content-based model that captured user interests could impact the overall effectiveness of the recommender system. The addition of content provided only limited improvements. The work in [649] discussed how to handle cold-start in location-aware recommender systems by combining content and collaborative systems, and also incorporating community endorsement.

9.7 Exercises

1. Design a method for performing collaborative filtering with the Bayes model, while incorporating time decay. Refer to Chapter 3 for collaborative filtering with the Bayes algorithm.

2. Design a latent factor model that incorporates time-decay in the factorization process.

3. Implement the *time-SVD++* algorithm.

4. Suppose that you want to design an order-k Markov model on a set of actions Σ, so that $|\Sigma| = n$. Furthermore, we are assured that there is never any repetition of an action in a window of size $(k + 1)$. What is the maximum number of states and transition edges in such a model, assuming that we do not keep any states or edges with probability 0?

5. Implement a sequential pattern mining algorithm for making temporal recommendations. You have wide leeway in making appropriate design choices for your algorithm.

6. Suppose you have a large log containing sequences of actions from various users. The discussion in the chapter shows how one might perform recommendations with *item-based* rules. Show how to design an analogous approach with user-based rules. How well do you think such an approach will work in practice?

7. Discuss why an R-Tree might not be as suitable as the pyramid tree for the preference-locality technique of collaborative filtering.

Chapter 10

Structural Recommendations in Networks

"In nature, we never see anything isolated, but everything in connection with something else which is before it, beside it, under it and over it." – Johann Wolfgang von Goethe

10.1 Introduction

The growth of various Web-enabled networks has enabled numerous models of recommendation. For example, the Web itself is a large and distributed repository of data, and a search engine such as Google can be considered a keyword-centric variation of the notion of recommendation. In fact, a major discourse in the recommendation literature is to distinguish between the notions of search and recommendations. While search technologies also recommend content to users, the results are often not personalized to the user at hand. This lack of personalization has traditionally been the case because of the historical difficulty in tracking large numbers of Web users. However, in recent years, many personalized notions of search have arisen, where the Web pages recommended to users are based on personal interests. Many search engine providers, such as Google, now provide the ability to determine personalized results. This problem is exactly equivalent to that of *ranking* nodes in networks with the use of personalized preferences.

Networks have become ubiquitous as a modeling tool in many applications, such as social and information networks. Therefore, it is particularly useful to discuss the various structural elements of a network that can be recommended in different scenarios. Each of these different types of structural recommendation may have a different set of applications in different scenarios. Some key examples of these different variations are as follows:

© Springer International Publishing Switzerland 2016   309
C.C. Aggarwal, *Recommender Systems: The Textbook*,
DOI 10.1007/978-3-319-29659-3_10

1. *Recommending nodes by authority and context:* In this case, the quality of nodes is judged by their incoming links, and the personalized relevance of nodes is judged by their context. Nodes of high quality have many incoming links. This problem is very closely related to that of search engines. A major observation is that the traditional notion of search in such engines does not distinguish between various users, and is therefore not personalized to a specific user. In search engines, Web pages (or nodes in the Web graph) are ranked on the basis of their *authority* and their content. Little emphasis is placed on the identity of the user performing the search. However, notions such as personalized *PageRank* were eventually developed that can tailor the results to various interests. These forms of personalization incorporate context into the ranking by modifying the traditional notion of *PageRank* with context-specific personalization. As will be examined in the next chapter, notions such as *FolkRank*, which are closely related to *PageRank*, are used in social tagging settings.

2. *Recommending nodes by example:* In many recommendation applications, one may want to recommend nodes that are similar to other example nodes. This problem is that of collective classification of nodes. Interestingly, personalized *PageRank* methods are often used in the problem of collective classification. Therefore, these two types of recommendation are closely related. Such applications may also be useful in an information network of users and other types of nodes in which some of the nodes may be tagged with specific properties.

3. *Recommending nodes by influence and content:* In many Web-centric applications, users may propagate knowledge about products of various types. This problem is referred to as *viral marketing*. In these cases, a merchant is looking for *users* who are most likely to propagate views about their specific product. In *topic-sensitive* influence analysis, users who are most likely to propagate specific topics are sought out. The problem of influence analysis can be viewed as that of recommending users to merchants on the basis of their viral potential in influencing others, as well as their topical specificity.

4. *Recommending links:* In many social networks, such as Facebook, it is in the interest of the social network to increase the connectivity of the network. Therefore, users are often recommended potential friends. This problem is equivalent to that of recommending potential links in a network. Interestingly, many ranking methods are used for link prediction. Many matrix factorization methods can also be adapted to link prediction. Furthermore, some of the link prediction methods are used for collective classification. Some of these interrelationships will be noted in this chapter where applicable.

The applications of these structural recommendation methods extend beyond the social network domain. Such structural recommendation methods can be used to recommend elements in any system, which can be modeled as a Web-centric network. Examples might include news, blog posts, or other Web-enabled content.

Furthermore, even the traditional product recommendation problem can be addressed with these methods. This is because any user-product recommendation problem can be modeled as a user-item graph. In Chapters 2 and 3, we have provided specific examples of how user-item graphs are used for such product recommendations. The next chapter provides an even more detailed view of how various forms of content in social systems can be used to enhance recommendations. While this chapter is closely related to the material in the following chapter, our study in this chapter is more deeply focused on the structural aspects of networks without being explicitly focused on sociocentric aspects such as *trust* or *user tagging behavior*. Furthermore, the methodologies discussed in this chapter can be used in applications beyond social network analysis. The next chapter focuses on socially sensitive methods of improving recommendations, whether they are network-centric or not.

This chapter is organized as follows. The next section studies the problem of node ranking in networks and its use in personalized ranking applications. Section 10.3 reviews the problem of collective classification and its use in various forms of recommendations. Section 10.4 examines the problem of link prediction. The problem of influence analysis is studied in section 10.5. The problem of topic-sensitive influence analysis is also studied in the same chapter. A summary is given in section 10.6.

10.2   Ranking Algorithms

The *PageRank* algorithm was first proposed in the context of Web search. The main motivation of the algorithm was to improve search quality. Since the Web allows open publishing, a problem faced by the earliest search engines was that the use of purely content-centric matching of Web pages with keywords for ranking provided results of poor quality. In particular, users can often publish spam, misleading information, or other incorrect content on pages, and a purely content-centric matching is unable to distinguish between results of varying quality. Therefore, a mechanism was needed to determine the *reputation* and quality of Web pages. This is achieved using the *citation structure* of the Web. When a page is of high quality, many other Web pages point to it. A citation can be logically viewed as a vote for the Web page. While the number of in-linking pages can be used as a rough indicator of the quality, it does not provide a complete view because it does not account for the quality of the pages pointing to it. To provide a more holistic citation-based vote, an algorithm referred to as *PageRank* is used. The *PageRank* algorithm generalizes the notion of citation-based ranking in a recursive way.

While the *PageRank* algorithm is not directly a recommendation approach, it is, nevertheless, closely related to the topic of recommendation analysis. Many variations of *PageRank* are used for personalized ranking mechanisms. This is because many settings for recommendations can be expressed as linked networks, including the traditional user-item recommendation scenario. This section will therefore explore the relationship between the two closely related problems of search and recommendation and the application of the *PageRank* algorithm in numerous recommendation scenarios. First, we will introduce the generic *PageRank* algorithm in the context of traditional Web ranking.

10.2.1   PageRank

The *PageRank* algorithm models the importance of nodes with the use of the citation (or linkage) structure in the Web graph. In the context of the Web graph, the nodes correspond to Web pages and the edges correspond to hyper-links. The basic idea is that highly reputable documents are more likely to be cited (or in-linked) by other reputable Web pages. Similarly, in a social network such as Twitter, highly reputable users are likely to be followed by other reputable users. For the purpose of the following discussion, we will assume a directed graph (like the Web), although the notion can be easily extended to undirected graphs by replacing each undirected edge with two directed edges. For many recommendation applications, the undirected representation is usually sufficient.

A random surfer model on the Web graph is used to achieve the goal of ranking pages. Consider a random surfer who visits random pages on the Web by selecting random links on a page. The long-term relative frequency of visits to any particular page is clearly influenced by the number of in-linking pages to it. Furthermore, the long-term frequency of visits to any page will be higher if it is linked to by other frequently visited (or *reputable*) pages. In other words, the *PageRank* algorithm models the reputation of a Web page in terms of its long-term frequency of visits by a random surfer. This long-term frequency is also referred to as the *steady-state probability*, and the model is also referred to as the *random-walk model*.

The basic random surfer model does not work well for all possible graph topologies. A critical issue is that some Web pages may have no outgoing links, which may result in the random surfer getting trapped at specific nodes. In fact, a probabilistic transition is not even meaningfully defined at such a node. Such nodes are referred to as *dead ends*. An example of a dead-end node is illustrated in Figure 10.1(a). Clearly, dead ends are undesirable because the transition process for *PageRank* computation cannot be defined at that node. To address this issue, two modifications are incorporated in the random surfer model. The first modification is to add links from the dead-end node (Web page) to all nodes (Web pages), including a self-loop to itself. Each such edge has a transition probability of $1/n$. This does not fully solve the problem, because the dead ends can also be defined on *groups of nodes*. In these cases, there are no outgoing links from *a group of nodes* to the remaining nodes in the graph. This is referred to as a *dead-end component*, or *absorbing component*. Figure 10.1(b) provides an illustration of a dead-end component.

Dead-end components are common in the Web graph (and other networks) because the Web is not strongly connected. In such cases, the transitions at individual nodes can be meaningfully defined, but the steady-state transitions will stay trapped in these dead-end components. All the steady-state probabilities will be concentrated in dead-end components because there can be no transition out of a dead-end component after a transition occurs into it. Therefore, as long as even a minuscule probability of transition into a dead-end component[1] exists, *all* the steady-state probability becomes concentrated in such components. This situation is not desirable from the perspective of *PageRank* computation in a large Web graph, where dead-end components are not necessarily an indicator of popularity. Furthermore, in such cases, the final probability distribution of nodes in various dead-end components is not unique and it is dependent on the state at which the random walk starts. This is easy to verify by observing that random walks starting in different dead-end components will have their respective steady-state distributions concentrated within the corresponding components.

While the addition of edges solves the problem for dead-end nodes, an additional step is required to address the more complex issue of dead-end components. Therefore, aside from the addition of these edges, a *teleportation* or *restart step* is used within the random surfer model. This step is defined as follows. At each transition, the random surfer may either jump to an arbitrary page with probability α, or follow one of the links on the page with probability $(1 - \alpha)$. A typical value of α used is 0.1. Because of the use of teleportation, the steady state probability becomes unique and independent of the starting state. The value of α may also be viewed as a *smoothing* or *damping probability*. Large values of α typically result in the steady-state probability of different pages becoming more even. For example, if the value of α is chosen to be 1, then all pages will have the same steady-state probability of visits.

---

[1] A formal mathematical treatment characterizes this in terms of the *ergodicity* of the underlying Markov chains. In ergodic Markov chains, a necessary requirement is that it is possible to reach any state from any other state using a sequence of one or more transitions. This condition is referred to as *strong connectivity*. An informal description is provided here to facilitate understanding.

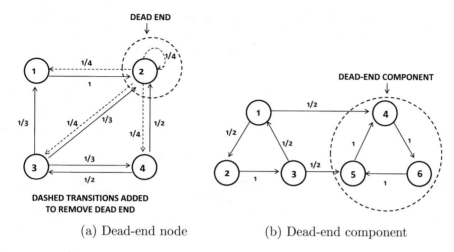

(a) Dead-end node   (b) Dead-end component

Figure 10.1: Transition probabilities for *PageRank* computation with different types of dead ends

How are the steady-state probabilities determined? Let $G = (N, A)$ be the directed network, in which nodes correspond to pages, and edges correspond to hyperlinks. The total number of nodes is denoted by n. It is assumed that A also includes the added edges from dead-end nodes to all other nodes. The set of nodes incident on i is denoted by $In(i)$, and the set of end points of the outgoing links of node i is denoted by $Out(i)$. The steady-state probability at a node i is denoted by $\pi(i)$. In general, the transitions of a Web surfer can be visualized as a *Markov chain*, in which an $n \times n$ transition matrix P is defined for a Web graph with n nodes. The *PageRank* of a node i is equal to the steady-state probability $\pi(i)$ for node i in the Markov chain model. The probability[2] p_{ij} of transitioning from node i to node j, is defined as $1/|Out(i)|$. Examples of transition probabilities are illustrated in Figure 10.1. These transition probabilities do not, however, account for teleportation, which will be addressed[3] separately below.

Let us examine the transitions into a given node i. The steady-state probability $\pi(i)$ of node i is the sum of the probability of a teleportation into it and the probability that one of the in-linking nodes directly transitions into it. The probability of a teleportation into the node is exactly α/n because a teleportation occurs in a step with probability α, and all nodes are equally likely to be the beneficiary of the teleportation. The probability of a transition into node i is given by $(1 - \alpha) \cdot \sum_{j \in In(i)} \pi(j) \cdot p_{ji}$, as the sum of the probabilities of transitions from different in-linking nodes. Therefore, at steady-state, the probability of a transition into node i is defined by the sum of the probabilities of the teleportation and transition events:

$$\pi(i) = \alpha/n + (1 - \alpha) \cdot \sum_{j \in In(i)} \pi(j) \cdot p_{ji} \qquad (10.1)$$

---

[2]In some applications such as bibliographic networks, the edge (i, j) may have a weight denoted by w_{ij}. The transition probability p_{ij} is defined in such cases by $\frac{w_{ij}}{\sum_{j \in Out(i)} w_{ij}}$.

[3]An alternative way to achieve this goal is to modify G by multiplying existing edge-transition probabilities by the factor $(1 - \alpha)$ and then adding α/n to the transition probability between each pair of nodes in G. As a result, G will become a directed clique with bidirectional edges between each pair of nodes. Such strongly connected Markov chains have unique steady-state probabilities. The resulting graph can then be treated as a Markov chain without having to separately account for the teleportation component. This model is equivalent to that discussed in the chapter.

For example, the equation for node 2 in Figure 10.1(a) can be written as follows:

$$\pi(2) = \alpha/4 + (1 - \alpha) \cdot (\pi(1) + \pi(2)/4 + \pi(3)/3 + \pi(4)/2)$$

There will be one such equation for each node, and therefore it is convenient to write the entire system of equations in matrix form. Let $\overline{\pi} = (\pi(1)\ldots\pi(n))^T$ be the n-dimensional column vector representing the steady-state probabilities of all the nodes, and let $\overline{e}$ be an n-dimensional column vector of all 1 values. The system of equations can be rewritten in matrix form as follows:

$$\overline{\pi} = \alpha\overline{e}/n + (1 - \alpha)P^T\overline{\pi} \tag{10.2}$$

The first term on the right-hand side corresponds to a teleportation, and the second term corresponds to a direct transition from an in-linking node. In addition, because the vector $\overline{\pi}$ represents a probability, the sum of its components $\sum_{i=1}^{n} \pi(i)$ must be equal to 1.

$$\sum_{i=1}^{n} \pi(i) = 1 \tag{10.3}$$

Note that this is a linear system of equations that can be easily solved using an iterative method. The algorithm starts off by initializing $\overline{\pi}^{(0)} = \overline{e}/n$, and it derives $\overline{\pi}^{(t+1)}$ from $\overline{\pi}^{(t)}$ by repeating the following iterative step:

$$\overline{\pi}^{(t+1)} \Leftarrow \alpha\overline{e}/n + (1 - \alpha)P^T\overline{\pi}^{(t)} \tag{10.4}$$

After each iteration, the entries of $\overline{\pi}^{(t+1)}$ are normalized by scaling them to sum to 1. These steps are repeated until the difference between $\overline{\pi}^{(t+1)}$ and $\overline{\pi}^{(t)}$ is a vector with a magnitude that is less than a user-defined threshold. This approach is also referred to as the *power-iteration method*.

The *PageRank* values can be shown to be the n components of the largest left eigenvector[4] of (a modified version of) the stochastic transition matrix P, for which the eigenvalue is 1. The modifications to the stochastic transition matrix directly incorporate the effect of restarts within the transition matrix by adding "restart" edges between every pair of nodes.

10.2.2   Personalized PageRank

The notion of personalized *PageRank* is also referred to as *topic-sensitive PageRank* in Web recommender systems. Although *PageRank* is an excellent mechanism to find popular nodes in terms of the linkage structure, it does little for finding items that are well-matched to the interests of specific users. The notion of personalized *PageRank* is designed to find popular nodes, which are also similar to specific nodes in the network. For example, consider an information network, such as Flickr, in which nodes might be users, image descriptions, or images. It is desirable to leverage the network structure to recommend popular content in the network to specific users. However, it is important to personalize this popular content to the user at hand. How do we recommend specific images or particular users, or vice versa? The key here is to understand that the teleportation mechanism provides a way of biasing the random walk towards specific nodes.

---

[4]The left eigenvector $\overline{X}$ of P is a row vector satisfying $\overline{X}P = \lambda\overline{X}$. The right eigenvector $\overline{Y}$ is a column vector satisfying $P\overline{Y} = \lambda\overline{Y}$. For asymmetric matrices, the left and right eigenvectors are not the same. However, the eigenvalues are always the same. The unqualified term "eigenvector" refers to the right eigenvector by default.

Another application of this approach is that of Web recommender systems, for which providing greater importance to some topics than others is desired in the ranking process. While personalization is less common in large-scale commercial search engines, it is more common is smaller scale site-specific search applications. Typically, users may be more interested in certain combinations of topics than others. The knowledge of such interests may be available to a personalized search engine because of user registration. For example, a particular user may be more interested in the topic of automobiles. Therefore, it is desirable to rank pages related to automobiles higher when responding to queries by this user. This can also be viewed as the *personalization* of ranking values. How can this be achieved?

Consider a Web recommender system in which users have the ability to express interest in specific topics. The first step is to fix a list of base topics, and determine a high-quality sample of pages from each of these topics. This can be achieved with the use of a resource such as the *Open Directory Project (ODP)*,[5] which can provide a base list of topics and sample Web pages for each topic. The *PageRank* equations are now modified, so that the teleportation is only performed on this sample set of Web documents, rather than on the entire space of Web documents.

Let $\overline{e_p}$ be an n-dimensional personalization (column) vector with one entry for each page. An entry in $\overline{e_p}$ takes on the value of 1, if that page is included in the sample set, and 0 otherwise. Let the number of nonzero entries in $\overline{e_p}$ be denoted by n_p. Then, the *PageRank* Equation 10.2 can be modified as follows:

$$\overline{\pi} = \alpha \overline{e_p}/n_p + (1 - \alpha)P^T\overline{\pi} \qquad (10.5)$$

The same power-iteration method can be used to solve the personalized *PageRank* problem. The selective teleportations bias the random walk, so that pages in the structural locality of the sampled pages will be ranked higher. As long as the sample of pages is a good representative of different (structural) localities of the Web graph, in which pages of specific topics exist, such an approach will work well. Therefore, for each of the different topics, a separate *PageRank* vector can be precomputed and stored for use during query time. The choice of α regulates the trade-off between the topical criterion and the popularity criterion. Larger values of α will make the approach more topic-sensitive, whereas smaller values of α will make the approach more sensitive to the structure of the network.

Consider the case where a user has specified an interest in specific *combinations of* topics such as sports and automobiles. Clearly, the number of possible combinations of interests can be very large, and it is not reasonably possible or necessary to prestore every personalized *PageRank* vector. In such cases, only the *PageRank* vectors for the base topics are computed. The final result for a user is defined as a weighted linear combination of the topic-specific *PageRank* vectors, where the weights are defined by the user-specified interest in the different topics.

The personalized *PageRank* approach can be viewed as an approach that provides similarity scores to nodes based on both their structural similarity to the restart nodes and their absolute level of connectivity to other nodes in the network. The precise importance to each of these factors depends on the value of α. There is, however, a limit to this control. By picking very large values of α, one also loses the sensitivity of the approach in computing similarity to nodes located at modest distances to the restart nodes, and only the restart nodes receive the lion's share of the probability. In some cases, it is desired to cancel out the popularity effects in a more meaningful way to make the numerical quantities reflect only

---

[5]http://www.dmoz.org

similarities. A method to reduce the impact of popularity is to perform standard *PageRank*, and subtract it from the personalized *PageRank*. By doing so, the ranking values can be both positive and negative, reflecting relative similarities or dissimilarities. A value of 0 would be considered the break-even point. This approach is related to the *FolkRank* method, which is commonly used in social tagging applications (cf. section 11.4.4.2 of Chapter 11).

10.2.3   Applications to Neighborhood-Based Methods

It is noteworthy that the teleportation mechanism of personalized *PageRank* methods increases the ranking of nodes which are structurally closer to the nodes at which restart is executed. This property is particularly useful in defining the neighborhoods of nodes in a network. When the personalized *PageRank* algorithm is used, the returned neighborhoods will also be of higher quality in terms of their citation ranking. The trade-off between quality and topical specificity can be regulated by modifying the restart probability. The basic question for neighborhood discovery is as follows:

Given a target node i_q and a subset of nodes $S \subseteq N$ from graph $G = (N, A)$, rank the nodes in S in their order of similarity to i_q.

Such a query is very useful in recommender systems in which users and items are arranged in the form of a bipartite graph of preferences, in which nodes correspond to users and items, and edges correspond to preferences. The node i_q may correspond to an item node, and the set S may correspond to user nodes. Alternatively, the node i_q may correspond to a user node, and the set S may correspond to item nodes. The use of personalized *PageRank* methods are discussed later in this chapter and also in the next chapter. Recommender systems are closely related to search in that they also perform ranking of target objects, but do so while taking user preferences into account.

   This problem can be viewed as a limiting case of topic-sensitive *PageRank*, in which the teleportation is performed to the *single node* i_q. Therefore, the personalized *PageRank* Equation 10.5 can be directly adapted by using the teleportation vector $\overline{e_p} = \overline{e_q}$, that is, a vector of all 0s, except for a single 1, corresponding to the node i_q. Furthermore, the value of n_p in this case is set to 1.

$$\overline{\pi} = \alpha \overline{e_q} + (1 - \alpha) P^T \overline{\pi} \tag{10.6}$$

The solution to the aforementioned equation will provide high ranking values to nodes in the structural locality of i_q. This definition of similarity is *asymmetric* because the similarity value assigned to node j starting from query node i is different from the similarity value assigned to node i starting from query node j. Such an *asymmetric* similarity measure is suitable for *query-centered* applications such as search engines and recommender systems. In typical collaborative filtering applications, one attempts to determine neighborhoods of *target* users or items. After these neighborhoods have been discovered, they can be used to make recommendations based on the content properties of these nodes. This approach can be utilized either for making recommendations in traditional social networks or for finding neighborhoods in network models of traditional collaborative filtering applications. We discuss both these cases below. As discussed at the end of the previous section, one can cancel out the effects related to popularity (if needed) by subtracting the unbiased *PageRank* values from the personalized *PageRank* computation. This approach, referred to as *FolkRank*, is also discussed in more detail in section 11.4.4.2 of Chapter 11.

10.2.3.1 Social Network Recommendations

Consider the case where the underlying network is a social network in which users have explicitly specified interests, whereas the linkages represent friendship relationships. In such cases, it may be desirable to leverage the neighborhood profile of the user for the purpose of recommendations. The neighborhood of a user can be discovered in a social network by using the personalized *PageRank* algorithm, which restarts at that user node. The social profile of the neighborhood can be retrieved in terms of the specified keywords, likes, or explicitly specified ratings. The social profile in the neighborhood of a target node can be aggregated, and the most liked items in these profiles can be recommended to the target. Therefore, this approach can be viewed as a type of hybrid recommender system, in which the structural data is used to determine the neighbors, but the user-specified interests are used to make the final recommendations.

This approach leverages the notion of *homophily* in social networks. The basic idea is that connected users in social networks often have similar properties. Therefore, the properties, profiles, and ratings of the neighborhood of a user can be leveraged in order to make recommendations. This problem is closely related to that of collective classification, which is discussed in section 10.3 of this chapter. In collective classification, the same goal is achieved with the use of machine learning models. Interestingly, random walk algorithms are among the most common methodologies used in collective classification models. This is because personalized *PageRank* methods are naturally designed for finding nodes, which are similar to pre-specified nodes in the network. These pre-specified nodes are the training data in collective classification algorithms.

10.2.3.2 Personalization in Heterogeneous Social Media

The personalized *PageRank* approach can be used to determine popular content, which is relevant to a specific node or query in the network. Such scenarios are common in the context of various forms of content recommendations, product recommendations, or question-answering systems in which the relevant nodes to a query are naturally embedded in a linked network structure [16, 81, 602, 640, 663]. In heterogeneous social media, the same network may contain users, media content, and text descriptions. An example of such a scenario is the Flickr network [700], in which users, nodes, and text content are connected with various types of links. A conceptual illustration of a heterogeneous social network with text, users, and images is illustrated in Figure 10.2. The personalized *PageRank* approach can be used to determine high-ranking nodes relevant to specific queries and users. The main idea in these methods is that high-quality users and content are naturally connected within the network structure. This notion is similar to the principle used by *PageRank* algorithms. Therefore, by using the mutually reinforcing nature of the underlying linkage structure, it is possible to simultaneously discover relevant users and content. At the same time, a personalized ranking methodology needs to be used because the results may be tailored to specific users or queries. It is important to note that the query to such a network can be general and can comprise any combination of social (actor), keyword, and content information. Similarly, the recommendations can also be provided from any one (or more) of these different modalities.

A heterogeneous ranking approach [602], referred to as *SocialRank*, has been designed to provide personalized recommendations in response to user queries. For example, consider a scenario in which the user enters the keyword *"birds"* in a social media network such as Flickr in order to determine images of interest. A personalized *PageRank* mechanism

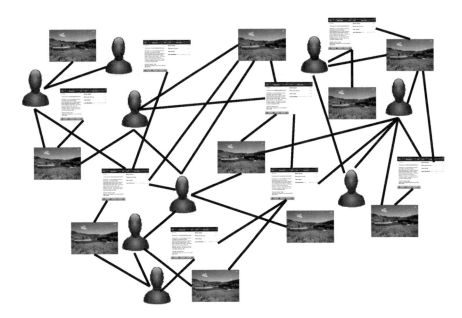

Figure 10.2: Heterogeneous social media network with users, images, and text

is used in which text nodes containing that keyword are weighted to a greater degree for teleportation purposes. Furthermore, if needed, specific user nodes can also be assigned greater weights in order to bias the random walk in the vicinity of the nodes. The choice of the teleportation probability α regulates the trade-off between the importance given to the personalization process and the citation-based popularity of a particular node in the network.

The main challenge with the use of the approach in a heterogeneous network is that a particular modality of the network (e.g., users, images, or text) can overwhelmingly dominate the entire ranking process if a significantly larger number of nodes is present in any particular modality. This is particularly common in many real settings. Therefore, it is important to perform the ranking process in such a way that each modality obtains hints from other modalities, but the ranking process for each class of objects is kept separate. Therefore, an iterative approach is used in [602], in which a separate ranking process is performed within each modality, and then the ranks from other modalities are used to modify the similarity matrix in the next iteration within each modality. Therefore, the approach starts by constructing a node-node similarity matrix within each modality, and uses the following two-step iterative process until convergence:

1. Use *PageRank* separately on the similarity matrix within each modality (e.g., text, images, actors) to create a ranking for each node.

2. Use ranks to readjust the similarity matrix. The similarity between a pair of nodes is increased if they are connected to the same node or to highly interconnected nodes of high rank in a different modality.

The reader should refer to [602] for details of the second step of readjusting the similarity matrix. It has been shown in [602] how this approach can yield personalized ranking results by giving greater importance to the underlying social cues.

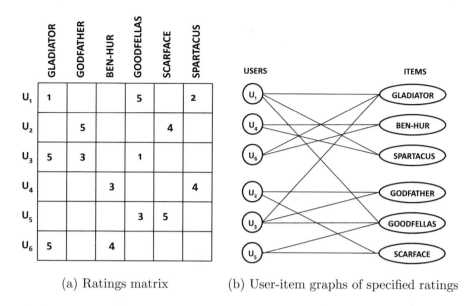

(a) Ratings matrix   (b) User-item graphs of specified ratings

Figure 10.3: A ratings matrix and corresponding user-item graph (Revisiting Figure 2.3 of Chapter 2)

10.2.3.3 Traditional Collaborative Filtering

The personalized *PageRank* approach can also be used to discover the neighborhoods in user-item graphs or user-user graphs in traditional collaborative filtering applications. A discussion of the use of graph models in traditional collaborative filtering applications is provided in section 2.7 of Chapter 2. An undirected user-item graph is constructed on the basis of the specified entries in the ratings matrix. The example from Chapter 2 is replicated in Figure 10.3. By performing random walks starting from a given user, it is possible to discover the other users in its neighborhood. This is a direct application of the *personalized PageRank* approach. If needed, the unbiased *PageRank* of nodes can be subtracted to cancel out the effect of popularity according to the discussion above. After the neighborhood of the user has been discovered, the specified ratings of the neighborhood can be used to make predictions. The methodology for constructing the user-item graph from a ratings matrix is discussed in more detail in section 2.7 of Chapter 2.

Instead of user-item graphs, it is also possible to work with user-user graphs or item-item graphs. In the following, we will describe the use of item-item graphs. The case of user-user graphs is similar. The item-item graph is also referred to as the *correlation graph* [232] because it defines the correlations between items. In this case, a weighted and directed network $G = (N, A)$ is constructed, in which each node in N corresponds to an item, and each edge in A corresponds to a relationship between items. The weight w_{ij} is associated with each edge (i, j). If items i and j have been rated by at least one user, then both the directed edges (i, j) and (j, i) exist in the network. Otherwise, no edges exist between nodes i and j. The directed network is, however, asymmetric because the weight of edge (i, j) is not necessarily the same as that of edge (j, i). Let U_i be the set of users that have specified ratings for item i and U_j be the set of users that have specified ratings for item j. The weight of the edge (i, j) is set as follows. First, we set the weight w_{ij} of edge (i, j) to $|U_i \cap U_j|$. Then, the weights of the edges are normalized, so that the sum of the weights of

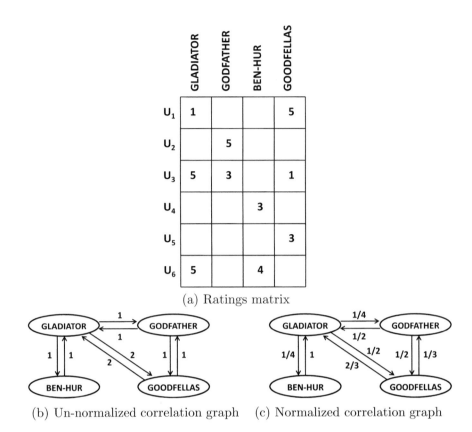

(a) Ratings matrix

(b) Un-normalized correlation graph   (c) Normalized correlation graph

Figure 10.4: A ratings matrix and its correlation graphs (Revisiting Figure 2.5 of Chapter 2)

the outgoing edges of a node is equal to 1. This normalization step results in asymmetric weights, because each of the weights w_{ij} and w_{ji} are divided by different quantities. This results in a graph in which the weights on edges correspond to random-walk probabilities. An example of the correlation graph for a ratings matrix is illustrated in Figure 10.4. It is clear that the weights on the *normalized* correlation graph are not symmetric because of the scaling of the weights to transition probabilities. Furthermore, it is noteworthy that the ratings values are not used in the construction of the correlation graph. Only the *number of* mutually specified ratings in common between two items are used. It is sometimes not desirable to ignore the ratings in creating the correlation graph. It is, of course, possible to define the correlation graph in other ways, such as the use of the cosine function, where the ratings are used as well.

A variety of personalized *PageRank* methods can be used to perform the recommendation. The following two methods are most commonly used:

1. One can perform a random-walk with restart at a particular item node in order to determine the relevant neighborhood items. Traditional item-based neighborhood algorithms (cf. section 2.3.2 of Chapter 2) can be used to predict the rating of that item.

2. One can also directly perform the recommendations with the use of an approach referred to as *ItemRank*. In this case, the *PageRank* bias vector is further influenced

by the ratings given by users to various items. For each user i, a different *PageRank* restart vector is used. Therefore, the system of *PageRank* equations is specific to user i and one needs to solve this system m times in order to determine the preferences of all the users. In practice, however, one is usually looking to make a recommendation for a specific user; therefore, the system needs to be solved only once. For each node (item) j in the correlation graph, the restart probability is set to be proportional to the rating r_{ij} of user i for item j. The resulting *PageRank* values over the different nodes yield the preference of user i for each item. The top-k values are returned as the corresponding recommendations.

The main criticism of the *ItemRank* approach is that is still provides importance to nodes that have been rated poorly by restarting at each node where the user has specified ratings. In general, it makes sense only to restart at nodes where the user has provided positive ratings rather than restart at nodes where the user has provided both positive and negative ratings. The ranking methods are particularly effective in the context of *unary* ratings, where there is a mechanism to specify a liking for an item, but no mechanism to specify a dislike. In such cases, the *ItemRank* approach will work quite well.

10.2.4 SimRank

In some applications, symmetric pairwise similarity between nodes is required. While it is possible to average the two topic-sensitive *PageRank* values in opposite directions to create a symmetric measure, the *SimRank* method provides an elegant and intuitive solution. The approach can be used to determine *reputable* neighborhoods of specific query nodes. The notion of *SimRank* was defined to compute the structural similarity between nodes. *SimRank* determines *symmetric* similarities between nodes. In other words, the similarity between nodes i and j is the same as that between j and i. Obviously, such a measure is intended only for undirected networks.

The *SimRank* approach works as follows. Let $In(i)$ represent the in-linking nodes of i. The *SimRank* equation is naturally defined in a recursive way as follows:

$$SimRank(i,j) = \frac{C}{|In(i)| \cdot |In(j)|} \sum_{p \in In(i)} \sum_{q \in In(j)} SimRank(p,q) \qquad (10.7)$$

Here C is a constant in $(0,1)$ that can viewed as a kind of decay rate of the recursion. As the boundary condition, the value of $SimRank(i,j)$ is set to 1 when $i = j$. When either i or j do not have in-linking nodes, the value of $SimRank(i,j)$ is set to 0. To compute $SimRank$, an iterative approach is used. The value of $SimRank(i,j)$ is initialized to 1 if $i = j$, and 0 otherwise. The algorithm subsequently updates the *SimRank* values between all node pairs, iteratively using Equation 10.7 until convergence is reached.

The notion of *SimRank* has an interesting intuitive interpretation in terms of random walks. Consider two random surfers walking *in lockstep* backwards from node i and node j till they meet. The number of steps taken by each of them is a random variable $L(i,j)$. Then, $SimRank(i,j)$ can be shown to be equal to the expected value of $C^{L(i,j)}$. The decay constant C is used to map random walks of length l to a similarity value of C^l. Note that because $C < 1$, smaller distances will lead to higher similarity and larger distances will lead to lower similarity.

One shortcoming of the *SimRank* method is that the path from each user to the common node must be of the same length. As a result, it is possible for the *SimRank* value between two directly connected nodes to be 0, when no path of the same length exists to a common

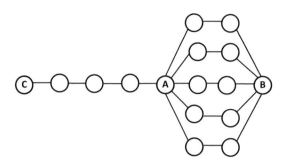

Figure 10.5: A bad case for *SimRank*

node. This is likely to occur when only paths of odd length occur between a pair of connected nodes. For example, in Figure 10.5, nodes A and B are connected only by paths of length 3. Therefore, the *SimRank* between nodes A and B is always 0, even though these nodes are well connected. On the other hand, even though the nodes A and C are not as well connected, the *SimRank* between nodes A and C is nonzero. Therefore, it is important to be aware of cases where[6] the *SimRank* approach is not applicable. For example, the *SimRank* value between a user and item node will always be 0 in a bipartite user-item graph. This is because all paths between user and item nodes are of odd length. On the other hand, the *SimRank* method can be used to effectively compute similarity between pairs of users, or between pairs of items. Therefore, such an approach can be used for neighborhood-based methods in traditional collaborative filtering applications by computing user peers or item peers.

10.2.5   The Relationship Between Search and Recommendation

The discussion in this section shows the close relationship between the two problems of search and recommendation. The main difference between these two problems is the personalization aspect. When users search for documents on the Google search engine, they are not necessarily expecting to discover results that are tailored to their tastes. The only expectation is to see high-quality content, which is tailored to the search phrase. However, in a personalized search application, the user expects to discover new items *that they will like*. Some applications, such as Google news, have both personalized and non-personalized versions of search. The main difference is that the latter is agnostic to previous user behavior, whereas the former directly incorporates user interests into the search process. Nevertheless, the goals of search algorithms are desired even in personalized applications. For example, search result relevance and quality are important in both cases. This is the reason that many variations of random walk algorithms are used in both cases for the ranking process. In fact, the problems of search and recommendation have become increasingly integrated in recent years. For example, Google search results may often depend on a user's location or browsing history, depending[7] on the settings of their browser or sign-in status of their Google accounts.

---

[6]It is possible to ameliorate this problem to some extent by making minor modifications such as adding self-loops to the graph. However, such methods are not a formal part of the original *SimRank* algorithm.

[7]http://googleblog.blogspot.com/2009/12/personalized-search-for-everyone.html

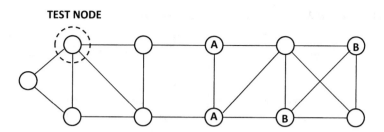

Figure 10.6: Label sparsity issues in collective classification

10.3   Recommendations by Collective Classification

Collective classification methods are particularly effective for incorporating content into the recommendation process. For example, consider the case of a social networking application in which a golf-equipment manufacturer wishes to determine all individuals interested in *"golf."* It is assumed that the manufacturer might already have several examples of individuals interested in golf. This might be achieved with several mechanisms in social networks, such as the utilization of user profile, or specification of a Facebook *"like"* button on a golf-related post. Furthermore, in some cases, where customer feedback is available, it may be possible for the manufacturer to have both like and dislike information for various nodes in the network. These categories of specific actors in the network can be specified with the use of labels. Therefore, a subset of the nodes are associated with labels. It is desired to use these labels as training data to determine the labels of the other nodes where they are unspecified. It is assumed that for labeled nodes, the index of the label is drawn from $\{1 \ldots r\}$. Like the collaborative filtering problem, this is also an incomplete data estimation problem, except that it is done in the context of network structures.

The solution to this problem is dependent on the notion of *homophily*. This notion can be viewed as the social network analog of using neighborhoods. The solution to this model is crucially dependent on the notion of homophily. Because nodes with similar properties are usually connected, it is reasonable to assume that this is also true of node labels. A simple solution to this problem is to examine the k labeled nodes in the proximity of a given node and report the majority label. This approach is, in fact, the network analog of a nearest neighbor classifier. However, such an approach is generally not possible in collective classification because of the sparsity of node labels. An example of a network is illustrated in Figure 10.6, in which the two classes are labeled A and B. The remaining nodes are unlabeled. For the test node in Figure 10.6, it is evident that it is generally closer to instances of A in the network structure, but there is no labeled node *directly* connected to the test instance. Thus, the problem of label sparsity arises in the context of network-based prediction, just as it does in the case of ratings-based data. How can one address these sparsity issues? In order to handle sparsity, one must not only use the direct connections to labeled nodes, but also use the indirect connections through unlabeled nodes. In this chapter, we will briefly discuss two algorithms, one of which is an iterative classification algorithm, and the other is a random walk-based method.

10.3.1 Iterative Classification Algorithm

The *Iterative Classification Algorithm (ICA)* is one of the earliest classification algorithms in the literature and has been applied to a wide variety of data domains. Consider the (undirected) network $G = (N, A)$ in which class labels are drawn from $\{1 \ldots r\}$. Each edge $(i, j) \in A$ is associated with the weight w_{ij}. Furthermore, the content $\overline{X}_i$ is available at the node i in the form of a multidimensional feature vector. The total number of nodes is denoted by n, from which n_t nodes are unlabeled test nodes.

An important step of the *ICA* algorithm is to derive a set of *link features* in addition to the available content features in $\overline{X}_i$. The most important link features correspond to the distribution of the classes in the immediate neighborhood of the node. Therefore, a feature is generated for each class, containing the fraction of its incident nodes belonging to that class. For each node i, its adjacent node j is weighted by w_{ij} for computing its credit to the relevant class. In principle, it is also possible to derive other link features based on structural properties of the graph such as the degree of the node, *PageRank* values, number of closed triangles involving the node, or connectivity features. Such link features can be derived on the basis of an application-specific understanding of the network data set.

The basic iterative classification algorithm is structured as a meta-algorithm. A base classifier $\mathcal{A}$ is leveraged within an iterative framework. Many different base classifiers have been used in different implementations, such as the naive Bayes classifier, a logistic regression classifier, and a neighborhood voting classifier. The main requirement is that these classifiers should be able to output a numeric score that quantifies the likelihood of a node belonging to a particular class. While the framework is independent of the specific choice of classifier, the use of the naive Bayes classifier is particularly common because of the interpretation of its numeric score as a probability. Therefore, the following discussion will assume that the algorithm $\mathcal{A}$ is instantiated to the naive Bayes classifier.

The link and content features are used to train the naive Bayes classifier. For many nodes, it is difficult to robustly estimate important class-specific features, such as the fractional presence of the different classes in their neighborhood. This is a direct result of label sparsity, and it makes the class predictions of such nodes unreliable. Therefore, an iterative approach is used for augmenting the training data set. In each iteration, n_t/T (test) node labels are made "certain" by the approach, where T is a user-defined parameter controlling the maximum number of iterations. The test nodes for which the Bayes classifier exhibits the highest class membership probabilities are selected to be made final. These labeled test nodes can then be added to the training data, and the classifier is retrained by extracting the link features again with the augmented training data set. The approach is repeated until the labels of all nodes have been made final. Because the labels of n_t/T nodes are finalized in each iteration, the entire process terminates in exactly T iterations. The overall pseudocode is illustrated in Figure 10.7.

One advantage of the iterative classification algorithm is that it can seamlessly use content and structure in the classification process. For example, if a node contains features corresponding to interests in other related products, then these features can also be used in the labeling process. The classifier can automatically select the most relevant features using off-the-shelf feature-selection algorithms. On the other hand, the errors in the earlier phases of iterative classification can propagate and multiply in later phases because of augmented training examples with incorrect labels. This can increase the cumulative error in noisy training data sets.

Algorithm *ICA*(Graph $G = (N, A)$, Weights: $[w_{ij}]$, Node Class Labels: $\mathcal{C}$,
　　　　Base Classifier: $\mathcal{A}$, Number of Iterations: T)
begin
　repeat
　　Extract link features at each node with current training data;
　　Train classifier $\mathcal{A}$ using both link and content features of
　　　current training data and predict labels of test nodes;
　　Make (predicted) labels of most "certain" n_t/T
　　　test nodes final, and add these nodes to training
　　　data, while removing them from test data;
　until T iterations;
end

Figure 10.7: The iterative classification algorithm

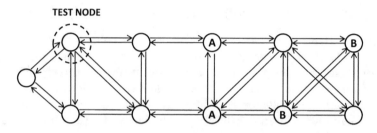

Figure 10.8: Creating directed transition graphs from undirected graph of Figure 10.6

10.3.2　Label Propagation with Random Walks

The label propagation method directly uses random walks on the undirected network structure $G = (N, A)$. The weight of edge (i, j) is denoted by $w_{ij} = w_{ji}$. To classify an unlabeled node i, a random walk is executed starting at node i and terminated at the first labeled node encountered. The class at which the random walk has the highest probability of termination is reported as the predicted label of node i. The intuition for this approach is that the walk is more likely to terminate at labeled nodes in the proximity of node i. Therefore, when many nodes of a particular class are located in its proximity, node i is more likely to be labeled with that class.

An important assumption is that the graph must be *label connected*. In other words, every unlabeled node needs to be able to reach a labeled node in the random walk. For undirected graphs $G = (N, A)$, this means that every connected component of the graph needs to contain at least one labeled node. In the following discussion, it will be assumed that the graph $G = (N, A)$ is undirected and label-connected.

The first step is to model the random walks in such a way that they always terminate at their *first arrival* at labeled nodes. This can be achieved by removing outgoing edges from labeled nodes and replacing them with self-loops. Furthermore, to use a random-walk approach, we need to convert the undirected graph $G = (N, A)$ into a directed graph $G' = (N, A')$ with an $n \times n$ transition matrix $P = [p_{ij}]$. For each undirected edge $(i, j) \in A$, directed edges (i, j) and (j, i) are added to A' between the corresponding nodes. The transition probability p_{ij} of edge (i, j) is defined as follows:

$$p_{ij} = \frac{w_{ij}}{\sum_{k=1}^{n} w_{ik}} \tag{10.8}$$

The transition probability p_{ji} of edge (j, i) is defined as follows:

$$p_{ji} = \frac{w_{ji}}{\sum_{k=1}^{n} w_{jk}} \qquad (10.9)$$

For example, the directed transition graph created from the undirected graph of Figure 10.6 is illustrated in Figure 10.8.

A wide variety of random-walk methods are available for propagating the labels using this transition graph. Consider the case where the labels are drawn from $\{1 \ldots k\}$. The idea is to execute the personalized *PageRank* algorithm k times, where the personalization vector for the cth execution restarts at labeled nodes belonging to the cth class. Each class-specific personalized *PageRank* probability is multiplied with the prior probability of that class, or, equivalently, the number of labeled training nodes in that class. For each node, the class index that yields the highest (prior-scaled) personalized *PageRank* probability is reported.

10.3.3 Applicability to Collaborative Filtering in Social Networks

Collective classification techniques can also be used for collaborative filtering of users in social networks. Consider a scenario in which we have the ratings for various products specified by different users. Furthermore, we also have the data corresponding to the social connections of various users. Therefore, this problem can be viewed as a generalization of the traditional problem of collaborative filtering. In this case, a pure neighborhood-based algorithm for collaborative filtering will take the similarity in ratings into account, but it will not take the homophily of different users into account. Clearly, it can be advantageous to take the homophily between users into account in order to perform collaborative filtering. As discussed in Chapters 1 and 3, collaborative filtering methods are generalizations of traditional classification problems. This analogy continues to be true even in the social network setting.

Certain versions of this problem can be easily handled using collective classification methods. Consider the case where the ratings are unary, in which users have a mechanism to specify a liking for an item but no mechanism to specify a dislike. In such cases, the specification of a liking for a product can be included as a keyword at that node. The label of a node is defined by the particular product of interest. The labels for the other products are treated as content-centric keywords. The problem now reduces to that of collective classification with content at the nodes. This variation of the problem can be handled easily by the *ICA* algorithm.

In cases where the ratings are not unary, the problem can be modeled as a *multilabel* collective classification problem, in which the rating of each product is treated as a separate label [306]. When the number of possible ratings is small, each value of the rating can be treated as a discrete value. Since all the items are handled in one shot, a single node may have multiple nodes corresponding to the ratings of various items. The goal is to use the specified ratings at the nodes together with the network structure to predict the values of the ratings for the various items. In such cases, the technique in [306] can be applied directly.

10.4 Recommending Friends: Link Prediction

In many social networks, it is desirable to predict future links between pairs of nodes in the network. For example, commercial social networks, such as Facebook, often recommend users as potential friends. As we will see later, such methods also have direct applicability to collaborative filtering techniques. In this section, we will discuss various techniques that are commonly used for link prediction.

10.4.1   Neighborhood-Based Measures

Neighborhood-based measures use the number of common neighbors between a pair of nodes i and j in different ways to quantify the likelihood of a link between them in the future. For example, in Figure 10.9(a), Alice and Bob share 4 common neighbors. Therefore, it is reasonable to conjecture that a link might eventually form between them. In addition to their common neighbors, they also have their own disjoint sets of neighbors. There are different ways of normalizing neighborhood-based measures to account for the number and relative importance of different neighbors. These are discussed below.

Definition 10.4.1 (Common Neighbor Measure) *The common-neighbor measure between nodes i and j is equal to the number of common neighbors between nodes i and j. In other words, if S_i is the neighbor set of node i, and S_j is the neighbor set of node j, the common-neighbor measure is defined as follows:*

$$CommonNeighbors(i, j) = |S_i \cap S_j| \qquad (10.10)$$

The major weakness of the common-neighbor measure is that it does not account for the *relative* number of common neighbors between them as compared to the number of other connections. In the example of Figure 10.9(a), Alice and Bob each have a relatively small node degree. Consider a different case in which Alice and Bob are either spammers or very popular public figures who were connected to a large number of other actors. In such a case, Alice and Bob might easily have many neighbors in common, *just by chance*. The *Jaccard measure* is designed to normalize for varying degree distributions.

Definition 10.4.2 (Jaccard Measure) *The Jaccard-based link prediction measure between nodes i and j is equal to the Jaccard coefficient between their neighbor sets S_i and S_j, respectively.*

$$JaccardPredict(i, j) = \frac{|S_i \cap S_j|}{|S_i \cup S_j|} \qquad (10.11)$$

The Jaccard measure between Alice and Bob in Figure 10.9(a) is $4/9$. If the degrees of either Alice or Bob were to increase, it would result in a lower Jaccard coefficient between them. This kind of normalization is important, because of the power-law degree distributions of nodes.

    The Jaccard measure adjusts much better to the variations in the degrees of the nodes *between which* the link prediction is measured. However, it does not adjust well to the degrees of their *intermediate* neighbors. For example, in Figure 10.9(a), the common neighbors of Alice and Bob are Jack, John, Jill, and Mary. However, all of these common neighbors could be very popular public figures with very high degrees. Therefore, these nodes are statistically more likely to occur as common neighbors of many pairs of nodes. This makes them *less* important in the link prediction measure. The Adamic-Adar measure is designed to account for the varying importance of the different common neighbors. It can be viewed as a weighted version of the common-neighbor measure, where the weight of a common neighbor is a decreasing function of its node degree. The typical function used in the case of the Adamic-Adar measure is the inverse logarithm. In this case, the weight of the common neighbor with index k is set to $1/\log(|S_k|)$, where S_k is the neighbor set of node k.

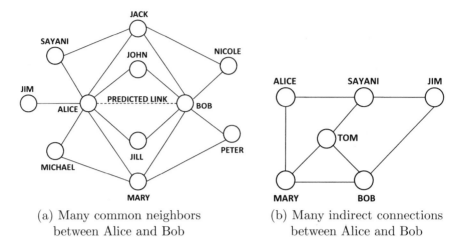

(a) Many common neighbors   (b) Many indirect connections
between Alice and Bob   between Alice and Bob

Figure 10.9: Examples of varying effectiveness of different link-prediction measures

Definition 10.4.3 (Adamic-Adar Measure) *The common-neighbor measure between nodes i and j is equal to the weighted number of common neighbors between nodes i and j. The weight of node k is defined is $1/log(|S_k|)$.*

$$AdamicAdar(i,j) = \sum_{k \in S_i \cap S_j} \frac{1}{log(|S_k|)} \tag{10.12}$$

The base of the logarithm does not matter in the previous definition, as long as it is chosen consistently for all pairs of nodes. In Figure 10.9(a), the Adamic-Adar measure between Alice and Bob is $\frac{1}{log(4)} + \frac{1}{log(2)} + \frac{1}{log(2)} + \frac{1}{log(4)} = \frac{3}{log(2)}$.

10.4.2   Katz Measure

While the neighborhood-based measures provide a robust estimation of the likelihood of a link forming between a pair of nodes, they are not quite as effective when the number of shared neighbors between a pair of nodes is small. For example, in the case of Figure 10.9(b), Alice and Bob share one neighbor in common. Alice and Jim also share one neighbor in common. Therefore, neighborhood-based measures have difficulty in distinguishing between different pairwise prediction strengths in these cases. Nevertheless, there also seems to be a significant *indirect* connectivity in these cases through longer paths. In such cases, walk-based measures are more appropriate. A particular walk-based measure that is used commonly to measure the link-prediction strength is the *Katz* measure.

Definition 10.4.4 (Katz Measure) *Let $n_{ij}^{(t)}$ be the number of walks of length t between nodes i and j. Then, for a user-defined parameter $\beta < 1$, the Katz measure between nodes i and j is defined as follows:*

$$Katz(i,j) = \sum_{t=1}^{\infty} \beta^t \cdot n_{ij}^{(t)} \tag{10.13}$$

The value of β is a discount factor that de-emphasizes walks of longer length. For small enough values of β, the infinite summation of Equation 10.13 will converge. If A is the symmetric adjacency matrix of an undirected network, then the $n \times n$ pairwise Katz coefficient matrix K can be computed as follows:

$$K = \sum_{i=1}^{\infty}(\beta A)^i = (I - \beta A)^{-1} - I \tag{10.14}$$

The eigenvalues of A^k are the kth powers of the eigenvalues of A. The value of β should always be selected to be smaller than the inverse of the largest eigenvalue of A to ensure convergence of the infinite summation. A weighted version of the measure can be computed by replacing A with the weight matrix of the graph. The Katz measure often provides prediction results of excellent quality.

It is noteworthy that the sum of the Katz coefficients of a node i with respect to other nodes is referred to as its *Katz centrality*. Other mechanisms for measuring centrality, such as closeness and *PageRank*, are also used for link prediction in a modified form. The reason for this connection between centrality and link-prediction measures is that highly central nodes have the propensity to form links with many nodes.

10.4.3 Random Walk-Based Measures

Random walk-based measures are a different way of defining connectivity between pairs of nodes. Two such measures are *PageRank* and *SimRank*. These methods are described in detail in section 10.2 of this chapter.

The first way of computing the similarity between nodes i and j is with the use of the personalized *PageRank* of node j, where the restart is performed at node i. The idea is that if j is the structural proximity of i, it will have a very high personalized *PageRank* measure, when the restart is performed at node i. This is indicative of higher link prediction strength between nodes i and j. The personalized *PageRank* is an asymmetric measure between nodes i and j. Because the discussion in this section is for the case of undirected graphs, one can use the average of the values of $PersonalizedPageRank(i, j)$ and $PersonalizedPageRank(j, i)$. Another possibility is the *SimRank* measure that is already a symmetric measure. This measure computes an inverse function of the walk length required by two random surfers moving backwards to meet at the same point. The corresponding value is reported as the link prediction measure.

10.4.4 Link Prediction as a Classification Problem

The aforementioned measures are *unsupervised* heuristics. For a given network, one of these measures might be more effective, whereas another might be more effective for a different network. How can one resolve this dilemma and select the measures that are most effective for a given network?

The link prediction problem can be viewed as a classification problem by treating the presence or absence of a link between a pair of nodes as a binary class indicator. Thus, a multidimensional data record can be extracted for *each pair of nodes*. The features of this multidimensional record include all the different neighborhood-based, Katz-based, or walk-based similarities between nodes. In addition, a number of other preferential-attachment features, such as node-degrees of each node in the pair, are used. Thus, for each node pair, a multidimensional data record is constructed. The result is a positive-unlabeled classification problem, where node pairs with edges are the positive examples, and the remaining pairs are unlabeled examples. The unlabeled examples can be approximately treated as negative examples for training purposes. Because there are too many negative example pairs in large and sparse networks, only a sample of the negative examples is used. Therefore, the supervised link prediction algorithm works as follows:

1. *Training phase:* Generate a multidimensional data set containing one data record for each pair of nodes with an edge between them, and a sample of data records from pairs of nodes without edges between them. The features correspond to extracted similarity and structural features between node pairs. The class label is the presence or absence of an edge between the pair. Construct a training model on the data.

2. *Testing phase:* Convert each test node pair to a multidimensional record. Use any conventional multidimensional classifier to make label predictions.

Logistic regression [22] is a common choice for the base classifier. Cost-sensitive versions of various classifiers are commonly used because of the imbalanced nature of the underlying classification problem.

One advantage of this approach is that content features can be used in a seamless way. For example, the content similarity between a pair of nodes can be used. The classifier will automatically learn the relevance of these features in the training process. Furthermore, unlike many link prediction methods, the approach can also handle *directed networks* by extracting features in an asymmetric way. For example, instead of using node degrees, one might use in-degrees and out-degrees as features. Random walk features can also be defined in an asymmetric way on directed networks, such as when computing the *PageRank* of node j with restart at node i and vice versa. In general, the supervised model is more flexible because of its ability to *learn* relationships between links and features of various types.

10.4.5 Matrix Factorization for Link Prediction

Like collaborative filtering, link prediction methods can be viewed as matrix completion problems with implicit feedback matrices. Let A be the $n \times n$ adjacency matrix of the underlying graph. We assume that the matrix A is binary, where the presence and absence of edges is indicated by 1s and 0s, respectively. Note that the matrix A is asymmetric for directed graphs and symmetric for undirected graphs. The matrix factorization methodology can be used in two different ways depending on whether the graph is directed or undirected. For directed graphs, the factorization is very similar to that in collaborative filtering:

$$A \approx UV^T \tag{10.15}$$

Here, A is the adjacency matrix of the graph. Furthermore, $U = [u_{is}]$ and $V = [v_{js}]$ are both factor matrices of size $n \times k$. After U and V have been learned, one can recommend the edges with largest predicted weight in UV^T.

The matrix A can be viewed in a similar way to an implicit feedback matrix, in which we need a sample of both positive and negative entries (cf. section 3.6.6.2 of Chapter 3). Note that we could use all the elements in the matrix A as observed elements but such an approach would be computationally very expensive when the number of nodes n is large. Furthermore, the sparsity of the adjacency matrix ensures that the factorization is dominated by the less important zero entries. Therefore, we work with only a sample of "observed" elements from the matrix. We define the positive and negative entries S_P and S_N as follows:

$$S_P = \{(i,j) : a_{ij} \neq 0\}$$
$$S_N = \{\text{Random sample of } (i,j) : a_{ij} = 0\}$$

All positive elements are included because they are rare and therefore too valuable to be discarded. Then, we define the "observed" elements of A as $S = S_P \cup S_N$ for the optimization process. Only the set S is used for training, and therefore the choice of S_N has an impact on the results obtained by the algorithm. The relative sizes of S_P and S_N will control the relative importance of the two types of entries. Note that if S_N were selected to be equal to all the zero entries in the matrix, the factorization would be dominated by the zero entries, and it may sometimes not be as effective in capturing the all-important edges. Differential importance to the two types of entries is appropriate, as in all rare-class detection problems. For example, the size of S_N can be set to be equal to S_P.

For any element $(i,j) \in S$, it can be predicted to the value $\hat{a}_{ij}$ as follows:

$$\hat{a}_{ij} = \sum_{s=1}^{k} u_{is} v_{js} \tag{10.16}$$

The error of predicting an entry is given by $e_{ij} = a_{ij} - \hat{a}_{ij}$. We want to minimize this error over the observed entries. The regularized objective function is as follows:

$$\text{Minimize } J = \frac{1}{2} \sum_{(i,j) \in S} e_{ij}^2 + \frac{\lambda}{2} \sum_{i=1}^{n} \sum_{s=1}^{k} u_{is}^2 + \frac{\lambda}{2} \sum_{j=1}^{n} \sum_{s=1}^{k} v_{js}^2$$

$$= \frac{1}{2} \sum_{(i,j) \in S} \left(a_{ij} - \sum_{s=1}^{k} u_{is} \cdot v_{js} \right)^2 + \frac{\lambda}{2} \sum_{i=1}^{n} \sum_{s=1}^{k} u_{is}^2 + \frac{\lambda}{2} \sum_{j=1}^{n} \sum_{s=1}^{k} v_{js}^2$$

Here, λ is the regularization parameter. It is noteworthy that this objective function is virtually identical to that discussed in section 3.6.4.2 of Chapter 3, except that the matrix A is a square $n \times n$ matrix. However, the solution methodology and gradient-descent updates are exactly identical. One can use either vectorized gradient-descent in which the gradient is computed with respect to the errors over all entries, or stochastic gradient descent in which the derivative is stochastically approximated using the errors over randomly chosen edges. In regular gradient-descent, the matrices U and V are randomly initialized, and the following updates are executed repeatedly for each entry (i,q) of U and each entry (j,q) of V:

$$u_{iq} \Leftarrow u_{iq} - \alpha \frac{\partial J}{\partial u_{iq}} = u_{iq} + \alpha \left(\sum_{j:(i,j) \in S} e_{ij} \cdot v_{jq} - \lambda \cdot u_{iq} \right)$$

$$v_{jq} \Leftarrow v_{jq} - \alpha \frac{\partial J}{\partial v_{jq}} = v_{jq} + \alpha \left(\sum_{i:(i,j) \in S} e_{ij} \cdot u_{iq} - \lambda \cdot v_{jq} \right)$$

Here, $\alpha > 0$ is the learning rate. One can perform these updates within the framework of Figure 3.8 of Chapter 3. The updates can also be performed using sparse matrix operations.

Algorithm *LinkPrediction*(Adjacency Matrix: A, Regularization: λ, Step Size: α)
begin
 Randomly initialize matrices U and V;
 $S_P = \{(i,j) : a_{ij} \neq 0\}$;
 $S_N = \{\text{Random sample of } (i,j) : a_{ij} = 0\}$;
 $S = S_P \cup S_N$;
 while not(convergence) **do**
 begin
  Randomly shuffle observed entries in S;
  for each $(i,j) \in S$ in shuffled order **do**
  begin
   $e_{ij} \Leftarrow a_{ij} - \sum_{s=1}^{k} u_{is} v_{js}$;
   $\overline{u_i}^{(+)} \Leftarrow \overline{u_i} + \alpha \left(e_{ij} \overline{v_j} - \dfrac{\lambda \overline{u_i}}{\text{OutDegree}_{(i)}} \right)$
   $\overline{v_j}^{(+)} \Leftarrow \overline{v_j} + \alpha \left(e_{ij} \overline{u_i} - \dfrac{\lambda \overline{v_j}}{\text{InDegree}_{(j)}} \right)$
   $\overline{u_i} = \overline{u_i}^{(+)}; \overline{v_j} = \overline{v_j}^{(+)}$;
  end
  Check convergence condition;
 end
end

Figure 10.10: Stochastic gradient descent for directed link prediction

The first step is to compute an error matrix $E = [e_{ij}]$ in which the unobserved entries of E (i.e., entries not in S) are set to 0. Note that E is a very sparse matrix, and it makes sense to compute the value of e_{ij} for only the observed entries $(i,j) \in S$ and store the matrix using a sparse data structure. Subsequently, the updates can be computed as follows:

$$U \Leftarrow (1 - \alpha \cdot \lambda)U + \alpha E V$$
$$V \Leftarrow (1 - \alpha \cdot \lambda)V + \alpha E^T U$$

Next, we describe stochastic gradient descent. The basic idea is to approximate the gradient stochastically with respect to the error component contributed by a single entry (including a "zero" edge in S_N). The edges in the set S are processed in randomly shuffled order, and the latent factors are updated based on the error gradient with respect to that edge. Starting with a random initialization of U and V, one can use the following updates with respect to the randomly chosen entry $(i,j) \in S$:

$$u_{iq} \Leftarrow u_{iq} - \alpha \cdot \left[\frac{\partial J}{\partial u_{iq}} \right]_{\text{Portion contributed by } (i,j)} \qquad \forall q \in \{1 \ldots k\}$$

$$v_{jq} \Leftarrow v_{jq} - \alpha \cdot \left[\frac{\partial J}{\partial v_{jq}} \right]_{\text{Portion contributed by } (i,j)} \qquad \forall q \in \{1 \ldots k\}$$

One can expand the aforementioned expression and consolidate the updates over different values of $q \in \{1 \ldots k\}$ into a single vectorized update of the corresponding row of U (or V). Let $\overline{u_i}$ be the ith row of matrix U and $\overline{v_j}$ be the jth row of matrix V. Then, the stochastic gradient descent updates may be written as follows:

$$\overline{u_i} \Leftarrow \overline{u_i} + \alpha \left(e_{ij}\overline{v_j} - \frac{\lambda \overline{u_i}}{\text{OutDegree}(i)} \right)$$

$$\overline{v_j} \Leftarrow \overline{v_j} + \alpha \left(e_{ij}\overline{u_i} - \frac{\lambda \overline{v_j}}{\text{InDegree}(j)} \right)$$

Here, $\alpha > 0$ is the learning rate. We continue to cycle over the various edges in S until convergence is reached. The overall framework of the stochastic gradient descent method is illustrated in Figure 10.10.

We have used a slightly more refined regularization term here than used in Chapter 3. Here, OutDegree(i) and InDegree(j) denote the out-degree and in-degree of nodes i and j, respectively. Note that the out-degree and in-degree of the nodes need to be computed with respect to $S_P \cup S_N$ rather than only S_P.

It is possible to further improve the accuracy of the method with an ensemble approach. The matrix is factorized multiple times with different draws of the negative sample S_N. Each factorization might provide a slightly different prediction of an edge. The different predictions of a particular entry in the matrix are then averaged to create the final result. Instead of sampling, it is also possible to include all zero entries in S_N, and then define a weighted optimization problem in which the nonzero entries are given a larger weight $\theta > 1$ than zero entries. The actual value of the weight parameter θ is learned using cross-validation. In such cases, stochastic gradient descent is no longer viable because of the large number of (specified) entries of the matrix. However, since most of the entries are zeros, it is possible to use some tricks [260] to leverage weighted ALS methods efficiently.

This approach is quite general because it can be applied to directed and/or signed networks. In the case of unsigned networks, one can impose non-negativity constraints on the latent factors to avoid overfitting. The only change to the update equations is that any negative factor values after an iteration are set to 0. Undirected networks can be addressed by replacing each undirected edge with two directed edges. Furthermore, the set S_N in undirected networks should be constructed by first sampling pairs of nodes (without an edge between them) and then including edges in both directions in S_N. In the next section, we will propose a method that is specifically optimized to undirected networks by reducing the number of learned parameters.

10.4.5.1 Symmetric Matrix Factorization

For undirected graphs, we do not need two separate factor matrices U and V because the matrix A is symmetric. Using fewer optimization parameters has the advantage of reducing overfitting. In such cases, we can use a single factor matrix U and represent the factorization[8] as follows:

$$A \approx UU^T \tag{10.17}$$

Here, $U = [u_{is}]$ is an $n \times k$ factor matrix. As in the previous case, the observed entries in $S = S_P \cup S_N$ include the existing edges in S_P and also some "zero" edges in S_N. For each edge (i, j) in the undirected graph, both (i, j) and (j, i) are included in S_P. The zero-edges are chosen from the pairs of nodes between which an edge does not exist, and both directions

---

[8]An implicit assumption here is that the matrix A is positive semi-definite. However, by setting the (unobserved) diagonal entries of A to the node degrees, it can be shown that A is positive semi-definite. These unobserved diagonal entries do not affect the final solution because they are not a part of the optimization problem.

of the edge are included in S_N. In other words, if (i,j) is included in S_N, then (j,i) is also included in it. Because of the nature of condition $A \approx UU^T$, each observed entry $(i,j) \in S$ can be predicted as follows:

$$\hat{a}_{ij} = \sum_{s=1}^{k} u_{is} u_{js} \qquad (10.18)$$

The corresponding error of the prediction is given by $e_{ij} = a_{ij} - \hat{a}_{ij}$. We want to minimize this error over the observed entries. The regularized objective function is as follows:

$$\text{Minimize } J = \frac{1}{2} \sum_{(i,j) \in S} e_{ij}^2 + \frac{\lambda}{2} \sum_{i=1}^{n} \sum_{s=1}^{k} u_{is}^2$$

$$= \frac{1}{2} \sum_{(i,j) \in S} \left(a_{ij} - \sum_{s=1}^{k} u_{is} \cdot u_{js} \right)^2 + \frac{\lambda}{2} \sum_{i=1}^{n} \sum_{s=1}^{k} u_{is}^2$$

Upon taking the partial derivative of J with respect to each of the decision variables, one obtains the following results:

$$\frac{\partial J}{\partial u_{iq}} = \sum_{j:(i,j) \in S} \left(a_{ij} + a_{ji} - 2 \sum_{s=1}^{k} u_{is} \cdot u_{js} \right)(-u_{jq}) + \lambda u_{iq}$$

$$\forall i \in \{1 \dots n\}, q \in \{1 \dots k\}$$

$$= \sum_{j:(i,j) \in S} (e_{ij} + e_{ji})(-u_{jq}) + \lambda u_{iq} \quad \forall i \in \{1 \dots n\}, q \in \{1 \dots k\}$$

$$= \sum_{j:(i,j) \in S} 2(e_{ij})(-u_{jq}) + \lambda u_{iq} \quad \forall i \in \{1 \dots n\}, q \in \{1 \dots k\}$$

$$\frac{\partial J}{\partial u_{jq}} = \sum_{i:(i,j) \in S} 2(e_{ij})(-u_{iq}) + \lambda u_{jq} \quad \forall j \in \{1 \dots n\}, q \in \{1 \dots k\}$$

Note that the value of $e_{ij} + e_{ji}$ is replaced with $2e_{ij}$ because the original matrix A, the predicted matrix $\hat{A}$, and the error matrix $[e_{ij}]$ are all symmetric. The steps for performing the gradient descent remain similar to those discussed in the previous case. Let $E = [e_{ij}]$ be the error matrix in which only the observed entries in S are set to the value of $a_{ij} - \sum_{s=1}^{k} u_{is} u_{js}$ and unobserved entries are set to 0. This matrix can be computed entry-by-entry for all elements in S and stored in sparse form. Subsequently, the updates can be performed using sparse matrix multiplication as follows:

$$U \Leftarrow U(1 - \lambda \alpha) + 2\alpha EU \qquad (10.19)$$

Here, $\alpha > 0$ represents the step-size. Note that the constant factor 2 in $2\alpha EU$ can be ignored by adjusting the step-size and regularization parameters appropriately.

The stochastic-gradient descent method may be used for faster convergence, although the quality of the resulting solution is typically lower. In the case of stochastic gradient-descent, the derivative is decomposed into the error components over individual entries (edges), and the update is specific to the error in each entry (edge). In this case, the following $2 \cdot k$ updates may be executed for each observed entry $(i,j) \in S$:

$$u_{iq} \Leftarrow u_{iq} + \alpha \left(2e_{ij} \cdot u_{jq} - \frac{\lambda \cdot u_{iq}}{\text{Degree}(i)} \right) \quad \forall q \in \{1 \ldots k\}$$

$$u_{jq} \Leftarrow u_{jq} + \alpha \left(2e_{ij} \cdot u_{iq} - \frac{\lambda \cdot u_{jq}}{\text{Degree}(j)} \right) \quad \forall q \in \{1 \ldots k\}$$

Here Degree(i) represents the number of edges incident on i, including the "edges" in S_N. One can also write these updates in terms of the ith row $\overline{u_i}$ and jth row $\overline{u_j}$ of U:

$$\overline{u_i} \Leftarrow \overline{u_i} + \alpha \left(2e_{ij}\overline{u_j} - \frac{\lambda \overline{u_i}}{\text{Degree}(i)} \right)$$

$$\overline{u_j} \Leftarrow \overline{u_j} + \alpha \left(2e_{ij}\overline{u_i} - \frac{\lambda \overline{u_j}}{\text{Degree}(j)} \right)$$

The value of λ is typically selected with the use of cross-validation methods or by trying various values of λ on a hold-out set. A nice characteristic of the matrix factorization methodology is that it works seamlessly for signed and unsigned networks. Furthermore, modest variants of the approach can be used for directed and undirected networks. This is not true for many of the other link prediction methods, which are inherently designed for undirected and unsigned networks.

It is also possible to incorporate bias variables within the matrix factorization process as in the case of traditional collaborative filtering (cf. section 3.6.4.5 of Chapter 3). The incorporation of bias variables within the matrix factorization framework for link prediction is intuitively equivalent to using preferential attachment principles [22] in networks. In cases where the graphs are unsigned, it is possible to use non-negative matrix factorization methods. Some of these methods have dual use in collaborative filtering when the ratings matrices can be represented as user-item graphs [235]. The following section will discuss these connections in detail.

10.4.6 Connections Between Link Prediction and Collaborative Filtering

Both link prediction and collaborative filtering attempt to estimate missing values. There-fore, it is natural to explore the connection between them. Link prediction is very similar to the implicit feedback setting of collaborative filtering in which the presence of a link is similar to a unary rating. The notion of user-item graphs provides a natural connection between link prediction and collaborative filtering. A detailed discussion of the process of creating user-item graphs is provided in section 2.7 of Chapter 2. For unary ratings matri-ces (or implicit feedback data sets), conventional link-prediction methods can be applied to the user-item graph in order to predict the affinities (links) between users and items. Each user corresponds to a user node in the user-item graph, and each item corresponds to an item node. All the 1s in the matrix correspond to edges between user nodes and item nodes. Examples of cases in which the ratings are unary are illustrated in Figures 10.11(a) and (b), respectively. Note that the predicted strength of links between user nodes and item nodes provides predictions of how much the corresponding users like the corresponding items. Because of this connection, link-prediction methods can be used to perform collaborative filtering. Furthermore, the converse is also true wherein collaborative filtering algorithms can be adapted to link prediction.

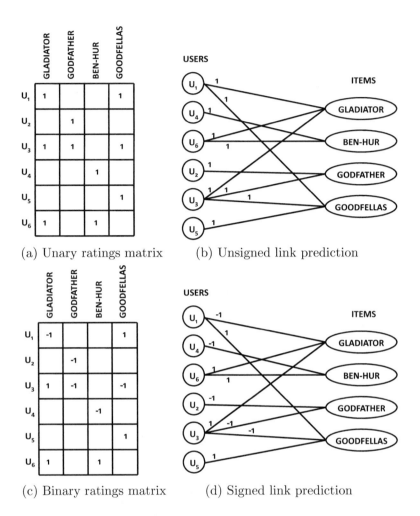

Figure 10.11: Link prediction for collaborative filtering

10.4.6.1   Using Link Prediction Algorithms for Collaborative Filtering

By predicting the top-k user-item links that are likely to be formed at a user node in the user-item graph, one can predict the top-k items for the user. Furthermore, by determining the top-k user-item links that are likely to be formed at an *item* node, the merchant can determine the top-k users to which she can promote a particular item. It is noteworthy that this approach can be used even in cases when the social network structure of the users is known. In such cases, the edges between users are included within the link prediction process. The inclusion of such edges will result in the incorporation of the homophily effects of social links within the recommendation process. These methods will be discussed in greater detail in section 11.3.7 of Chapter 11.

The case of explicitly specified ratings is somewhat more challenging because ratings might indicate either like or dislike of the item at hand. The conventional link-prediction problem is inherently designed to handle the notion of positive relationships, rather than that of negative relationships. However, a number of recent advances in link prediction can handle these cases as well. For ease in discussion, consider the case where the ratings are drawn from $\{-1, +1\}$, corresponding to the user liking or disliking the item. In this case, the edges are labeled with the sign of the rating. Examples of cases where the ratings are binary are illustrated in Figures 10.11(c) and (d), respectively. The resulting network is a *signed network*, and it is desired to predict the top-k positive links incident at a user in order to determine the items that a user likes the most. By predicting the top-k negative links, one can even discover the top-k items that a user might *dislike* the most. This problem is that of positive or negative link prediction in *signed* networks. Although the signed link prediction problem has not been discussed in this chapter, it has been shown in the literature [324–326, 346, 591] how methods for unsigned link prediction can be extended to the case of signed networks. The link prediction approach is most effective for unary or binary ratings data, although arbitrary ratings can be used as well. In that case, the ratings need to be mean-centered for each user, and then either a positive or negative weight is associated with the links corresponding to the mean-centered value of the ratings. This process results in a signed network in which links are weighted, and many of the methods for signed link prediction can handle such settings. The work in [324, 325] also shows how to use signed networks in the context of collaborative filtering applications, although the approach used is different from the one discussed here.

10.4.6.2 Using Collaborative Filtering Algorithms for Link Prediction

Both collaborative filtering and link prediction are missing value estimation problems. The only difference is that collaborative filtering is performed on user-item matrices whereas link prediction is performed on node-node matrices. Although the difference in matrix dimensions can affect the performance of the algorithms, a relatively unappreciated fact is that virtually all collaborative filtering methods can be used for link prediction. However, some amount of adaptation of the collaborative filtering algorithms is required.

For example, one can use almost all neighborhood-based methods, sparse linear models, and matrix factorization methods for link prediction. A user-based neighborhood method maps to a row-wise method on the adjacency matrix, and an item-based neighborhood method maps to a column-wise method on the adjacency matrix. However, because adjacency matrices of undirected networks are symmetric, one cannot distinguish between user-based and item-based methods (see Exercises 8 and 9). An important observation is that these methods can be used for both undirected and directed link prediction, whereas many other link prediction methods are applicable only to undirected networks. In the context of directed networks, user-based and item-based methods would map to outgoing edge-based and incoming edge-based methods, respectively. Recently, this relationship between link prediction and collaborative filtering has become increasingly appreciated; the work by [432] is particularly instructive, as it adapts matrix factorization methods for link prediction. Nevertheless, significant scope still exists in leveraging collaborative filtering methods for link prediction. Most of the neighborhood methods and linear regression models have not been explored significantly in the context of link prediction.

10.5   Social Influence Analysis and Viral Marketing

All social interactions result in varying levels of influence between individuals. In traditional social interactions, this is sometimes referred to as "word-of-mouth" influence. This general principle is also true for online social networks. For example, when an actor tweets a message in Twitter, the followers of the actors are exposed to the message. The followers may often retweet the message in the network. This results in the spread of information, ideas, and opinions in the social network. Many companies view this kind of information spread as a valuable advertising channel. By tweeting a popular message to the right participants, millions of dollars worth of advertising can be generated, if the message spreads through the social network as a *cascade*. This type of approach allows the rapid spread of information in a network in much the same way as a virus would spread in a biological epidemic or computer network. In fact, the models used in both cases share many similarities. Therefore, this methodology of influencing market participants is also referred to as *viral marketing*.

Different actors have different abilities to influence their peers in the social network. The two most common factors that regulate the influence of an actor are as follows:

1. The centrality of a actor within the social network structure is a crucial factor in her influence level. For example, actors with high levels of centrality are more likely to be influential. In directed networks, actors with high prestige are more likely to be influential. Centrality and prestige measures are discussed in [22]. *PageRank* can also be used as a measure of centrality and prestige.

2. The edges in the network are often associated with weights that are dependent on the likelihood that the corresponding pair of actors can be influenced by each other. Depending on the diffusion model used, these weights can sometimes be directly interpreted as *influence propagation probabilities*. Several factors may determine these probabilities. For example, a well-known individual may have higher influence than lesser known individuals. Similarly, two individuals, who have been friends for a long time, are more likely to influence one another. It is often assumed that the influence propagation probabilities are already available for analytical purposes, although a few recent methods show how to estimate these probabilities in a data-driven way.

An *influence propagation model* is used to quantify the precise impact of the aforementioned factors. These models are also known as *diffusion models*. The main goal of influence propagation models is to determine a set of seed nodes that maximize influence with the dissemination of information. In this sense, *influence maximization models can be viewed as recommenders of valuable social actors to merchants*. Therefore, the influence maximization problem is as follows:

Definition 10.5.1 (Influence Maximization) *Given a social network $G = (N, A)$, determine a set of k seed nodes S, influencing which will maximize the overall spread of influence in the network.*

The value of k can be viewed as a budget on the number of seed nodes that one is allowed to initially influence. This is consistent with real-life models in which advertisers are faced with budgets on initial advertising capacity. The goal of social influence analysis is to extend this initial advertising capacity with word-of-mouth methods.

Each model or heuristic can quantify the influence level of a node with the use of a function of S that is denoted by $f(\cdot)$. This function maps subsets of nodes to real numbers representing influence values. Therefore, after a model has been chosen for quantifying the influence $f(S)$ of a *given* set S, the optimization problem is that of determining the set S that maximizes $f(S)$. An interesting property of a very large number of influence analysis models is that the optimized function $f(S)$ is *submodular*.

What does submodularity mean? It is a mathematical way of representing the natural law of diminishing returns, as applied to sets. In other words, if $S \subseteq T$, then the *additional* influence obtained by adding an individual to set T cannot be larger than the additional influence of adding the same individual to set S. Thus, the *incremental* influence of the same individual diminishes, as larger supersets of cohorts are available as seeds. The submodularity of set S is formally defined as follows:

Definition 10.5.2 (Submodularity) *A function $f(\cdot)$ is said to be submodular, if for any pair of sets S, T satisfying $S \subseteq T$, and any set element e, the following is true:*

$$f(S \cup \{e\}) - f(S) \geq f(T \cup \{e\}) - f(T) \tag{10.20}$$

Virtually all natural models for quantifying influence turn out to be submodular. Submodularity is algorithmically convenient because a very efficient greedy optimization algorithm exists for maximizing submodular functions, as long as $f(S)$ can be evaluated for a given value of S. This algorithm starts by setting $S = \{\}$ and incrementally adds nodes to S that increase the value of $f(S)$ as much as possible. This procedure is repeated until the set S contains the required number of influencers k. The approximation level of this heuristic is based on a well-known classical result on optimization of submodular functions.

Lemma 10.5.1 *The greedy algorithm for maximizing submodular functions provides a solution with an objective function value that is at least a fraction $\left(\frac{e-1}{e}\right)$ of the optimal value. Here, e is the base of the natural logarithm.*

Thus, these results show that it is possible to optimize $f(S)$ effectively, as long as an appropriate submodular influence function $f(S)$ can be defined for a given set of nodes S.

Two common approaches for defining the influence function $f(S)$ of a set of nodes S are the *Linear Threshold Model* and the *Independent Cascade Model*. Both these diffusion models were proposed in one of the earliest works on social influence analysis. The general operational assumption in these diffusion models is that nodes are either in an active or inactive state. Intuitively, an *active* node is one which has already been influenced by the set of desired behaviors. Once a node moves to an active state, it never deactivates. Depending on the model, an active node may trigger activation of neighboring nodes either for a single time, or over longer periods. Nodes are successively activated until no more nodes are activated in a given iteration. The value of $f(S)$ is evaluated as the total number of activated nodes at termination.

10.5.1 Linear Threshold Model

In this model, the algorithm initially starts with an active set of seed nodes S and iteratively increases the number of active nodes based on the influence of neighboring active nodes. Active nodes are allowed to influence their neighbors over multiple iterations throughout the execution of the algorithm until no more nodes can be activated. The influence of neighboring nodes is quantified with the use of a linear function of the edge-specific weights b_{ij}. For each node i in the network $G = (N, A)$, the following is assumed to be true:

$$\sum_{j:(i,j)\in A} b_{ij} \leq 1 \tag{10.21}$$

Each node i is associated with a random threshold $\theta_i \sim U[0, 1]$ that is fixed up front and stays constant over the course of the algorithm. The total influence $I(i)$ of the *active* neighbors of node i on i, at a given time-instant, is computed as the sum of the weights b_{ij} of all *active* neighbors of i.

$$I(i) = \sum_{j:(i,j)\in A, j \text{ is active}} b_{ij} \tag{10.22}$$

The node i becomes active in a step when $I(i) \geq \theta_i$. This process is repeated until no further nodes can be activated. The total influence $f(S)$ may be measured as the number of nodes activated by a given seed set S. The influence $f(S)$ of a given seed set S is typically computed with simulation methods.

10.5.2   Independent Cascade Model

In the aforementioned linear threshold model, once a node becomes active, it has multiple chances to influence its neighbors. The random variable θ_i was associated with a *node* in the form of a threshold. On the other hand, in the independent cascade model, after a node becomes active, it obtains only a *single chance* to activate its neighbors, with *propagation probabilities* associated with the *edges*. The propagation probability associated with an edge is denoted by p_{ij}. In each iteration, only the *newly* active nodes are allowed to influence their neighbors, that have not already been activated. For a given node j, each of the edges (i, j) joining it to its newly active neighbors i flips a coin independently with success probability p_{ij}. If the coin toss for edge (i, j) results in a success, then the node j is activated. If node j is activated, it will get a single chance in the next iteration to influence its neighbors. In the event that no nodes are newly activated in an iteration, the algorithm terminates. The influence function value is equal to the number of active nodes at termination. Because nodes are allowed to influence their neighbors only once over the course of the algorithm, a coin is tossed for each edge at most once over the course of the algorithm.

10.5.3   Influence Function Evaluation

Both the linear threshold model and the independent cascade model are designed to compute the influence function $f(S)$ with the use of a model. The estimation of $f(S)$ is typically accomplished with simulation.

For example, consider the case of the linear threshold model. For a given seed node set S, one can use a random number generator to set the thresholds at the nodes. After the thresholds have been set, the active nodes can be labeled using any deterministic graph-search algorithm starting from the seed nodes in S and progressively activating nodes when the threshold condition is satisfied. The computation can be repeated over different sets of randomly generated thresholds, and the results may be averaged to obtain more robust estimates.

In the independent cascade model, a different simulation may be used. A coin with probability p_{ij} may be flipped for each edge. The edge is designated as *live* if the coin toss was a success. It can be shown that a node will eventually be activated by the independent cascade model, when a path of live edges exists from at least one node in S to it. This can be used to estimate the size of the (final) active set by simulation. The computation is repeated over different runs, and the results are averaged.

The proof that the linear threshold model and the independent cascade model are sub-modular optimization problems can be found in pointers included in the bibliographic notes. However, this property is not specific to these models. Submodularity is a very natural consequence of the laws of diminishing returns, as applied to the incremental impact of individual influence in larger groups. As a result, most reasonable models for influence analysis will satisfy submodularity.

10.5.4 Targeted Influence Analysis Models in Social Streams

The aforementioned models for influence analysis are highly *static* and are completely agnostic to the specific topic of interest. Consider a scenario in which a dealer in baseball equipment wishes to influence interested customers using the Twitter stream. The most influential actors on the network are typically *topic-agnostic* and may not be interested in baseball at all. For example, if one uses the number of followers of an actor in Twitter as a rough proxy for their influence, it is easy to see that such individuals are often famous actors, politicians, or sportsmen. Targeting famous politicians with tweets or promotions about baseball equipment is not necessarily the most efficient way for the dealer to increase the reach of her product. However, it would certainly be useful for the dealer to influence famous sportsmen, *specific to* baseball. Clearly, the influence-mining approach in the previous section will not achieve these goals. Furthermore, it is assumed in the previous section that influence propagation probabilities of edges are available. The determination of such probabilities also requires a separate model because such information is not directly available from the Twitter stream. Therefore, the influence analysis models discussed in the previous section are incomplete, because they assume more inputs than are truly available from the base data. In fact, the only data available to the user is the Twitter stream, which contains a large volume of tweets. In general, streams such as that of Twitter are referred to as *social streams*. In such streams, the trends in the network may evolve over time, and the most relevant influencers may also change over time.

In the context of social streams, it is important to make the influence analysis models *data-driven* or *content-centric*. In the method discussed in [573], the approach is made topically sensitive by selecting a set of relevant keywords, in terms of which the social stream is expressed. The flow of these keywords can then be tracked in the network to determine how the various actors influence one another, *specific to the topic at hand*. For example, a baseball manufacturer would select a set of keywords, which are relevant to the topic of baseball. The initial stage of feature selection is, therefore, a crucial one. For example, in the context of a Twitter stream, it may be possible to use *hashtags* belonging to specific topics for tracking purposes.

After these keywords have been selected, their propagation through the network structure is analyzed in terms of the underlying *flow paths*. A valid flow path is a sequence of actors that tweet (or post) the same keyword in sequence, and the sequence of actors are also connected by social network links. For example, consider the case where we have a hashtag related to baseball games. The *path* of propagation of this hashtag through the social network of actors provides very useful information about the *topic-specific* influence related to the topic of baseball. For example, in the network shown in Figure 10.12, the flow of the hash tags[9] *#baseball* and *#sammysosa* along various paths is the result of re-tweets (or copying behavior) in the social network. In this case, it is evident that Sayani is an influential tweeter in the specific topic of baseball, and her tweets on this topic are often considered authoritative enough to be picked up by other participants. However, if the hashtags were related to other subjects, unrelated to baseball, then even with the same pattern of propagation Sayani would not be considered influential in the *specific* context of baseball.

Influential actors will frequently occur in the early portions of such paths. Therefore, by determining *frequently* occurring paths, one can also determine important *epicenters* of various cascades. The work in [573] uses a constrained sequential pattern-mining model [23] in order to determine the most commonly occurring flow paths in the stream. The early

---

[9]Sammy Sosa is a retired Major League baseball player.

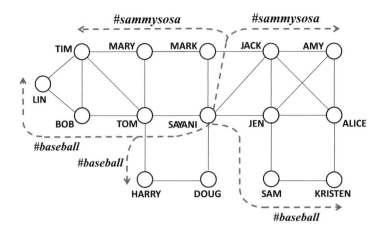

Figure 10.12: Examples of propagation paths of tweets related to baseball

points of these flow paths are declared as influential actors specific to the topic at hand. Other recent methods explicitly use topic modeling techniques to discover such influential actors. Refer to the bibliographic notes for more details of these methods.

10.6   Summary

Many recommendation problems in practical scenarios can be modeled as structural recommendation problems in networks. For example, traditional collaborative filtering problems can be modeled as user-item graphs. Various methods such as ranking techniques, collective classification methods, and link prediction techniques can be used on these user-item graphs.

Ranking and search are problems that are closely related to recommendation analysis. The main difference is that the results in the latter are personalized to specific users. In recent years, the methods for search and recommendation have become increasingly integrated, as search providers have started using user-specific information in order to personalize search results. Many variants of structural recommendation algorithms can be used in conjunction with social network analysis or user-item graphs. For example, both collective classification and link prediction can be used in conjunction with user-item graphs.

In collective classification, the goal is to infer labels at the remaining vertices from the pre-existing labels at a subset of the vertices. Collective classification techniques are particularly useful in the context of content-centric recommendation analysis in social networks.

In the link-prediction problem, the goal is to predict the links from the currently available structure in the network. The structural methods use local clustering measures such as the Jaccard measure or personalized *PageRank* values for making predictions. Supervised methods are able to discriminatively determine the most relevant features for link prediction. Link prediction methods are used to predict friends in social networks.

Social networks are often used for influencing individuals using "word-of-mouth" techniques. These methods can be viewed as techniques for recommending users to merchants for viral marketing. Typically, centrally located actors are more influential in the network. Diffusion models are used to characterize the flow of information in social networks. Two examples of such models include the linear threshold model and the independent cascade model. In recent years, such methods have been extended to use topic-specific techniques in the context of social streams.

10.7 Bibliographic Notes

The *PageRank* algorithm is described in [104, 465]. The *HITS* algorithm is also used for topic-sensitive search [302]. The topic-sensitive *PageRank* algorithm is described in [243], and the *SimRank* algorithm is described in [278]. Methods for leveraging the *personalized PageRank* algorithm in various forms of social recommender systems are discussed in [16, 81, 350, 602, 640, 663]. The work in [350] shows how to use random walks for grocery shopping recommendations. Temporal recommendations with the use of random walks on graphs are discussed in [639].

The iterative classification algorithm (ICA) has been presented in the context of many different data domains, including document data [143] and relational data [453]. Several base classifiers have been used within this framework, such as logistic regression [379] and a weighted voting classifier [387]. The discussion in this chapter is based on [453]. Many different variations of random-walk methods [56, 674, 678] have also been proposed. Collective classification of directed graphs is discussed in [675]. Detailed surveys on node classification methods may be found in [77, 375]. A toolkit for collective classification may be found in [388].

The link-prediction problem for social networks was proposed in [354]. The measures discussed in this chapter are based on this work. The merits of supervised methods are discussed in [355], and matrix factorization methods are discussed in [432]. Matrix factorization for link prediction can be viewed as analogous to similar methods that are used for collaborative filtering. A survey on link-prediction methods for social network analysis may be found in [42]. Methods for signed link prediction are discussed in [157, 324–326, 346, 591]. Other signed network techniques for collaborative filtering are discussed in [324, 325]. The work in [157] is notable because it shows the connections between the matrix factorization methods for link prediction and for collaborative filtering. A large and growing area of research is in the context of heterogeneous networks [36, 576, 577] in which links of multiple types are predicted from one another. In other related work, links from *multiple networks* are predicted from one another [488].

The problem of influence analysis has been studied both in the context of viral marketing and social networks. This problem was first studied in the context of viral marketing in [176, 510]. Subsequently, the problem was also studied in the context of social networks [297]. The linear threshold and independent cascade models are presented in this work. A degree-discount heuristic was proposed in [152]. A discussion of the submodularity property may be found in [452]. Other recent models for influence analysis in social networks are discussed in [153, 154, 369, 589]. One of the main problems in social influence models is a difficulty in learning the influence propagation probabilities, though there has been some recent focus on this issue [234]. Recent work has also shown how influence analysis can be performed directly from the social stream [80, 233, 573]. The method in [573] also shows how this approach may be made topic-sensitive. A survey on models and algorithms for social influence analysis is provided in [575].

10.8 Exercises

1. Apply the *PageRank* algorithm to the graph of Figure 10.1(b), using teleportation probabilities of 0.1, 0.2, and 0.4, respectively. What is the impact on the dead-end component (probabilities) of increasing the teleportation probabilities?

2. Repeat the previous exercise, except that the restart is performed from node 1. How are steady-state probabilities affected by increasing the teleportation probability?

3. Show that the transition matrix of the graph of Figure 10.1(b) will have more than one eigenvector with an eigenvalue of 1. Why is the eigenvector with unit eigenvalue not unique in this case?

4. Implement the personalized *PageRank* approach for collaborative filtering on a implicit feedback matrix. Your implementation should automatically construct the user-item graph.

5. Implement the Jaccard and Adamic-Adar measures for link prediction.

6. Create a link prediction measure that can perform the degree normalizations performed both by the Jaccard measure and the Adamic-Adar measure.

7. Implement the linear threshold and independent cascade model for influence analysis.

8. Describe the adaptation of user-based neighborhood models in collaborative filtering for undirected link prediction. Does it make a difference whether one adapts user-based methods or item-based methods in undirected networks? How about directed networks?

9. Describe the adaptation of sparse linear models in Chapter 3 to directed link prediction.

Chapter 11

Social and Trust-Centric Recommender Systems

"Society is like a large piece of frozen water; and skating well is the great art of social life."– Letitia Elizabeth Landon

11.1 Introduction

With increasing access to social information about users, merchants can directly incorporate social context in collaborative filtering algorithms. Although some of these methods are discussed in Chapter 10, the focus of this chapter is primarily on recommending nodes and links in network settings. Social context is a much broader concept, not only including social (network) links, but also various types of side information, such as tags or folksonomies. Furthermore, the social context can also be understood in a network-agnostic way, as a special case of context-sensitive recommender systems (cf. Chapter 8). The social setting results in a number of human-centric factors, such as *trust*. When users are aware of the identity of the actors who participate in the feedback process, the trust factor plays an important role. Therefore, the material in this chapter is closely related to that in Chapter 10, but nevertheless it is distinct enough a merit a separate chapter in its own right. In particular, we will study the following aspects of social context in recommender systems:

1. *Social context as a special case of context-aware recommender systems:* Context-aware recommender systems are discussed in Chapter 8. An important framework for contextual recommendations is that of the multidimensional model [6]. One of the possible forms of context is the *social context* in which the social information is used as side information to improve the effectiveness of the recommendation process. For example, the choice of movie that a user might watch depends on the companion with whom she chooses

© Springer International Publishing Switzerland 2016
C.C. Aggarwal, *Recommender Systems: The Textbook*,
DOI 10.1007/978-3-319-29659-3_11

to watch the movie. In other words, she would often choose a different film depending on whether she was watching it with her friends, parents, or significant other. Such recommendations can be handled directly with the multidimensional model, without the need to use the structure of the social network in the recommendation process.

2. *Social context from a network-centric and trust-centric perspective:* In these cases, it is assumed that the merchant is aware of the social structure of the user. Users are often likely to ask their friends for suggestions about movies, restaurants, or other items. Therefore, the social structure of a user can be viewed as a *social trust network*, which is useful for the recommendation process. For example, if a user is friends with many people who have watched a particular movie, then she is more likely to watch that movie. Furthermore, if the user is densely connected with a community of users interested in the movie, this provides further evidence of user interest. Therefore, the network structure and interests in the vicinity of the user play a key role in the recommendation process.

   In some networks, such as Epinions.com [705], trust networks are created between users, which provide feedback on how much users can rely on each other's opinions in the recommendation process. The trust factor is particularly important because a user's personalized interests can be better predicted from the rating patterns of other users whom she has trusted in the past. It has now been conclusively shown in the literature that the incorporation of trust has a significant and positive effect on the recommendation process. Such methods are closely related to some of the network-centric methods discussed in Chapter 10. Here, we discuss these methods in further detail, especially in the context of trust-centric systems.

3. *User interaction perspective:* The user interactions with social networks create many forms of feedback, such as comments or tags. These tags can be viewed as *folksonomies*, which collaboratively annotate and classify content. Such folksonomies are very informative and can be used to improve the recommendation process. These methods are closely related to content-centric recommendations except that a combination of a collaborative and content-centric approach is used. This is particularly natural because sufficient data is available in such cases to leverage both collaborative and content-centric factors.

It is noteworthy that these methods apply to completely different recommendation settings and input data. Furthermore, the social information is used in a completely different way in each of these settings. Therefore, social recommender systems can be understood from many different perspectives, depending on whether social participants serve as the context, as the peer recommenders or as the providers of interaction data.

In this chapter, we will discuss all the aforementioned scenarios for social network recommendations. We will discuss the key settings in which each of these methods apply, and also the settings in which they work most effectively. We will also discuss how many of these techniques relate to the methods discussed in earlier chapters. The use of multidimensional context to address the social setting is closely related to the techniques introduced in Chapter 8. On the other hand, the use of network-centric methods is closely related to the techniques introduced in Chapter 10. The discussion in this chapter expands on these themes as they relate to the social context.

This chapter is organized as follows. In section 11.2, we will discuss the use of social context as a special case of social recommender systems. In other words, we will discuss the use of the multidimensional model [6] to address social context. Network-centric methods for social recommendations are discussed in section 11.3. The utilization of user interaction in social recommendations is discussed in section 11.4. A summary is given in section 11.5.

11.2 Multidimensional Models for Social Context

The multidimensional model of Chapter 8 is the simplest method for incorporating social information within the recommendation process. This approach has the merit that we can reuse traditional collaborative filtering models by using the reduction-based approach of Chapter 8. The use of ratings associated with a social context is one instance where this approach is applicable. Data about social context may either be directly collected or inferred from other information sources. Some typical modes of collecting data about social context are as follows:

1. *Explicit feedback:* While rating items, such as a movie, the system can be designed to capture various types of information, such as details about who the movie was watched with. Similarly, the destination of a tourist might depend on their travel companion. For example, a tourist is far more likely to travel to Disneyland than to Las Vegas when her children accompany her. The main challenge with this approach is that users are generally not very willing to spend too much effort in specifying such contextual details while providing ratings. Therefore, it becomes more difficult to collect sufficient data. Nevertheless, when it is possible to collect such data through explicit feedback, it is generally of high quality. Therefore, it should be considered the first choice where possible.

2. *Implicit feedback:* The social context of a user can often be inferred from where, when, and how the item was bought, or her other social activities. For example, if a tourist uses the same credit card to book a set of tickets for herself and her travel companions, this provides useful contextual information to the tour operator for future recommendations. In some cases, the collection of contextual data might require the use of machine learning techniques. With the increasing availability of mobile phones and the ability to perform online user activity analysis, it has become increasingly easy to collect such information in an automated way.

Let U be the set of users, I be the set of items, and C be the set of alternatives representing the social context. The ratings can then be viewed as a mapping g_R on a 3-dimensional ratings cube R. The domain of the mapping is defined by $U \times I \times C$, and the range corresponds to the values of the ratings. This mapping may be written as follows:

$$g_R : U \times I \times C \rightarrow rating$$

For example, consider a travel recommendation application in which the context is the travel companion. Figure 11.1 illustrates an example of a 3-dimensional ratings matrix with social context. Here, the items correspond to the tourist locations and the contexts correspond to the travel companions. Each entry in the cube corresponds to the rating of a specific travel location of a user in a particular context. It is noteworthy that this example is a simple adaptation of Figure 8.3 in Chapter 8 to fit the social context. It is also possible to have multiple social contexts. In such a case, the dimensionality of the underlying cube will increase accordingly and one can work with a w-dimensional cube of ratings. It is evident

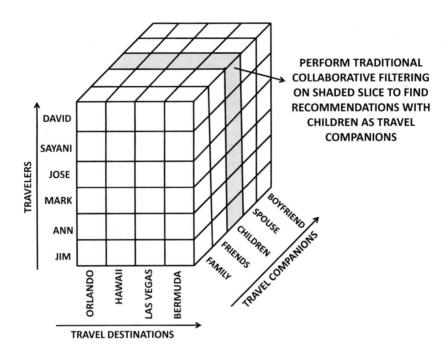

Figure 11.1: Travel recommendations with varying social context (Adaptation of Figure 8.3 in Chapter 8)

that the multidimensional model in the case of social context is not very different from that in the case of other types of context. Therefore, the algorithms discussed in Chapter 8 can be generalized to this scenario in a relatively painless way.

Queries can be posed in a way that is similar to multidimensional context by partitioning the dimensions into "what" dimensions and "for whom" dimensions. A typical query is of the following form:

Determine the top-k possibilities in the "what" dimensions for a particular set of specified values in the "for whom" dimensions.

In the aforementioned example, some possibilities for the various queries include:

1. Determine the top-k destinations for a particular user.

2. Determine the top-k destination-companion pairs for a particular user.

3. Determine the top-k destinations for a particular user-companion pair.

4. Determine the top-k companions for a particular user-destination pair.

The reduction approach of section 8.2 in Chapter 8 can be used to provide responses to such queries. Let $f_{R'} : U \times I \rightarrow rating$ be a conventional collaborative filtering algorithm on a 2-dimensional ratings matrix R'. Then, each of the aforementioned queries can be reduced to a standard collaborative filtering problem. For example, in order to determine the best destinations to visit with children, one can extract the corresponding 2-dimensional slice $R'(children)$ from the original 3-dimensional ratings matrix R. This slice is shaded in

Figure 11.1. Then, a standard collaborative filtering algorithm can be applied to this 2-dimensional matrix. In the event that there are multiple travel companions, this results in context set V. The data cube slices for each of the contexts in V can be extracted and the ratings can be averaged over the different contextual values for a particular user-item combination. This process is similar to that of Equation 8.2 in Chapter 8:

$$g_R(User, Item, V) = \text{AVERAGE}_{[y \in V]} \; g_R(User, Item, y) \qquad (11.1)$$

Therefore, the problem can again be reduced to the 2-dimensional case with the use of the averaged slice over the social context set V. A similar approach can be used for the problem of determining the top-k destinations of a particular user without any specific context in mind. In such a case, the ratings can be averaged over all the different contexts rather than a specific social context set V. This approach is referred to as prefiltering. However, other methods, such as postfiltering, latent factor models, or other machine learning models, are also discussed in Chapter 11. All these methods can be easily generalized for providing recommendations in these settings.

11.3 Network-Centric and Trust-Centric Methods

The basic idea in network-centric methods is that the friendship structure of a user has a profound influence on her tastes, choices, or consumption patterns. Users often ask their friends for suggestions regarding movies, travel, or other items. Furthermore, social connections exhibit the well-known principle of *homophily*, in which connected users often have similar interests and tastes. This similarity in tastes often leads a user to trust the recommendations from their connected users more than others. Numerous methods can be designed to incorporate such links into the recommendation process. Although such links may have varying effectiveness depending on the application domain, they are usually particularly helpful in cases of cold-start, when there is little information available about a particular user's ratings. In such cases, the knowledge embedded in the user's social connections can be particularly helpful in identifying her most relevant peers. In the following, a number of key methods will be discussed for incorporating social knowledge into the recommendation process. First, we will discuss the two important concepts of trust and homophily, which are related but not quite the same.

11.3.1 Collecting Data for Building Trust Networks

Trust and homophily both play an important role in the social recommendation process. These concepts are related but they are not quite the same. *Homophily* refers to the fact that linked users in social networks are likely to be similar to one another in terms of their tastes and interests. *Trust* refers to the fact that users are more likely to trust the tastes and recommendations of their friends. In some cases, trust is the consequence of homophily. As linked users tend to be similar to one another, they tend to trust each other's tastes and recommendations. The strong correlation between trust and homophily has been shown in [224, 681].

Either homophily or trust or both may be relevant in a given network. In some social networks, such as Facebook, both homophily and trust are relevant because links typically represent friendship relationships. In fact, trust relationships can often be computationally inferred from such Web-based social networks [226]. Many characteristics, such as feature similarity and email exchanges can be used to infer the trust links. For example, one might use the following user-to-user similarity [588] to determine the trust t_{ij} between user i and j.

$$t_{ij} = \begin{cases} \text{Cosine}(i, j) & \text{if } i \text{ and } j \text{ are connected} \\ \text{Undefined} & \text{otherwise} \end{cases} \tag{11.2}$$

The cosine similarity is calculated on the ratings of users i and j. It is noteworthy that when i and j are not connected, the trust between them is undefined. As we will see later, this undefined value can also be inferred with the use of trust propagation methods. Therefore, such methods infer the trust values between connected users in a different way from how they infer trust values between unconnected users.

The aforementioned methodology can be viewed as an implicit way of inferring the trust. In some networks such as Epinions [705], the trust links are explicitly specified by the users. Some examples of such networks are as follows:

1. In Golbeck's Filmtrust system [225], users are asked to evaluate their trust in their acquaintances' ratings in addition to providing ratings. This data is then used to make recommendations.

2. In the Epinions site [705], users are explicitly asked to specify the other users that they trust or distrust.

3. In the Moleskiing site [461], the inter-user trust information is obtained via explicit feedback. Users are allowed to rate how useful they found the comments by other users. This can help in inferring trust links between users. When a user frequently expresses a positive opinion about the comments of another user, a directed edge can be added from the former to the latter. A modeling approach may be used to relate this frequency to an explicit trust value. An example of such a modeling approach is provided in [591], although this work is focused on distrust relations rather than trust relations. The ability to leave feedback about reviews is also available on sites such as Amazon.com.

4. Trust and distrust relationships are also available in the Slashdot network [706], which is a technology blog. In this case, the relevant trust relationships are directly specified by the users.

In all cases, whether the trust relationships are implicitly inferred or explicitly specified by the users, a *trust network* can be created. This trust network is also referred to as the *Web of trust*. For the purpose of this chapter, we will assume that trust is specified as an $m \times m$ user-user matrix $T = [t_{ij}]$, in which each value of t_{ij} is drawn from the range $(0, 1)$. Large values of t_{ij} indicate user i trusts user j to a greater degree. The case where $t_{ij} \in (0, 1)$ represents the *probabilistic* model of trust representation. This representation provides a way of modeling trust but not distrust. In general, the value of t_{ij} may not be the same as that of t_{ji}, although some implicit inference models may make this assumption.

In some cases, distrust relationships are also available. For example, Epinions provides users the option to specify *block lists* corresponding to distrusted users. Ideally, distrust relations should be negative values and one can extend the model to use values in $[-1, +1]$. However, it is often challenging to generalize inference algorithms to trust networks with both trust and distrust relationships. Most of the work in the literature has focused only on using the trust relationships while ignoring the distrust relations. Therefore, most of the discussion in this section will be based on positive trust relations between nodes. The bibliographic notes contain more details about methods that use distrust relations.

Trust-aware recommender systems are able to use the knowledge in the Web of trust in order to make personalized and accurate recommendations. Such recommender systems are also referred to as *trust-enhanced recommender systems*. Many of these methods use specialized operators, referred to as *trust aggregation* and *trust propagation*. These are mechanisms to estimate the unknown trust level between two users with the use of *transitivity* in the trust network. In other words, once it is known how much A trusts B and how much B trusts C, it can also be estimated how much A trusts C. *Trust metrics* estimate how much one user should trust another based on existing trust relationships in the network [682].

Trust networks are directed, especially when they are explicitly specified by users. This is because trust relations are asymmetric. The level of trust of A for B may be different from that of B for A. Most trust-based algorithms take the edge directions into account during the computation. However, in some cases, the simplifying assumption of undirected networks is made, especially when the trust relations are implicitly inferred from Web-based social networks. For example, the trust relation of Equation 11.2 is symmetric.

11.3.2 Trust Propagation and Aggregation

Trust propagation and aggregation play an important role in the design of social recommender systems. These operators are motivated by the fact that trust networks are sparsely specified, in which all pairs of users do not necessarily have trust relationships between them. Therefore, the transitivity in trust relations needs to be used to infer the missing trust relations with the use of operators like propagation and aggregation.

What does transitivity mean? For example, if Alice trusts John and John trusts Bob, then one can infer the fact that Alice might trust Bob. This fact is, in turn, useful for Alice to make recommendations based on the items that Bob might have liked. In other words, one needs to determine *paths* in the trust network in order to make such inferences. The determination of the unknown value of the trust between two nodes at the end points of a path is referred to as trust propagation. However, there are typically multiple paths in a trust network between a pair of users. For example, in the simple trust network of Figure 11.2, the trust values on the edges are assumed to be drawn from $(0, 1)$. The value on the directed edge from any user A to any other user B indicates how much A trusts B. There are two paths between Alice and Bob, and the (propagated) trust values between Alice and Bob need to be aggregated over these two paths. When quantifying Alice's trust in Bob, Alice is the *source* and Bob is the *sink*. The trust propagation and trust aggregation operators are computed as follows:

1. *Trust propagation along a single path:* A multiplicative approach is commonly used for trust propagation [241, 509]. In this case, the trust values on the edges are multiplied in order to obtain the trust between the two endpoints. For example, consider the path Alice $\rightarrow$ John $\rightarrow$ Bob in Figure 11.2. In this case, multiplying the trust values on the path yields a propagated trust value of $0.7 \times 0.6 = 0.42$. Similarly, for the path Alice $\rightarrow$ Mary $\rightarrow$ Tim $\rightarrow$ Bob, the multiplicatively propagated trust value is $0.3 \times 0.4 \times 1 = 0.12$. Many methods also use trust decay to deemphasize long paths, or simply use shortest paths. For example, a user-defined decay factor $\beta < 1$ is used to multiply the computed trust value with β^q, where q is the path length of the propagation. In Figure 11.2, the decay approach would multiply the propagated results with β^2 for the upper path and β^3 for the lower path. The resulting computed values

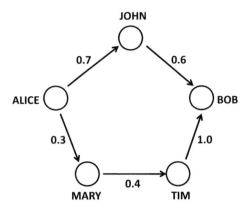

Figure 11.2: A simple trust network

are $0.42 \times \beta^2$ and $0.12 \times \beta^3$, respectively. More sophisticated methods for incorporating decay include the *Appleseed* algorithm [682], in which a spreading activation model is used.

These types of multiplicative propagation algorithms are only designed for non-negative trust values in the range $(0, 1)$. Distrust relations pose significant challenges because a sequence of two distrusts does not always imply a trust relationship [241, 590, 591]. Therefore, a multiplicative approach cannot be directly used in case of negative trust values. Refer to the bibliographic notes for pointers on propagation methods that are designed for distrust relations.

2. *Trust aggregation across multiple paths:* In trust aggregation, the propagated values over various paths are aggregated into a single value. Common aggregation operators include the use of the minimum, maximum, average, weighted average, or weighted sum. In weighted averages, some propagation paths are considered more important than others. For example, shorter paths or recommendations from closer friends might be considered more important. Such weighting can also be handled within the trust propagation operator with the use of decay function.

Consider the example of Figure 11.2. The use of the average operator in the afore-mentioned example leads to an estimated trust value between Alice and Bob of $(0.42 + 0.12)/2 = 0.27$, whereas the summation operator leads to an estimated value of $(0.42 + 0.12) = 0.54$. The bibliographic notes contain pointers related to various trust aggregation methods.

Trust propagation and aggregation are unsupervised methods for performing recommendations in trust-centric systems because they use fixed heuristics, irrespective of the underlying data. Supervised methods use low rank representations, such as matrix factorization, to learn these dependencies. In a later section, we will also discuss supervised methods, in which the algorithms learns the importance of different paths. It is noteworthy that some of the decay-based propagation algorithms with summation-based aggregation are very similar to the unsupervised Katz measure used in link prediction. The use of the Katz measure for link prediction is discussed in Chapter 10. As we will see later in section 11.3.7, the problem of trust-aware recommendations can be directly transformed to an instance of the link prediction problem.

11.3.3 Simple Recommender with No Trust Propagation

Consider a scenario in which a trust network is available, but only directly observed trust values (e.g., by user feedback in Epinions) are used. Furthermore, propagation and aggregation are not used to infer trust values between indirectly connected users. In other words, if user i has not *directly* provided feedback about user j, then no trust value between i and j is available. We have an $m \times n$ ratings matrix $R = [r_{ij}]$ for m users and n items, and an $m \times m$ trust matrix $T = [t_{ij}]$ representing the trust relationships. In other words, t_{ij} represents the degree to which user i trusts user j.

A simple approach to predict the rating $\hat{r}_{ij}$ of user i for item j is to define the peer group of user i as all users $N(i, \theta)$ who have rated the item j and are trusted by user i above a given threshold θ. Then, we can use the formula that is commonly used in neighborhood-based methods:

$$\hat{r}_{ij} = \frac{\sum_{k \in N(i,\theta)} t_{ik} r_{kj}}{\sum_{k \in N(i,\theta)} t_{ik}} \tag{11.3}$$

This approach can be viewed as a user-based version of neighborhood methods, in which trust values are used instead of the Pearson correlation coefficient. The formula is also referred to as the *trust weighted mean*. An alternative is to use the mean rating μ_k of each user k for centering the ratings, as in traditional collaborative filtering:

$$\hat{r}_{ij} = \mu_i + \frac{\sum_{k \in N(i,\theta)} t_{ik} (r_{kj} - \mu_k)}{\sum_{k \in N(i,\theta)} t_{ik}} \tag{11.4}$$

This approach can lead to predictions that do not lie within the specified rating scale. In such cases, one can adjust the rating to the nearest rating within the specified scale.

11.3.4 TidalTrust Algorithm

The *TidalTrust* algorithm is based on the observation that shorter paths are more reliable for propagation. Therefore, one should use the shortest path between a source-sink pair for the trust computation. For the purpose of further discussion, assume that the trust needs to be computed from source i to sink j. The algorithm derives its name from the fact that it first has a forward phase in which nodes are explored from source i to sink j in breadth-first order in order to discover all the shortest paths from i to j and also set a trust threshold $\beta(i, j)$. Then, the algorithm uses a backward phase in which recursive trust computations are made in reverse order from that in which nodes were explored in the forward phase (i.e., from sink to source). Only edges lying on the shortest paths (discovered in the forward phase) with trust at least equal to $\beta(i, j)$ are used in the backward phase. Therefore, the algorithm may be summarized as follows:

1. *Forward phase:* The goal of the forward phase is to determine a minimum threshold $\beta(i, j)$ on the trust values for them to be considered relevant in the trust computation between source i and sink j. The approach for computing $\beta(i, j)$ will be discussed later. Furthermore, all shortest paths from source to sink are determined during this phase with breadth-first search. Note that the subgraph $\mathcal{G}(i, j)$ of (all) shortest paths from source i to sink j is always a directed acyclic graph with no cycles. The *children* $C(q)$ of each node q are defined as all nodes to which node q points in this directed acyclic graph $\mathcal{G}(i, j)$ of shortest paths. Only the edges in this subgraph are relevant for the backward phase. The forward phase will be described in more detail later.

2. *Backward phase:* In the backward phase, starting from the sink node j, nodes are processed in reverse order of their distance to source node s using the edges of $\mathcal{G}(i,j)$. In other words, nodes closest to the sink are processed first. Let the currently processed node be denoted by q. If the edge (q,j) is already present in the trust network, then we can trivially set the predicted trust value $\hat{t}_{qj}$ to the observed trust value t_{qj}. Otherwise, if the edge (q,j) is not present in the trust network, then the predicted trust value $\hat{t}_{qj}$ between user node q and sink node j is recursively computed using only the edges in $\mathcal{G}(i,j)$ with observed trust values of at least $\beta(i,j)$:

$$\hat{t}_{qj} = \frac{\sum_{k \in C(q), t_{qk} \geq \beta(i,j)} t_{qk} \hat{t}_{kj}}{\sum_{k \in C(q), t_{qk} \geq \beta(i,j)} t_{qk}} \tag{11.5}$$

It is noteworthy that computing the trust value $\hat{t}_{qj}$ according to Equation 11.5 always requires the computed values $\hat{t}_{kj}$ for all children $k \in C(q)$. The value of $\hat{t}_{kj}$ is always available at the time of computing the trust value $\hat{t}_{qj}$ because k is a child of q, and all computations are performed in the backward direction. Even though the approach computes many intermediate values $\hat{t}_{kj}$, the computed source-sink value $\hat{t}_{ij}$ is the only relevant one for a particular source-sink pair (i,j) and the other intermediate values are discarded. The approach, therefore, needs to be repeated over various source-sink pairs.

It remains to describe the forward phase in more detail. In the forward phase, a modified version of breadth-first search is used starting from node i in order to compute the directed acyclic graph of shortest paths $\mathcal{G}(i,j)$. Standard breadth-first search only discovers the first shortest path between i and j (depending on node exploration order), whereas we would like to find all of them. The main difference from standard breadth-first search is that previously visited neighbors of a node are also checked to see if they might be children of a given node. The source i is labeled with a distance value $d(i)$ of 0. All other distances are labeled as ∞. All outgoing neighbors of i are then labeled with a distance value of 1 and added to a list L. In each iteration, the node q with the smallest distance label $d(q)$ from L is selected. The labels of each its neighbors k on outgoing edges are modified as follows:

$$d(k) = \min\{d(k), d(q) + 1\} \tag{11.6}$$

The node k is added to the children $C(q)$ of q if and only if $d(k) = d(q) + 1$ after the update. The node q is deleted from the list L after updating the labels of all its neighbors (including those that were visited earlier). The algorithm terminates when the node with the smallest distance label in L is the sink j. At this point, all nodes in the graph with distance labels greater than or equal to that of the sink node j are deleted from the network. Furthermore, any edge (q,k) not satisfying the condition $d(k) = d(q) + 1$ is deleted. The remaining subgraph $\mathcal{G}(i,j)$ contains all the shortest paths from node i to node j. For example, the shortest path subgraph $\mathcal{G}(i,j)$ for the trust network in Figure 11.3(a) is shown in Figure 11.3(b). Note that node 6 is missing in Figure 11.3(b) because it is irrelevant to any of the paths between the source node 1 and sink node 8. Several edges have also been dropped from the original graph because they do not lie on any shortest path. For each source to sink path in $\mathcal{G}(i,j)$, the minimum weight edge is determined. The value of $\beta(i,j)$ is set to the maximum of these various minima. A dynamic programming approach can also be used to efficiently compute $\beta(i,j)$ by keeping track of intermediate values of $\beta(i,k)$ during the forward phase. We initialize $\beta(i,i) = \infty$ and $\beta(i,k) = 0$ for each $k \neq i$. Whenever, the label

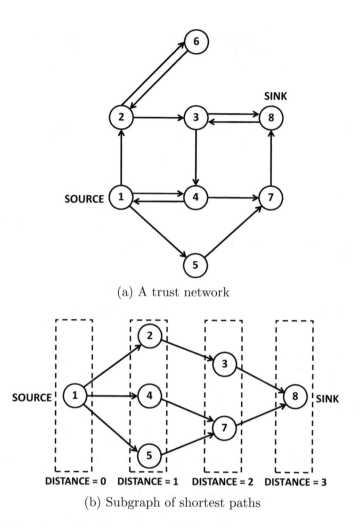

(a) A trust network

(b) Subgraph of shortest paths

Figure 11.3: The subgraph of shortest paths found by *TidalTrust* for a trust network

of node k strictly reduces because of incoming edge (q, k) (based on Equation 11.6), the following update is also executed:

$$\beta(i, k) = \max\{\beta(i, k), \min\{t_{qk}, \beta(i, q)\}\} \tag{11.7}$$

As a result, the end of the forward phase also yields the value of $\beta(i, j)$.

So far, we have only discussed user-to-user trust computation in *TidalTrust*. How can this computation help in recommendation of *items*? The final rating of an item is computed using the trust-weighted mean, in a manner similar to Equation 11.3. The main difference is that predicted trust values $\hat{t}_{ik}$ can also be used on the right-hand side of Equation 11.3, rather than only the observed trust values of the neighbors of node i. Let I_i be the indices of the items rated by users i. Therefore, Equation 11.3 is modified as follows:

$$\hat{r}_{ij} = \frac{\sum_{k:k \in I_i \hat{t}_{ik} \geq \theta} \hat{t}_{ik} r_{kj}}{\sum_{k:k \in I_i \hat{t}_{ik} \geq \theta} \hat{t}_{ik}} \tag{11.8}$$

As before, θ is a user-defined threshold on the inferred trust value, for it to be used in the computation. These methods have a particularly beneficial effect on recommendations to controversial users whose ratings on items differ significantly from other users [223].

11.3.5   MoleTrust Algorithm

The *MoleTrust* algorithm shares a number of conceptual similarities with the *TidalTrust* algorithm, but is quite different in terms of how it is implemented. The *TidalTrust* algorithm uses a forward phase followed by a backward phase for each source-sink pair, whereas the *MoleTrust* algorithm uses two forward phases for each source node. Note that a single application of the forward and backward phase in the *TidalTrust* algorithm is able to compute the trust from a particular source to a particular sink, whereas the *MoleTrust* algorithm is able to compute the trust from the source i to *all* other nodes within a maximum distance threshold in two forward phases. As a sink is not specified in *MoleTrust*, a different criterion (in terms of maximum path length δ) is used to terminate the shortest path computation. Furthermore, a user-defined trust threshold α is used across all source-sink pairs, rather than one that is computed for each source-sink pair. Therefore, the two phases are as follows:

1. *Forward phase 1:* Determine all shortest paths starting from source node i with length at most δ. As in *TidalTrust*, the modified breadth-first approach is used, except that the termination criterion is based on maximum path length rather than on reaching the sink node. We determine the directed acyclic graph $\mathcal{G}(i, \delta)$ in which all edges lie on one of these shortest paths. The *predecessors* $P(q)$ of each node are the nodes that point to q in the graph $\mathcal{G}(i, \delta)$. Note that the notion of predecessor in *MoleTrust* is exactly the converse of the notion of children in *TidalTrust*.

2. *Forward phase 2:* The algorithm starts by setting $\hat{t}_{ik} = t_{ik}$ for all nodes k such that the edge (i, k) is present in the graph $\mathcal{G}(i, \delta)$. These represent nodes at distance 1 from source node i. Then, the trust value between the source and the nodes at higher distances are computed. For any node q at a distance 2 or more from the source node i in $\mathcal{G}(i, \delta)$, the trust $\hat{t}_{iq}$ is computed as follows:

$$\hat{t}_{iq} = \frac{\sum_{k \in P(q), t_{kq} \geq \alpha} \hat{t}_{ik} \cdot t_{kq}}{\sum_{k \in P(q), t_{kq} \geq \alpha} t_{kq}} \qquad (11.9)$$

Note the similarity with the *TidalTrust* computation. The main difference is that this computation is in the forward direction and the threshold α is user-defined. and it is invariant across all source-sink pairs. Unlike *TidalTrust*, in which a *source-sink specific* threshold $\beta(\cdot, \cdot)$ (computed during the forward phase) is used, the threshold α of the *MoleTrust* algorithm is invariant across all source-sink pairs.

The final approach for item recommendation is similar to that of *TidalTrust*. After all the trust values have been computed, one can use Equation 11.8 to make ratings predictions.

The directed acyclic subgraph for a maximum horizon of length 2, for the graph of Figure 11.3(a) is shown in Figure 11.4. As in the case of Figure 11.3, node 1 is used as the source node. Note that unlike Figure 11.3(b), node 6 is present in Figure 11.4, but node 8 is absent. In the *TidalTrust* algorithm, the trust values of nodes that are outside the distance horizon of the source node cannot be the computed. Therefore, $\hat{t}_{18}$ cannot be computed by *MoleTrust*. The assumption is that the trust computation $\hat{t}_{18}$ is too unreliable to be used in the recommendation process. Therefore, such trust values are implicitly set to 0. *MoleTrust* is more efficient than *TidalTrust* because it needs only be applied once for each source node, rather than for each source-sink pair.

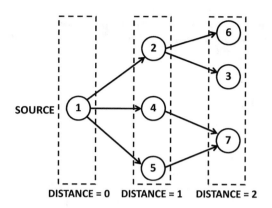

Figure 11.4: The subgraph of shortest paths found by *MoleTrust* at a maximum horizon of 2 for the trust network in Figure 11.3(a)

11.3.6 TrustWalker Algorithm

The TrustWalker algorithm [269] is based on the observation that social network links provide an independent source of information from ratings [172]. Therefore, a random-walk approach is used to discover the similar users. However, a major dilemma is that if one goes too far in the random walk, then irrelevant users might be used. An important observation in this context is that the ratings of strongly trusted friends on similar items are better predictors than the ratings of weakly trusted friends on the same item. Therefore, the *TrustWalker* approach combines trust-based user similarity and item-based collaborative filtering models in a unified random-walk framework.

The *TrustWalker* algorithm uses a random-walk approach on the social network of users. The algorithm starts with the source user i in order to determine the rating $\hat{r}_{ij}$ for item j. At each step of the random walk, it checks whether the visited user k in the random walk has rated item j. If this is indeed the case, then the observed rating r_{kj} is returned. Otherwise, the algorithm has two choices, which can be viewed as a modified version of the restart method in random walks:

1. At step l of the random walk, the algorithm can terminate at node k with probability ϕ_{kjl}. In such a case, the rating of the user k on a random item similar to j is returned. Among all items rated by user k, this random item is chosen with a probability proportional to its item-item similarity with target item j. Note that the returned rating can be viewed as a randomized and trust-based version of item-based collaborative filtering algorithms.

2. With probability $(1 - \phi_{kjl})$, the random walk is continued to the neighbors of k.

The random walk is repeated multiple times, and the ratings are averaged in a probabilistic way over the various walks. This weighting is based on the probability of termination at various random walks and the probability of selecting the specific item used for making the prediction. Refer to [269] for details.

It is noteworthy that the restart probability ϕ_{kjl} is dependent on the currently visited user k, item j, and number of steps l. The intuition for deciding this value is as follows. The value of the termination probability ϕ_{kjl} increases with the number of steps l to avoid the use of weakly trusted users that are far away from the source user. This is consistent with all trust-based algorithms that avoid using longer paths for trust propagation. Furthermore, the probability of termination should also be high, if we are confident that similar items rated by k will provide a reliable prediction. This is achieved by increasing the probability of termination when the similarity value of target item j to the closest item rated by k is high. Let this maximum similarity value be $\Delta_{kj} \in (0, 1)$. Therefore, the overall probability of termination is set as follows:

$$\phi_{kjl} = \frac{\Delta_{kj}}{1 + \exp(-l/2)} \tag{11.10}$$

The aforementioned computation requires the determination of item-item similarity. To compute the similarities between two items, a discounted version of the Pearson correlation coefficient is used. First, only items with positive correlation are considered. Second, the discount factor is set in such a way that the similarity value is reduced when the number of common users rating the items is small. Therefore, for two items j and s with N_{js} raters in common, we have:

$$\text{Sim}(j, s) = \frac{\text{Pearson}(j, s)}{1 + \exp(-N_{js}/2)} \tag{11.11}$$

Therefore, the *TrustWalker* algorithm is able to combine the notions of user trust and item-item similarity in a seamless way, within a single random-walk framework.

11.3.7   Link Prediction Methods

Most of the aforementioned methods are designed to work with trust propagation and aggregation *heuristics*. The effectiveness of a particular heuristic might depend on the data set at hand. This is because such methods are *unsupervised*, and they do not always adapt well to the particular structure of the network at hand. A natural question arises whether one can *directly* learn the relevance of different parts of the trust network in a data-driven manner while performing the propagation and aggregation. Link prediction methods are useful if only a ranked list of recommended items is required rather than the prediction of exact values of the ratings. This caveat is primarily because most link prediction methods are good at recommending ranked lists of edges, but do not work very well for predicting weights on the edges exactly.

As discussed in section 10.4.6 of Chapter 10, traditional collaborative filtering problems can be posed as link prediction problems on user-item graphs. Refer to sections 10.2.3.3 and 10.4.6 for a detailed discussion of how user-item graphs can be used for traditional collaborative filtering. A detailed discussion of the process of user-item graph construction is also provided in section 2.7 of Chapter 2. In this case, the user-item graphs need to be augmented with social links corresponding to the links between various users. The augmentation of user-item graphs with social links allows the use of social information in the collaborative filtering process.

Consider an $m \times n$ ratings matrix with m users and n items. It is assumed that the users are arranged in the form of a social network $G_s = (N_u, A_u)$. Here, N_u denotes the set of nodes representing users, and A_u denotes the set of social links between the users. A one-to-one correspondence exists between users and nodes in N_u. Since the number of users is m, we have $|N_u| = m$. An example of a toy social network between a set of users is illustrated in Figure 11.5(a).

The user-item graph can be viewed as an augmentation of the social network graph with item nodes. Let N_i be the set of nodes representing items. As in the case of user nodes, a

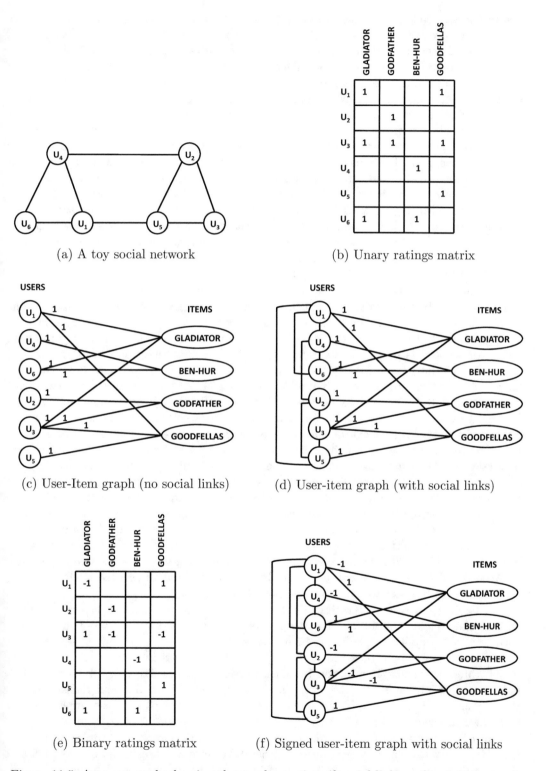

Figure 11.5: A toy example showing the amalgamation of social links and user-item graphs

one-to-one correspondence exists between items and nodes in N_i. Since we have n items, we have $|N_i| = n$. We construct the graph $G = (N_u \cup N_i, A_u \cup A)$. Here, A is a set of edges that exist between user nodes in N_u and item nodes in N_i. Note that the nodes and edges of this graph are supersets of those in the original social network G_s. The edges in A correspond to the relationships in the user-item graph (cf. section 2.7 of Chapter 2). Specifically, an edge exists between a user node in N_u and an item node in N_i, if the user has rated that item. The weight on that edge is equal to the mean-centered rating of the user for that item. This will often result in negative weights on the edges. In the case of implicit feedback data sets, the feedback is not mean-centered but the corresponding weight (e.g., a 0-1 value or the number of items bought) is used. The reason for mean-centering in the former case is that ratings are supposed to indicate both likes and dislikes, whereas implicit feedback provides a form of unary rating with no explicit mechanism to specify a dislike. In the case of implicit feedback, the resulting network is a conventional network with only non-negative weights on the links. In the case of explicit feedback, the resulting network is a *signed* network with positive and negative edge weights. It is noteworthy that the resulting network can be viewed as a union of the nodes and edges in the original social network, and the user-item graphs discussed in section 2.7 of Chapter 2.

In order to illustrate this point, we show an example of a unary ratings matrix in Figure 11.5(b). This matrix is the same as that shown in Figure 10.11(a) of Chapter 10. The corresponding user-item graph (without social connections) is illustrated in Figure 11.5(c). This graph is identical to that in Figure 10.11(b) of Chapter 10. The user-item graph with social connections is illustrated in Figure 11.5(d). Note that the graph in Figure 11.5(d) is a union of the graphs in Figures 11.5(a) and (c). Furthermore, social links can also have weights depending on the strength of the social ties or the level of trust between the corresponding social actors. As discussed in section 10.4.6, link prediction methods can be used to determine user affinity for items. Most link prediction methods also return a quantification of the strength of the predicted link. The strengths of predicted links of users for items can be ranked in order to create a ranked item list for the user. Link prediction methods are discussed in section 10.4 of Chapter 10. In the case of implicit ratings, conventional link prediction methods may be used because all link weights are non-negative. The only difference from the approach in section 10.4.6 is that the user-item graphs are enhanced with social connections. One challenging issue with the use of this approach is that the social links and the user-item links may not be equally important for the particular application at hand. In order to address this issue, the weights of all the social links are multiplied with the parameter λ. The value of λ regulates the relative importance between social (trust) links and user-item links. The optimal value of λ is chosen using cross-validation in order to maximize the prediction accuracy.

For explicit feedback, the ratings need to be mean-centered, which will result in edges with signed weights. In the special case of binary ratings, the values of -1 and $+1$ are used to retain simplicity. A value of $+1$ indicates a "like," whereas a value of -1 indicates a "dislike." An example of a binary ratings matrix is shown in Figure 11.5(e), and its socially augmented user-item graph is shown in Figure 11.5(f). For such problems, *signed* link prediction methods [346, 591] can be used to predict both like and dislike ratings. Furthermore, it is also possible to include distrust relationships in these predictions by using negative social links.

One of the nice aspects of link prediction methods is that they do not require the explicit use of trust propagation and aggregation heuristics because the transitivity of user trust and corresponding preferences is already learned in a data-driven manner with the use of machine learning algorithms. In fact, one can even use link prediction methods to infer the trust values between pairs of users in the social network, rather than directly inferring user-item affinities. In other words, the machine learning techniques of link prediction can automatically propagate and aggregate trust in a data-driven manner. It is particularly helpful to use *supervised* methods (cf. section 10.4.4 of Chapter 10) for link prediction because such methods can learn the importance of the trust network in a data-driven manner. In fact, many of the trust propagation methods can be viewed as *unsupervised* heuristics, whereas link prediction provides a route for incorporating supervision into the computation. Indeed, unsupervised measures for link prediction, such as the Katz measure (cf. section 10.4.2 of Chapter 10), are very similar in principle to some of the decay-based trust propagation heuristics. It is well known [355] that supervised methods for link prediction generally outperform unsupervised methods.

Many link prediction methods are designed for undirected networks. While we have assumed an undirected trust network in the aforementioned example for simplicity, it could very easily have been used in a directed manner. In the aforementioned user-item graph model, the user-user links can be assumed to be asymmetric and directed according to the specified trust relationships, whereas the user-item links are always directed from users to items. Therefore, directed paths from users to items imply trust-based affinities from users to items. Either supervised methods or matrix factorization methods [432] can be used to perform directed link prediction. Therefore, link prediction methods provide a very general framework that can be used in a variety of scenarios. The bibliographic notes contain pointers to some recent methods that have use link prediction for recommendations.

11.3.8 Matrix Factorization Methods

Matrix factorization methods are closely related to link prediction [432]. Although one can use matrix factorization methods within the link prediction framework of the previous section by using the approach in [432] as the base algorithm, it is more fruitful to design and optimize matrix factorization methods directly for trust networks.

Let R be an $m \times n$ ratings matrix with m users and n items. Let us assume that the social trust matrix is given by an $m \times m$ matrix $T = [t_{ip}]$. Note that both R and T are incomplete matrices that are highly sparse. Let S_R and S_T be the observed indices in these matrices:

$$S_R = \{(i,j) : r_{ij} \text{ is observed}\}$$
$$S_T = \{(i,p) : t_{ip} \text{ is observed}\}$$

In cases where all observed values of t_{ip} are strictly positive, it is helpful to set a sample of the unobserved values of t_{ip} to 0, and include the corresponding indices within S_T. Such an approach can help in avoiding overfitting because it compensates for the lack of negative feedback (cf. section 3.6.6.2 of Chapter 3).

We start by introducing the *SoRec* algorithm. The *SoRec* algorithm [381] can be viewed as an extension of the matrix factorization methods in Chapter 3 to include social information. We emphasize that the presentation here is a simplified version of the *SoRec* algorithm, which is originally presented as a probabilistic factorization algorithm. The simplified presentation helps in understanding the key ideas behind the algorithm, by abstracting out the less important but complex details. Readers are referred to [381] for the exact description.

In Chapter 3, a matrix factorization model of rank-k is proposed to create an $m \times k$ user-factor matrix $U = [u_{ij}]$ and an $n \times k$ item-factor matrix $V = [v_{ij}]$, so that the following condition is satisfied as closely as possible over the observed entries:

$$R \approx UV^T \tag{11.12}$$

In order to incorporate social information, we introduce a second $m \times k$ user factor matrix $Z = [z_{ij}]$, so that the following condition is satisfied as closely as possible over the observed trust values:

$$T \approx UZ^T \tag{11.13}$$

Two user-factor matrices are employed here because matrix U is for the initiator and matrix Z is for the receiver. Furthermore, as T might not be symmetric, U and Z need not be the same. Intuitively, the initiator is the participant who decides whether or trust or not to trust (i.e., source), and the receiver is the participant who is the recipient of this trust/distrust (i.e., sink). Note that the user matrix U, which is the initiator, is shared in both factorizations. The initiator is shared rather than the receiver because the trust opinions of sources for sinks are used to predict ratings in such systems. It is this sharing of U that results in a factorization with the incorporation of the social trust information. Therefore, a joint factorization objective function is set up in which the errors in factorizing each of R and T are added. How much should the errors in each of the two factorizations be weighted? This is achieved with the use of a balance parameter β. Then, the overall objective function may be stated as follows:

$$\text{Minimize } J = \underbrace{||R - UV^T||^2}_{\text{Observed entries in } R} + \underbrace{\beta \cdot ||T - UZ^T||^2}_{\text{Observed entries in } T} + \underbrace{\lambda \left(||U||^2 + ||V||^2 + ||Z||^2 \right)}_{\text{Regularizer}}$$

The parameter λ controls the level of regularization. Note that this objective function is computed only over the observed entries and the unspecified entries are ignored in the computation of the Frobenius norm. This is consistent with the approach used in Chapter 3. The resulting objective function is, therefore, a straightforward extension of the matrix factorization methods in Chapter 3 with an additive term for the social contribution. We rewrite the objective function in terms of observed entries in S_R and S_T:

$$\text{Min. } J = \underbrace{\sum_{(i,j) \in S_R} \left(r_{ij} - \sum_{s=1}^{k} u_{is} v_{js} \right)^2}_{\text{Observed entries in } R} + \beta \underbrace{\sum_{(i,p) \in S_T} \left(t_{ip} - \sum_{s=1}^{k} u_{is} z_{ps} \right)^2}_{\text{Observed entries in } T} + \underbrace{\lambda \left(||U||^2 + ||V||^2 + ||Z||^2 \right)}_{\text{Regularizer}}$$

A gradient-descent approach can be used to determine the factor matrices U, V, and Z. A gradient vector of J with respect to all parameters in U, V, and Z is used to update the current vector of parameters representing all entries in U, V, and Z. The gradient descent steps depend on the errors $e_{ij}^{(r)}$ and $e_{ip}^{(t)}$ in the matrices between the observed and predicted values in the two matrices:

$$e_{ij}^{(r)} = r_{ij} - \hat{r}_{ij} = r_{ij} - \sum_{s=1}^{k} u_{is} v_{js}$$

$$e_{ip}^{(t)} = t_{ip} - \hat{t}_{ip} = t_{ip} - \sum_{s=1}^{k} u_{is} z_{ps}$$

The error matrix on the ratings can be written as $E_r = [e_{ij}^{(r)}]$ in which unobserved entries (i.e., entries not in S_R) are set to 0. The error matrix on the trust entries can be written as $E_t = [e_{ij}^{(t)}]$ in which unobserved entries (i.e., entries not in S_T) are set to 0. Then, the gradient-descent steps can be written in the form of matrix updates as follows:

$$U \Leftarrow U(1 - \alpha \cdot \lambda) + \alpha E_r V + \alpha \cdot \beta E_t Z$$
$$V \Leftarrow V(1 - \alpha \cdot \lambda) + \alpha E_r^T U$$
$$Z \Leftarrow Z(1 - \alpha \cdot \lambda) + \alpha \cdot \beta E_t^T U$$

Here, $\alpha > 0$ represents the step-size. The details of the derivation of the gradient-descent method are left as an exercise for the reader. Note that only the observed entries of E_r and E_t need to be computed in each iteration, and it makes sense to use a sparse data structure to represent these matrices because unobserved entries are set to 0. Although we have used a single regularization parameter λ and update step-size α for all updates, it often makes sense to use different regularization parameters and step-sizes for the different matrices U, V, and Z.

Next, we describe the *stochastic* gradient descent method in which the error is approximated in a randomized way with that over a single entry. This entry is selected in random order and might belong to either the ratings matrix or the trust matrix. Then, the stochastic gradient-descent approach first iterates through each observed entry $(i, j) \in S_R$ in the ratings matrix in random order and makes the following updates:

$$u_{iq} \Leftarrow u_{iq} + \alpha \left(e_{ij}^{(r)} \cdot v_{jq} - \frac{\lambda \cdot u_{iq}}{2 \cdot n_i^{user}} \right) \quad \forall q \in \{1 \ldots k\}$$

$$v_{jq} \Leftarrow v_{jq} + \alpha \left(e_{ij}^{(r)} \cdot u_{iq} - \frac{\lambda \cdot v_{jq}}{n_j^{item}} \right) \quad \forall q \in \{1 \ldots k\}$$

Here, $\alpha > 0$ represents the step-size. Furthermore, n_i^{user} represents the number of observed ratings for user i, and n_j^{item} represents the number of observed ratings for item j. Note that this set of updates is identical to that used in matrix factorization for collaborative filtering without the trust matrix (cf. section 3.6.4.2 of Chapter 3). One difference is that we have respectively normalized[1] the two regularization components with the number of observed ratings for users and items.

Subsequently, the stochastic gradient-descent approach iterates through each observed entry $(i, p) \in S_T$ in the trust matrix in random order and makes the following update steps:

$$u_{iq} \Leftarrow u_{iq} + \alpha \left(\beta \cdot e_{ip}^{(t)} \cdot z_{pq} - \frac{\lambda \cdot u_{iq}}{2 \cdot n_i^{out}} \right) \quad \forall q \in \{1 \ldots k\}$$

$$z_{pq} \Leftarrow z_{pq} + \alpha \left(\beta \cdot e_{ip}^{(t)} \cdot u_{iq} - \frac{\lambda \cdot z_{pq}}{n_p^{in}} \right) \quad \forall q \in \{1 \ldots k\}$$

Here n_i^{out} denotes the number of observed entries in S_T, for which i is the origin of the edge, and n_p^{in} denotes the number of observed entries in S_T for which p is the destination of the edge. We alternately cycle through the observed entries in the ratings matrix and the trust matrix with these updates, until convergence is reached. During a particular

---

[1]Strictly speaking, such a normalization should also be used in traditional matrix factorization, but it is often omitted on a heuristic basis. In the particular case of trust-centric systems, normalization becomes more important because of the varying sizes of the ratings matrix and trust matrix.

cycle, the entries are processed in random order, corresponding to the "stochastic" nature of this gradient-descent approach. The parameters β and λ can be selected using cross-validation or by simply trying various values of these parameters on a hold-out set. Different regularization parameters can be used for different matrices for better results, although such an approach increases the complexity of parameter tuning.

As in the case of all matrix factorization methods, the ratings matrix can be reconstructed as $\hat{R} = UV^T$. Note that one can also perform full reconstruction of the trust matrix T as $\hat{T} = UZ^T$. In fact, the reconstruction of the trust matrix can be viewed as a data-driven method for trust propagation and aggregation in which the rating information is used in addition to the existing trust relations.

11.3.8.1   Enhancements with Logistic Function

The aforementioned description provides a simplified version of *SoRec* in order to align it more closely with the discussion in Chapter 3. The actual *SoRec* algorithm uses a somewhat more sophisticated objective function. Matrix factorization methods have the drawback of predicting values that are outside the range of item ratings in R or trust values in T. One way of forcing the factorization to produce range-bound ratings is to use a logistic function $g(x) = 1/(1 + \exp(-x))$ within the factorization. The logistic function always maps values to the range $(0, 1)$. Without loss of generality, the ratings in R and trust values in T can be assumed[2] to be drawn from the range $(0, 1)$. In other words, the ratings R and trust matrix T should be reconstructed as $R \approx g(UV^T)$ and $T \approx g(UZ^T)$. The expression $g(UV^T)$ denotes the fact that the function $g(\cdot)$ is applied to each matrix element of UV^T. Then, the aforementioned objective function also needs to be modified as follows:

$$\text{Minimize } J = \underbrace{||R - g(UV^T)||^2}_{\text{Observed entries in } R} + \underbrace{\beta \cdot ||T - g(UZ^T)||^2}_{\text{Observed entries in } T} + \underbrace{\lambda \left(||U||^2 + ||V||^2 + ||Z||^2 \right)}_{\text{Regularizer}}$$

Note the use of the logistic function within the objective function. Correspondingly, the gradient descent method will multiplicatively incorporate the derivative of the logistic function in one of its terms. It is noteworthy that the logistic function-based enhancement is really an optimization and can be used within the context of any matrix factorization method in Chapter 3, not just in trust-based methods.

11.3.8.2   Variations in the Social Trust Component

Many variations of the aforementioned matrix factorization method exist, especially in terms of how the social (trust) part of the objective function is formulated.

1. Instead of using an $m \times k$ social factor matrix Z to impose $T \approx UZ^T$, one might use a $k \times k$ matrix H to impose $T \approx UHU^T$. The corresponding social term in the objective function will be modified to $||T - UHU^T||^2$. Intuitively, the matrix H captures the pairwise correlations between the various latent components of the users. The approach is referred to as *LOCALBAL* [594]. It is also possible to use a logistic function within the objective function as in the case of *SoRec*, although the original work does not use this approach.

   Note that this approach has a similar form as *SoRec*, except that it parameterizes $Z = UH^T$. The matrix H has only k^2 variables, whereas Z has $m \cdot k$ variables. Therefore,

---

[2]Ratings do not always lie in $(0, 1)$. If needed, the ratings matrix can be scaled using its ranges to $(r_{ij} - r_{min})/(r_{max} - r_{min})$, so that all its entries lie in $(0, 1)$.

LOCALBAL makes stronger assumptions about the social correlation structure of the users, as compared to *SoRec*. Fewer variables reduce the likelihood of overfitting, at the expense of incorporating some bias.

2. The *SocialMF* [270] algorithm imposes the constraint $U \approx TU$. Note that TU is not defined because some of the entries in T might be unspecified. Such entries are set to 0 for the purpose of computing TU. The corresponding social term in the objective function is $||U - TU||^2$. It is assumed that each row of T is normalized to sum to 1. The logistic function is used only within the term $||R - g(UV^T)||^2$ involving the ratings matrix. Note that the number of factor variables in this case is even fewer because the matrix Z is missing. In fact, the number of factor variables is exactly the same as in conventional matrix factorization. Reducing the number of factor variables will help in avoiding overfitting, but comes at the expense of greater bias.

   The approach sets the preference vector of each user to her trust-weighted average preference vector over all her neighbors. This is a direct result of the normalization of each row of T to sum to 1. The basic assumption is that the behavior of a user is affected by the behavior of her direct neighbors due to social influence.

3. *Social regularization:* In this approach [382], the user factors are forced to be more similar across links, and the difference in similarity is weighted with the trust values in the objective function. In other words, if $\overline{u_i}$ is the ith row of U, then the social part of the objective function is $\sum_{(i,j):t_{ij}>0} t_{ij}||\overline{u_i} - \overline{u_j}||^2$. This approach can be viewed as an indirect way of forcing homophily, and works best with implicitly inferred trust values t_{ij}. An example of such an implicitly inferred trust value is provided in Equation 11.2. Many variations of this approach, such as average-based regularization, are also discussed in the same work. The average-based regularization approach is somewhat similar to the *SocialMF* algorithm.

The bibliographic notes also provide pointers related to several other variations of the basic objective function.

11.3.9 Merits of Social Recommender Systems

Social recommender systems have a number of merits because they incorporate additional trust information into the recommendation process. This is particularly useful for improving the recommendation quality of items, addressing cold-start issues, and making the approach attack-resistant.

11.3.9.1 Recommendations for Controversial Users and Items

The greatest advantages in incorporating trust lie in the improvement of the recommendation quality for controversial users and items. Controversial users are those that disagree with the other users about the ratings of specific items [223]. Controversial items are those that receive diverse or polarized reviews. In such cases, the use of trust metrics generally enhances the user-specific or item-specific accuracy significantly [223, 406, 617] because the opinions of users are highly personalized in such cases. For example, users who are more similar and trust one another are more likely to provide similar ratings for controversial items.

11.3.9.2   Usefulness for Cold-Start

Social links are particularly useful to handle the cold-start problem for new *users*. Consider the case where a link prediction system is used for recommendation. A new user comes into the system and therefore has no ratings associated with any of the items covered by the recommender system; no user-item links are incident on that user. On the other hand, if social links are incident on that user, the link prediction method can still be used in order to predict the top matching items. This observation is true for other recommendation methods, such as matrix factorization. The main assumption is that the social links for the users are often available even before the user actively starts using the system. This is especially true for implicitly inferred trust networks. In any case, social links do add more data, which is helpful for alleviating the sparsity issue in the recommendations.

11.3.9.3   Attack Resistance

In general, there are significant commercial motivations for merchants to try to "cheat" recommender systems hosted by third parties. For example, the manufacturer of an item might try to post fake reviews for his item on Amazon.com. In many cases, such reviews are posted with the use of fake profiles created by the manufacturer. Trust-based recommender systems are more resistant to such attacks because these algorithms rely on the trustworthy peers of a user for predicting ratings. For example, Equations 11.3 and 11.4 explicitly weight the trust of a user for other users in the prediction process. A user is highly unlikely to specify a trust relation with a fake profile. As a result, such an approach is less likely to use the ratings posted from fake profiles in the prediction process. The topic of attack-resistant recommendation systems will be discussed in more detail in the next chapter.

11.4   User Interaction in Social Recommenders

The next-generation Web, which is also referred to as Web 2.0, has supported the development of a number of open systems in which users can actively participate and leave feedback. In particular, the development of *social tagging systems* allows the user to create and share meta-data about media objects. Such meta-data are also referred to as *tags*. Users may tag any form of object supported by the social network, such as an image, a document, music, or video. Virtually all social media sites allow some form of tagging. Some examples of such tagging systems are as follows:

- Flickr [700] allows users to tag images with keywords. For example, a keyword might describe the scenery or object in a specific image.

- The site last.fm [692] hosts music and allows users to tag music.

- Delicious [702] promotes the sharing of bookmarks and online links.

- The Bibsonomy [256, 708] system allows the sharing and tagging of publications.

- For a while, Amazon.com allowed its customers to tag products [709].

It is instructive to examine the nature of the tags created by a social tagging site such as last.fm. For the well-known musical album *Thriller* by Michael Jackson, the top tags at last.fm are as follows:

1001 albums you must hear before you die, 1980s, 1982, 1983, 80s pop, albums, albums I own, albums I own on vinyl, beat it, classic, classic pop, classic rock, crates of vinyl, dance-pop, disco, epic, thirller ...

These tags are quite informal because they are created by users in an open and participatory environment, rather than by specialists. Note that the tag "thirller" is a misspelling, which is quite common in such settings. Furthermore, all songs that were tagged by various users with a particular tag are indexed by it. For example, by clicking on the tag "classic rock," one can access various resources (artists, albums, or events) related to this tag because the corresponding albums and songs were tagged by various users. In other words, the tag "classic rock" serves as a bookmark or index to other relevant resources.

This process, therefore, results in the organization of content and the creation of a knowledge resource referred to as a *folksonomy*. The term "folksonomy" derives its roots from "folk" and "taxonomy," and it therefore intuitively refers to the classification of Web objects by non-specialist, voluntary, participants on the World Wide Web (i.e., common folk). This term was coined by Thomas Vander Wal, who defined it as follows [707]:

> "Folksonomy is the result of personal free tagging of information and objects (anything with a URL) for one's own retrieval. The tagging is done in a social environment (usually shared and open to others). Folksonomy is created from the act of tagging by the person consuming the information.
>
> The value in this external tagging is derived from people using their own vocabulary and adding explicit meaning, which may come from inferred understanding of the information/object. People are not so much categorizing, as providing a means to connect items (placing hooks) to provide their meaning in their own understanding."

Other terms used to describe social tagging include *collaborative tagging, social classification*, and *social indexing*. Tags provide an understanding of the topic of the object, and they often use vocabulary that are commonly used and understood by other participants. Therefore, the non-specialist nature of the participants is actually an asset, and it contributes to the collaborative power of such a system. Tags are also referred to as *social indexes* because they serve the dual role of organizing items. For example, by clicking on a tag, a user might be able to browse items related to that tag.

Folksonomies have numerous applications, including recommender systems [237]. In the particular context of recommender systems, folksonomies are valuable because they contribute to the available knowledge about the object at hand. At the very least, each tag can be considered a feature describing an object, although the underlying description may sometimes be noisy and irrelevant. In spite of their noisy nature, it has been observed that such social tagging methods can be used to improve the effectiveness of recommender systems significantly by complementing the knowledge available in ratings and other sources.

11.4.1 Representing Folksonomies

In tagging systems, *users* annotate *items* (or *resources*) with *tags*. The nature of the resource depends on the underlying system at hand. For example, the resource might be an image for Flickr, or a song for last.fm. Therefore, a 3-way relationship exists between users, items, and tags. Correspondingly, it can be represented as a hypergraph, in which each hyper-edge connects three objects. One can also represent it as a 3-dimensional cube (or tensor) containing unary bits with information about whether a user has tagged a specific resource (e.g., image) with a particular tag (e.g., "landscape"). The action of a user tagging a resource sets the corresponding bit to 1, and it is unspecified otherwise. In many cases, the unspecified values are approximated as 0 for analytical purposes. Figure 11.6 shows a toy example of 6 users with 4 items (images) and 5 tags both in the hypergraph and in the tensor representation. Figure 11.6(a) shows the hypergraph representation, whereas the Figure 11.6(b) shows the hypergraph representation. For example, Ann has tagged item 2 with the tag "flower." This results in a hyper-edge between these three entities in Figure 11.6(a), whereas the corresponding bit is set to 1 in Figure 11.6(b). Formally, we define a folksonomy as follows:

Definition 11.4.1 (Folksonomy) *A folksonomy is defined over m users, n items, and p tags as a 3-dimensional array $F = [f_{ijk}]$ of size $m \times n \times p$. The element f_{ijk} is a unary value indicating whether user i has tagged item j with the kth tag. In other words, the value of f_{ijk} is defined as follows:*

$$f_{ijk} = \begin{cases} 1 & \text{if user } i \text{ has tagged the } j\text{th resource with the } k\text{th tag} \\ \text{unspecified} & \text{otherwise} \end{cases} \tag{11.14}$$

In practical settings, the unspecified values are set to 0 by default, as is common in highly sparse implicit-feedback settings. Henceforth, we will refer to F as the *tag-cube*. It is immediately evident from Figure 11.6 that folksonomies have much in common with the multidimensional representation of context-sensitive recommender systems (see Chapter 8). As we will see later, this similarity is very useful because many of the methods of Chapter 8 can be used to resolve some of the queries.

Although Figure 11.6 illustrates a small toy example, the number of users and items may be on the order of hundreds of millions in a social platform like Flickr, and the number of tags may be on the order of millions. Therefore, such systems face challenges of scalability in a data-rich environment. This is both a challenge and an opportunity for research in the field of social-tagging recommender systems.

11.4.2   Collaborative Filtering in Social Tagging Systems

The nature of the recommendation formulation depends on the type of the underlying application. In some sites, such as Flickr, tagging information is available but rating information is not. In such cases, one can still develop a social tagging recommender system in which either tags or items are recommended based on the underlying patterns in the tag cube. In other cases, a separate $m \times n$ ratings matrix R is available along with the $m \times n \times p$ tag cube F. The ratings matrix is defined over the same set of users and items as the tag cube. For example, the *MovieLens* site contains both ratings and tagging information. The resulting collaborative filtering system is said to be a tag-*aware* recommender system in which the ratings matrix is the primary data, and the tagging information provides additional side information to improve the accuracy of rating prediction. Note that the ratings matrix could be an implicit-feedback matrix, such as in the case of last.fm, where accesses to resources are logged. In fact, implicit feedback is more common in social tagging sites. From an algorithmic perspective, implicit-feedback matrices are easier to use because they generally do not contain negative preferences, and missing entries can often be treated as 0 as an approximation. In the following, we will assume an explicit ratings matrix, unless otherwise specified.

When the ratings matrix is available, richer varieties of collaborative filtering queries can be formulated than in the case where only tagging information is available. In such cases,

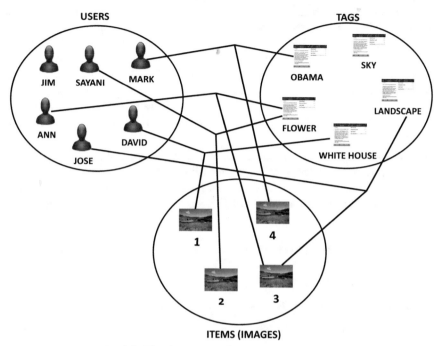

(a) The hypergraph representation

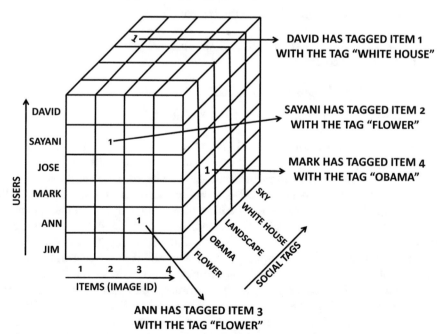

(b) The multidimensional cube representation

Figure 11.6: Representations of a folksonomy

the tag-cube and the ratings matrix share the user and item dimensions in common, but the ratings matrix does not contain the tag dimension. The information in these two sources of information can then be integrated in order to provide recommendations. It is noteworthy that this scenario can be viewed as a generalization of content-based collaborative filtering applications. In content-based collaborative filtering, keywords are associated only with items, whereas in the tag-cube keywords are associated with user-item *combinations*. One can view content-based recommender systems as a special case of the tag-cube, in which the 2-dimensional item-tag slice for each user is identical. Therefore, many of the methods discussed in the following sections can also be used for content-based collaborative filtering applications.

Because of this wide diversity in the underlying applications, the problem of collaborative filtering can be posed in a variety of ways, not all of which have been fully explored in the literature. In fact, much remains to be done to advance the field with respect to collaborative filtering, as it is a relatively recent area of interest. Some examples of possible queries are as follows:

1. **(Tagging data only)** Given an $m \times n \times p$ tagging cube F, recommend

   (a) a ranked list of tags to user i,

   (b) a ranked list of other users with similar interests (tagging patterns) to user i,

   (c) a ranked list of items to user i,

   (d) a ranked list of tags to user i for item j, and

   (e) a ranked list of items to user i for tag context k.

2. **(Tagging data and ratings matrix)** Given a ratings matrix R and an $m \times n \times p$ tagging cube F, recommend

   (a) a ranked list of items to user i, and

   (b) a ranked list of items to user i for tag context k.

The aforementioned queries can be partitioned into two categories. The first set of queries does not use a ratings matrix. In such queries, the recommendation of tags and users is generally more important than the recommendation of items, although one might also use the approach for item recommendation. Because tags serve as bookmarks and indexes for resources (items), finding relevant tags is a way of finding relevant items. The second set of queries is more closely aligned with traditional recommender systems because they are primarily based on the ratings matrix R. The only difference from traditional recommender systems is that the tag-cube is used as side information, and it plays a secondary role because of the amount of noise in it. Such methods are also referred to as *tagommenders* [535] or *tag-informed collaborative filtering* [673]. The primary strength of these systems lies in their ability to integrate the best of both worlds in user ratings and tag activity. There are generally fewer methods of the second type, but an increasing number of methods are able to integrate the knowledge in ratings matrices and the tag-cube. It is important to note that the matrix R might be unary in cases where explicit ratings are not available, and only implicit feedback (e.g., buying behavior) is available. It is, nevertheless, important to understand that the matrix R is an independent source of information from the tag-cube even when it is implicitly derived.

Table 11.1: A list of features [536] used to evaluate tag quality

Feature	Specificity	Criterion by which tags are ordered
num-item-apps	per item-tag	Number of times tag has been applied to a particular item
num-apps	per tag	Number of times tag has been applied across all items
num-users	per tag	Number of users who have applied the tag across all items
num-searches	per tag	Number of searches for the tag
num-search-users	per tag	Number of users who have searched for the tag
tag-share	per item-tag	Fraction of item tags for a particular tag value
avg-fraction-items-tagged	per tag	Average across all users, the fraction of all the items tagged by the user that have the tag
apps-per-item	per tag	Average number of times they are applied to their items
num-tag-words	per tag	Number of words in the tag
tag-length	per tag	Number of letters in the tag

11.4.3 Selecting Valuable Tags

Tags are generally quite noisy because of the open way in which they are contributed and used. In many cases, users might use non-standard vocabulary or misspellings to tag items. This can result in a large fraction of noisy and irrelevant tags. If irrelevant tags are used, then it can have a detrimental effect on many recommendation applications. Therefore, it is helpful to preselect a smaller number of tags. The pre-selection of tags also helps reduce the complexity of the mining process from a computational point of view. Therefore, tag-selection algorithms generally order the tags based on simple criteria and then preselect the top-ranked tags based on these criteria.

Many tagging sites use a simple methodology, referred to as *num-item-apps*, in which the number of people who have added a particular tag to an item is used as an estimate of how much other people would like to see that tag in the future. This can also be considered a proxy for the value of the tag. There are other intuitive features that are commonly used to estimate the quality of tags. For example, some tags may be globally valuable, whereas the value of other tags may be specific to particular items. A number of such features have been proposed in [536] for evaluating the quality of tags. A list of some of these features is provided in Table 11.1. In each case, the specificity of the tag (global or local) is also indicated. It is noteworthy that some of the features in [536] assume that users have rated the tags themselves with thumbs-up or thumbs-down ratings. Such information may not always be available in all systems, and therefore these features are not included in Table 11.1. An experimental methodology for evaluating these features is discussed in [535, 536]. It was found that features such as *num-item-apps*, *tag-share*, and *avg-fraction-items-tagged*, provided good performance. On the other hand, some features, such as *num-apps*, *num-users*, and *tag-length* did not provide the best performance. Furthermore, combining the five best features into a single feature called *all-implicit* provided better performance than any of the individual features. More details are provided in [535, 536] on the inference of this particular feature.

Aside from these methods, it is also possible to use the feature selection methods in section 4.3.4 of Chapter 4. The first step is to convert the tag-cube into a 2-dimensional item-tag slice by aggregating all the item-tag frequencies of the various users. By treating each tag as a "term," this approach results in a term-document matrix. Any of the methods discussed in section 4.3.4 may be used to select the most discriminative tags.

11.4.4   Social-Tagging Recommenders with No Ratings Matrix

This case can also be viewed as a special case of the multidimensional model in context-sensitive recommender systems. The tag-cube can be viewed as a multidimensional cube, in which the tags represent the context. Therefore, the context-sensitive model can be used to resolve these queries. In fact, the tensor factorization models used for context-sensitive ranking [495, 496] are not very different in principle from those used in tag recommendation [497, 498]. A detailed discussion of the multidimensional model for context-sensitive systems is provided in section 8.2 in Chapter 8.

As discussed earlier, queries in social-tagging recommenders can be formulated in a variety of ways, where one might recommend item, tags, or users. The tagging cube is 3-dimensional, and one might recommend along any of the dimensions. Among these various forms, the recommendation of tags is the most common. The reason for this is that the recommendation of tags has a benefit both to the user and to the platform hosting the tagging system:

1. *Utility to hosting platform:* As tags are non-standard, different users may describe the same resource using different keywords. Recommending tags for a specific item helps in consolidating the descriptions. Such a consolidation of the underlying description helps the system to collect better tags, and therefore improve the quality of the recommendations.

2. *Utility to user:* Users may either be recommended a tag specific to an item, or they may be recommended a tag specific to their own interests. The recommendation of item-specific tags is motivated by the fact that users may find it burdensome to assign tags to items. When relevant tags are recommended for a given item, it makes their job easier and also makes it more likely that they will participate in the tagging process. This is, in turn, beneficial in collecting more tagging data. User-specific recommendation of tags is beneficial because tags often serve the goal of organizing the items in a personalized way for various users. For example, Figure 11.6 might represent an image-browsing environment such as Flickr. If Ann is recommended the tag *sky* based on her other tags, then by clicking on this tag she might be able to discover other items of interest to her. It is also possible to combine tagging data with ratings matrices to make high-quality recommendations.

The following section reviews the variety of methods that have been proposed for recommendations in social-tagging systems.

11.4.4.1   Multidimensional Methods for Context-Sensitive Systems

The multidimensional methods discussed in section 8.2 of Chapter 8 can be used to build social-tagging recommenders. The basic idea is to project the data along a particular pair of dimensions for queries along two dimensions and use prefiltering methods for contextual queries along three dimensions.

For example, in order to recommend the best tags to a particular user, one can aggregate the frequencies of the tags over various items. In other words, one determines the number of times that a user used a particular tag over all items. This results in a 2-dimensional user-tag matrix of non-negative frequencies. Any traditional collaborative filtering algorithm can be used on this matrix in order to recommend tags to a user. Such an approach is best for recommending tags to users, but they do not use the item context; nonetheless, this approach is quite useful in real-life settings. Since tags serve the dual function of indexes to resources, the tags can be used by users to discover resources they might be interested in. Similarly, aggregating frequencies along the tag dimension leads to a user-item matrix. This matrix can be used to recommend items to users.

One disadvantage of using these aggregation methods is that the information along one of the dimensions is ignored. It is also possible to combine the information from all the dimensions during recommendation. Suppose that we want to recommend the best tags or the best items to a particular target user. One way of doing this would be to compute the similarity of users to the target user based on the aggregated user-tag matrix. This computation can also be performed with the aggregated user-item matrix. A linear combination of these two criteria is used to generate the most similar users to the target user. Then, the standard prediction method (cf. Equation 2.4 of Chapter 2) for user-based prediction can be leveraged to recommend either the most relevant items or the most relevant tags to the target. A similar approach can be used for item-based collaborative filtering methods by starting with a target item and finding the most similar items on the basis of either the aggregated user-item matrix or the aggregated tag-item matrix.

Another useful query is the recommendation of items to a user for a particular tag context. The prefiltering and postfiltering methodologies (cf. sections 8.3 and 8.4 of Chapter 8) for context-sensitive systems can be used to achieve this goal. For example, if one wanted to recommend movies related to the tag "animation," then the slice of the tag-cube corresponding to "animation" can be extracted. This process results in a 2-dimensional user-item matrix that is specific to animation movies. Traditional collaborative filtering algorithms can be applied to this matrix in order to make recommendations. One challenge with the use of the approach is that the extracted user-item slice might be too sparse. In order to address the sparsity issue, one can group related tags with the use of *tag clustering*. For example, a tag cluster might contain "animation," "children," "for kids," and so on. The user-item tag frequencies over these related tags can be added together to create an aggregated user-item matrix, which is less sparse. Recommendations can be performed over this aggregated matrix. Tag-clustering methods are proposed in [70, 215, 542]. Although the works in [70, 215, 542] explore the use of tag clustering for content-based methods, such techniques can also be used to improve the effectiveness of collaborative filtering applications.

Finally, the class of tensor factorization methods have found increasing popularity in social tagging. These methods are discussed in section 8.5.2 of Chapter 8, as a special case of context-sensitive systems. A particularly popular method, which is discussed in that section, is the *Pairwise Interaction Tensor Factorization (PITF)* method. In addition, these methods have been generalized to the notion of factorization machines, which can be viewed as generalizations of large classes of latent factor models. Refer to section 8.5.2.1. ·

11.4.4.2 Ranking-Based Methods

Ranking-based methods use the *PageRank* methodology in order to make recommendations in the presence of tags. A detailed description of ranking methods is provided in section 10.2 of Chapter 10. The two notable methods in this regard are *FolkRank* [256], and *SocialRank* [602]. The main difference between *SocialRank* and *FolkRank* is that *SocialRank* also uses content-centric similarity between the objects in the ranking process. For example, links might be added between pairs of images based on their content-centric similarities. Furthermore, *SocialRank* can be applied to arbitrary social media networks, rather than the tagging hypergraph. Therefore, *SocialRank* makes significant changes to the *PageRank* algorithm in order to balance the effects of the different modalities. The method can, nevertheless, be applied to a folksonomy as well. *FolkRank* is specifically designed to work with the tagging hypergraphs created in folksonomies. As *SocialRank* is already discussed in section 10.2.3.2 of Chapter 10, we will focus only on the *FolkRank* method in this description.

FolkRank is a simple adaptation of personalized *PageRank* (cf. 10.2.2 of Chapter 10). The first step in applying *FolkRank* is to extract a tripartite graph from the tag hypergraph. The tripartite graph $G = (N, A)$ is extracted from the tag hypergraph as follows:

1. Each tag, user, and item becomes a node in graph G. In other words, each $i \in N$ is a user, tag, or item. Therefore, for m users, n items, and p tags, the graph G contains $(m + n + p)$ nodes.

2. For each hyperedge between a tag, user, and item, undirected edges are added between each pair of entities. Therefore, three edges are added for each hyperedge.

The personalized *PageRank* method is then applied directly to this network. The personalization vector of section 10.2.2 is set in such a way that preferred items, users, or tags have a higher probability of restart. By setting the restart probability in different ways, one can query for specific users, tags, items, user-item pairs, user-tag pairs, or tag-item pairs. The responses to the queries can also be obtained in all modalities.

As a result of the process, highly ranked tags, users, and items provide different views of relevant nodes in the network. An important aspect of *FolkRank* is that it takes global popularity (reputation) into account in addition to the user-specific relevance. This is because all ranking mechanisms tend to favor highly connected nodes. For example, a tag that is used heavily will always be ranked highly even in personalized *PageRank* mechanisms. The value of the restart probability regulates the trade-off between specificity and popularity. Therefore, a *differential* version of *FolkRank* has also been developed, in which these effects are canceled out. The basic idea of the differential version is to perform the following steps:

1. *PageRank* is performed on the extracted tripartite graph with no bias. In other words, the restart probabilities of all nodes have the same value of $1/(m+n+p)$. Recall that the tag-cube is of size $m \times n \times p$ and the number of nodes in the network is $(m+n+p)$. Let the resulting probability vector be given by $\overline{\pi_1}$.

2. Personalized *PageRank* is performed by setting an increased bias value for the specific user-item combination being queried. For example, consider the case where a particular user-item combination is queried. The restart probability for the queried user node can be set to be proportional to $(m + 1)/(2m + 2n + p)$, the restart probability for the queried item node can be set to be proportional to $(n + 1)/(2m + 2n + p)$, and the restart probability of the remaining nodes can be set to be proportional to $1/(2m + 2n + p)$. Let the resulting probability vector be given by $\overline{\pi_2}$.

3. The relevance of the various nodes in all modalities can be extracted from the vector $\overline{\pi_2} - \overline{\pi_1}$. The values may be either positive or negative, depending on the level of similarity or dissimilarity.

The main advantage of such an approach is that it cancels out the global popularity effects to a large extent.

11.4.4.3   Content-Based Methods

Content-based methods can be used in order to make the recommendations of both items and tags to users. In order to recommend items to users, a user-specific training data set can be created in which each item is described by its tag frequencies over the m users. These frequencies can be represented in tf-idf format. For a given user, the training data contains all the items that she has tagged, as well as a negative sample of items she has not tagged. These are objects for which the tagging frequency needs to be learned. The feature variables and the dependent variable (for the learning process) correspond to the tf-idf representation of each item, and the number of tags the user has placed on each item. Note that the dependent variable is 0 for negative samples. A regression-based model is applied to this training data in order to make predictions.

A similar approach can be used to recommend tags to users instead of recommending items to users. The main difference is that tags are represented as tf-idf vectors of items instead of the other way around. The training data is now generated using tags as objects that need to be classified. Therefore, tags have labels attached to them, based on the number of times that the user has used them on different items. This training model is used to predict the interest of the user in the tags for which the user interest is unknown. A comparison of various content-based methods for tag recommendation is provided in [264].

An item recommendation algorithm, based on tag clustering, was presented in [542]. The clusters are created using the tf-idf representation of the tags in terms of items. In other words, each tag is treated as a vector of item frequencies; then these vectors are used to create m clusters. The clustering process provides an intermediate representation in terms of which the user interest and item relevance are measured and integrated.

Let the interest of the ith user in the sth cluster be denoted by $ucW(i, s)$, and the relevance of the jth item (resource) to the sth cluster be denoted by $rcW(j, s)$. The value $ucW(i, s)$ is computed as the fraction of the tags of user i belonging to the sth cluster, and the value of $rcW(j, s)$ is computed as the fraction of the tags of item j belonging to the sth cluster. Then, the overall interest $I(i, j)$ of user i in item j is computed as follows:

$$I(i, j) = \sum_{s=1}^{m} ucW(i, s) \times rcW(j, s) \tag{11.15}$$

The computation of the interest with the use of clusters as an intermediate step is shown in Figure 11.7. Note that this interest can be used to rank items for users. The basic idea is that clusters provide a robust summary of the sparse user-item tagging behavior, which can be used to make high-quality interest computation.

In addition, the work in [542] uses the approach to provide personalized item responses to user tag queries. For example, if Mary searches for "animation," she might not be recommended the same movies that Bob would be suggested for the same query. For a given queried tag q, its similarity $S(j, q)$ to item j is defined in terms of the relative frequency f_{jq} with which the item j is tagged with q, in comparison with the frequency of other tags on item j:

$$S(j, q) = \frac{f_{jq}}{\sqrt{\sum_s f_{js}^2}} \tag{11.16}$$

Although the value $S(j, q)$ can directly be used to rank items in response to the search of a particular user i, the results are personalized by using the user interest $I(i, j)$ of the searcher. The value of $I(i, j)$ is computed using Equation 11.15. The query results are, therefore, ordered by $S(j, q) \times I(i, j)$ instead of ordering them by $S(j, q)$. It is noteworthy that recommending items for tag queries does not necessarily require the use of user-specific personalization, because one can simply use $S(j, q)$ to rank items. Furthermore, recommending tags for items does not require the use of personalization either. One can simply use the tagging characteristics of the items in order to make recommendations of tags to users. In such

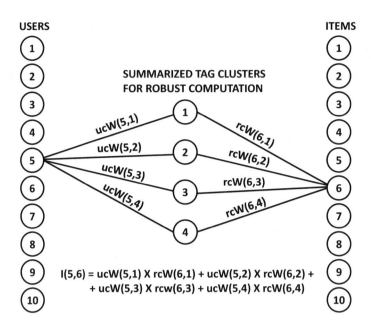

Figure 11.7: Using clusters as a bridge to compute the interest of users for items. The example illustrates the computation of the interest of user 5 for item 6. Such a computation can be performed for any user-item pair.

cases, the recommended tag will not depend on the querying user, but rather will depend on the queried item. In fact, the earliest works on tag recommendation use co-occurrence statistics between tags and items to make recommendations. The results are, therefore, not dependent on the user making the query.

Similarly, an approach proposed in [316] uses latent Dirichlet allocation (LDA) to make recommendations based on content-centric topic modeling by treating each item as a "document" containing tag (or "word") frequencies. Similar to traditional topic modeling on documents, the approach shows that the qth tag is related to item j by the following relationship:

$$P(\text{Tag } = q|\text{Item } = j) = \sum_{s=1}^{K} P(\text{Tag } = q|\text{Topic } = s) \cdot P(\text{Topic } = s|\text{Item } = j) \quad (11.17)$$

Here, K represents the total number of topics, which is a user-defined parameter. Note that the left-hand side of Equation 11.17. provides the recommendation probability for ranking purposes, whereas the quantities on the right-hand side can be estimated from the parameter learning process of LDA methods. It is not necessary to use LDA for topic modeling. For example, the simpler Probabilistic Latent Semantic Analysis (PLSA) model may be used in lieu of LDA. Note that it is also possible to personalize this recommendation by treating the sets of tags of a *user* as "documents" and clustering these users into topics with the use of topic modeling After these topics have been determined, we can compute the relevance of the various tags to each user as follows:

$$P(\text{Tag } = q|\text{User } = i) = \sum_{s=1}^{K} P(\text{Tag } = q|\text{Topic } = s) \cdot P(\text{Topic } = s|\text{User } = i) \quad (11.18)$$

Note that Equation 11.18 uses a different set of topics than Equation 11.17; the former clusters the users, whereas the latter clusters the items. A linear combination of Equations 11.17 and 11.18 can be used to determine the relevance of tag q to item j, given the personalized context of user i. The weights of the linear combination will decide the trade-off between user-specificity and item-specificity.

There are other ways of combining user-specificity and item-specificity by directly adapting some of the Bayesian ideas in topic modeling [315]. Specifically, we can directly compute the *personalized* and *item-specific* recommendation probability $P(\text{Tag} = q|\text{User} = i, \text{Item} = j)$. This probability can be simplified with the use of the naive Bayes rule as follows:

$$P(\text{Tag} = q|\text{User} = i, \text{Item} = j) = \frac{P(\text{User} = i, \text{Item} = j|\text{Tag} = q) \cdot P(\text{Tag} = q)}{P(\text{User} = i, \text{Item} = j)} \quad (11.19)$$

$$\approx \frac{P(\text{User} = i|\text{Tag} = q) \cdot P(\text{Item} = j|\text{Tag} = q) \cdot P(\text{Tag} = q)}{P(\text{User} = i, \text{Item} = j)} \quad (11.20)$$

$$\propto P(\text{User} = i|\text{Tag} = q) \cdot P(\text{Item} = j|\text{Tag} = q) \cdot P(\text{Tag} = q) \quad (11.21)$$

Note that we have ignored the term $P(\text{User} = i, \text{Item} = j)$ in the denominator to a constant of proportionality. This is because we wish to *rank* the different tags in order of their recommendation probability for a *specific* user and item, which are already fixed. Therefore, this constant term can be ignored for ranking purposes.

Now, the terms $P(\text{User} = i|\text{Tag} = q)$ and $P(\text{Item} = j|\text{Tag} = q)$ in the right-hand side of the aforementioned equations can be expressed in terms of user recommendation and item recommendation probabilities using Bayes rule:

$$P(\text{User} = i|\text{Tag} = q) = \frac{P(\text{User} = i)P(\text{Tag} = q|\text{User} = i)}{P(\text{Tag} = q)} \quad (11.22)$$

$$P(\text{Item} = j|\text{Tag} = q) = \frac{P(\text{Item} = j)P(\text{Tag} = q|\text{Item} = j)}{P(\text{Tag} = q)} \quad (11.23)$$

Therefore, on substituting these terms in Equation 11.21, we obtain the following:

$$P(\text{Tag} = q|\text{User} = i, \text{Item} = j) \propto \frac{P(\text{Tag} = q|\text{User} = i) \cdot P(\text{Tag} = q|\text{Item} = j)}{P(\text{Tag} = q)} \quad (11.24)$$

The terms on the right-hand side can be estimated easily in a data-driven manner, as in any Bayes classifier. For example, the value of $P(\text{Tag} = q)$ can be estimated as the fraction of non-empty cells in the tag-cube for which the qth tag has been specified. The value of $P(\text{Tag} = q|\text{User} = i)$ can be estimated as the fraction of non-empty cells of the slice of the tag-cube for user i that correspond to the qth tag. The value of $P(\text{Tag} = q|\text{Item} = j)$ can be estimated as the fraction of the non-empty cells of the slice of the tag-cube for item j that correspond to the qth tag. Laplacian smoothing is often used to avoid overfitting.

The probabilities in Equation 11.24 are used to rank the tags for the specific user-item combination. The work in [315] also discusses a simpler frequency-based model for performing the recommendation.

11.4.5 Social-Tagging Recommenders with Ratings Matrix

Tags have significant potential in improving the quality of recommendations when they are used in addition to item ratings. For example, consider a scenario where Mary has watched many movies such as *Shrek* and *Lion King*, which are annotated with the tag "animated" on a rating site such as IMDb. However, Mary might not have tagged any of these movies in the tag-cube, and these preferences are derived from the ratings matrix.

Now consider the case where a movie such as *Despicable Me* is also tagged as "animated," although Mary has not watched this movie yet. In such a case, it is reasonable to assume that Mary might also be interested in watching the movie *Despicable Me*. Although the ratings matrix might also provide the same prediction, the chances of prediction error are reduced when tag information is incorporated because it provides an independent source of information. In particular, this is true when a movie is new and there are too few ratings or tags to make reliable predictions about user preference. In such cases, the ratings and tags can complement each other to make more robust decisions. In most cases, tagging systems contain *implicit* ratings (e.g., whether or not a user has viewed an item) in the ratings matrix. This is because sites such as last.fm automatically log data about the items that a user might have consumed. Note that the implicit ratings are an independent source of information because a user might view an item, even though they might not tag it. In this section, we will study both the case of implicit and explicit ratings.

The most straightforward approach is to use hybrid recommender systems to combine the predictions based on tags and ratings. For example, any of the methods discussed in section 11.4.4 can be used to make item predictions based purely on tags. Furthermore, any traditional collaborative filtering algorithm can be used to make predictions based on ratings. A weighted average of the two ratings can be used to make the final prediction. The weights can be learned using the method discussed in section 6.3 of Chapter 6 on hybrid recommender systems. However, such an approach does not integrate the two sources of prediction very tightly. Better results may be obtained with algorithms that tightly integrate the various sources of data in the recommendation process.

11.4.5.1 Neighborhood-Based Approach

The method in [603] works with implicit feedback data sets in which the ratings matrices are assumed to be unary. This is quite commonly the case in social-tagging systems. For example, in a site such as last.fm, user accesses of items are available, but explicit ratings are not available. The paper treats missing entries as 0 values. Therefore, the ratings matrix R is treated as a binary matrix rather than as a unary matrix.

The approach of [603] augments the $m \times n$ ratings matrix R with data from the $m \times n \times p$ tag cube F, by creating additional pseudo-users or pseudo-items. For example, user-based collaborative filtering can be performed on a ratings matrix with an extended set of items. In order to create a ratings matrix R_1 with the item dimension extended, each tag is treated as a pseudo-item. Furthermore, the value of a user-tag combination is assumed to be 1, if the user has used that tag at least once (possibly over multiple items). Otherwise, the value is equal to 0. Note that there are $m \times p$ user-tag combinations. One can then append these $m \times p$ combinations to the $m \times n$ ratings matrix R by treating the tags as new pseudo-items. This results in an extended matrix R_1 of size $m \times (n + p)$. The similarity between a user i and other users is computed using this extended matrix. The similarity computation is enriched because of the additional columns containing user-tag activity information. The item ratings of the user i are computed using the number of 1 values in the peer group of i. These predicted ratings $\hat{r}_{ij}^{user}$ are normalized to sum to 1 over the different values of item index j so that they represent probabilities of accessing (or buying) various items. Note that ratings represent frequency of activity in these implicit feedback settings.

The item-based approach can be extended in a similar way. In this case, a $p \times n$ matrix corresponding to tag-item combinations is created. A value in this matrix is 1 if the item is tagged at least once. The tags are now treated as pseudo-users, and appended as rows to the original ratings matrix R. This results in an extended matrix R_2 of size $(m + p) \times n$. This extended matrix is used to perform similarity computations in item-based collaborative filtering. The predicted ratings $\hat{r}_{ij}^{item}$ for a given user i are then normalized so that they sum to 1 over all j. Therefore, the predicted ratings represent probabilities of accessing or buying items in this case as well.

After performing user-based and item-based collaborative filtering, the ratings predictions of the two cases are fused using a parameter $\lambda \in (0, 1)$:

$$\hat{r}_{ij} = \lambda \cdot \hat{r}_{ij}^{user} + (1 - \lambda) \cdot \hat{r}_{ij}^{item} \tag{11.25}$$

The optimal value of λ can be chosen using cross-validation. The results in [603] showed improvements over traditional collaborative filtering when tag information was also used. The fusion of user-based and item-based methods was necessary to achieve the improvements from the incorporation of tagging.

11.4.5.2 Linear Regression

The method in [535] uses linear regression to integrate the tags into the recommendation process. As tags are generally less statistically precise in identifying user preference than are ratings, it is important to select only valuable tags for the recommendation process. To achieve this goal, the methodology described in section 11.4.3 may be used. The basic approach in [535] fuses the information in the user ratings in order to enrich the information about the tag preferences for various items. For example, if a user has rated *Lion King* and *Shrek* highly, and both movies are tagged as "animation," it can be inferred that the user is likely to be interested in movies with this tag. The first step is to determine the relevance weighting between an item and a tag. For example, any of the item-tag specific quantifications in Table 11.1 may be used. Then, if q_{jk} is the relevance of item j to tag k, then the item-preference value is further transformed with the sigmoidal function:

$$v_{jk} = \frac{1}{1 + \exp(-q_{jk})}$$

Then, the user preference u_{ik} of user i for tag k is computed by combining the tag-item relevance with the user interest in the items. The user interest in the items may be inferred using the ratings matrix $R = [r_{ij}]$. The preference u_{ik} of user i for tag k may be inferred as follows:

$$u_{ik} = \frac{r_{ij} \cdot v_{jk}}{\sum_{s=1}^{n} r_{is} \cdot v_{sk}} \tag{11.26}$$

Items that are not rated by user i are ignored in the numerator and denominator. When ratings are not available, the value of u_{ik} can also be indirectly inferred from the frequency of the user's visits, clicks, buys, or tags on items. For example, the value of r_{ij} in Equation 11.26 can be replaced with the number of times the user has tagged item j (not necessarily with the tag k).

A simple approach to predict a preference score p_{ij} of an item j for user i is to determine all the tags T_j of that item and average the value of u_{ir} over all tags $r \in T_j$:

$$p_{ij} = \frac{\sum_{r \in T_j} u_{ir} \cdot v_{jr}}{\sum_{r \in T_j} v_{jr}} \tag{11.27}$$

Note that the value p_{ij} might not lie in the range of ratings. Nevertheless, it can still be used to *rank* items for a user.

A more effective approach to predict ratings is to use *linear regression*. The basic idea in linear regression is to assume that the rating r_{ij} of user i for item j is based on a linear relationship, which is true for fixed j and varying values of i:

$$r_{ij} = \sum_{r \in T_j} u_{ir} \cdot w_{jr} \quad \forall i : r_{ij} \text{ is observed} \tag{11.28}$$

The (unknown) coefficient w_{jr} represents the importance of tag r for item j, and it can be learned using regression over all the ratings that are observed for the item j. The main difference from Equation 11.27 is that instead of using a heuristic value of v_{jr} as the weight of tag r (specific to item j), we are *learning* w_{jr} using linear regression on the ratings matrix. The resulting approach is generally superior because of the greater level of supervision. As the regression training process includes all users that have rated item j, the approach does use the collaborative power of the ratings in different users. Furthermore, this method also provides superior results to conventional collaborative filtering algorithms because of its use of side information available in the tags. Combining this methodology with a simple matrix factorization method into a hybrid system provided even better results [535]. It was shown that regression support vector machines provide the best results for the training process, although least-squares regression provides a simpler alternative. Linear regression methods are discussed in section 4.4.5 of Chapter 4.

11.4.5.3   Matrix Factorization

A matrix factorization approach, referred to as *TagiCoFi* [673], uses a variation of the methods discussed in Chapter 3 to approximately factorize the ratings matrix R into two matrices, an $m \times q$ matrix U and an $n \times q$ matrix V. This condition can be expressed as follows:

$$R \approx UV^T \;\forall \text{ observed entries of } R \tag{11.29}$$

This condition can be imposed by approximately minimizing the Frobenius norm $g(U, V, R) = ||R - UV^T||^2$ over the observed entries of R.

In addition, a similarity constraint is imposed over the user factor matrices U, so that users with similar tagging behavior have similar factors. Let S_{ij} be the similarity between users i and j and let $\overline{u_i}$ be the ith row of U. The computation of S_{ij} from the tagging behavior will be described later. Then, in order to ensure that users with similar tagging behavior have similar factors, we would like to also minimize the following factor similarity objective $f(U)$:

$$f(U) = \sum_{i=1}^{m} \sum_{j=1}^{m} S_{ij} ||\overline{u_i} - \overline{u_j}||^2 \tag{11.30}$$

As we have two different criteria, defined by the objective functions $g(U, V, R)$ and $f(U)$, the balancing parameter β can be introduced to minimize $g(U, V, R) + \beta f(U)$. In addition, we have the standard regularization term in matrix factor factorization, which is given by the sum of the Frobenius norms of the factor matrices. This regularization term is

$\lambda \left(||U||^2 + ||V||^2 \right)$, where λ is the regularization parameter. Summing up these different terms, we derive the following objective function:

$$\text{Minimize } J = \underbrace{||R - UV^T||^2}_{\text{Observed entries in } R} + \underbrace{\beta \cdot \sum_{i=1}^{m} \sum_{j=1}^{m} S_{ij} ||\overline{u_i} - \overline{u_j}||^2}_{\text{Tagging similarity objective}} + \underbrace{\lambda \left(||U||^2 + ||V||^2 \right)}_{\text{Regularizer}}$$

As in the case of all matrix factorization methods, a gradient descent method is used to determine the factor matrices U and V. The values of β and λ can be computed using cross-validation methods.

It is noteworthy that this approach is technically similar to a social regularization approach [382] discussed in section 11.3.8.2 on trustworthy recommender systems. In that approach, a trust matrix T is used to add the similarity term $\sum_{i,j:t_{ij}>0} t_{ij} ||\overline{u_i} - \overline{u_j}||^2$ to the objective function. Here, the tagging similarity matrix is used to add the term $\sum_{i=1}^{m} \sum_{j=1}^{m} S_{ij} ||\overline{u_i} - \overline{u_j}||^2$. In other words, the trust/homophily t_{ij} between users i and j is replaced with the tagging similarity S_{ij} between users i and j. Thus, minor variations of the same technical model can be used to address diverse social recommendation scenarios. Furthermore, instead of forcing the user factors to be more similar, one can also force the item-factors to be more similar based on tagging behavior (see Exercise 5).

Computing Tagging Similarity

The aforementioned approach requires the computation of the tagging similarity S_{ij} between the users i and j. First, the tf-idf matrix is generated from the tag-cube F in which the number of times the user has used a particular tag is computed. In other words, the number of 1s of a particular user-tag combination over all items are summed. Thus, a frequency vector is generated for each of the m users. This frequency is then normalized with the standard tf-idf normalization used in information retrieval. The work in [673] proposes two different methods for computing the similarity:

1. *Pearson similarity:* The Pearson correlation coefficient ρ_{ij} is computed over all tags used by user i and user j. Tags not used by either user are ignored. The sigmoidal function is used to transform the correlation coefficient into a non-negative similarity value S_{ij} in $(0, 1)$:

$$S_{ij} = \frac{1}{1 + \exp(-\rho_{ij})} \tag{11.31}$$

2. *Cosine similarity:* The standard cosine similarity between the frequency vectors is used as the similarity value. Refer to Chapter 4 for a discussion of the similarity function.

3. *Euclidean similarity:* The Euclidean distance d_{ij} is computed between the similarity vectors, and then a Gaussian kernel is applied to the distance to transform it into a similarity value in $(0, 1)$:

$$S_{ij} = \exp\left(-\frac{d_{ij}^2}{2\sigma^2} \right) \tag{11.32}$$

Here, σ is a user-controlled parameter, which can be chosen using cross-validation.

In the results reported in [673], the Pearson similarity provided the best performance, whereas the Euclidean similarity provided the worst.

11.4.5.4   Content-Based Methods

Social tagging methods provide a straightforward way of using content-based methods. The frequency vector of the tags on a movie can be viewed as its description. The movies that the user has rated are treated as the training examples on the feature space defined by the tags. The ratings are treated as the class labels. A training model, which is specific to the user, is constructed using this training data. The model is used to predict the ratings of the other movies. Either a classification or a regression model may be used depending on whether the ratings are unary or interval-based. Such content-based models can also be combined with any of the aforementioned collaborative systems.

A simple content-based model for recommendations on the IMDb data set was presented in [584]. It uses the notion of tag clouds to represent the tag-based descriptions of movies. The various keywords are weighted according to their relevance and then combined with ratings to make a final prediction. One challenge with the use of content-based methods is that the tags are very noisy with a significant amount of synonymy. Linguistic methods were used in [178] for disambiguation, and then combined with a naive Bayes classifier. It is also useful to leverage the feature selection methods discussed in Chapter 4 to improve the representation quality.

11.5   Summary

Social information can be used in a wide variety of ways in recommender systems. The standard multidimensional model can be used to incorporate social information in a limited way. Trust-centric methods can be used to create robust recommender systems. Unsupervised methods use trust propagation and aggregation methods to incorporate trust into the recommender system. Supervised methods use link prediction and matrix factorization for more effective performance. Supervised methods are generally considered the state-of-the-art today. Incorporating trust knowledge can make the systems attack resistant and also help in avoiding cold-start issues.

In recent years, social tagging systems have become an avenue for users to collaboratively tag resources on the Web with free-form descriptions. These descriptions are also referred to as folksonomies, which are represented as tag cubes. Such descriptions are useful in terms of the rich content-centric knowledge they contain about user interests. Tag cubes can either be used on a stand-alone basis, or they can be combined with ratings matrices to make recommendations. The former class of methods shares similarities with the multidimensional model for recommendations. The latter class of methods can be based on either collaborative or content-based methods. A variety of techniques, such as neighborhood methods, linear regression, and matrix factorization, have been developed for this scenario.

11.6   Bibliographic Notes

Overviews of trust-based recommender systems may be found in [221, 588, 616, 646]. The doctoral dissertation of Jennifer Golbeck [222] provides several seminal algorithms on the topic. The correlation between homophily in social networks and the notion of trust has been shown in [224, 681]. In these cases, trust relationships can be computationally inferred from Web-based social networks. The work in [187] showed how to infer trust relationships directly from ratings data [187], although there is some debate on whether this notion of trust is generally accepted as it is conventionally used in the literature. One of the earliest works in applying trust-based methods to such networks was proposed in the context of movie recommendations [223, 225]. The Filmtrust system [225] showed how to use trust metrics for movie recommendations. The work in [592] studies the predictability of distrust relationships from interaction data. Other sites that collect trust-based information include Epinions [705], Moleskiing [461], and Slashdot [706].

Trust metrics play a key role in recommendations in trust networks [344, 680]. The work of [680] provides a good overview of the relevant trust metrics. Although much of the work on trust networks focuses on trust (positive) relationships only, some recent work also discusses the use of both trust and distrust relationships [241, 287, 590, 593, 614, 680]. Furthermore, most of these methods have only discussed methods for (positive) trust propagation, with the exception of the work in [287], which has proposed methods for distrust aggregation as well. The interaction between trust and distrust concepts is studied in [590, 591] in the context of the recommendation and link prediction problems. Methods for trust propagation with multiplicative methods are discussed in [241, 509]. A variety of other trust propagation methods include the use of decay factors along paths [240], using only shortest paths [222], distance from a fixed propagation horizon [403], spreading factors [682, 683], rules [345, 597], and semantic distances [1]. It was shown in [227] through experiments that transitively propagated trust values are more accurately inferred with the use of shortest paths, rather than using all paths. This observation formed the basis of the *TidalTrust* algorithm. Sophisticated methods for de-emphasizing shorter paths include the *Appleseed* algorithm [682], in which a spreading activation model is used. Trust is modeled as energy, which is injected from the source node. The energy is divided among subsequent nodes based on the trust scores along edges. The amount of energy reaching the sink provides the total amount of trust. Clearly, if the sink is connected to the source with many short paths, then more energy will reach the sink. The *EigenTrust* algorithm [292] uses the principal eigenvectors of the trust network to calculate trust values of the source node for all other nodes. However, the approach provides a ranking of the trustworthiness rather than actual trust values. The exploitation of homophily effects for trust propagation is discussed in [594], in which a matrix factorization model is introduced.

A second important aspect of trust computation is that of aggregation. Aggregation rules in social networks are discussed in [1, 221, 222, 287, 449, 615]. The work in [405] discusses methods for weighting the different components of the aggregation based on path length or based on closeness of friendships.

The combination of propagation and aggregation leads to the creation of trust metrics [221, 344]. The *Advogato* trust metric is discussed in [344], and it is one of the classical metrics used in the literature for many applications beyond recommender systems. The trust metrics discussed in this chapter are specialized to recommendation algorithms. The best description of the *TidalTrust* algorithm together with a pseudocode may be found in [222]. The *MoleTrust* algorithm is described in [406]. The effectiveness of *MoleTrust* in the presence of cold-start is shown in [403, 404]. The *TrustWalker* approach is presented in [269], and an axiomatic approach to trust-based recommendation is provided in [48]. The use of link prediction in signed and unsigned networks for recommendation is studied in [157, 324, 325, 580, 581]. The work in [157] is notable because it shows the connections between the matrix factorization methods for link prediction, and the matrix factorization methods for collaborative filtering. The *SoRec* algorithm is proposed in [381], and the *LOCALBAL* algorithm in [594]. The use of both trust and distrust relationships in matrix factorization methods was explored in [383]. The *SocialMF* algorithm was discussed in [270], whereas the similarity-based regularization approach was proposed in [382]. An ensemble method using matrix factorization, known as *social trust ensemble (STE)*, was presented in [384].

The utility of recommender systems for controversial items and users has been studied in several works [222, 406, 617]. It is generally accepted that trust-based methods are particularly useful in such cases. The effectiveness of such systems in the presence of cold-start is shown in [403, 404]. The attack-resistant nature of trust-aware systems is discussed in [344].

A general survey on social-tagging techniques may be found in [237]. A survey of tagging recommender systems is provided in [671], although most of the works discussed in this survey do not use a ratings matrix for the recommendation process. Finally, the recommender systems handbook contains an overview of social-tagging recommender systems [401]. One of the earliest works on tag recommendations was provided in [553], in which simple methods such as co-occurrence, voting, and summing are used to perform recommendations. A hierarchical clustering method for content-based recommendation was proposed in [542]. Probabilistic latent factor models are presented in [316]. Some of the works [135, 179, 584] focus primarily on content-based systems.

Tensor-based methods for tag recommendation are presented in [497, 498, 582, 583]. The notion of factorization machines has found significant popularity in these cases [493, 496]. A particularly notable pairwise approach is the PITF method [496]. A method has been proposed in [487] to leverage latent factor models for application to mining algorithms in the presence of tags. Although this work is not specifically focused on recommender systems, the underlying latent factor models can be used in virtually any application, including recommender systems. Machine learning methods for tag recommendation algorithms are discussed in [250, 555, 556]. Among these works, the techniques in [556] are designed to perform the tag recommendation in real time. Tag clustering methods [70, 215, 542] are often used to alleviate the sparsity problem in collaborative filtering applications. A weighted hybrid method for social tagging methods is discussed in [216].

Various evaluations of tag recommendation methods are provided in [264, 277]. Methods for evaluating tag quality are discussed in [536]. Only a small number of systems today combine the power of ratings matrices with that of social tags [535, 603, 673]. Content-based methods for combining ratings with tagging data are discussed in [179, 584]. For specific data domains, such as music, valuable insights can sometimes be gleaned from the music files for the recommendation process [191]. A solution to the cold-start problem with social tags is discussed in [672].

11.7   Exercises

1. Implement the neighborhood-based method for recommending tags for an item to a user with the use of a linear combination of results obtained on the user-tag matrix and the item-tag matrix.

2. Discuss the relationship of the Katz measure for link prediction with trust propagation and aggregation methods.

3. Implement the gradient descent method of section 11.3.8.

4. The method in section 11.4.5.3 forces the user factors to be more similar based on user-tagging similarity.

    (a) Design a method that forces item factors to be more similar based on item-tagging similarity.

    (b) Design a method that forces both user and item factors to be more similar based on corresponding similarities in user and item tagging.

Chapter 12

Attack-Resistant Recommender Systems

"The truth is incontrovertible. Malice may attack it, ignorance may deride it, but in the end, there it is." – Winston Churchill

12.1 Introduction

The input to recommender systems is typically provided through open platforms. Almost anyone can register and submit a review at sites such as Amazon.com and Epinions.com. Like any other data-mining system, the effectiveness of a recommender system depends almost exclusively on the quality of the data available to it. Unfortunately, there are significant motivations for participants to submit incorrect feedback about items for personal gain or for malicious reasons:

- The manufacturer of an item or the author of a book might submit fake (positive) reviews on Amazon in order to maximize sales. Such attacks are also referred to as *product push* attacks.

- The competitor of an item manufacturer might submit malicious reviews about the item. Such attacks are also referred to as *nuke* attacks.

It is also possible for an attack to be designed purely to cause mischief and disrupt the underlying system, although such attacks are rare relative to attacks motivated by personal gain. This chapter studies only attacks that are *motivated to achieve a particular outcome* in the recommendation process. The person making the attack on the recommender system is also referred to as the *adversary*.

By creating a concerted set of fake feedbacks from many different users, it is possible to change the predictions of the recommender system. Such users become *shills* in the attack process. Therefore, such attacks are also referred to as *shilling attacks*. It is noteworthy that the addition of a single fake user or rating is unlikely to achieve the desired outcome. In most

© Springer International Publishing Switzerland 2016
C.C. Aggarwal, *Recommender Systems: The Textbook*,
DOI 10.1007/978-3-319-29659-3_12

cases, an adversary would need to create a large number of fake users (or fake *profiles*) to achieve the desired outcome. For the purpose of this chapter, a profile refers to a set of ratings corresponding to a fake user created by the adversary. Of course, the number of injected profiles may depend on the specific recommendation algorithm being attacked, and the specific approach used to attack it. An attack that requires a smaller number of injected profiles is referred to as an *efficient* attack because such attacks are often difficult to detect. On the other hand, if an attack requires a large number of injected profiles, then such an attack is *inefficient* because most systems should be able to detect a sudden injection of a large number of ratings about a small number of items. Furthermore, the effectiveness of the attacks may depend on the specific recommendation algorithm being used. Some recommendation algorithms are more robust to attacks than others. Furthermore, different attacks may work more or less efficiently with different algorithms.

Attacks can also be classified based on the amount of knowledge required to mount them successfully. Some attacks require only limited knowledge about the ratings distribution. Such attacks are referred to as *low-knowledge* attacks. On the other hand, attacks that require a large amount of knowledge about the ratings distribution are referred to as *high-knowledge attacks*. As a general rule, a trade-off exists between the amount of knowledge required to make an attack and the efficiency of the attack. If adversaries have more knowledge about the ratings distribution, then they can generally make more efficient attacks.

This chapter is organized as follows. In the next section, we will discuss the nature of the trade-offs between the required knowledge and the efficiency of the attack. We will also discuss the impact of using a specific recommendation algorithm on the effectiveness of the attack. The various types of attacks are discussed in section 12.3. The problem of attack detection in recommender systems is discussed in section 12.4. The design of robust recommender systems is discussed in section 12.5. A summary is given in section 12.6.

12.2   Understanding the Trade-Offs in Attack Models

Attack models have a number of natural trade-offs between the efficiency of the attack and the amount of knowledge required to mount a successful attack. Furthermore, the effectiveness of a particular attack may depend on the specific recommendation algorithm being used. In order to understand this point, we will use a specific example.

Consider the toy example illustrated in Table 12.1 where we have 5 items and 6 (real) users. The ratings are all drawn from the range of 1 to 7, where 1 indicates extreme dislike and 7 indicates extreme like. Furthermore, an attacker has injected 5 fake profiles, which are denoted with the labels *Fake-1*, *Fake-2*, *Fake-3*, *Fake-4*, and *Fake-5*. The goal of this attacker is to inflate the ratings of item 3. Therefore, this attacker has chosen a rather naive attack, in which they have inserted fake profiles containing a single item corresponding to the item 3. However, such an attack is not particularly efficient. It is highly detectable because only a single item is included in every injected profile with a very similar rating. Furthermore, such ratings injections are unlikely to have a large impact on most recommendation algorithms. For example, consider the *non-personalized* recommendation algorithm in which the highest rated item is recommended. In such a case, the *naive* attack algorithm will increase the predicted rating of item 3, and it will be more likely to be recommended. The attack might also increase the predicted rating of item 3 in cases where item bias is explicitly used as a part of the model construction. There is, however, little chance that such an attack will significantly affect a neighborhood-based algorithm. Consider, for example, a user-based neighborhood algorithm in which the profiles are used to make predictions for Mary. None of the injected profiles will be close to the rating profile of Mary; therefore, Mary's predicted

Table 12.1: A naive attack: injecting fake user profiles with a single pushed item

Item ⇒	1	2	3	4	5
User ⇓					
John	1	2	1	6	7
Sayani	2	1	2	7	6
Mary	1	1	?	7	7
Alice	7	6	5	1	2
Bob	?	7	6	2	1
Carol	7	7	6	?	3
Fake-1	?	?	7	?	?
Fake-2	?	?	6	?	?
Fake-3	?	?	7	?	?
Fake-4	?	?	6	?	?
Fake-5	?	?	7	?	?

Table 12.2: Slightly better than naive attack: injecting fake user profiles with a single pushed item and random ratings on other items

Item ⇒	1	2	3	4	5
User ⇓					
John	1	2	1	6	7
Sayani	2	1	2	7	6
Mary	1	1	?	7	7
Alice	7	6	5	1	2
Bob	?	7	6	2	1
Carol	7	7	6	?	3
Fake-1	2	4	7	6	1
Fake-2	7	2	6	1	5
Fake-3	2	1	7	6	7
Fake-4	1	7	6	2	4
Fake-5	3	5	7	7	4

ratings of item 3 will not be affected by the injection of the fake profiles. This particular injection of ratings is, therefore, not particularly efficient because it is hard to affect the predicted ratings, even when a large number of fake users are injected. Furthermore, such a profile injection can be detected in most cases because of the injection of ratings involving a single item.

Consider a second example of an attack, shown in Table 12.2, in which the attacker is trying to promote item 3, but he or she also adds random ratings to other items in order to reduce the detectability. Note that the genuine ratings in this second example are the same as the first, but the fake profiles are different. Such an attack is more effective than the one shown in Table 12.1. Consider the case where a user-based neighborhood is used to perform the recommendation. When only genuine profiles are used, John and Sayani are among Mary's neighbors and item 3 will have a low predicted rating for Mary. When fake profiles are also injected before user-based recommendation, most of the fake profiles are not close to Mary because the ratings are randomly chosen. However, the profile *Fake-3* happens

Table 12.3: Highly knowledgeable attacker injects fake user profiles

Item ⇒	1	2	3	4	5
User ⇓					
John	1	2	1	6	7
Sayani	2	1	2	7	6
Mary	1	1	?	7	7
Alice	7	6	5	1	2
Bob	?	7	6	2	1
Carol	7	7	6	?	3
Fake-1	6	7	7	2	1
Fake-2	7	7	6	1	1
Fake-3	1	1	7	6	7
Fake-4	1	1	6	7	6
Fake-5	2	1	7	7	7

to be close to Mary by chance. As a result, the predicted rating of Mary for item 3 will increase. Therefore, from the point of view of an adversary, this type of attack does a better job than a completely naive attack with a single item. Nevertheless, this attack is quite inefficient because a large number of injected profiles would be required to affect the results of a neighborhood-based algorithm. In general, it is hard to ensure that randomly injected ratings will be close enough to a particular target user to whom the recommendation is to be made. After all, it is important for fake profiles to be close to target users in order to significantly affect the recommendation in any way.

In order to understand the impact of greater knowledge on the attack process, consider the example of an attacker who has a significant knowledge of the underlying ratings distributions. Therefore, we have illustrated an example in Table 12.3 in which the genuine ratings are the same as those in Table 12.1. However, the injected ratings are designed to reflect the underlying item correlations and also push the ratings for item 3 higher. For example, the attacker is aware that the ratings of items 1 and 2 are positively correlated in the ratings database, and the ratings of items 4 and 5 are also positively correlated. Furthermore, these two groups of items are negatively correlated with one another. Therefore, the attacker injects the ratings, while respecting these correlations. Correspondingly, it is evident that these correlations are respected in the fake profiles of Table 12.3. In this case, it is evident that the predicted ratings of Mary for item 3 are affected more significantly, than in the earlier examples of Tables 12.1 and 12.2. This is because the three profiles *Fake-3*, *Fake-4*, and *Fake-5* are all very close to Mary, and they may be included among her peers in a neighborhood-based algorithm. Therefore, this attack is very efficient because it requires a small number of profiles in order to cause a significant shift in the underlying ratings. On the other hand, such an attack requires a significant amount of knowledge, which may not always be available in practical settings.

It is noteworthy that the efficiency of a particular attack also depends on the particular algorithm being attacked. For example, user-based and item-based neighborhood algorithms have very different levels of propensity of being attacked. If an item-based algorithm were to be applied to the high-knowledge case of Table 12.3, then the predicted ratings of Mary for item 3 would not be affected very significantly. This is because the item-based algorithm uses the ratings of other users only in the item-item similarity computation. The fake

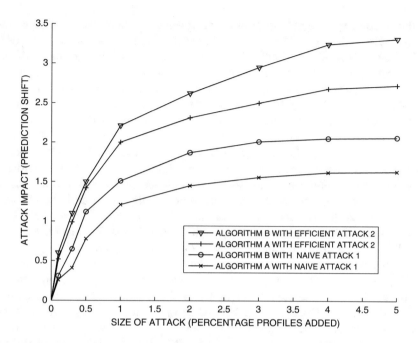

Figure 12.1: Typical examples of the effect of the combination of the specific push attack algorithm and the specific recommendation algorithm

profiles affect the similarity computation used to discover the most similar items to item 3; subsequently Mary's *own ratings* on these items are used to make the prediction. In order to change the most similar items to item 3, one must typically inject a large number of ratings, which makes the attack more detectable. Furthermore, it is much harder to change the predictions in a particular direction for the target item by changing its similar items; after all, Mary's *own ratings* on these items are used to make the prediction rather than those in the fake profiles. Algorithms that are less prone to attacks are referred to as *robust* algorithms. It is one of the goals of recommender systems to design algorithms that are more robust to attacks.

The aforementioned examples lead us to the following observations:

1. Carefully designed attacks are able to affect the predictions with a small number of fake profile insertions. On the other hand, a carelessly injected attack may have no effect on the predicted ratings at all.

2. When more knowledge about the statistics of the ratings database is available, an attacker is able to make more efficient attacks. However, it is often difficult to obtain a significant amount of knowledge about the ratings database.

3. The effectiveness of an attack algorithm depends on the specific algorithm being attacked.

In order to understand the nature of these trade-offs, consider a recommendation algorithm A that is robust to attacks, and another algorithm B that is not robust. Similarly, consider a naive attack (labeled 1), and an efficient attack (labeled 2), both of which are push attacks. Therefore, there are four different combinations of algorithm and attack-type. In Figure 12.1, we have shown some typical examples of how a specific recommender system might respond to a particular attack type. The X-axis illustrates the fraction of fake profiles that are inserted in the attack, and the Y-axis illustrates the amount of shift in the predicted ratings. In each case, the shift in predicted ratings is positive because it is a push attack. Intuitively, the shift is defined as the amount by which the predicted ratings have moved on the average over all users. The shift may be computed over a particular (pushed) item, or it may be computed over a subset of (pushed) items. More details of the methodology for shift computation are provided in section 12.2.1.

The higher the curve is, the more efficient the attack will be. Efficient attacks are more desirable for an adversary because they are more difficult to detect. It is evident that the combination of recommendation algorithm B and attack type 2 leads to the highest curve, both because of the weakness of the recommendation algorithm and the efficiency of the attack. It is also possible to measure the impact of an attack in terms of other evaluation metrics such as the hit-ratio rather than through the prediction shift. In all cases, the impact of adding the fake profiles to a specific evaluation metric is quantified.

However, it is sometimes not possible to easily extrapolate the effectiveness of a particular attack into a concrete statement about the robustness of the recommendation algorithm at hand. This is because adversaries might tailor the attack to a specific recommendation algorithm, and therefore the robustness of the recommendation algorithm depends on the type of attack. For example, an attack algorithm that works well for a user-based neighborhood algorithm might not work well for an item-based neighborhood technique, and vice versa. By tailoring the attack to the specific recommendation algorithm at hand, more efficient attacks can be constructed. Fortunately, it is often difficult for an adversary to achieve this goal, unless she is aware of the specific recommendation algorithm being used.

An adversarial relationship perpetually exists between recommender systems and attackers. Attackers try to design increasingly clever algorithms to influence the recommender systems, whereas the designers of recommender systems try to propose more robust algorithms. Although the goal of this chapter is to learn how to design robust algorithms, it is important to understand attack strategies in order to be able to design robust algorithms. Therefore, we will first introduce the various types of attacks before discussing the design of robust algorithms.

12.2.1 Quantifying Attack Impact

In order to analyze the impact of various types of attacks, it is important to be able to quantify their impact. For example, the respective impacts of the attacks shown in Figures 12.1, 12.2, and 12.3 are quantified with a measure abstractly referred to as the "prediction shift." This measure is shown on the Y-axis in Figure 12.1. It is useful to examine in greater detail how the prediction shift is actually computed.

Consider a ratings matrix R with user set U and item set I. The first step is to select a subset $U_T \subseteq U$ of test users. Furthermore, let $I_T \subseteq I$ be the set of test items pushed in the testing process. Then, the attack is performed one at a time for each item $j \in I_T$, and the effect on predicted rating of users in U_T on item j is measured. The average prediction shift over all users and items is measured. Therefore, the attack needs to be performed $|I_T|$ times in order to measure the prediction shit over all test items.

Let $\hat{r}_{ij}$ be the predicted rating of user $i \in U_T$ for item $j \in T_T$ before the attack, and $\hat{r}'_{ij}$ be the corresponding predicted rating after the attack on item j. Then, the prediction shift of user i for item j is given by $\delta_{ij} = \hat{r}'_{ij} - \hat{r}_{ij}$. Note that δ_{ij} can be either positive or negative. A positive value indicates that the push attack has been successful, and therefore

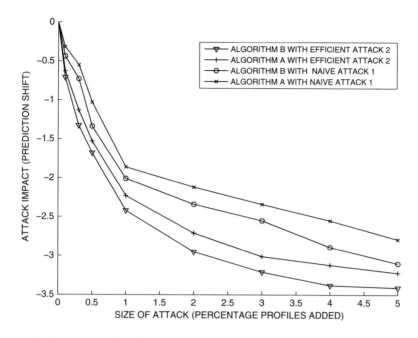

Figure 12.2: Typical examples of prediction shifts in case of nuke attack (Compare with push attacks of Figure 12.1)

the item j is more positively rated. If the attack is a nuke attack, then negative values of the prediction shift are indicative of success. Then, the average shift $\Delta_j(U_T)$ for test-user set U_T and item j is computed as follows:

$$\Delta_j(U_T) = \frac{\sum_{i \in U_T} \delta_{ij}}{|U_T|} \tag{12.1}$$

Then, the overall prediction shift $\Delta^{all}(U_T, I_T)$ over all items in I_T is equal to the average value of the per-item shift over all test items:

$$\Delta^{all}(U_T, I_T) = \frac{\sum_{j \in I_T} \Delta_j(U_T)}{|I_T|} \tag{12.2}$$

The prediction shift is a way of quantifying how well a pushed (or nuked) item has been shifted in a direction favoring its goal. Note that δ_{ij} can be either positive or negative; therefore, shifts in a direction opposite to the desired outcome are penalized by this measure. Furthermore, the prediction shift curves will be upward sloping in the case of a push attack, whereas they will be downward sloping in the case of a nuke attack. For example, typical curves for prediction shifts in the case of nuke attacks are illustrated in Figure 12.2. It is evident that these plots have trends opposite to those shown in Figure 12.1.

Although the prediction shift is a good way of quantifying the changes in the ratings, it may often not measure the true impact from the perspective of the end user. The end user only cares about whether her pushed item made it to the top-k list (or was removed from the top-k list). In many cases, a large prediction shift may not be sufficient to move an item into the top-k list. Therefore, a more appropriate measure is the *Hit-Ratio* $h_j(U_T)$, which is defined for item j and test user set U_T. The hit-ratio $h_j(U_T)$ is defined as the fraction of

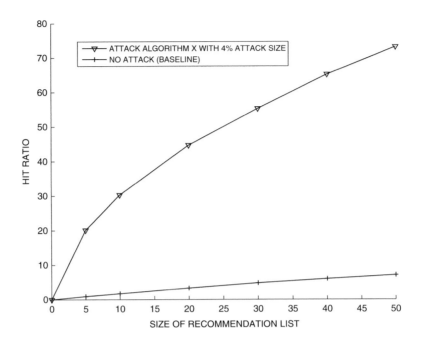

Figure 12.3: Effect of push attack on hit-ratio

users in U_T for which item j appears among the top-k recommendations. Then, the overall hit-ratio $h^{all}(U_T, I_I)$ over all test users and items is averaged over all the test items in I_T:

$$h^{all}(U_T, I_T) = \frac{\sum_{j \in I_T} h_j(U_T)}{|I_T|} \tag{12.3}$$

It is noteworthy that the hit-ratio is not a differential measure, as it does not compute the shift in the ratings. Therefore, unlike the prediction shift, one needs to plot the hit ratio both before and after the attack. In this type of plot, the X-axis depicts the size of the recommendation list, and the Y-axis depicts the hit-ratio; the size of the attack (i.e., the number of injected profiles) is fixed. An example of such a plot is illustrated in Figure 12.3 in which the hit-ratio of both the original algorithm and attacked algorithm is shown. The distance between these two curves provides an idea of the level of success of an adversary in making the pushed items appear on the recommendation list. It is also possible to fix the size of the recommendation list and plot the hit-ratio with the size of the attack. Such a plot would be somewhat similar to Figure 12.1 because it provides an idea of how the hit-ratios are affected with increasing attack size.

12.3 Types of Attacks

Although the rating of a particular item may be targeted in an attack, it is important to inject ratings of other items in order to make the attack effective. If fake profiles of only a single (pushed or nuked) item are inserted, they generally do not significantly affect the outcome of many recommendation algorithms. Furthermore, such attacks are generally easy to detect using automated methods. Therefore, the ratings of additional items are included in the injected profile. Such items are referred to as *filler* items. The importance of including filler items is particularly emphasized by the example of Table 12.1, where the addition of only a single item containing a rating is not sufficient to create an effective attack.

The ratings of most of the items will not be specified in the fake user profiles, just as in the case of genuine user profiles. Such unspecified items are also referred to as *null* items. It is also evident from the example of Table 12.3 that the attacks are most effective when the filler items are correlated with the target item in terms of the underlying rating patterns. For example, if a target movie, such as *Gladiator,* is rated frequently with another movie, such as *Nero,* then it is generally beneficial to add filler ratings of *Nero* when trying either to use either a push attack or a nuke attack on *Gladiator*. It would not be quite as beneficial to add filler ratings for a completely unrelated item like *Shrek*. However, such attacks require a greater knowledge of the ratings distribution because correlated sets of items need to be identified. Therefore, there is a natural trade-off between the efficiency and knowledge requirements of different types of attacks.

Some attacks are specifically designed to be push attacks or to be nuke attacks. Although many attacks can be used in both capacities, each attack is generally more effective in one of the two settings. There are also subtle differences in the evaluation of these two types of attacks. The two types of attacks often show very different behavior in terms of the prediction shift and the hit-ratio. For example, it is much easier to nuke an item with a few bad ratings, given that only a few top items are recommended in any given setting. In other words, the effects on the hit-ratio can be more drastic than the effects on the prediction shift in the case of a nuke attack. Therefore, it is important to use multiple measures while evaluating push attacks and nuke attacks.

In the following, we will discuss the various types of attacks that are commonly used, and also discuss their common use case as either push attacks or as nuke attacks. These attacks require various levels of knowledge from the adversary. We will study these different attacks, starting with the ones that require the least knowledge.

12.3.1 Random Attack

In the random attack, the filler items are assigned their ratings from a probability distribution that is distributed around the *global* mean of all ratings across all items. Because the global mean is used, the ratings of the various filler items are drawn from the same probability distribution. The filler items are chosen randomly from the database, and therefore the selection of items to rate is also not dependent on the target item. However, in some cases, the same set of filler items may be used for each profile. There is no advantage in choosing the same set of filler items for each profile because it does not reduce the required level of knowledge to mount the attack, but it only makes the attack more conspicuous.

The target item is either set to the maximum possible rating value r_{max} or the minimum possible rating value r_{min}, depending on whether it is a push attack or a nuke attack. The main knowledge required to mount this attack is the mean values of all ratings. It is not very difficult to determine the global mean of the ratings in most settings. The limited knowledge required for a random attack comes at a disadvantage for the attacker, because such attacks are often not very efficient.

12.3.2 Average Attack

The average attack is similar to the random attack in terms of how the filler items are selected for rating. The same set of filler items are selected for each profile. However, the average attack differs from the random attack in terms of how the ratings are *assigned* to the selected items. In the average attack, the ratings that are assigned to the filler items have values that are specified at or approximately at the average of the *specific* item. The target item is assigned either the maximum rating or the minimum rating depending on whether the attack is a push attack or a nuke attack. Note that the average attack requires a greater amount of knowledge than the random attack, because knowing the global mean is not sufficient. One also needs to know the mean of *each* filler item. Furthermore, the attack is somewhat conspicuous because the same set of filler items is used for each fake profile.

In order to reduce the possibility of detection, one can also use randomly selected filler items for each injected user profile. The drawback of doing this is that a greater amount of knowledge will be required for making the attack. For example, the global mean of *each* of the injected filler items will be required. However, this can sometimes be reasonable in settings where the ratings are public. For example, the ratings on Amazon.com are public, and the average values can be computed easily. In other systems, such as IMDb, the average rating of each item is often directly advertised. Alternatively, one can randomly select items out of a small set of candidate items in order to determine the fillers for each fake profile. Such a strategy requires much less knowledge. Furthermore, it has been shown [123] that it does not lose a significant amount of knowledge.

12.3.3 Bandwagon Attack

The main problem with many of the aforementioned attacks is the inherent sparsity of the ratings matrix, which prevents the injected profiles from being sufficiently similar to the existing profiles. When too many items are selected as filler items, the attack becomes conspicuous. On the other hand, when a small number of filler items are selected randomly for a fake profile, then it might not have an inadequate number of observed ratings in common with other users. In user-based collaborative filtering, a fake profile has no impact when it does not have any rated items in common with the target user to whom the recommendation is being made. The efficiency of the attack reduces as a result.

The basic idea of the bandwagon attack is to leverage the fact that a small number of items are very popular in terms of the number of ratings they receive. For example, a blockbuster movie or a widely used textbook might receive many ratings. Therefore, if these items are always rated in the fake user profile, it increases the chance of a fake user profile being similar to the target user. In such cases, the predicted ratings of the target user are more likely to be affected by the attack. Therefore, the knowledge about the popularity of the items is used to improve the efficiency of the attack. In addition to the popular items, a set of random items is used as additional filler items.

In the bandwagon attack, the ratings of popular items are set to their maximum possible rating value r_{max}. The other filler items are rated randomly. The reason for assigning the maximum rating value to the most popular items is to the increase the chances that more users will be found within the fake profiles, which are close to any particular target user to whom the recommendation is being made. This is because the popular items are more likely to be assigned positive ratings in real settings. The target item is set to the maximum possible rating r_{max} or the minimum possible rating r_{min}, depending on whether it is a push attack or a nuke attack.

It is noteworthy that the notion of "popular" items in this particular case does not necessarily refer to the most frequently rated items, but rather refers to widely liked items. Such items are likely to be frequently rated *in a positive way* in the ratings database. One does not need to use the ratings matrix in order to determine the most popular items. It is usually easy to determine the most popular products of any type from sources independent of the ratings matrix. This is the main reason that the bandwagon attack requires much less knowledge as compared to the average attack. Bandwagon attacks can often perform almost as well as the average attack, in spite of their smaller knowledge requirements. In general, bandwagon attacks can influence user-based collaborative filtering algorithms significantly, but they have greater difficulty in influencing item-based algorithms.

12.3.4 Popular Attack

The popular attack shares a number of similarities with the bandwagon attack in that it also uses popular items in order to create the filler items. However, the popular items might be either widely liked or widely disliked items, but they must have many ratings. The popular attack also assumes more knowledge about the ratings database to set the ratings of these popular items. Furthermore, it does not assume the existence of an additional set of filler items. Therefore, more popular items have to be used in this attack than in the case of the bandwagon attack.

In order to set the ratings on the popular items in an intelligent way, more knowledge needs to be assumed about the underlying rating database. In particular, it is assumed that average values of the ratings of the popular items are known. In order to achieve a push attack, the ratings of the various filler items in a fake user profile are set as follows:

1. If the average rating of a filler item in the ratings matrix is less than the global rating average over all items, then the rating of that item is set to its minimum possible value r_{min}.

2. If the rating of a filler item is greater than the overall average rating of all items, then the rating of the item is set to $r_{min} + 1$.

3. The rating of the target item is always set to r_{max} in the fake user profile.

The reason for setting the ratings in this unusual way is (a) to increase the likelihood of finding a profile similar to that of the target user within the fake profiles by choosing differential ratings of r_{min} and $r_{min} + 1$ for filler items; and (b) to increase the ratings gap between the target item and the filler items to push the item more effectively. This attack can also be used for the case of a nuke attack with minor modifications. In a nuke attack, the ratings of filler items are set to $r_{max} - 1$ for low-rated popular items, r_{max} for highly rated popular items, and r_{min} for the target item.

As in the case of the bandwagon attack, one does not need to assume that the popular items need to be inferred from the ratings database. Such information can be easily inferred from other data sources. However, one does need to know the average values of the ratings, albeit only for the popular items. It is also possible to use external sources to *estimate* the popular items with lower or higher ratings. For example, one might use the text of the reviews to determine the items with positive or negative sentiment. Nevertheless, the knowledge requirements of a popular attack are always greater than those of a bandwagon attack.

12.3.5 Love/Hate Attack

The love/hate attack is specifically designed to be a nuke attack, and its main advantage is that it requires very little knowledge to mount this attack. In the love/hate attack, the nuked item is set to the minimum rating value r_{min}, whereas the other items are set to the maximum rating value r_{max}. In spite of the minimal knowledge requirements, this attack is very effective. As discussed earlier, nuke attacks are generally easier to mount than push attacks. Therefore, such low-knowledge attacks often have a better chance of being successful in the case of nuke attacks as compared to push attacks. For example, a symmetrically designed attack in which the ratings of the filler items are set to r_{min} and the rating of the target item is set to r_{max}, is not quite as successful for pushing items. The love/hate attack is highly specific to user-based collaborative filtering algorithms, and it is almost completely ineffective with item-based collaborative filtering algorithms.

12.3.6   Reverse Bandwagon Attack

This attack is specifically designed to nuke items. The reverse bandwagon attack is a variation of the bandwagon attack, in which widely disliked items are used as filler items to mount the attack. The fact that such items are "widely disliked" means that they have received many ratings. For example, if a movie is highly promoted before its release but then turns out to be a box-office failure, then it will receive many low ratings. These items are selected as filler items. Such filler items are assigned low ratings together with the nuked item. As in the case of the bandwagon attack, it is often not very difficult to discover such items from other channels. This attack works very well as a nuke attack when an item-based collaborative filtering algorithm is used for recommendation. Although it can also be used in the case of user-based collaborative filtering algorithms, many other attack methods, such as the average attack, are generally more effective.

12.3.7   Probe Attack

An important aspect of many of the aforementioned methods is that the ratings are often artificially set to values, such as r_{min} and $r_{min}+1$, in an identical way across many profiles. The use of such ratings tends to make the attack rather conspicuous, and therefore easily detectable. The probe attack tries to obtain more realistic ratings for items *directly from a user-based recommender system* in order to use these values in the attack. In other words, the operation of a recommender system is probed to mount the attack.

In the probe attack, a seed profile is created by the attacker, and the predictions generated by the recommender system are used to learn related items and their ratings. Since these recommendations have been generated by user-neighbors of this seed profile, they are highly likely to be correlated with the seed profile. One can also use this approach to learn the ratings of items within a specific genre. For example, in a movie recommendation scenario, consider the case where the target item to be pushed or nuked corresponds to action movies. The seed profile might contain the ratings of a set of popular action movies. The seed profile can then be extended further by observing the operation of a user-based collaborative filtering algorithm when the seed profile is used as the target user. The recommended items and their predicted ratings can be used to augment the seed profile in a realistic way. The rating of the target item is set to r_{max} or r_{min}, depending on whether it is pushed or nuked, respectively. The ratings of other filler items learned from the probing approach are set to the average values predicted by the recommender system.

12.3.8   Segment Attack

Almost all the aforementioned attack methods work effectively with user-based collaborative filtering algorithms, but they do not work quite as effectively with item-based algorithms. The only exception is the reverse bandwagon attack, which is designed only to nuke items, but not to push items. It is generally harder to attack item-based collaborative filtering algorithms. One of the reasons for this phenomenon is that an item-based algorithm leverages the target user's *own* ratings in order to make attacks. The target user is always an authentic user. Obviously, one cannot use fake profile injection to manipulate a genuine user's specified ratings.

However, it is possible to change the *peer* items with the use of fake profiles. Changing the peer items has an effect on the quality of the predicted ratings. In the segment attack, the attacker uses their domain knowledge to identify a targeted set of users (i.e., users with specific interests) to which they push the item. For example, the attacker might decide to push a historical movie, such as *Gladiator*, to users who have liked historical movies in the past. Note that the relevant genre for a particular movie is often common knowledge, and it does not require any specific information from the ratings matrix. Therefore, the first step for the attacker is to determine which *segment* (i.e., category or genre) of items are the closest to a given item. Such items are assigned the maximum possible rating together with the pushed item. An additional set of sampled filler items are assigned the minimum rating. This maximizes the variations in the item similarities towards items of the same genre. The basic idea is for the attacker to make sure that only very similar items are used in the item-based recommendation process. It is generally assumed that increasing the likelihood of using items of a similar genre in the prediction process for the target item will give it an additional advantage over other items in the recommendation process. After all, users tend to rate similar items in a similar way. Therefore, for users who have liked movies of this genre in the past, the predicted ratings of the targeted item will be pushed up more than other items of the same genre because of greater relevance. Therefore, it is more likely that such users will be recommended the targeted item. Although one might also use a variation of the segment attack for nuke attacks, it is mostly effective for push attacks. Furthermore, the segment attack may also be used effectively in the context of user-based collaborative filtering algorithms.

The segment attack is a generalization of the notion of the *favorite item* attack [123]. The favorite item attack is designed only with a specific *user* in mind. Filler items are selected to be a set of items, such that their ratings are greater than the average user rating. In such a case, the ratings of these items and the pushed item are set to their maximum value, and the ratings of filler items are set to the minimum value. Although the favorite item attack works well for both user-based and item-based collaborative filtering algorithms, the attack is restricted to a specific user. Furthermore, the attack requires a significant amount of knowledge of the values of the ratings. These characteristics tend to make this attack rather impractical. Its main utility is in establishing an upper bound on the effectiveness of other attacks.

12.3.9 Effect of Base Recommendation Algorithm

As discussed earlier, the choice of attack is highly specific to the particular recommendation algorithm at hand. In general, user-based recommendation algorithms are more susceptible to attacks as compared to item-based algorithms. Only a few attacks, such as the reverse bandwagon attack and the segment attack are specifically designed for item-based algorithms. Most of the other attack methods are effective for user-based algorithms, but are able to affect the item-based algorithms only to a limited degree. Some attack methods such as the love/hate attack are completely ineffective against item-based algorithms.

Interestingly, much of the work on attack algorithms is largely focussed on neighborhood-based methods and there are only a few studies on the effectiveness with respect to model-based algorithms. Some recent work [446, 522] has analyzed the vulnerability of model-based algorithms against attacks. Examples of the analyzed algorithms included clustering-based algorithms, PCA-based methods, LSA-based methods, and association rule methods. The experiments showed that model-based algorithms are generally more robust to attacks as compared to user-based collaborative filtering algorithms, although there were some variations among different algorithms. Hybridizing the algorithm tends to make the approach more robust, especially when external domain knowledge is used. This is because domain knowledge cannot be influenced by profile injection mechanisms. A summary of the effects of various attacks on various model-based collaborative filtering algorithms may be found in [523].

Although this chapter is primarily focussed on explicit ratings, a few attack methods have also been designed for implicit feedback data sets [79]. Just as explicit data sets require the injection of fake *profiles*, implicit feedback data sets require the injection of fake *actions*. The basic idea is to correlate the fake actions with other popular actions so as to give the impression that the fake actions are similar to these popular actions. Consider a Web site that wants to increase the likelihood of recommendation a particular page by injecting fake actions in the click-stream. The mechanism for injecting a fake action is to use an automated crawler that simulates Web browsing sessions. The crawler visits carefully selected Web pages in combination, so that the target item is pushed effectively. An example of such an attack is the *popular page* attack [79] in which the target page is crawled together with other popular pages. Such an attack can be viewed as an implicit version of the bandwagon attack.

12.4 Detecting Attacks on Recommender Systems

An adversarial relationship exists between attackers and the designers of recommender systems. From the point of view of maintaining a robust recommender system, the best way to thwart attacks is to detect them. Detection allows corrective measures (such as the removal of fake user profiles) to be taken. Accordingly, the detection of fake user profiles is a pivotal element in the design of a robust recommender system. However, the removal of fake profiles is a mistake-prone process, in which genuine profiles might be removed. It is important not to make too many mistakes, because the removal of authentic profiles can be counter-productive. On the other hand, the inability to remove fake profiles is also undesirable. This results in a natural trade-off between the precision and recall of fake profile removal. Correspondingly, attack detection algorithms are often measured in terms of this precision and recall. In fact, one can also use[1] the receiver operating characteristic (ROC) curve (see Chapter 7), which plots the trade-off between the true positive rate (TPR) and the false positive rate (FPR). An alternative way of evaluating the effectiveness of attack

---

[1]The ROC curve is used in a different context here than in Chapter 7. In Chapter 7, the ROC curve measures the effectiveness of ranking items for recommendations. Here, we measure the effectiveness of ranking user profiles based on their likelihood of being fake. However, the general principle of using the ROC curve is similar in both cases, because a ranking is compared with the binary ground-truth in both cases.

removal is by measuring the impact of profile removal on recommender system accuracy. For example, one can measure the mean absolute error both before and after the filtering of the profiles. The various detection algorithms can be compared in terms of this measure.

Almost all attacks use multiple profiles in order to undermine the recommender system. Therefore, the profiles may be removed either individually or as a group. Different attack algorithms have been designed for each case. Furthermore, attack detection algorithms may be either supervised or unsupervised. The difference between these two types of detection algorithms is as follows:

1. *Unsupervised attack detection algorithms:* In this case, ad hoc rules are used to detect fake profiles. For example, if a profile (or significant portion of it) is identical to many other profiles, then it is likely that all these profiles have been injected for the purpose of creating an attack. The basic idea in this class of algorithms is to identify the key characteristics of attack profiles that are not similar to genuine profiles. Such characteristics can be used to design unsupervised heuristics for fake profile detection.

2. *Supervised attack detection algorithms:* Supervised attack detection algorithms use classification models to detect attacks. Individual user profiles or groups of user profiles are characterized as multidimensional feature vectors. In many cases, these multidimensional feature vectors are derived using the same characteristics that are leveraged for the unsupervised case. For example, the number of profiles to which a given user profile is identical can be used as a feature for that user profile. Multiple features can be extracted corresponding to various characteristics of different types of attacks. A binary classifier can then be trained in which known attack profiles are labeled as $+1$, and the remaining profiles are labeled as -1. The trained classifier is used to predict the likelihood that a given profile is genuine.

Supervised attack detection algorithms are generally more effective than unsupervised methods because of their ability to learn from the underlying data. On the other hand, it is often difficult to obtain examples of attack profiles.

Attack detection methods are either *individual* profile detection methods or *group* profile detection methods. When detecting individual attack profiles, each user profile is assessed independently to determine whether or not it might be an attack. In the case of group detection, a set of profiles is assessed as a group. Note that both the unsupervised and supervised methods can be applied to either individual or group profile detection. In the following, we will discuss various methods for detecting attack profiles as individuals, and for detecting attack profiles as groups.

12.4.1 Individual Attack Profile Detection

Individual attack-profile detection is also referred to as *single attack-profile detection*. An unsupervised method for individual attack-profile detection is discussed in [158]. In this technique, a set of features is extracted from each user profile. The features are such that unusually high or unusually low values are indicative of an attack, depending on the feature at hand. In many cases, these features measure the *consistency* of a particular profile with other profiles in the system. Therefore, the fraction of features that take on abnormal values can be used as a measure to detect attacks. Other heuristic functions can also be used in conjunction with these features, which can enumerated as follows:

1. *Number of prediction differences (NPD):* For a given user, the *NPD* is defined as the number of prediction changes after removing that user from the system. Generally,

attack profiles tend to have larger prediction differences than usual, because the attack profiles are designed to manipulate the system predictions in the first place.

2. *Degree of disagreement with other users (DD):* For the ratings matrix $R = [r_{ij}]_{m \times n}$, let ν_j be the mean rating of item j. Then, the degree to which the user i differs from other users on item j is given by $|r_{ij} - \nu_j|$. This value is then averaged over all the $|I_i|$ ratings observed for user i to obtain the degree of disagreement $DD(i)$ of user i:

$$DD(i) = \frac{\sum_{j \in I_i} |r_{ij} - \nu_j|}{|I_i|} \tag{12.4}$$

Users with a larger degree of disagreement with other users are more likely to be attack profiles. This is because attack profiles tend to be different from the distribution of the other ratings.

3. *Rating deviation from mean agreement (RDMA):* The rating deviation from mean agreement is defined as the average absolute difference in the ratings from the mean rating of an item. The mean rating is biased with the inverse frequency if_j of each item j while computing the mean. The inverse frequency if_j is defined as the inverse of the number of users that have rated item j. Let the biased mean rating of an item j be ν_j^b. Let I_i be the set of items rated by user i. Then, the value $RDMA(i)$ for user i is defined as follows:

$$RDMA(i) = \frac{\sum_{j \in I_i} |r_{ij} - \nu_j^b| \cdot if_j}{|I_i|} \tag{12.5}$$

Note the presence of the inverse frequency if_j in the aforementioned equation, so that rare items are given greater importance. It is instructive to compare this equation with Equation 12.4, which does not use such weightings at any stage of the computation. Larger values of this metric indicate the possibility that the user profile might represent an attack.

4. *Standard deviation in user ratings:* This is the standard deviation in the ratings of a particular user. If μ_i is the average rating of user i, and I_i is the set of items rated by that user, then the standard deviation σ_i is computed as follows:

$$\sigma_i = \frac{\sum_{j \in I_i} (r_{ij} - \mu_i)^2}{|I_i| - 1} \tag{12.6}$$

Even though the ratings of fake profiles differ significantly from *other* users, they are often quite *self*-similar because many filler items are set to the same rating value. As a result, the standard deviation σ_i tends to be small for fake profiles.

5. *Degree of similarity with top-k neighbors (SN):* In many cases, attack profiles are inserted in a coordination fashion, with the result being that the similarity of a user with her closest neighbors is increased. Therefore, if w_{ij} is the similarity between the users i and j, and $N(i)$ is the set of neighbors of user i, then the degree of similarity $SN(i)$ is defined as follows:

$$SN(i) = \frac{\sum_{j \in N(i)} w_{ij}}{|N(i)|} \tag{12.7}$$

The value of w_{ij} can be computed with any standard user-user similarity metric, such as the Pearson correlation coefficient.

It is noteworthy that most of these metrics, with the exception of *RDMA*, have also been proposed [43] in the context of finding influential users in a recommender system. This coincidence is because fake profiles are designed by attackers to manipulate the predicted ratings as unusually influential entities in the recommender system. Furthermore, all these metrics, with the exception of the standard deviation, take on larger values in the case of an attack profile. The algorithm in [158] declares a profile to be an attack when all these metrics take on abnormal values in the direction indicative of an attack. Many variations of these basic principles are also possible for designing attack detection methods. Other features may be extracted as well. For example, the presence of an unusually large number of ratings in a profile may be considered suspicious [630].

The aforementioned features are useful not only for unsupervised attack detection algorithms, but also for supervised methods. The main difference between supervised and unsupervised methods is that examples of previous attacks are available. In such cases, these features are used to create a multidimensional representation and a classification model is constructed. For a given user profile for which the attack behavior is unknown, these features can be extracted. The classification model, which is built on the training data of example attacks, can be used on these features to assess the likelihood that it is indeed an attack.

An example of such a supervised attack detection algorithm is discussed in [124]. The metrics discussed above are used as features for the attack detection algorithm. In addition to these features, a number of generic and model-specific features were introduced. *Model-specific features* are designed to detect a specific type of attack, such as an average attack or segment attack. The generic features introduced in [124] are as follows:

1. *Weighted deviation from mean agreement (WDMA):* The *WDMA* metric is similar to the *RDMA* metric, but it places greater weight on the ratings of rare items. Therefore, the square of the inverse frequency is used instead of the inverse frequency in the *WDMA* computation. Therefore, using the same notations as Equation 12.5, the *WDMA* feature is computed as follows:

$$WDMA(i) = \frac{\sum_{j \in I_i} |r_{ij} - \nu_j| \cdot if_j^2}{|I_i|} \qquad (12.8)$$

2. *Weighted degree of agreement (WDA):* The second variation of the *RDMA* metric uses only the numerator of the *RDMA* metric, defined by the right-hand side of Equation 12.5:

$$WDA(i) = \sum_{j \in I_i} |r_{ij} - \nu_j| \cdot if_j \qquad (12.9)$$

3. *Modified degree of similarity:* The modified degree of similarity is computed in a similar way to the degree of similarity defined by Equation 12.7. The main difference is that the similarity value w_{ij} in Equation 12.7 is proportionally discounted by the number of users who rate both items i and j. This discounting is based on the intuition that the computed similarity is less reliable when the number of items in common between users i and j is small.

In addition, a number of model-specific features have been used in [124]. Readers are referred to [124] for details of these features. Three different algorithms corresponding to the k-nearest neighbor classifier, the C4.5 decision tree, and the support vector machine were tested. It was found that these different classifiers had different trade-offs between precision and recall of fake profile detection, with the support vector machine providing the best overall performance.

12.4.2   Group Attack Profile Detection

In these cases, the attack profiles are detected as groups rather than as individuals. The basic principle here is that the attacks are often based on groups of related profiles, which are very similar. Therefore, many of these methods use clustering strategies to detect attacks. Some of these methods perform the detection at recommendation time [397], whereas others use more conventional preprocessing strategies [427] in which detection is performed *a priori*, with the fake profiles are removed up front.

12.4.2.1   Preprocessing Methods

The most common approach is to use clustering to remove fake profiles. Because of the way in which attack profiles are designed, authentic profiles and fake profiles create separate clusters. This is because many of the ratings in fake profiles are identical, and are therefore more likely to create tight clusters. In fact, the relative tightness of the clusters containing the fake profiles is one way of detecting them. The method proposed in [427] uses *PLSA* to perform the clustering of the user profiles. Note that *PLSA* already creates a soft clustering, in which each user profile has a particular probability of belonging to an aspect. This soft clustering is converted to a hard clustering by assigning each user profile to the cluster with which it has the largest probability of membership. Although the *PLSA* approach is used for clustering in this case, virtually any clustering algorithm can be used in principle. After the hard clusters have been identified, the average Mahalanobis radius of each cluster is computed. The cluster with the smallest Mahalanobis radius is assumed to contain fake users. This approach is based on the assumption of the relative tightness of the clusters containing fake profiles. Such an approach works well for relatively overt attacks, but not necessarily for subtle attacks.

A simpler approach uses only principal component analysis (*PCA*) [425]. The basic idea is that the covariance between fake users is large. On the other hand, fake users often exhibit very low covariances with other users, when the users are treated as dimensions. How can one identify such highly inter-correlated dimensions with *PCA*, which are not correlated with the normal users? This problem is related to that of *variable selection* in *PCA* [285]. Let us examine the transpose of the ratings matrix in which users are treated as dimensions. According to the theory of variable selection in principal component analysis [427], this problem amounts to that of finding the dimensions (users in the transposed ratings matrix) with small coefficients in the small eigenvectors. Such dimensions (users) are likely to be fake profiles.

The ratings matrix is first normalized to zero mean and unit standard deviation and then the covariance matrix of its transpose is computed. The smallest eigenvector of this matrix is computed. Those dimensions (users) with small contributions (coefficients) in the eigenvector are selected. A slightly more enhanced approach is discussed in [427]. In this case, the top (smallest) 3 to 5 eigenvectors are identified instead of using only the smallest eigenvector. The sum of the contributions over these 3 to 5 eigenvectors is used in order to determine the spam users.

Another algorithm to detect the group profiles is the *UnRAP* algorithm [110]. In the *UnRAP* algorithm, a measure called the H_v-score is used. This measure is adapted from the bioinformatics area, where it is used in the context of biclustering of gene clusters. Let μ_i be the mean rating of user i, ν_j be the mean rating of item j, γ be the mean over all ratings, and I_i be the set of items rated by user i. Then, the H_v-score of user i is defined as follows:

$$H_v(i) = \frac{\sum_{j \in I_i}(r_{ij} - \mu_i - \nu_j + \gamma)^2}{(r_{ij} - \mu_i)^2} \qquad (12.10)$$

Larger values of the H_v-score are more indicative of an attack profile. The basic idea is that fake profiles tend to be self-similar in rating values, but they tend to be different from other users. This is captured by the H_v-score because of the way in which the numerator and denominator are constructed. When the ratings are random, the H_v-score will be close to 1. The algorithm first determines the top-10 users with the largest H_v-scores. This set of users is then used to identify the target item that deviates the most from the mean user rating.

The identification of the target item then sets the stage for the next phase of the algorithm. The criterion for considering users to be candidates for being fake is then relaxed, and more than 10 user profiles are considered as candidates for being fake. However, such candidates will contain many false positives. The *UnRAP* algorithm also discusses methods to remove those users that have not rated the target item, or who have rated the target item in the "wrong" direction. Refer to [110] for details of how the larger candidate set is computed with the use of a sliding-window method.

12.4.2.2 Online Methods

In these methods, the fake profiles are detected during recommendation time. Consider a scenario in which a user-based neighborhood algorithm is used during recommendation time. The basic idea is to create two clusters from the neighborhood of the active user [397]. Note that the main goal of the attacker is to either push or nuke a particular item. Therefore, if a sufficiently large difference exists in the average ratings of the active items in the two clusters, it is assumed that an attack has taken place. The cluster in which the active item has the smaller variance of ratings is assumed to be the attack cluster. All profiles in this attack cluster are removed. This detection method has the merit of being able to be directly integrated into attack-resistant recommendation algorithms during the process of neighborhood formation. Therefore, this approach is not just a method to remove fake profiles, but also an online method for providing more robust recommendations. If desired, the fake profiles can be removed incrementally during the operation of the system.

12.5 Strategies for Robust Recommender Design

A variety of strategies are available for building recommenders in a more robust way. These strategies range from the use of better recommender-system design to better algorithmic design. In the following sections, we will discuss the use of some of these strategies.

12.5.1 Preventing Automated Attacks with CAPTCHAs

It is noteworthy that it requires a significant number of fake profiles in order to result in a significant shift in the predicted ratings. It is not uncommon for the adversary to require between 3% to 5% of the number of authentic profiles to be fake profiles to initiate an attack. For example, consider a ratings matrix containing over a million authentic users.

Figure 12.4: An example of a CAPTCHA from the official CAPTCHA site (http://www.captcha.net)

In such a case, as many as 50,000 fake profiles may be required. It is hard to insert so many fake profiles manually. Therefore, attackers often resort to automated systems to interact with the Web interface of the rating system, and insert the fake profiles.

How can one detect such automated attacks? CAPTCHAs are designed [619] to tell the difference between humans and machines in the context of Web interaction. The acronym CAPTCHA stands for "Completely Automated Public Turing test to tell Computers and Humans Apart." The basic idea is to present a human with distorted text, which is hard for a machine to decipher but can still be read by a human. This distorted text serves as a "challenge" text or word that needs to be entered into the Web interface in order to allow further interaction. An example of a CAPTCHA is illustrated in Figure 12.4. The recommender system can prompt for CAPTCHAs to allow the entry of ratings, especially when a large number of them are entered from the same IP address.

12.5.2   Using Social Trust

The previous chapter reviewed methods for using social trust in the context of a recommender system. In these methods, the social trust between participants is used to influence the ratings. For example, users may specify trust relationships based on their experience with the ratings of other users. These trust relationships are then used to make more robust recommendations. Such methods are able to reduce the effectiveness of attacks, because users are unlikely to specify trust relationships towards fake profiles, which are rather contrived. Detailed discussions of how social trust is used for more effective recommendations are provided in Chapter 11.

The work in [502, 503] proposes an algorithm, referred to as the *influence limiter*, to build trustworthy recommender systems. A global measure of the reputation of each user is used in the recommendation process. Each user is weighted with her reputation score while making the recommendation. The reputation is itself learned on the basis of the accuracy of a user predicting the rating of her neighbors. A theoretical bound on the impact of a negative attack was also shown by this work.

12.5.3   Designing Robust Recommendation Algorithms

It is evident from the discussion in this chapter that different algorithms have different levels of susceptibility to attacks. For example, user-based algorithms are generally much easier to attack than item-based algorithms. Therefore, a number of algorithms have specifically been designed with attack resistance in mind. This section will discuss some of these algorithms.

12.5.3.1 Incorporating Clustering in Neighborhood Methods

It has been shown in [446], how clustering can be used in the context of neighborhood-based methods. This work clusters the user profiles with the use of *PLSA* and k-means techniques. An aggregate profile is created from each cluster. The aggregate profile is based on the average rating of each item in the segment. Then, a similar approach to user-based collaborative filtering is used, except that the aggregate (clustered) profiles are used instead of the individual profiles. For each prediction, the closest aggregated profiles to the target users are used to make recommendations. It was shown in [446] that the clustering-based approach provides significantly more robust results than a vanilla nearest-neighbor method. The main reason for the robustness of this approach is that the clustering process generally maps all the profiles to a single cluster, and therefore limits its influence on the prediction when alternative clusters are available.

12.5.3.2 Fake Profile Detection during Recommendation Time

The attack detection algorithms discussed in the earlier sections can also be used to make robust recommendations, particularly when the detection is done during recommendation time. Such a method is discussed in section 12.4.2.2. In this approach, the neighborhood of the active user is partitioned into two users. An attack is suspected when the active item has very different average values in the two clusters. The cluster that is the most self-similar (i.e., smaller radius) is considered the attack-cluster. The profiles from this cluster are then removed. The recommendations are then performed using the profiles from the remaining cluster. This approach has the dual purpose of being both an attack-detection method and a robust recommendation algorithm.

12.5.3.3 Association-Based Algorithms

Rule-based collaborative filtering algorithms are discussed in section 3.3 of Chapter 3. It was shown in [522] that such algorithms are robust to the average attack when the maximum attack size is less than 15%. The reason for this phenomenon is that there is generally not sufficient support for the attack profiles in order to mount a successful attack. However, such an algorithm is not immune to the segment attack.

12.5.3.4 Robust Matrix Factorization

Matrix factorization methods are generally more robust to attacks because of their natural ability to treat the attack profiles as noise. It has been shown in [424, 427], how *PLSA* methods can be used to detect and remove attacks. Note that many matrix factorization recommenders are themselves based on *PLSA*. Therefore, if the attack profiles are removed in the intermediate step and the probabilistic parameters are renormalized, they can directly be used for recommendation.

Another approach [428] is to modify the optimization function used for matrix factorization to make it more robust to attacks. In matrix factorization, the $m \times n$ ratings matrix R is factorized into user factors and item factors as follows:

$$R \approx UV^T \tag{12.11}$$

Here $U = [u_{is}]$ and $V = [v_{js}]$ are $m \times k$ and $n \times k$ matrices. The predicted value $\hat{r}_{ij}$ of an entry is as follows:

$$\hat{r}_{ij} = \sum_{s=1}^{k} u_{is} v_{js} \tag{12.12}$$

Therefore, the error of predicting an observed entry is given by $e_{ij} = r_{ij} - \hat{r}_{ij}$. As discussed in Chapter 3, the matrix entries of U and V are determined by minimizing the sum of squares of e_{ij} over all the *observed* entries in the matrix R, along with some regularization terms.

How can one change the objective function to de-emphasize the contribution of attacking profiles? The main insight here is attacking profiles often cause outlier entries with large absolute values $|e_{ij}|$ in the *residual matrix* $(R - UV^T)$. Therefore, if one simply used the Frobenius norm of the observed portion of $(R - UV^T)$, the presence of fake profiles would significantly change the user factors and item factors. The natural solution is to de-emphasize the contribution of entries in the residual matrix with large absolute values. Let S be the set of observed entries in the ratings matrix R. In other words, we have:

$$S = \{(i,j) : r_{ij} \text{ is observed}\} \tag{12.13}$$

As discussed in Chapter 3, the objective function of matrix factorization is defined as follows:

$$\text{Minimize } J = \frac{1}{2} \sum_{(i,j) \in S} e_{ij}^2 + \frac{\lambda}{2} \sum_{i=1}^{m} \sum_{s=1}^{k} u_{is}^2 + \frac{\lambda}{2} \sum_{j=1}^{n} \sum_{s=1}^{k} v_{js}^2$$

In order to de-emphasize the impact of very large absolute values of e_{ij}, a new set of error terms is defined:

$$\epsilon_{ij} = \begin{cases} e_{ij} & \text{if } |e_{ij}| \le \Delta \\ f(|e_{ij}|) & \text{if } |e_{ij}| > \Delta \end{cases} \tag{12.14}$$

Here Δ is a user-defined threshold, which defines the case when an entry becomes large. $f(|e_{ij}|)$ is a *damped* (i.e., sublinear) function of $|e_{ij}|$ satisfying $f(\Delta) = \Delta$. This condition ensures that ϵ_{ij} is a continuous function of e_{ij} at $e_{ij} = \pm\Delta$. The damping ensures that large values of the error are not given undue importance. An example of such a damped function is as follows:

$$f(|e_{ij}|) = \sqrt{\Delta(2|e_{ij}| - \Delta)} \tag{12.15}$$

This type of damped function has been used in [428]. The objective function for robust matrix factorization then replaces the error values e_{ij} with the adjusted values ϵ_{ij} as follows:

$$\text{Minimize } J^{robust} = \frac{1}{2} \sum_{(i,j) \in S} \epsilon_{ij}^2 + \frac{\lambda}{2} \sum_{i=1}^{m} \sum_{s=1}^{k} u_{is}^2 + \frac{\lambda}{2} \sum_{j=1}^{n} \sum_{s=1}^{k} v_{js}^2$$

An iterative re-weighted least-squares algorithm, which is described in [426], is used for the optimization process. Here, we describe a simplified algorithm. The first step is to compute the gradient of the objective function J^{robust} with respect to each of the decision variables:

$$\frac{\partial J^{robust}}{\partial u_{iq}} = \frac{1}{2} \sum_{j:(i,j) \in S} \frac{\partial \epsilon_{ij}^2}{\partial u_{iq}} + \lambda u_{iq}, \quad \forall i \in \{1 \ldots m\}, \forall q \in \{1 \ldots k\}$$

$$\frac{\partial J^{robust}}{\partial v_{jq}} = \frac{1}{2} \sum_{i:(i,j) \in S} \frac{\partial \epsilon_{ij}^2}{\partial v_{jq}} + \lambda v_{jq} \quad \forall j \in \{1 \ldots n\}, \forall q \in \{1 \ldots k\}$$

Note that the aforementioned gradients contain a number of partial derivatives with respect to the decision variables. The value of $\frac{\partial \epsilon_{ij}^2}{\partial u_{iq}}$ can be computed as follows:

$$\frac{\partial \epsilon_{ij}^2}{\partial u_{iq}} = \begin{cases} 2 \cdot e_{ij}(-v_{jq}) & \text{if } |e_{ij}| \le \Delta \\ 2 \cdot \Delta \cdot \text{sign}(e_{ij})(-v_{jq}) & \text{if } |e_{ij}| > \Delta \end{cases}$$

Here, the sign function takes on the value of $+1$ for positive quantities and -1 for negative quantities. The case-wise description of derivative can be consolidated to simplified form as follows:

$$\frac{\partial \epsilon_{ij}^2}{\partial u_{iq}} = 2 \cdot \min\{|e_{ij}|, \Delta\} \cdot \text{sign}(e_{ij}) \cdot (-v_{jq})$$

It is noteworthy that the gradient is damped when the error is larger than Δ. This damping of the gradient directly makes the approach more robust to a few large errors in the ratings matrix. Similarly, we can compute the partial derivative with respect to v_{jq} as follows:

$$\frac{\partial \epsilon_{ij}^2}{\partial v_{jq}} = \begin{cases} 2 \cdot e_{ij}(-u_{iq}) & \text{if } |e_{ij}| \le \Delta \\ 2 \cdot \Delta \cdot \text{sign}(e_{ij})(-u_{iq}) & \text{if } |e_{ij}| > \Delta \end{cases}$$

As before, it is possible to consolidate this derivative as follows:

$$\frac{\partial \epsilon_{ij}^2}{\partial v_{jq}} = 2 \cdot \min\{|e_{ij}|, \Delta\} \cdot \text{sign}(e_{ij}) \cdot (-u_{iq})$$

One can now derive the update steps as follows, which need to be executed for each user i and each item j:

$$u_{iq} \Leftarrow u_{iq} + \alpha \left(\sum_{j:(i,j) \in S} \min\{|e_{ij}|, \Delta\} \cdot \text{sign}(e_{ij}) \cdot v_{jq} - \lambda \cdot u_{iq} \right) \quad \forall i, \ \forall q \in \{1 \ldots k\}$$

$$v_{jq} \Leftarrow v_{jq} + \alpha \left(\sum_{i:(i,j) \in S} \min\{|e_{ij}|, \Delta\} \cdot \text{sign}(e_{ij}) \cdot u_{iq} - \lambda \cdot v_{jq} \right) \quad \forall j, \ \forall q \in \{1 \ldots k\}$$

These updates are performed to convergence. The aforementioned steps correspond to global updates. These updates can be executed within the algorithmic framework of gradient descent (cf. Figure 3.8 of Chapter 3).

One can also isolate the gradients with respect to the errors in individual entries and process them in random order. Such an approach corresponds to *stochastic* gradient descent. For each observed entry $(i, j) \in S$, the following update steps are executed:

$$u_{iq} \Leftarrow u_{iq} + \alpha \left(\min\{|e_{ij}|, \Delta\} \cdot \text{sign}(e_{ij}) \cdot v_{jq} - \frac{\lambda \cdot u_{iq}}{n_i^{user}} \right) \quad \forall q \in \{1 \ldots k\}$$

$$v_{jq} \Leftarrow v_{jq} + \alpha \left(\min\{|e_{ij}|, \Delta\} \cdot \text{sign}(e_{ij}) \cdot u_{iq} - \frac{\lambda \cdot v_{jq}}{n_j^{item}} \right) \quad \forall q \in \{1 \ldots k\}$$

Here n_i^{user} denotes the number of observed ratings for user i and n_j^{item} denotes the number of observed ratings for item j. One cycles through the observed entries in the matrix in random order and performs the aforementioned update steps until convergence is reached. This is based on the framework of Figure 3.9 (cf. Chapter 3) with the modified set of update steps discussed above. These update steps are different from traditional matrix factorization only in terms of capping the absolute values of the gradient components, when the error is larger than Δ. This is consistent with the stated goals of a robust matrix factorization approach, where large errors might be the result of anomalies in the ratings matrix structure. These anomalies might be indicative of attacks.

It is important to note that this approach will work only when the number of attack profiles is small compared to the correct entries in the ratings matrix. On the other hand, if the number of attack profiles is very large, it will significant affect the factor matrices, and the damping approach will not work. Robust matrix factorization and PCA has a rich history in the context of the recovery of the structure of corrupted matrices. Refer to the bibliographic notes for pointers to work in this area.

Intuitively, the notion of robust matrix factorization is not very different from that of robust regression, which is used commonly to reduce the impact of outliers in regression modeling [512]. In this case, the least-squares optimization function is modified in a similar way to robust matrix factorization. Indeed, robust regression modeling can be used to make many of the collaborative filtering methods in section 2.6 of Chapter 2 more robust. Although there are no existing experimental results on such methods, it is reasonable to assume that robust regression modeling methods are likely to be attack-resistant. This would be an interesting direction of future research in the field.

12.6   Summary

Shilling attacks can significantly reduce the effectiveness of recommender systems because of the presence of fake profiles, which distort the recommendations provided to real users. A variety of push attack methods have been designed in an attempt to influence recommender systems. Some of these include the random attack, the average attack, the bandwagon attack, and the segment attack. Another set of tactics exist, such as the reverse bandwagon attack and the love-hate attack, that are designed to "nuke" items (lower their ratings in the system). Nuke attacks are generally easier to carry out than are push attacks. Attack detection methods use a variety of common characteristics of attacks. These characteristics include the self-similarity of injected profiles, and the differences of these profiles from those of other users. Attack detection methods can be used to design robust recommendation systems. Many robust recommendation systems directly incorporate the fake-profile removal process deep into the recommender system. Other techniques use trustworthy recommender systems or increase fake-profile injection costs. The design of robust recommender systems is a perpetual game between attackers and recommender designers, in which increasingly clever measures and countermeasures are developed by both parties over time.

12.7   Bibliographic Notes

Surveys on shilling attacks and attack-resistant recommender systems may be found in [119, 236]. Attack-resistant methods for collaborative filtering are surveyed in [424]. The idea of using fake user profiles for attacking recommendation algorithms was used in [394]. Some of the earliest methods, such as the average attack and random attack, were proposed and evaluated in [122, 329]. The differential behavior of various recommendation algorithms was discussed in [329]. For example, it was shown that item-item recommendation algorithms are more robust to attack than user-user recommendation algorithms. A related problem is that of asking users to re-rate items to reduce the effect of *noise* [44] in recommender systems. However, noisy ratings are not necessarily the same as fake profiles, which are created intentionally to mislead the recommender system. The approach in [44], therefore, addresses a different scenario from attack-resistant models.

The bandwagon attack works effectively for user-user collaborative filtering algorithms, but it is not quite as effective for item-based algorithms [246, 329, 445]. The main advantage of the bandwagon attack is that it is nearly as effective as the average attack method, but it requires much less knowledge [329]. A discussion of the popular item attack, along with explanations of the prediction shift, is provided in [395]. The effectiveness of this attack is also studied in [396]. The segment attack was proposed in [445], and it was shown to be effective for item-item collaborative filtering algorithms. The segment attack is a generalization of the favorite item attack [123]. The two nuke attack models, namely the reverse bandwagon attack and the love/hate attack, were proposed in [444]. In group-shilling attacks [572], several human agents cooperate together to either push or nuke an item.

Most of the aforementioned attack systems are designed for the case of explicit ratings. Attack systems for implicit ratings require injection of fake *actions* rather than fake *profiles*. Such systems can be implemented with an automated crawler that simulates Web browsing sessions. The crawler visits carefully selected Web pages in combination, so that the target item is pushed effectively. An example of such an attack is the *popular page* attack in which the target page is crawled together with other popular pages. Such an attack can be viewed as an implicit version of the bandwagon attack. Refer to [79] for a discussion of these strategies.

An unsupervised algorithm for individual/single profile attack detection is discussed in [158]. This algorithm is based on the fact that users with undue influence on the ratings are suspicious. The approach uses a number of metrics discussed earlier for detecting influential users [43]. The presence of an unusually large number of ratings for a user profile may also be considered suspicious [630]. These methods were combined with the *RDMA* metric for unsupervised attack detection. These features were further combined with other features for supervised attack detection [124]. An attack-detection algorithm, which monitors the changes in the ratings over time, is proposed in [668]. The basic idea of this approach is that sudden fake profile injections often lead to anomalous temporal changes in the ratings over time, and they can therefore be detected with time-series monitoring. A related method [78] uses anomaly detection to detect attacks. A method for detection of group shilling attacks is discussed in [572]. In this approach, clusters of users are detected who have co-rated many items and provided atypical ratings compared to other ratings in the database; these clusters are generally fake profiles.

A number of methods for group-based attack detection have also been proposed [110, 425, 427]. The use of principal component analysis (*PCA*) for spam detection is discussed in [425]. The work in [427] discusses the use of *PLSA*-based clustering for group-attack detection. Enhancements of the *PCA* approach, originally discussed in [425], are presented in [427]. The *UnRAP* algorithm is discussed in [110].

A variety of methods can be designed to build attack-resistant recommender systems. CAPTCHAs have been designed [619] to tell humans and computers apart. Such CAPTCHAs can be used to increase the costs of injecting fake profiles into the system. The notion of social trust can also be used to reduce the effectiveness of attacks. Such systems are discussed in detail in Chapter 11. The notion of an influence limiter in order to build attack-resistant recommendation algorithms was proposed in [502, 503]. The integration of attack detection into attack-resistant recommendation algorithms is discussed in [397]. The use of association methods for building robust algorithms is discussed in [522]. A variety of robust matrix factorization methods for designing attack-resistant recommender systems are discussed in [424, 426–428, 609]. Methods for robust PCA and matrix factorization have also been proposed in the traditional machine-learning literature in other settings, where the low rank structure of corrupted data needs to be recovered [132]. A possible avenue for future research in this area is robust regression in order to reduce the impact of outliers on the recommendation process [512].

One of the challenges with attack-resistant recommender systems is that attackers continue to devise ever more sophisticated methods for attacking the recommender system. For example, attackers might use knowledge of the criteria for detecting attack profiles [397], use obfuscated methods to mount attacks [631], or design attack methods targeting specific collaborative filtering models [522]. Therefore, it is important for research to keep up with the advances in attack algorithms in a perpetual game between the attacker and recommendation system designer.

12.8 Exercises

1. For each of the attack methods discussed in this chapter, write a computer program to implement it.

2. Suppose that you are aware that an average attack has been mounted on your recommender system. Discuss a method to remove the fake profiles.

3. Suppose you had perfect knowledge available about the ratings in the recommender system. In other words, all the ratings in the recommender system are available to you. Show how to design an attack that would be hard to detect. [The answer to this question is not unique.]

4. Implement the online neighborhood method for attack detection (see section 12.4.2.2). Refer to the original publication [397] if needed.

Chapter 13

Advanced Topics in Recommender Systems

"In the last fifty years, science has advanced more than in the two thousand previous years and given mankind greater powers over the forces of nature than the ancients ascribed to their gods." – John Boyd Orr

13.1 Introduction

Recommender systems are often used in a number of specialized settings that are not covered in previous chapters of this book. In many cases, the recommendations are performed in settings where there might be multiple users or multiple evaluation criteria. For example, consider a scenario where a group of tourists wish to take a vacation together. Therefore, they may want to obtain recommendations that match the overall interests of the various members in the group. In other scenarios, users may use multiple criteria to provide ratings to items. These variations in the problem formulation can sometimes make the prediction problem more challenging. In particular, we will study the following advanced variations of recommender systems in this chapter:

1. *Learning to rank:* Most of the models discussed in the previous chapters treat the recommendation problem as a rating prediction problem in which the squared error of prediction is minimized. However, in real-life settings, users are presented only with the top-k recommendations, and the other predictions are ignored. Therefore, it makes sense to explore whether one can directly optimize ranking-based evaluation criteria, such as the mean reciprocal rank or the area under curve of the receiver operating characteristic curve.

© Springer International Publishing Switzerland 2016
C.C. Aggarwal, *Recommender Systems: The Textbook*,
DOI 10.1007/978-3-319-29659-3_13

411

2. *Online learning with multi-armed bandits:* In many recommendation domains, such as that of recommending news articles, the cold-start problem is pervasive. New articles and stories appear all the time, and the effectiveness of various algorithms may also vary with time. In such cases, it is crucial for the approach to continuously explore the space of various choices as new data are received. At the same time, the learned data are exploited in order to optimize the payoff in terms of the conversion rate. This type of trade-off between exploration and exploitation is managed with the help of multi-armed bandit algorithms.

3. *Group recommender systems:* In many settings, the recommendations may not be made to individuals, but to groups of users. Such recommendations are typically associated with activities undertaken by groups of users. Examples include a visit to the movies by a group, travel services bought by a group, the choice of music or television programs played or watched by a group, and so forth. In these cases, users may have varying tastes and interests that are reflected in their different choices. Group recommender systems are designed to work with these various trade-offs in order to make meaningful recommendations.

4. *Multi-criteria recommender systems:* In multi-criteria systems, ratings might be specified on the basis of different criteria by a single user. For example, a user might rate movies based on the plot, music, special effects, and so on. Such techniques often provide recommendations by modeling the user's utility for an item as a vector of ratings corresponding to various criteria. In fact, it has been shown [271, 410] that some of the methods for group recommender systems can also be adapted to multi-criteria recommender systems. However, the two topics are generally considered different because they emphasize different aspects of the recommendation process.

5. *Active learning in recommender systems:* Active learning is a well-known technique that is used in classification to acquire labels of training examples so as to maximize classification accuracy. The acquisition of labels is generally expensive; therefore, one must choose the training examples judiciously in order to maximize the accuracy of the classifier for a given cost budget. As the problem of recommendation can be viewed as a generalization of classification, the available methods for active learning can also be generalized to recommendations. Active learning provides methods of acquiring ratings within a given budget in order to maximize prediction accuracy.

6. *Privacy preservation in recommender systems:* Recommender systems are deeply dependent on the ability of users to voluntarily provide information about their interests. Such information is quite sensitive because it may reveal information about political opinions, sexual orientation, and so on. Therefore, it is crucial to develop privacy-preserving methods for the recommendation process. When there is a risk of public disclosure, owners of ratings data become less likely to release it. An example is the case of the Netflix Prize, in which a sequel to the contest was not pursued because of privacy concerns [714].

In addition to the aforementioned topics, this chapter will also study the application of recommender systems technology to a variety of application domains, such as news recommendations, computational advertising, and reciprocal recommender systems. The idea of studying these topics is to provide an understanding of how the methods discussed in various chapters are applied to these different domains. In some cases, the techniques discussed in these chapters cannot be applied directly, and therefore new techniques must be developed. Therefore, one of our goals is to provide an understanding of the limitations of the various methods used in current settings.

This chapter is organized as follows. The next section will introduce the problem of learning to rank. Multi-armed bandit algorithms are introduced in section 13.3. Various techniques for designing group recommender systems will be discussed in section 13.4. Multi-criteria recommender systems are discussed in section 13.5. Active learning methods are introduced in section 13.6. Methods for privacy in collaborative filtering are discussed in section 13.7. A number of interesting application domains are discussed in section 13.8. A summary is given in section 13.9.

13.2 Learning to Rank

Most of the models discussed in previous chapters treat the recommendation problem as a prediction problem in which the squared error of rating prediction is optimized. However, in practice, recommender systems rarely present all the ratings to the user. In practice, only the top-k items are presented to the user as a ranked list. Furthermore, the user is more likely to pay attention to the items at the top of the list than the lower-ranked items. The predicted values for the items not included in the list are irrelevant from the user perspective. In many cases, optimizing predicted values of the ratings may not provide the best recommendation lists to the end user. For example, if all the low-ranked ratings are predicted very accurately, but significant errors are made on the higher-ranked ratings, the resulting solution will not provide a high-quality recommendation list to the end-user. On the other hand, a prediction-based objective function may report it as a high-quality solution because the lower-ranked items are given equal importance. This problem arises because the objective functions of prediction-based methods are not fully aligned with the end-user experience.

The classical objective function used in optimization models for recommender systems (such as matrix factorization) is the aggregate squared error. This type of objective function is optimized to the RMSE measure used for evaluating recommender systems. It is also particularly easy to optimize from an algorithmic standpoint. This is one of the reasons that such prediction-based objective functions dominate the recommendation modeling landscape. However, as discussed in Chapter 7 on evaluating recommender systems, there are many rank-centric measures used for evaluating recommender systems. Such rank-centric measures can also be directly optimized in the context of collaborative filtering (or even content-based) models. As discussed in the chapter on evaluating recommender systems (cf. Chapter 7), there are two primary types of ranking measures:

1. *Global ranking measures:* Such measures evaluate the entire ranked lists of all the items. Examples, include the Kendall coefficient, Spearman coefficient, and the area under curve (AUC) of the receiver operating characteristic (ROC).

2. *Top-heavy ranking measures:* These are typically utility-based measures in which the top-ranked items are given much more importance. Examples of such measures include the normalized cumulative discounted gain (NDCG) and mean reciprocal rank (MRR). Such measures are often the most realistic from the perspective of the end-user because they ignore the lower-ranked items. Such items are not visible to the end-user in the recommended list.

Many of the ranking-based measures are used for evaluating implicit data settings. Correspondingly, many of the ranking-based learning methods are also designed in the context of implicit data settings.

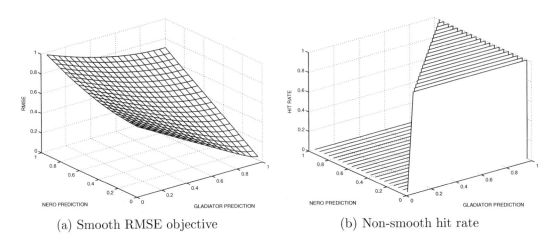

(a) Smooth RMSE objective   (b) Non-smooth hit rate

Figure 13.1: Ranking objective is not a smooth function of predicted ratings (and underlying model parameters)

For example, consider the problem of factorizing the ratings matrix R into user- and item-factors, U and V, respectively. One would like to determine U and V such that a specific ranking objective is optimized. Then, one might pose the optimization problem as follows:

Optimize $J = [$Objective function quantifying *ranking* evaluation between R and $UV^T]$
    subject to:
        Constraints on U and V

As in traditional matrix factorization, it is possible to add a regularizer to improve the generalization power of the objective function. The constraints on U and V might depend on the specific application setting. For example, in an implicit feedback setting, one might impose nonnegativity constraints on U and V. The optimization objective function might be derived from ranking-based measures such as the NDCG, MRR, AUC, and so on. A specific example of such a matrix factorization method, which optimizes the AUC, is discussed in [432]. In this work, the link recommendation problem is solved with the use of the AUC-based objective.

The main challenge in ranking-based methods is that the underlying objective functions are often *non-smooth* [490], which can be hard to optimize with off-the-shelf gradient-descent techniques. Tiny changes in the predicted ratings can change the item rankings and the corresponding objective functions suddenly. For example, consider a setting where there are two movies *Nero* and *Gladiator*, with true ratings 0 and 1, respectively. The predicted ratings can be converted to ranks, and the top-1 recommended movie can be reported. The (smooth) RMSE for various combinations of predicted ratings is shown in Figure 13.1(a), whereas the (non-smooth) hit-rate of the top-1 predicted rating is shown in Figure 13.1(b). Note the sudden jump in objective function in the case of Figure 13.1(b) at particular values of the predicted ratings. In the case of ranking-based objective functions, such non-smooth jumps or drops can occur with small changes in not just the predicted values but also the underlying model parameters. For example, in matrix factorization methods, tiny changes in the parameters of the user and item factors can cause sudden jumps or drops in ranking-based objectives. Such non-smooth changes are not observed with conventional measures such as the squared error, which are much easier to optimize. For example, a gradient-descent method would have difficulty in determining the correct descent direction with a non-smooth objective function because important changes in the objective function might occur at non-differentiable points in the parameter space. To get around this problem, smooth approximations of the underlying objective functions are often used. For each individual ranking-based objective, a specific lower-bound or approximation is used to design a smooth variation of the underlying objective function. Since these smooth variations are only approximations, the quality of the algorithm will often depend on that of the underlying approximation. In the following, we provide a brief discussion of some of the common ranking-based methods.

The traditional approach to ranking is to first predict the ratings with a loss function and then rank the items using the predicted ratings. One can view this approach as a *pointwise* methodology. Many of these methods are not specifically optimized to ranking because they focus on predicting the values of the ratings. A particularly notable work in this category is *OrdRec* [314], which treats ratings as ordinal values rather than as numerical values. There are two other primary types of methods that are specifically optimized to rank-centric learning, and they are referred to as *pairwise* or *listwise* rank-learning methods [128]. In the following, we will discuss these different types of rank learning methods.

13.2.1 Pairwise Rank Learning

In pairwise rank learning, pairs of items for which the users have provided preferences are used as the training data. Each pair contains only information about whether the first item of the pair is preferred to the second one or not, with a +1 or a −1, respectively. For example, consider a scenario where John has provided ratings for *Terminator*, *Alien*, and *Gladiator*, as 4, 3, and 5, respectively. Then, one can create the following pairs of training points:

John, *Terminator*, *Alien*, +1
John, *Terminator*, *Gladiator*, −1
John, *Alien*, *Gladiator*, −1

One can generate similar pairs for Peter, Bob, Alice, and so on, to create the training data across all users. For implicit feedback data sets, one can treat unobserved values as 0s. With this training data, one can now try to learn the relative item preferences such as the following:

Alice, *Terminator*, *Gladiator*, ?
Bob, *Terminator*, *Gladiator*, ?
John, *Nero*, *Cleopatra*, ?

Note that this transformation essentially creates a binary classification problem, and the learning method implicitly tries the minimize the number of pairwise inversions in the training data. This objective is intimately related to the Kendall rank correlation coefficient. It is also possible to optimize other measures such as the AUC in this setting. One can use any off-the-shelf ranking classifier (such as ranking SVMs) to learn an appropriate ranking objective. The main challenge in doing so is that the underlying representation is very sparse, since each training example contains only three nonzero elements of the form $\langle User, Item1, Item2 \rangle$. Note that the base dimensionality might contain hundreds of thousands of users and items. Such a setting is particularly well-suited to factorization machines (cf. section 8.5.2.1 of Chapter 8). With m users and n items, one can create a $p = (m+2 \cdot n)$-dimensional binary representation $x_1 \ldots x_p$, such that exactly three elements of the representation are set to 1. The remaining elements are set to 0. The m elements in the representation correspond to users and the $2 \cdot n$ elements correspond to pairs of items. The predicted value $y(\overline{x})$ is either $+1$ or -1 depending on whether or not the rating of the first item is larger than the second one. Then, the prediction function of Equation 8.9 is modified to a form used in logistic regression:

$$P(y(\overline{x}) = 1) = \frac{1}{1 + \exp(-[g + \sum_{i=1}^{p} b_i x_i + \sum_{i=1}^{p} \sum_{j=i+1}^{p} (\overline{v_i} \cdot \overline{v_j}) x_i x_j])} \tag{13.1}$$

The model parameters g, b_i, and $\overline{v_i}$ are defined in the same way as in section 8.5.2.1 of Chapter 8. A log-likelihood criterion can be optimized to learn the underlying model parameters with a gradient-descent approach. Factorization machines also provide the flexibility to do the feature engineering in other ways [493]. For example, one can use the $(m + n)$-dimensional binary representation $x_1 \ldots x_{m+n}$, in which two entries are nonzero (corresponding to user-item combination) and assume that the prediction $y(\overline{x})$ is equal to the value of the rating. Then, one can directly optimize a ranking objective function over pairs of predictions $(y(\overline{x_i}), y(\overline{x_j}))$, depending on which one is larger in the observed data. The main difference between this approach and the previous one is that the current approach optimizes pairwise ranking over all pairs $(y(\overline{x_i}), y(\overline{x_j}))$ (irrespective of whether $\overline{x_i}$ and $\overline{x_j}$ correspond to the same user), whereas the previous one does not allow the ratings of a particular pair to belong to different users.

Other well-known models used to learn these predictions include the Bayesian personalized ranking model (BPR) [499], EigenRank model [367], pLPA [368], and CR [59]. Many of these methods use ranking-based measures in the underlying objective function for learning.

13.2.2   Listwise Rank Learning

In listwise rank learning, the quality of the entire list is evaluated by using a ranking-based objective function. Examples of such objective functions include the normalized cumulative discounted gain (NDCG), mean reciprocal rank (MRR), and so on. One can view an ordered list as a permutation of items with a specific objective function value, depending on the ranking measure. Therefore, the key is to devise an optimization model that can determine this permutation directly. These methods generally tend to be more focussed on implicit feedback matrices because of the natural importance of ranking-based methods in these methods. Some examples of listwise methods are as follows:

1. *CoFiRank:* This approach [624, 625] is optimized for maximizing the NDCG measure with the use of *structured estimation* methods. A structured estimation method is designed to work for complex output domains such as sequences. One can view the output of a listwise method to belong to a structured output domain because a list is also an ordered sequence. The idea is to define a structured loss function that works on lists rather than individual points, and whose optimization results in the best possible ranking. The basic idea is that the dot product of a permutation of the predicted ratings of all items with the sorted vector $\overline{c} = (\frac{1}{\sqrt[4]{2}}, \frac{1}{\sqrt[4]{3}}, \ldots \frac{1}{\sqrt[4]{n+1}})$ is maximized when the predicted ratings are in decreasing order (based on the Polya-Littlewood-Hardy inequality). In other words, the dot product $\overline{c} \cdot \overline{r_u}^\pi$ of $\overline{c}$ with the corresponding permutation $\overline{r_u}^\pi$ of the estimated ratings $\overline{r_u}$ is maximized when the ratings in $\overline{r_u}^\pi$

are in decreasing order. The overall loss function is defined by maximizing a sum of $1 - NDCG(\pi)$ and $\bar{c} \cdot (\overline{r_u}^\pi - \overline{r_u})$ over all possible values of π. An upper bound on the loss function can be shown because of the Poly-Littlewood-Hardy inequality. The loss function is summed over all users, and a maximum margin optimization problem is defined in order to determine the optimal value of the predicted ratings.

2. *CLiMF:* This approach [545, 546] optimizes the mean-reciprocal rank (MRR), which has the tendency to obtain at least a few interesting items at the top of the list. The basic idea is to determine a smoothed version of the MRR for optimization and determine a lower bound on this smoothed version. Note that this approach is designed for implicit feedback data sets because of its use of the MRR. A related method, referred to as xCLiMF, is designed for explicit ratings.

Numerous other methods have been proposed for incorporating context into such methods [549].

One can further improve the quality of ranking methods with ensemble learning. Multiple techniques are used to learn the rankings, and the different sets of ranks are aggregated into a single ranked list. This problem is that of *rank aggregation* [190]. For example, one might use the average or median rank across different ensemble learners to re-rank the items. However, other sophisticated methods are possible, such as the use of the best rank across different learners or combining the two methods in some way. The median rank is known to have several desirable theoretical properties in terms of the quality of the aggregation. This area remains relatively unexplored, and is a good candidate for future research.

13.2.3 Comparison with Rank-Learning Methods in Other Domains

An excellent tutorial on ranking methods for recommendations may be found in [323]. It is noteworthy that the dichotomy between prediction-based and ranking-based models also exists in classification and regression modeling. For example, ranking support vector machines were introduced in [284] in the context of an internet search engine. Gradient-descent methods for ranking were discussed in [115] with a neural network model. Neural networks have the advantage that they are universal function approximators, and therefore multi-layer neural networks can often be quite effective with ranking-based cost functions. An elaborate tutorial on the ranking problem in the context of machine learning may be found in [15]. The typical application discussed in this class of works is that of internet search, which can also be viewed as a kind of recommendation. Since the recommendation problem can be viewed as a generalization of classification and regression modeling, it is natural to also design ranking variations of recommendation algorithms. In fact, ranking variations are much more important in the context of recommender design because most users are presented only with restricted sets of ranked lists rather than predicted values. Such methods have also been explored extensively in the context of ranking methods in the information retrieval domain. A tutorial on such methods may be found in [370], and the methods strongly overlap with those used in the machine learning literature for internet search [15, 115, 284]. The methods from information retrieval can be used to directly improve the effectiveness of content-based methods in the recommendation domain.

13.3   Multi-Armed Bandit Algorithms

An important challenge in many recommendation settings is that new users and items constantly appear in the system and it is important for the recommender system to constantly adapt to the changing patterns in the data. Therefore, unlike offline recommendation algorithms, the approach needs to simultaneously *explore* and *exploit* the search space of recommendations. Each time an opportunity arises to show a recommendation to a user, the recommender system has to choose between a number of strategies, objects, or algorithms that decide what is shown to the user. These choices may be different, depending on the application domain at hand. Some examples are as follows:

1. The system might use a number of different recommendation algorithms, which might be more or less effective with different users. For example, a user who prefers a high level of customization might be better served with a knowledge-based recommender system, whereas a "lazy" user might be better served with a collaborative recommender that does most of the work for her. Therefore, the approach may constantly need to learn the best choice of strategy for each user.

2. A special (and important) case of the aforementioned setting is one where each strategy corresponds to recommending a specific item. For example, a news portal might show articles from various topics to a particular user over a period of time, and then bias the article presentation depending on the previous history of interest (i.e., clicks) on the various articles. In the context-free case, the recommendation is independent of the user. However, in practice, a feature vector is associated with each user, which characterizes the interest of the user in a specific topic. This provides the means to incorporate personalization in multi-armed bandit algorithms. If a user is more interested in sports and entertainment, then the recommender system needs to learn this fact during the operation of the system, and frequently show recommendations belonging to these topics to that individual.

The main challenge in these systems is that new users and new articles constantly enter the system; therefore, one must *simultaneously* learn the user interests and exploit these interests *during the operation of the system.* This is different from the offline setting discussed in this book. This problem is related to that of *reinforcement learning*, in which exploration and exploitation of the search space are performed simultaneously. One such important class of reinforcement learning algorithms is that of the multi-armed bandit algorithms.

This class of algorithms derives its name from the fact that one can view the recommender system in a manner similar to a gambler in a casino, who is faced with a choice of a number of slot machines (recommendation algorithms or strategies). This scenario is illustrated in Figure 13.2. By pulling the arms of each of these machines, the gambler will receive a payoff with a specific probability distribution. The gambler suspects that one of these slot machines might have higher (expected) payoff than the others, although it is impossible for the gambler to identify this machine without playing all the machines. Playing these machines for learning purposes can be viewed as an *exploration* of the search space of strategies. Of course, this learning phase is likely to waste trials because it is not optimized to the best paying machine. However, once the gambler learns that one of these machines has a better payoff, he or she can play that machine to achieve a better payoff. Like all reinforcement learning algorithms, multi-armed bandit algorithms are faced with a natural trade-off between exploration and exploitation of the search space.

Let us explain this scenario in the context of Web page recommender systems. Whenever a recommender system has to decide on the recommendation of a Web page to a user, it is faced with a number of different choices of strategies. For example, the recommender system

Figure 13.2: The multi-armed bandit analogy

may have to decide on the choice of Web pages to recommend. These choices correspond to the arms of various slot machines. When a user clicks on the link of a recommended page, the recommender system receives a payoff in terms of the success of the recommendation. In the simplest case, the click-through problem is modeled with binary payoffs, where a click amounts to a payoff of 1 unit. This payoff can be viewed in an analogous way to that received by a gambler from the slot machine. In most practical settings, additional contextual information may be available to the recommender system about the user or the context of the recommendation. Some examples of such contextual information are as follows:

1. A set of features describing the profile of the user or the item-context may be available. Examples of item context might include the content of the Web page on which a recommendation is displayed. For example, a recommendation on a Web page describing the movie *Terminator* might be very different from that on a page describing the movie *Nero*. This type of context is particularly common in settings such as computational advertising.

2. The users may be clustered into groups, and the cluster identifier of the group may be used as semantic knowledge about the user. This is because similar users might have similar payoffs, and therefore the analysis can be segmented in group-wise fashion.

In cases where contextual information is available about the users, it is often assumed that user identification mechanisms are available. In order to explain the use of multi-armed bandit algorithms, we will first discuss the traditional setting where no contextual information is available. We will then provide a basic understanding of how contextual information may be incorporated within multi-armed bandit algorithms.

There are a number of strategies that the gambler can use to regulate the trade-off between exploration and exploitation of the search space. In the following, we will briefly describe some of the common strategies used in multi-armed bandit systems.

13.3.1 Naive Algorithm

In this approach, the gambler plays each machine for a fixed number of trials in the exploration phase. Subsequently, the machine with the highest payoff is used forever in the exploitation phase. This strategy shares a number of similarities with A/B-testing used for online evaluation of recommender systems. The difference is that A/B-testing uses only the exploration phase for evaluation purposes, whereas the bandit algorithm has an additional exploitation phase.

Although this approach might seem reasonable at first sight, it has a number of drawbacks. The first problem is that it is hard to determine the number of trials at which one can confidently predict whether a particular machine is better than the other. The process of estimation of payoffs might take a long time, especially in cases where the payoff events and non-payoff events are very unevenly distributed. For example, in a Web recommendation algorithm, the percentage of click-throughs might be low, as a result of which many trials will be required before one can confidently state whether one recommendation algorithm is better than the other. Using many exploratory trials will waste a significant amount of effort on suboptimal strategies. Furthermore, if the wrong strategy is selected in the end, the gambler will use the wrong slot machine forever. In practice, the payoffs of various machines (recommendation algorithms) might evolve over time. This is particularly true of the type of dynamic recommendation settings addressed by multi-armed bandit methods. Therefore, the approach of fixing a particular strategy forever is unrealistic in real-world problems.

13.3.2 ϵ-Greedy Algorithm

The ϵ-greedy algorithm is designed to use the best strategy as soon as possible, without wasting a significant number of trials. The basic idea is to choose a random slot machine for a fraction ϵ of the trials. These exploratory trials are also chosen at random (with probability ϵ) from all trials, and are therefore fully interleaved with the exploitation trials. In the remaining $(1 - \epsilon)$ fraction of the trials, the slot machine with the best average payoff so far is used. An important advantage of this approach is that one is guaranteed not be trapped in the wrong strategy forever. Furthermore, since the exploitation stage starts early, one is often likely to use the best strategy a large fraction of the time.

The value of ϵ is an algorithm parameter. For example, in practical settings, one might set $\epsilon = 0.1$, although the best choice of ϵ will vary with the application at hand. It is often difficult to know the best value of ϵ to use in a particular setting. Nevertheless, the value of ϵ needs to be reasonably small in order to gain significant advantages from the exploitation portion of the approach. Selecting a small value of ϵ, however, poses a significant challenge in settings where new slot machines (items) constantly enter the system. In such cases, one would explore this new slot machine only occasionally and miss an opportunity to obtain a better payoff.

To provide a specific example of the challenge posed by this situation, consider a setting in which the slot machines correspond to the different items, and the users are clustered into similar groups based on their specified profiles. The ϵ-greedy strategy is executed independently for each group of similar users. Whenever an opportunity arises to serve a recommendation to a user, the accumulated statistics of that user's group are used to choose the item with the use of the ϵ-greedy algorithm. At some point, a new item enters the system, which would be of great interest to John's group. However, at small values of ϵ, this item will be shown to John's group *very* occasionally, especially when the number of other items is very large. In a system with $10{,}000$ items and $\epsilon = 0.1$, the new item would be shown to John's group approximately once every $100{,}000$ trials of that group. This means that a large number of trials would be wasted before the relevance of this item to John's group is learned.

13.3.3 Upper Bounding Methods

Even though the ϵ-greedy strategy is better than the naive strategy in dynamic settings, it is still quite inefficient at learning the payoffs of new slot machines. In a dynamic recommendation setting, this problem is pervasive because new items enter the system all the time. In upper bounding strategies, the gambler does not use the mean payoff of a slot machine. Rather, the gambler takes a more optimistic view of slot machines that have not been tried sufficiently, and therefore uses a slot machine with the best *statistical upper bound* on the payoff. Note that rarely tested slot machines will tend to have larger upper bounds (because of larger confidence intervals) and will therefore be tried more frequently. Furthermore, one no longer needs to explicitly use a parameter ϵ to divide the trials into two categories; the process of selecting the slot machine with the largest upper bound has the dual effect of encoding both the exploration and exploitation aspects within each trial.

An important problem here is that of determining a statistical upper bound on the payoff of each machine. This can often be achieved with the help of the central limit theorem, which states that the sum of a large number of i.i.d. random variables (payoffs) converges to the normal distribution. One can estimate the mean and standard deviation of the normal distribution over various trials, and then set the upper bound of each slot machine at the required level of statistical confidence. Note that new slot machines will have large confidence intervals, and therefore the upper bounds will also be correspondingly large. Increasing the number of trials reduces the width of the confidence interval and therefore the upper bounds will tend to reduce over time. When a new slot machine enters the system, it will often be tried repeatedly, until its upper bound falls below that of one of the existing slot machines. One can regulate the trade-off between exploration and exploitation by using a specific level of statistical confidence. For example, an algorithm at 99% level of statistical confidence will perform a larger proportion of exploration as compared to an algorithm at 95% level of statistical confidence.

Such upper bounding strategies have been used recently for designing recommendation algorithms [348]. Many of these algorithms use the contextual features of the users and recommendation setting to design the various multi-arm bandit strategies for exploration and exploitation of the search space. The basic idea is that the gambler is shown a feature vector relevant to that trial (e.g., user or item profile in recommender system), and the gambler makes decisions on the slot machine (choice of recommendation strategy or choice of item) based on the knowledge of the feature vector. Such algorithms are also referred to as *contextual bandit algorithms*. The main goal of the gambler is to learn how the contextual features and the rewards on the arms relate to one another based on previous experience. The contextual feature vectors can be extracted from side-information such as user-profiles or the Web page on which the recommendation is shown. Therefore, contextual features provide a useful tool to incorporate various types of personalization in multi-armed bandit algorithms.

Consider a setting where the arms of the slot machines correspond to recommending different items. The basic idea of these algorithms is to use the following steps repeatedly:

1. **(Incremental) training:** Train a classification or regression learning model based on past history of feature-payoff pairs to learn the expected payoff of each arm. In most cases, this phase is executed incrementally, as new feature-payoff pairs enter the system over the time. Whenever a particular arm is selected by the recommender system, its feature attributes and payoff value is added to a training data set *that is specific to the corresponding arm*. Therefore, there are as many training data sets (and incrementally updated models) as the number of arms. The number of training

examples for each arm is equal to the number of times that the arm was played in the past. A separate model is constructed for each arm using its training data. It is desired to use a probabilistic or statistical learning algorithm that outputs the expected payoff and an estimated standard deviation (or maximum deviation) measure of the payoff of each arm (item) for a particular feature vector (context). Note that arms corresponding to newly added items will have smaller training data sets. Smaller training data sets will lead to larger estimated deviation of prediction. In general, there are two criteria to keep in mind while selecting the base model for payoff prediction:

- The base model should be incrementally updatable because new feature-payoff pairs are continually added to the training data.

- The base model should have the ability to output some measure of (or tight upper bound on) the expected error of prediction.

2. **Upper-bound estimation:** For the current contextual profile being shown to the recommender system, use the learned model to construct an upper bound on the expected payoff of each arm. The upper bound is computed as a linear sum of the expected payoff and an appropriate multiple of the standard deviation. In some cases, a tight upper bound on the maximum deviation is used instead of the standard deviation. The choice of deviation measure often depends on the ease of computing such measures with the model at hand.

3. **Recommendation:** Select the arm with the largest upper bound. Recommend the corresponding item to the user.

These steps are executed continuously over time, as recommendations are made and additional examples are added to the training data. In cases, where the payoff is a binary value (e.g., clicking or not clicking a link), a classification model may be used instead of a regression model.

The LinUCB algorithm is an upper bounding algorithm, which is based on a similar approach [348]. This approach uses a linear regression algorithm to learn the expected payoff. Consider a setting, where the ith arm has been played n_i times so far. In particular, if $\overline{X}$ is a d-dimensional (row) vector corresponding to the current context, D_i is the $n_i \times d$ feature matrix of the training data set of the ith arm, and $\overline{y_i}$ is the n_i-dimensional payoff (column) vector of the ith arm, then one can use ridge regression to predict the expected payoff of $\overline{X}$ with the ith arm as follows:

$$\text{Payoff}_i = \underbrace{\overline{X}}_{d \text{ features}} \underbrace{\left[(D_i^T D_i + \lambda I)^{-1} D_i^T \overline{y_i} \right]}_{d \text{ coefficients}} \tag{13.2}$$

Here, $\lambda > 0$ is the regularization parameter and I is a $d \times d$ identity matrix. Furthermore, a tight upper bound on the expected deviation can be quantified under conditional independence assumptions on the payoff (response) variables with respect to the feature variables. In particular, it can be shown [348] that with probability at least $(1 - \delta)$, the following is true for the binary payoff[1] setting:

$$\text{Deviation}_i \leq \left(1 + \sqrt{\ln(2/\delta)/2} \right) \cdot \sqrt{\overline{X}(D_i^T D_i + \lambda I)^{-1} \overline{X}^T} \tag{13.3}$$

---

[1]If the payoffs lie in the range $[0, \Delta]$, then the deviation also needs to be scaled up by Δ.

The deviation will reduce when D_i has a larger number of rows (training examples), because the entries in $(D_i^T D_i + \lambda I)^{-1}$ typically become smaller as the entries in $D_i^T D_i$ become larger. Furthermore, the deviation increases for smaller values of δ. The arm with the largest value of Payoff$_i$ + Deviation$_i$ is selected as the relevant one. By increasing or decreasing δ, one can select the desired point on the exploration-exploitation trade-off curve. In practice, one directly uses $\alpha = (1 + \sqrt{\ln(2/\delta)/2})$ as the relevant input parameter rather than δ, although the former's relationship to δ can be used to provide some intuitive guidance in selecting it. It is noteworthy that both $D_i^T D_i$ and $D_i^T \overline{y_i}$ can be maintained incrementally because they can expressed as linear sums of functions of the attributes/payoffs of individual training points. Nevertheless, it is still required to invert the $d \times d$ matrix $(D_i^T D_i + \lambda I)$ during each prediction. In cases where d is large, the inversion can be done periodically.

In practice, one can use virtually any probabilistic algorithm that outputs a robust measure of the expected payoff and maximum deviation for a given feature vector. It is noteworthy that LinUCB uses a tight upper bound on the deviation rather than the standard deviation because it is easier to estimate. In many settings, it may be desirable to present more than one recommendation at a time in the form of a ranked list. The simplest approach may be to use the top-k upper bounds as an approximation. A more sophisticated approach is to use the *slate* setting, and it is discussed in detail in [290].

13.4 Group Recommender Systems

Group recommender systems are designed to address scenarios in which items are consumed by groups of users, rather than a single user. Some examples of these scenarios and the systems developed to deal with them include the following:

1. *Movie domain:* In many scenarios, a group of users might wish to go out to see a set of movies. The recommendations must therefore be tailored to the composition of the group. An example of such a recommender system is *PolyLens* [168], which provides recommendations to groups of users. *PolyLens* can be viewed as an extension of the *MovieLens* system.

2. *Television domain:* Like movies, one might want to recommend programs to watch for groups of users. An example of such a television program recommender, which is based on user profile merging, is discussed in [653].

3. *Music domain:* Although it is less common for groups of users to hear music together, such scenarios arise when the music is to be played in a group setting, such as a fitness center or gym. An example of such a system is the *MusicFX* [412] group recommender system.

4. *Travel domain:* The travel domain is perhaps the most common one for group recommendations. This is because it is common for groups of tourists to make travel plans together. Some examples of such systems include *Intrigue* [52], *Travel Decision Forum* [272], and *Collaborative Advisory Travel System (CATS)* [413].

These processes lead to a natural question: why would one not use straightforward averaging to recommend items to a group in these situations? After all, if the goal is to maximize the overall utility, then using the average seems to be the most effective option. However, users can often influence one another based on social phenomena, such as *emotional contagion* and *conformity* [409]. These phenomena can be defined as follows:

1. *Emotional contagion:* The satisfaction of various users can have an impact on one another. For example, if a set of users are watching a movie together, and if some members of the group are not enjoying the movie, this can have a contagious effect on other users. In such cases, averaging does not work very well because the users infect one another with their tastes, and the final experience of the group may be very different from what the average rating might indicate.

2. *Conformity:* Conformity is closely related to the notion of emotional contagion, in that the *expressed* opinions of users have an impact on one another. However, the social phenomenon is slightly different, in that users either want to consciously have similar opinions as their peers (in spite of having a hidden difference of opinion), or their opinions unconsciously change because of peer influence. As a result, the final experience of the group may deviate significantly from what an average rating might indicate.

These two social phenomena, which are related to *social choice theory*, have significant effects on the performance of recommender systems. As a result, the averaging strategy will often not work well. For example, an evaluation of an averaging-based strategy for television recommendation service was performed in [654], and it was shown that the recommender performs well when the group had homogeneous tastes, but it does not perform quite as well when the tastes vary widely. Therefore, it is crucial to be able to use social phenomena in the modeling process. In addition, group recommenders are generally defined differently, depending on whether they are designed in the collaborative, content-based, or knowledge-based settings. Although the general principles of group recommendation in the collaborative and content-based settings are similar, the principles of knowledge-based systems are quite different. In the following, we will study these different settings.

13.4.1   Collaborative and Content-Based Systems

The collaborative and content-based systems are generally quite similar in terms of the approach used for creating the group recommendations. The general approach comprises the following two steps:

1. Perform the recommendation independently for each user as in any collaborative or content-based system. For a given group and a given universe of items, determine rating predictions for each user-item combination.

2. For each item, aggregate the ratings from the various members of the group into a single group rating by using a aggregation function of the ratings predicted for each member of the group. This function might use a simple weighted averaging over group members, an aggregation approach based on principles from social choice theory, or a combination of the two. All the items are then ranked for the group based on the predicted group rating of each item.

The main difference between the various methods is the implementation of the second aggregation step. A variety of different strategies are used to aggregate the diverse ratings into a single value in the second step. These strategies are as follows:

1. *Least misery strategy:* In the least misery strategy, the overall rating suggested to the group is the lowest rating of any member of the group. The basic idea of this approach is to prevent the negative effects of social contagion and conformity. An example a system using this approach is *PolyLens* [168].

2. *Weighted averaging:* This approach uses the average rating of the individual ratings, and a weight is associated with each individual. The weight is often used to model specific types of situations that prevent extreme dislike or infeasibility. For example, a casino resort should not be suggested as a tourist destination to a group containing a child, and a physically strenuous trip should not be suggested to groups containing one or more disabled individuals. Providing greater weights to the preferences of such individuals automatically increases the overall acceptability and feasibility of the group recommendation. A variation of such a strategy was used in *Intrigue* travel recommender [52]. It has also been suggested [168] that ratings from experts might be assigned greater weight. Finally, it is also possible to combine the least misery strategy with the averaging strategy by using a weighted summation of the least misery and averaging prediction over each item.

3. *Average without misery:* This approach averages the predicted ratings of the group members after excluding the ratings of individuals with the lowest ratings. Note that this approach tends to have an opposite focus to the least misery strategy, because it averages only over members who experience the greatest pleasure over a specific item. This type of approach was used in the *MusicFX* system [412]. When considering this approach, it is worth noting that pleasurable experiences can be emotionally contagious in the same manner as unhappy experiences.

A variation of the averaging approach is to use the median instead of the mean. The advantage of using the median is that it is less susceptible to noise and outliers. For example, a single highly negative rating may affect the mean significantly but it may not affect the median much. Such an approach is particularly useful when users are aware of the recommendations that other users are giving and respond by attempting to selectively provide highly positive or negative ratings that would have an outsized influence on the overall group recommendation. As a result, the average no longer remains representative of the group rating. Such an approach is used by *Travel Decision Forum* [272]. A variety of other aggregation strategies are suggested in [407]. Refer to the bibliographic notes.

13.4.2 Knowledge-Based Systems

The aforementioned systems are all based on ratings specifications. However, knowledge-based systems are not based on user ratings, but rather on the specification of user *requirements.* Therefore, the natural approach in such systems is to have each user specify his or her requirements, which are aggregated into a single set. Then, the item that fulfills most of these requirements is recommended. Such an approach is used by the *Collaborative Advisory Travel System (CATS)* [413]. Such systems also allow interactive feedback that allows the group to explore its interests in an interactive style. Knowledge-based systems are particularly well suited to group recommendations, as they allow the group to come to a consensus in an interactive way before actually consuming the item. This reduces the likelihood of dissatisfaction in the final recommendation. Although knowledge-based systems are designed for complex product domains, they are also useful in the context of complex *user* domains. A group recommendation setting can be viewed as a complex user domain. Knowledge-based recommender systems are discussed in Chapter 5.

Table 13.1: Effects of multiple criteria in defining similarity

Criterion ⇒ User ⇓	Visual Effects	Plot	Overall
Sayani	3	9	7
Alice	9	3	7
Bob	8	3	5

13.5 Multi-Criteria Recommender Systems

In many recommendation applications, users may be interested in items on the basis of different criteria. For example, in a movie recommender system, one user may be interested in visual effects, whereas another user may be interested in the plot. In such cases, the overall rating is often a poor reflection of the user's overall choices. Consider the hypothetical example illustrated in Table 13.1. In this case, three users have expressed their ratings for the movie *Gladiator* based on visual effects, plot, and overall rating. Note that the overall rating is specified directly by the users, and might not necessarily represent an average of all the ratings. Each rating value is specified on a scale from 1 to 10. It is clear that Alice and Sayani have exactly the same overall rating, but their patterns of ratings for the plot and visual effects are very different. On the other hand, Alice and Bob are slightly different in their overall ratings, but have similar ratings on the visual effects and plot. Therefore, for any peer-based prediction method, Alice and Bob should be considered more similar than Alice and Sayani. By using similarity computations based only on the overall rating, one can often obtain misleading predictions.

The overall rating in a multi-criteria system may be either explicitly specified by users, or it may be derived with the use of a global utility function (e.g., simple averaging). In cases where an overall rating is specified by users, it is possible to learn a user-specific utility function with the use of linear regression methods such as those discussed in Chapter 5 on knowledge-based recommender systems. For cases in which the overall rating is not specified by users, the items can be ranked directly by integrating the predictions from the various criteria without computing an overall rating. In other cases, one can implicitly average over various criteria in order to create the overall rating. If needed, the various criteria may be weighted using domain-specific knowledge (e.g., utility functions).

It should be pointed out that multi-criteria recommender systems are inherent to knowledge-based systems, which are designed for complex product domains such as cars. Such products have multiple criteria such as performance, interior design, luxury options, navigation, and so on. In such domains, users wish to rank items based on whether they satisfy certain user-specified criteria. As these methods are already discussed in Chapter 5, this chapter will primarily focus on content-based and collaborative filtering methods.

In the following, we will discuss some of the common methods used in multi-criteria recommender systems. Refer to the bibliographic notes for an up-to-date discussion of recent methods. For the purpose of the following discussion, we will assume that there are a total of c criteria, indexed by $\{1, 2, \ldots, c\}$. The $m \times n$ ratings matrix according to the kth criterion is denoted by $R^{(k)}$, and the rating of user i for item j in $R^{(k)}$ is denoted by $r_{ij}^{(k)}$. In the event that the user also specifies overall ratings, then the corresponding rating matrix is denoted by $R^{(0)}$, and the corresponding value of the overall rating of user i for item j is denoted by $r_{ij}^{(0)}$.

13.5.1 Neighborhood-Based Methods

Neighborhood-based methods can be easily adapted to work with multi-criteria systems because of the ease with which multiple criteria can be incorporated within the similarity function. Most of the existing neighborhood-based methods leverage *user-based* collaborative filtering methods rather than item-based collaborative filtering methods. However, it is possible, in principle, to generalize item-based methods to multi-criteria scenarios using similar techniques. In the following, we will discuss only user-based neighborhood methods because of its wider acceptance and available experimental results.

Let $\text{Sim}^k(i, j)$ represent the similarity between users i and j over criterion k, where $k \in \{1 \ldots c\}$. Furthermore, we will assume that the overall ratings matrix $R^{(0)}$ is available, and the corresponding similarity between users i and j is denoted by $\text{Sim}^0(i, j)$. Then, the neighborhood-based method can be implemented as follows:

1. Compute the similarity $\text{Sim}^k(i, j)$ between each pair of users i and j for each $k \in \{0 \ldots c\}$. Any of the methods introduced in Chapter 2, such as the Pearson correlation coefficient, may be used for computing $\text{Sim}^k(i, j)$.

2. Compute the aggregated similarity $\text{Sim}^{aggr}(i, j)$ between each pair of users i and j by aggregating the similarity values over the various criteria using an aggregation function $F(\cdot)$:

$$\text{Sim}^{aggr}(i, j) = F(\text{Sim}^0(i, j), \text{Sim}^1(i, j), \text{Sim}^2(i, j), \ldots \text{Sim}^c(i, j)) \tag{13.4}$$

   Determine the k-closest peers of each user with the aggregated similarity.

3. Use the similarity weighted values of the (overall) ratings of each peer of a user t for an item j in order to predict the rating of user t for item j. Typically, the approach is combined with row-wise mean-centering to prevent user-specific bias. Thus, this approach is equivalent to that of using Equation 2.4 of Chapter 2 on the overall ratings matrix $R^{(0)}$, except that aggregated similarities $\text{Sim}^{aggr}(\cdot, \cdot)$ are used for peer determination and weighting purposes within Equation 2.4.

It is noteworthy that the aggregation function of Equation 13.4 also uses $\text{Sim}^0(i, j)$ (similarity based on overall ratings) in the computation. The main differences among various methods arise in terms of how the aggregation of Equation 13.4 is computed. The common methods for aggregation are as follows:

1. *Average similarity:* This approach [12] is based on the averaging of the predictions of the $(c+1)$ different ratings (including the overall rating). Therefore, the function $F(\cdot)$ of Equation 13.4 is defined as follows:

$$\text{Sim}^{aggr}(i, j) = \frac{\sum_{k=0}^{c} \text{Sim}^k(i, j)}{c + 1} \tag{13.5}$$

2. *Worst-case similarity:* This approach [12] uses the smallest similarity across all the criteria (including the overall rating). Therefore, we have:

$$\text{Sim}^{aggr}(i, j) = \min_{k=0}^{c} \text{Sim}^k(i, j) \tag{13.6}$$

3. *Weighted aggregation:* This approach [596] is a generalization of the averaging technique and uses a weighted sum of the similarities across the different criteria.

Let $w_0 \ldots w_c$ be the weights of the various criteria. Then, the aggregated similarity is defined as follows:

$$\text{Sim}^{aggr}(i,j) = \sum_{k=0}^{c} w_k \cdot \text{Sim}^k(i,j) \tag{13.7}$$

The value of w_i determines the weight of criterion i, and the weights can be determined using straightforward parameter-tuning techniques such as cross-validation (cf. Chapter 7).

In addition to using similarities, it is also possible to use *distances* both for peer computation, and for the final step of weighted rating prediction. Note that similar items will have *smaller* distances, necessitating conversion of the distances into similarities in a heuristic way in order to perform the weighting. For any pair of users, the distances are computed based only on the items that the two users have rated in common. The distances are computed separately for each of the items by aggregating across various criteria. The distances across various items are averaged in a second aggregation step.

How is the first step of computing the distance $\text{ItemDist}^{aggr}(i,j,q)$ between users i and j with respect to a particular item q executed? Note that item q must be rated by both users i and j for this distance to be computed at all. A natural approach is to use the L_p-norm, which is defined as follows:

$$\text{ItemDist}^{aggr}(i,j,q) = \left(\sum_{k=0}^{c} |r_{iq}^k - r_{jq}^k|^p \right)^{(1/p)} \tag{13.8}$$

Commonly used values of p are $p = 1$ (Manhattan metric), $p = 2$ (Euclidean metric), and $p = \infty$ (L_∞-norm).

This approach is repeated over each of the items that the users i and j have rated in common. Let this set of items be denoted by $I(i,j)$. The overall distance $\text{Dist}^{aggr}(i,j)$ across all items is defined by the average distance over all items in $I(i,j)$:

$$\text{Dist}^{aggr}(i,j) = \frac{\sum_{q \in I(i,j)} \text{ItemDist}^{aggr}(i,j,q)}{|I(i,j)|} \tag{13.9}$$

One can convert the distances into similarity values with the use of simple kernel computations or inversion tricks:

$$\text{Sim}^{aggr}(i,j) = \frac{1}{1 + \text{Dist}^{aggr}(i,j)} \tag{13.10}$$

After the similarity values have been computed, one can leverage the user-based collaborative filtering methods as discussed above.

13.5.2   Ensemble-Based Methods

All the aforementioned methods make changes to a specific *algorithm*, such as the neighborhood algorithm, in order to perform the recommendations. However, it is possible to use ensemble-based methods, which can leverage any existing technique, to perform the recommendations [12]. The basic approach contains two steps:

1. For each value of $k \in \{1 \ldots c\}$, use any off-the-shelf collaborative filtering algorithm on ratings matrix $R^{(k)}$ to fill in the ratings for criterion k.

2. For each user i and item q, for which the ratings have been predicted, combine the predictions $\hat{r}(1)_{iq} \ldots \hat{r}_{iq}^{(c)}$ across the various criteria using an aggregation function $f()$ as follows:

$$\hat{r}_{iq}^{(0)} = f(r_{iq}^{(1)} \ldots r_{iq}^{(c)}) \tag{13.11}$$

The computed aggregation provides the overall predicted rating. The recommended items are then ranked for user i based on the overall predicted ratings.

The construction of the aggregation function $f()$ remains to be explained. There are three common techniques suggested in [12]:

1. *Domain-specific and heuristic methods:* In this case, the aggregation function is set by the domain expert depending on the perceived importance of the various criteria. The simplest possible approach is to use the average of the predicted ratings over the various criteria.

2. *Statistical methods:* These represent linear and non-linear regression methods. For example, the overall predicted rating can be expressed as a linear weighted sum of the predicted ratings over various criteria:

$$\hat{r}_{iq}^{(0)} = \sum_{k=1}^{c} w_k \cdot r_{iq}^{(k)} \tag{13.12}$$

The values of $w_1 \ldots w_c$ can be learned using linear regression techniques, as discussed in section 6.3 of Chapter 6. Note that the observed values of the ratings across various criteria can be used as the training data to learn the weights.

3. *Machine-learning methods:* This approach is not very different in principle from the second approach. Instead of using regression, any machine-learning method (such as a neural network) can be used. Note that simpler versions of neural networks can also be used to approximate linear regression. However, a neural network provides greater power in modeling arbitrarily complex functions.

The aforementioned discussion is based on the assumption of a global aggregation. However, it is also possible to learn user-specific or item-specific aggregation functions, if sufficient number of observed ratings about users and items are available. The ensemble-based approach is simple to implement because it provides the ability to use off-the-shelf tools in various phases of the process. This aspect of ensemble methods also provides it with greater flexibility in performing model selection, and tuning the system with an appropriate choice of learners.

13.5.3 Multi-Criteria Systems without Overall Ratings

The aforementioned methods require the availability of overall ratings in order to perform the recommendations. In cases where overall ratings are not available, the methods discussed in the previous sections cannot be used in their current form. However, one can still use the first step of the ensemble-based method discussed in the previous section. The main difference is that the second step of aggregating the predicted ratings needs to performed without any available learning data. Therefore, methods such as linear regression, nonlinear regression, neural networks, or other machine-learning methods are no longer possible. However, it is still possible to use heuristic and domain-specific combination functions in the aggregation step. The items can then be ranked on the basis of the aggregated value. A second approach for presenting the items to the user is to leverage the pareto-optimality of the predicted ratings across the various criteria. Only the pareto-optimal items are presented to the user along with an explanation of why they are presented. The bibliographic notes contain pointers to various multi-criteria systems that do not assume the availability of overall ratings.

13.6   Active Learning in Recommender Systems

Recommender systems are heavily dependent on historical data provided by the user. However, ratings matrices are sometimes excessively sparse, causing challenges in providing meaningful recommendations. This is especially true at start-up time, where *cold-start* problems are often encountered. In such cases, it is important to quickly acquire more ratings to build up the ratings matrix. The process of acquiring ratings is time-consuming and costly because users are often not willing to voluntarily provide ratings without a perceived benefit. Indeed, it has been argued [303] that users are willing to share private information in collaborative filtering applications only when they are fairly compensated. This implies that there is an inherent cost (often implicit) to acquiring ratings. An active learning system chooses specific user-item combinations for which to acquire ratings in order to maximize the accuracy of predicted ratings. For example, consider a scenario of a movie recommender system in which many action movies have already been rated, but no comedy movies have been rated. In such cases, it is intuitively fruitful to actively acquire ratings of comedy movies rather than action movies in order to maximize prediction accuracy. This is because the *incremental* improvement in accuracy by acquiring further ratings of other action movies is likely to be less than that obtained by acquiring ratings of comedy movies. After all, one can already predict the ratings of action movies reasonably well, whereas one cannot predict the ratings of comedy movies very well with the available ratings. The problem here is that one cannot acquire the rating of an arbitrary user-item combination. For example, a user who has not consumed an item cannot be reasonably expected to provide a rating.

Active learning is commonly used in classification applications [18]; therefore, the applicability of the approach to content-based methods is obvious. After all, content-based methods are essentially classification problems on user-specific training data. In the case of collaborative filtering applications, content or genre information is typically not specified, and one must make such predictions with the use of the currently available ratings matrix. In its simplest form, one can formulate the ratings acquisition problem as follows:

Given a ratings matrix R, a cost budget C, and a cost-per-acquisition c, determine the set of user-item combinations for which the ratings must be acquired in order to maximize the prediction accuracy.

It is evident that the active learning formulation for classification is similar to that of collaborative filtering. In the case of classification, *labels* of training points are queried. In collaborative filtering, the *ratings* of user-item combinations are queried. As collaborative filtering is a generalization of the classification problem (cf. Figure 1.4 of Chapter 1), the active learning methodologies of classification also generalize to the collaborative filtering scenario. However, there is one key difference between collaborative filtering and classification. In classification, it is assumed that an oracle exists that provides the label of any queried data point. This assumption cannot be made in collaborative filtering. For example, if a user has not consumed an item, she cannot be expected to provide a rating for it. Nevertheless, the principles of active learning in collaborative filtering applications are similar to those in classification, at least in terms of determining which user-item combinations are most valuable to acquire. In many cases, incentives can be provided to a user to rate a specific item. For example, the merchant might offer a free product in exchange for a specific number of ratings from a particular user.

The simplest approach to active learning is to query for items that have been rated sparsely by the users. This can naturally help in the cold-start setting. However, such an approach is useful only in the initial stages of the recommender system setup. In later stages, more refined techniques are required in which the entries of the matrix are selected on the basis of the particular *combination* of users and items. Such methods are based on ideas already available in the classification literature.

Active learning is still an emerging area in the topic of collaborative filtering, and there are relatively few methods proposed in this area. Therefore, this section will briefly discuss two common methodologies [18, 22] used in classification and their applicability to collaborative filtering applications. These two methods are *heterogeneity-based models* and *performance-based models*. In the former case, the data points (user-item combinations) are queried, for which their *predicted* rating values are the most *uncertain before* performing the query. In performance-based models, the data points are queried, so that the prediction accuracy on the *remaining entries* provide the *best expected performance or certainty* after incorporating the newly queried rating in the matrix.

13.6.1 Heterogeneity-Based Models

In heterogeneity-based models, the goal is to query for the rating of the user-item combination for which the predicted rating is the most uncertain *before* performing the query. The specific method for judging the level of uncertainty depends on the model at hand. For example, if a numeric rating is predicted with a specific variance, each user should be queried with the item with the largest predicted variance. In the case of a binary rating prediction with a Bayesian approach, the item q whose posterior probability p_q is closest to 0.5 (i.e., smallest value of $|p_q - 0.5|$) is queried. Some specific examples of how this approach may be used in the context of specific models are as follows:

1. In a user-based neighborhood approach, the variance of the prediction of user-item combination (i, q) can be computed as the sample variance of the ratings of the peer users of i for item q. If none of the peer users have rated item q, the sample variance is ∞.

2. In an item-based neighborhood approach, the variance of the prediction can be computed from the ratings of user i of the most similar items of q. If user i has not rated any of the most similar items of q, the sample variance is ∞. Therefore, the approach tends to guide the user towards rating different items and naturally increases the coverage of the recommender system. In this sense, the approach is also able to adjust well in the cold-start setting.

3. In a Bayesian model, a Bayes classifier (cf. Chapter 3) is used to predict ratings. Consider the case of binary ratings, in which the prediction of a value of 1 has posterior probability p_q. In this case, the uncertainty is quantified as $1 - |p_q - 0.5|$. The item with the largest uncertainty value is selected for querying.

4. One can use multiple models to predict the ratings. A rating is said to be uncertain when the different models have different predictions. The variance of the predictions over the different models can be used to quantify the uncertainty.

The aforementioned methods are simple adaptations of techniques in the classification literature. One can adapt most collaborative filtering algorithms in a natural way to compute the uncertainty. In collaborative filtering, some additional factors can be combined with the uncertainty level in a heuristic way (e.g., multiplicatively):

1. One can include a factor for the probability that a user is likely to rate an item. This is because users cannot provide ratings for items they have not consumed. Consider an implicit feedback matrix in which a value of the entry is 1 if a user has rated an item (irrespective of actual rating value) and 0, otherwise. The predicted "rating" with the use of any collaborative filtering algorithm provides a probability that a user will actually rate the item.

2. It is suggested in [513] that very popular items should not be queried because their ratings are often not representative of the other items.

Note that few experimental results exist on how active learning methods actually perform in the context of collaborative filtering. Therefore, this area is open to significant opportunities for further research.

13.6.2 Performance-Based Models

The goal of querying for ratings is to increase prediction accuracy and also reduce the uncertainty of prediction on the *currently available* entries. In performance-based models, the data points are queried so that the prediction accuracy on the *remaining entries* provides the *best expected performance or certainty after* incorporating the newly queried rating in the matrix. Note that uncertainty-based models focus on the prediction characteristics of the currently queried instance, whereas performance-based models focus on the *differential* impact of the *added* instance to the predictions of the *currently available* entries. Determination of what would happen after querying the rating of a user-item combination is challenging because the *expected performance* must be computed before actually querying the rating. Bayesian methods are used to compute this expected performance. The corresponding techniques are described in [18, 22].

13.7   Privacy in Recommender Systems

Collaborative filtering applications are heavily dependent on the collection of feedback from multiple users. In collaborative filtering applications, users need to specify ratings for items. These ratings reveal important information about user interests, their political opinions, sexual orientation, and so on. This revealing of private information that comes with rating items poses numerous challenges because it makes users less willing to contribute ratings.

All privacy-preservation methods change the data in some way so as to reduce its accuracy of representation. This is done in order to increase the privacy. The trade-off is that the data becomes less accurately represented. Therefore, mining algorithms are no longer as effective. Two classes of techniques are used to preserve privacy:

1. *Privacy at data collection time:* In these techniques, the data collection approach is modified so that individual ratings are not collected. Rather, distributed protocols [133] or perturbation techniques [35, 38, 484, 485] are used to collect the data *only in a perturbed way or in the aggregate.* Typically, specialized (secure) user-interfaces and data collection plug-ins are required in order to implement the approach. Furthermore, specialized data-mining methods are used on the collected data, because many of these techniques use aggregate distributions for mining, rather than individual data records.

   The advantage of such an approach is that users are assured that no single entity has access to their private data, at least in its exact form. Although privacy at data collection provides the strictest form of privacy, much of the work in this area is at the research stage only. To the best of our knowledge, there are no large-scale commercial implementations of such systems. This is, in part, because such systems typically require more effort from the user in terms of gaining access to specialized interfaces/infrastructures and more effort from the data miner after the aggregated data becomes available.

2. *Privacy at data publication time:* In most practical settings, a trusted entity (e.g, Netflix or IMDb) has access to all the ratings data it has collected over time. In such cases, the trusted entity might wish to *publish* the data to the broader technical community to enable further advancements in the field of collaborative filtering. A specific example of such a publication was the Netflix Prize data set, which was released after de-identification of the ratings. In such cases, models like k-anonymity [521] are used to preserve privacy. Typically, such methods use *group-based* anonymization techniques in which records belonging to groups of a minimum size become indistinguishable. This is achieved by carefully perturbing selected attributes of the data records so that one cannot join such records with publicly available information in order to exactly identify the subjects of the data records. Such systems are more common, and have wider applicability than the first scenario.

The two aforementioned models have different trade-offs. The first model provides stronger privacy guarantees because the individual's ratings are not stored anywhere, at least in their exact form. In some cases, the ratings are stored only in *an aggregate sense.* Therefore, the approach provides greater privacy guarantees. On the other hand, it is generally harder to use off-the-shelf collaborative filtering algorithms with such forms of data collection. This is because the data is either perturbed very highly, or the fundamental *representation* of the data has been changed to some aggregate form. In the case of methods using group-based anonymization, the privacy guarantees are typically weaker. On the other hand, the released data records are typically in the same format as the original data. Therefore, it is easier to make use of off-the-shelf collaborative filtering algorithms in these cases. The following passage provides a brief overview of group-based anonymization models.

Group-based anonymization methods are typically used by trusted entities at *data publication time.* The typical goal of publishing entities is to prevent identification of the subjects of the data records. For example, when Netflix released their ratings data set, the subjects of the data records were de-identified. In addition, the attributes are typically perturbed in such a way that groups of data records become indistinguishable. The basic idea in these methods is to perturb the data records sufficiently that attackers cannot match the records with other publicly available data sources in order to determine the identity of the subjects of the records. Some common models for perturbing the data records in group-wise fashion include k-anonymization [521], condensation [27], ℓ-diversity [386], and t-closeness [352].

The reader is encouraged to refer to the bibliographic notes for further details related to common privacy-preservation methods. In the following, we briefly discuss a condensation-based method that is easy to apply to the collaborative filtering setting. We will also discuss some challenges that surface when these methods are used for high-dimensional data.

13.7.1   Condensation-Based Privacy

The condensation-based approach was originally designed for multi-dimensional data records, which are *completely specified* [27]. However, the approach can easily be used for in-completely specified data records as well. One of the inputs to the algorithms is an anonymity level p, which defines the number of rows we wish to be indistinguishable from one another. Larger values of p result in a greater level of anonymity, but they reduce the accuracy of the modified data. Consider an $m \times n$ ratings matrix R, which is incompletely specified:

1. Partition the rows of R into clusters $\mathcal{C}_1 \ldots \mathcal{C}_k$, such that each cluster contains at least m records.

2. For each cluster $\mathcal{C}_r$, generate $|\mathcal{C}_r| > m$ synthetic data records matching the data distribution of the records in the cluster.

Both the two steps need to account for the fact that the rows in the matrix R are incompletely specified. Clustering methods can be modified relatively easily to work for incomplete data. For example, a k-medians algorithm can be modified by using only the specified values of the entries in the median computation. Similarly, the distances are computed using only the specified entries and then normalized by the number of observed dimensions. Similarly, while generating the synthetic data records from $\mathcal{C}_r$, one can use a simple multivariate Bernoulli distribution over the values of the ratings to model each item. This multivariate Bernoulli distribution is derived from the rating distribution of the records in the cluster. One must take care to generate the rating of an item the same number of times that it is present in that cluster.

This method of synthetic data generation comes with two primary advantages. The first advantage is that the data are generated in the same format as that of the original ratings matrix, allowing application of any off-the-shelf collaborative filtering algorithm; the second is that the anonymity of synthetic data is generally harder to compromise. This approach can also be generalized to dynamic settings [27].

13.7.2   Challenges for High-Dimensional Data

Ratings data is typically high-dimensional. For example, a typical ratings matrix may contain thousands of dimensions. Furthermore, some users might easily specify more than 10 or 20 ratings. In such cases, it is harder to preserve the privacy of such users with group-based anonymization methods, even when the data records are perturbed. For example, if a particular source releases a set of de-identified ratings, an attacker might use a different source of ratings that are not de-identified and match the two data sets in order to determine the subjects of the de-identified records. The larger the number of specified ratings is, the easier it is to de-identify the records. It has been shown in [30] that only about 10 to 20 specified values in a row are necessary in order to generate a powerful attack. The well-known Netflix Prize data set was attacked using this methodology [451]. The challenges for high-dimensional data are not trivial, and there are theoretical barriers [30] to the limits of anonymization. The development of new anonymization methods for high-dimensional and sparse data sets remains an open area of research.

13.8 Some Interesting Application Domains

In this section, we will study a number of interesting application domains for recommender systems. The goal of this section is to study the application of recommender systems to various application domains, and the specific challenges that arise in the context of each domain. Some examples are as follows:

1. *Query recommendation:* An interesting question is how Web logs can be used to recommend queries to users. It is not quite clear whether query recommendation should be considered a personalization application because the recommendations are typically *session-specific* (i.e., dependent on the history of user behavior in a short session) and do not use *long-term* user behavior. This is because queries are often issued in scenarios in which user re-identification mechanisms are not available over multiple sessions. We will not discuss this topic in detail, although relevant pointers are included in the bibliographic notes.

2. *Portal content and news personalization:* Many online portals have strong user identification mechanisms by which returning users can be identified. In such cases, the content served to the user can be personalized. This approach is also used by news personalization engines, such as Google News, in which Gmail accounts are used for user identification. News personalization is usually based on implicit feedback containing user behavior (clicks), rather than explicit ratings.

3. *Computational advertising:* Computational advertising is a form of recommendation, because it is desirable for companies to be able to identify advertisements for users based on a relevant context (Web page or search query). Therefore, many ideas from recommendation systems are directly used in the area of computational advertising.

4. *Reciprocal recommender systems:* In these cases, both the users and items have preferences (and not just the users). For example, in an online dating application, both parties (men and women) have preferences, and a successful recommendation can be created only by satisfying the preferences of both parties. Reciprocal recommender systems are closely related to the link-prediction methods discussed in Chapter 10.

This chapter will provide an overview of some of these different applications, with a specific focus on portal content personalization, computational advertising, and reciprocal recommender systems. The basic idea is to give the reader a sense of how recommendation technology can be used in diverse settings.

13.8.1 Portal Content Personalization

Many news portals personalize the news for their users by using their history of past accesses. An example of such a personalization system is the Google news engine. Google has strong user-identification mechanisms with the use of Gmail accounts. This mechanism is used to track the past history of user click behavior. This past history is used to recommend news of interest to users. Similar methods can be used to recommend content to users in many types of Web portals. The main assumption in all these cases is that a user log of their past actions is available.

13.8.1.1   Dynamic Profiler

Dynamic Profiler [636] is a portal content personalization engine that uses a combination of collaborative and content-based techniques. The system can be used for any form of portal content personalization, including news personalization. The approach contains several steps, most of which are periodically repeated to refresh summary statistics which need to be updated over time in order to prevent them from becoming stale. These statistics are used to make recommendations in real time. The overall approach contains the following broad steps:

1. **(Periodically updated)** A sample of documents from the portal is used to create a group of clusters. The clustering is done using a partially supervised clustering scheme [29]. The supervision of the clustering is done with the help of samples of documents belonging to semantically relevant topics. These samples are used as seeds for creating the clusters with a combination of an agglomerative and a k-means approach. As a result, the clusters contain the semantically important categories in the collection.

2. **(Periodically updated)** The user access log is used in conjunction with the aforementioned clusters to create user profiles. The user profile contains a count of a number of accesses of the user to documents belonging to each cluster. Therefore, the user profile is a multidimensional record with as many dimensions as the number of clusters.

3. **(Periodically updated)** The user profiles are then clustered into peer groups with the use of high-dimensional clustering methods. Several high-dimensional clustering methods are discussed in [19].

4. **(Online phase at recommendation time)** A neighborhood-based approach is used in conjunction with these peer groups to perform the recommendations. For any given target user, the frequent categories in the closest clusters form the relevant recommended categories. It is also possible to recommend individual documents to the target user using an approach described below.

It remains to be explained how the final step of performing the recommendations is executed. For a given user, the first step is to determine her closest peer group. This is achieved by computing the distance between her profile and the centroids of the various peer groups. The closest peer group is referred to as her *community*. The frequency of all the documents accessed by this community is efficiently determined from an indexed version of the logs. The most frequently accessed documents in this community, which have also not been accessed by the target user, are then presented as the relevant recommendations.

13.8.1.2   Google News Personalization

The Google news personalization engine [175] is based on a similar problem statement as the dynamic profiler model. Therefore, an implicit feedback data set of user clicks is available in this case. The Gmail accounts of users provide a strong identification mechanism in Google news. When users are signed in and access Web pages, their click behavior is stored. The goal is to use the stored statistics about user clicks to make recommendations to these users from a candidate list L of items. For the time being, we will assume that the candidate list L is given. Later, we will discuss how the candidate list can be generated.

The Google news system uses very different algorithms from *Dynamic Profiler*. Whereas the *Dynamic Profiler* is designed to work for individual Web sites, the Google news system is designed to work in a Web-scale environment. The basic idea of the approach is to use a similarity-based mechanism to make recommendations. As with user-based neighborhood algorithms, a weighted similarity of users to other users who have accessed a particular item is used to make the recommendations. Let r_{iq} be an indicator variable, which takes on the value 1, if user i has accessed item q, and 0, otherwise. Note that r_{iq} can be viewed as the implicit feedback version of a ratings matrix. Similarly, let w_{ij} be the computed similarity between users i and j based on the similarity of their access patterns to Web pages. Then, the *predicted* propensity p_{iq} of user i to access the news item q is defined as follows:

$$p_{iq} = \sum_{j \neq i} w_{ij} \cdot r_{jq} \tag{13.13}$$

Since the rating r_{jq} is assumed to be binary, the prediction propensity p_{iq} can also be binarized using an appropriate threshold. The similarity can be computed in a variety of ways. For example, one might compute the Pearson correlation coefficient or the cosine similarity between the item accesses of the two users.

The aforementioned formula is a straightforward generalization of the user-based collaborative filtering mechanism. Note that it is expensive to compute this predicted propensity in a Web-scale setting because the similarity w_{ij} between every pair of users needs to be pre-computed. The pairwise computation can be rather expensive, and the summation on the right-hand side will also contain as many terms as the number of users. Therefore, the work in [175] also proposes a number of more efficient model-based alternatives. These methods use clustering to speed up computation. Furthermore, clustering methods have some advantages in noise reduction for more effective collaborative filtering.

In model-based techniques, the users are either *probabilistically* or *deterministically assigned* to clusters with similar access behavior. In other words, users who have similar access patterns typically belong to similar clusters with high probability. Two clustering schemes are used, corresponding to *MinHash* and *PLSI*, and either of them can be used to implement the approach. The former users a *hard* assignment of users to clusters, and the latter uses a *soft assignment* to clusters. More details of these methods are discussed later in this section.

Assume that a total of m clusters are defined, and the *fraction* of user i assigned to cluster k is given by f_{ik}. In the case of deterministic clustering, the value of f_{ik} is either 0 or 1, whereas the value of f_{ik} lies in $(0, 1)$ in the case of soft clustering. Then, the propensity of user i to access item q is defined as follows:

$$p_{iq} = \sum_{k=1}^{m} f_{ik} \sum_{j:f_{jk}>0} r_{jq} \tag{13.14}$$

It is also possible to further refine this formula by incorporating f_{jk}, although this is not mentioned in [175]:

$$p_{iq} = \sum_{k=1}^{m} f_{ik} \sum_{j} f_{jk} r_{jq} \tag{13.15}$$

In the case where the clustering is a hard assignment, such as (*MinHash* scheme), this expression reduces to the following:

$$p_{iq} = \sum_{j} \text{CommonClusters}(i, j) \cdot r_{jq} \tag{13.16}$$

Here, CommonClusters(i, j) correspond to the number of common clusters in which users i and j co-occur. Furthermore, if the clustering is executed only once as a strict partitioning, the value of CommonClusters(i, j) is either 0 or 1. On the other hand, if the clustering is repeated several times with a fast randomized approach, the value of CommonClusters(i, j) is equal to the number of times that users i and j occur in the same cluster. For dynamic data sets, the value of the implicit feedback "rating" r_{jq} can be multiplied with a time-decay value.

In addition, a co-visitation score is added to the scores generated from the clustering-based computation. The co-visitation score is similar in principle to an item-based algorithm. Two items are co-visited when they are visited by the same user within a pre-defined span of time. For each item, the number of (time-decayed) co-visits to every other item is dynamically maintained. For the target user i and target item q, it is determined whether the frequent co-visits of item q are present in the recent item history of user i. For each such presence, a normalized value is added to the recommendation score of Equation 13.14. A specialized data structure is used to implement this operation efficiently.

Clustering Methods

As discussed earlier, *MinHash* and *PLSI* are used as the two clustering schemes. The *MinHash* scheme implicitly clusters users based on intra-similarity defined by the Jaccard coefficient of the sets of items they have visited in common. Although the *MinHash* scheme is a randomized clustering method, it creates deterministic clusters in which the probability of two users belonging to the same cluster is proportional to their Jaccard coefficient. The *PLSI* scheme, on the other hand, is a probabilistic clustering method in which each point is assigned to a cluster with a certain probability. Both the *MinHash* and *PLSI* methods are described in detail in [175]. The work in [175] describes *MapReduce* methods to implement these operations efficiently. The *MapReduce* approach is required to scale the approach to massive settings.

Candidate List Generation

So far, the generation of the candidate list L for a particular target user i has not been described in detail. The candidate list can be generated in one of two ways. The News Frontend can generate a list of candidates based on the news edition, language preferences of the user i, story freshness, customized sections selected by the user i, and so on. Alternatively, the candidates can also be generated as the union of (i) all stories that have been clicked by members of the same cluster as user i, and (ii) the set of stories that have been co-visited with the set of stories in the click-history of user i.

13.8.2 Computational Advertising versus Recommender Systems

In recent years, online computational advertising has received increasing attention because of the greater importance of the internet as a medium for content consumption, information search, and business transactions. These represent typical *activities* that users are often engaged in, and they also represent an opportunity for online advertisers, because the content consumed and the transactions completed can provide a context within which advertisements can be served. An activity that a user is engaged in typically reveals a lot about the user and can be leveraged to target the products specific to the activity at hand. For example, when a user queries a search engine such as Google or Bing with a keyword like "golf," it is common to see many "sponsored search results," in addition to the true search results. These sponsored search results are advertisements, which are placed by the search engine, and are typically related to the search engine query (i.e., "golf"). This advertisement methodology is referred to as *sponsored search*. In general, the two most common computational advertising models are as follows:

1. *Sponsored search:* In this case, the search engine serves as a *match-maker*, and it serves to place advertisements adjacent to the *query search results* posed by users. The query search results provide the *context* for the advertisements, because the goal of both advertisers and the match-maker is to display advertisements related to the returned search results. This is because the users are more likely to click on contextually relevant sponsored search results. This is helpful in increasing business revenue for the advertiser and also the advertising revenue for the match-maker because match-makers are often paid on the basis of successful click-throughs from the sponsored search result or the number of times the search result is shown. A combination of these payoffs may also be used.

2. *Display advertising:* In this case, publishers of content (e.g., news portals) physically place advertisements on the Web page corresponding to their content. Thus, the content publisher plays the role of the match-maker. The content of Web page serves as the context. For example, a news portal, which is displaying an article on a golf tournaments, might display an advertisement related to golf on the same page. The match-maker can paid by the advertiser with the use of a variety of metrics. For example, the match-maker might be paid for successful click-throughs on the advertisement, a successful transaction on the basis of the advertisement, or the number of times the advertisement is shown (i.e., number of *impressions*). A combination of these payoffs may also be used. Therefore, the model of display advertising shares many similarities with that of sponsored search.

In both cases, an *advertisement* (analogous to an item) is recommended to a *user*, in a specific *context* (defined by either the search results or topic of the page on which the display advertisement is placed). In both cases, the match-maker is a publisher of the content which provides the context for the advertisement. Note that a search query result is also a form of content publication, albeit it is *dynamically generated*, and it is *reactive* to a specific user query. Furthermore, it is in the interest of both the advertiser and the match-maker to ensure that the recommended advertisements are as relevant as possible. This relationship between the various entities in the online advertising scenario is illustrated in Figure 13.3.

There are several important similarities and distinctions between computational advertising methods and recommender systems. The advertisements are like items and the match-maker plays the role of the recommender to the users. However, before discussing ways in which recommendation technology can be used for computational advertising, we need to first understand the distinctions among them. This provides an understanding of the scenarios in which one can effectively use this approach, and the changes needed to achieve these goals. The specific distinctions between recommendations and computational advertising are as follows:

1. In traditional recommender systems, it is in the best interest of a recommender system, such as Amazon.com, to provide the most relevant recommendations to users. Therefore, the user and recommender system interests are perfectly aligned. In computational advertising, the match-maker is *paid* by the advertiser to recommend items to users. While this provides a motivation for publishers (match-makers) to increase

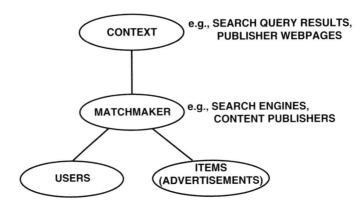

Figure 13.3: Relationships between various parties in the computational advertisement setting

the click-throughs to the advertisement, the interests of the advertiser, publisher, and user might not always be perfectly aligned. This is particularly true when publishers are paid by advertisers on the basis of the number of impressions. The cost models can be understood only in a game-theoretic sense, where the three entities try to maximize their own utility. However, in many cases, the interests of the three entities are more or less aligned.

2. Traditional recommender systems have strong user identification mechanisms. Even when users are anonymized, the *long-term history* of a returning user is known. This is not necessarily true in the case of computational advertising, where it is highly likely that no information is known about the long-term history of a user submitting a search on a search engine. In many cases, the data about past user interactions with advertisements (items) is not even available. This is particularly important because recommendations are all about *personalization*, whereas computational advertising is all about *immediate context*. Nevertheless, in some sites with strong user-identification mechanisms, both context and personalization are important. For example, if an online newspaper has a login mechanism, it can leverage the user identification to provide more relevant advertisement results. Similarly, Google does provide the ability to perform personalized search with the use of Gmail-based identification mechanisms.

3. Items have a long lifetime within a recommendation system. However, in a computational advertisement system, a particular advertisement campaign may have only a very short lifetime. Therefore, advertisements are inherently transient. However, it is possible to logically represent advertisements on the same subject as a "pseudo-item" in order to use recommendation technology.

It is clear from the aforementioned discussion that significant distinctions exist between the computational advertising and recommendation model. Nevertheless, there are a few scenarios in which one can adapt recommendation technology to computational advertising.

For cases in which strong user identification mechanisms are available and the advertiser interests can be properly aligned with publisher interests, the advertising model can be conceived as a recommendation process. The steps required to perform the modeling are as follows:

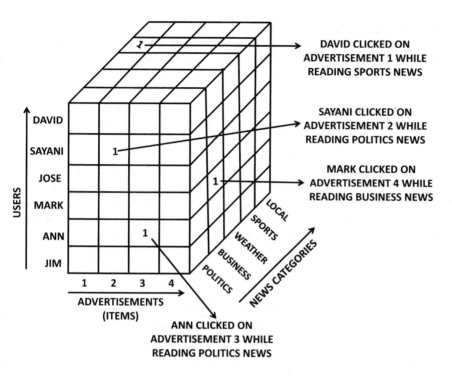

Figure 13.4: Representing advertisements as contextual recommendations for newspaper display advertisements (Note the similarity to Figures 8.1 and 11.6)

1. It is assumed that the set of (identified) users U participating in the system is known in advance for long-term tracking and analytical purposes.

2. Even though advertising campaigns are short-lived, they are all classified into sets of items. For example, two different advertisements on the same type of golf club are treated as a single item. The overall set of items is denoted by I.

3. The user *actions*, such as the act of clicking on an advertisement, are treated as implicit feedback. As advertisements have already been consolidated into items, the user actions can be used to create an implicit feedback between users and items. This implicit feedback could correspond to the frequency of user actions and can be effectively treated as "ratings."

4. All the publication sources (e.g., search phrases, or Web pages) are classified into a discrete set of categories at an appropriate level of granularity. These categories are treated as a fixed set of contexts, denoted by C. As discussed in Chapter 8, this additional set of contexts can be used to define a 3-dimensional ratings mapping function h_R:

$$h_R : U \times I \times C \to implicit\ feedback\ rating$$

This relationship is shown in Figure 13.4. In this case, a hypothetical example of a newspaper is shown, where all the articles have been categorized into specific topics. When users click on advertisements relating to a particular topic on that page, this information is recorded. The result is a multidimensional contextual representation, as discussed in Chapter 8. The level of similarity of Figure 13.4 to Figures 8.1 of Chapter 8 and Figure 11.6 of Chapter 11 is particularly striking. The use of multidimensional methods for context-sensitive recommendations [7] is a powerful technique, and it has recurred several times in this book in different scenarios.

Many of the same techniques of Chapter 8 may be used for recommending advertisements by treating them as items. However, the use of such techniques may need to be further enhanced with the cost information, such as the amount by which a publisher is paid for having a successful click-through on an advertisement. In other words, cost-sensitive variants of contextual collaborative filtering algorithms may be used, in which items with higher payoffs are prioritized over others. This can be achieved by ranking the predictions in terms of the expected payoffs, rather than in terms of the expected probability of a click. Content-based methods are particularly popular [105, 142, 327], and they use content similarity to match the context of the Web page with that in the advertisements.

13.8.2.1   Importance of Multi-Armed Bandit Methods

Multi-armed bandit methods are particularly useful for computational advertising. It is noteworthy that multi-armed bandit methods are particularly useful in settings where (a) new items enter the system all the time, and (b) the payoffs of selecting a particularly strategy can be precisely computed. Computational advertising in a domain in which the items are *extremely* transient and therefore it is particularly important to use exploration and exploitation simultaneously. Each arm of a slot machine can be viewed as one of the advertisements. Therefore, slot machines will constantly be added to and removed from the system. Also, since various types of context are associated with advertisements, it is particularly useful to leverage contextual bandit methods, where the context of the advertisement (e.g., search engine query keywords or the Web page on which an advertisement is displayed) is used in order to make decisions on whether to serve the advertisement. Refer to section 13.3 for a discussion of multi-armed bandit methods. A discussion of contextual bandit algorithms is also found in [348].

In many cases, the setting of computational advertising does not neatly fit into the traditional multi-armed bandit framework. For example, a publisher might present more than one advertisement at a time on a page, and a user might click on more than one advertisement presented to them. To handle this variation, the *slate problem* is proposed [290] for multi-armed bandits. In this variation of multi-armed bandits, the gambler is allowed to play more than one slot machine in a single try before he or she becomes aware of the rewards associated with that attempt. The simultaneous plays correspond to the different advertisements that are placed on a given page. The reward associated with a particular attempt is equal to the sum of the rewards obtained from the individual slot machines. In the advertisement setting, it translates to the placement of different advertisements (slot machine arms) on a Web page. In an ordered variation of this problem, different payoffs are associated with different placements of the advertisements on the Web page. For example, a higher placement in the ranked list will have a higher expected payoff than a lower placement. Refer to [290] for details of a randomized algorithm for computing the optimal policy.

13.8.3 Reciprocal Recommender Systems

The problem of computational advertising is related to the problem of *reciprocal recommen-dations* [481]. The basic idea is that the task of recommendation changes when one needs to consider the utility of the recommendation to multiple stakeholders with asymmetric interests. An example of such a scenario is that of online dating [480, 482], although the basic approach can be used in the context of various scenarios such as employer-employee matching [253] and mentor-mentee [103, 621] matching. Even the link-prediction problem discussed in Chapter 10 can be viewed as a form of reciprocal recommender system. A partic-ularly relevant variation of link prediction is that of *reciprocal relationship prediction* [254], in which one attempts to predict the likelihood of the occurrence of bidirectional "follower" links in a directed social-network setting. There are several key differences between tradi-tional recommender systems and reciprocal recommender systems. These differences [480] impact the nature of the algorithms that can be used in these settings:

1. In traditional recommender systems, the user receives recommendations about items and is the sole decider of the use or purchase of the items. On the other hand, in a reciprocal recommender system such as online dating, the user is aware that the success of the transaction depends on the agreement of the other party. In fact, the other party is the "item" in the reciprocal setting. Therefore, in traditional recommender systems, items are abundant and there is no need for the agreement of any other party to consume the item. This is not true in reciprocal recommender systems.

2. In traditional recommender systems, users and products constantly recur in the sys-tem. As a result, it is much easier to collect data about user preferences. In reciprocal recommender systems (such as online dating) users and items might occur only once in the system and they might never recur after a successful transaction. Therefore, the cold-start problem is much more significant in the reciprocal setting. However, this problem is not universal to all reciprocal domains. For example, in the link prediction problem for social networks, the nodes are typically persistent.

The term "reciprocal" is motivated by the fact that both users and "items" have preferences and successful transactions can be initiated only by satisfying both. Furthermore, one can view the problem in a symmetric sense. In an employer-employee matching, one can view the (potential) employer as the user and the (potential) employee as the item, or one can view the employer as the item and the employee as the user. Therefore, there are two different recommendations occurring in parallel, which need to be harmonized in order to maximize the likelihood of successful transactions. For example, if an employee is very interested in a specific employer, but the employer is not interested in the skill-set of that employee, it makes little sense to introduce them to one another.

Explicit ratings are less common in such systems as compared to implicit feedback caused by user actions. Therefore, most of these systems are based on implicit feedback data in which user actions are used in lieu of the ratings. For example, in an online dating application, the initiation of a contact, exchange of a message, or response to a message may be given varying levels of weight as an implicit indication of interest. The main challenge in such systems is the cold-start problem because successful transactions have a tendency to remove users and items from the system.

In cases where the cold-start problem is significant, content-centric methods may play a key role either directly or indirectly. In direct methods, content-centric methods can be used within the recommendation technique in order to compensate for the paucity of ratings.

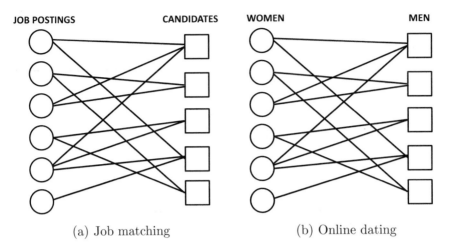

(a) Job matching   (b) Online dating

Figure 13.5: Relating link prediction to reciprocal recommendations

Content-centric methods are also facilitated by the fact that both[2] users and items are likely to have descriptive profiles in such systems. A second (indirect) method to handle the non-persistent nature of users and items is to create persistent *representatives*. For example, consider a job-matching application. For each posted job in the system, one might treat other similar jobs posted in the past as instantiations of this job. This "similarity" is defined on the basis of content-centric attributes. Similarly, for each candidate in the system, one might treat other similar candidates in the past as instantiations of this candidate. In an online dating application, one might treat (expired) users with similar profiles as instantiations of a current profile. Successful transactions between past representatives can be treated as pseudo-transactions between their current avatars. The weights of such pseudo-transactions can be computed as a function of the similarities between the current avatars and past instantiations of the representative users and "items." This augmented data set can be used in conjunction with various collaborative filtering and link-prediction methods to predict the most likely links between pairs of nodes. It is often possible to recommend a user and item pair to one another even when a pseudo-transaction already exists between. Note that some of the pseudo-transactions might be quite noisy and unreliable. However, since the underlying inference methods use the aggregate structure of the data set, the predictions are likely to be reasonably robust. In cases where the pseudo-transactions are noisy, the corresponding user-item pairs are less likely to be recommended by prediction algorithms such as robust matrix factorization.

In the following, we will give a brief description of two key methods that are common to reciprocal recommender systems. However, since this is an emerging area, we recognize that these methods only scratch the surface of what is really possible in this area. Significant opportunities exist for further research in this domain.

13.8.3.1   Leveraging Hybrid Methods

In these methods, two *traditional* recommender methods are constructed corresponding to the preferences of the two reciprocal parties. Then, the predictions from these two parties are combined. For example, in a job-matching application, a traditional recommender system

---

[2]In traditional recommender systems, items are more likely to have descriptive profiles than users.

$\mathcal{R}_1$ may be used to create a ranked list of potential employees for an employer. Then, a traditional recommender system $\mathcal{R}_2$ may be used to create a ranked list of potential employers for an employee. The results from these two recommendations are combined to maximize the likelihood of a successful transaction. The combination method may use the weighted hybrid methods discussed in Chapter 6. As discussed in Chapter 6, the weights can be learned using linear regression methods, where the observed data is defined by successful transactions in the past. In cases where sufficient observed data is not available because of cold-start issues, either simple averages or domain-specific weights may be used. In cases where the preference if one party is more important than that of the other, a cascade hybrid can be used. For example, in settings where the number of job-seekers is far greater than the number of job-postings, the recommender system can choose to prioritize employer interests over employee interests. In such a setting, the cascade hybrid is ideal because it naturally prioritizes the first cascade in the hybrid over the second one.

There are many other factors that can play an important role in deciding how to combine the recommendations. For example, it is possible that one of the two parties might be naturally proactive (i.e., initiating contact), and the other party might be naturally reactive (i.e., responding to initial contact). In such cases, the nature of the hybrid can depend on the system's relative interests in satisfying the proactive and reactive parties. For example, one might assume primacy of interests of the proactive party but only ensure that the reactive party does not reject the recommendations. Repeated rejections from the reactive party can be costly and can affect the popularity of the system. Therefore, two models can created: the first model $\mathcal{R}_1$ computes the "items" that the proactive party will like, and the second model $\mathcal{R}_2$ computes the users that reactive party (i.e., "items") will dislike. The idea of the second model is to prune the recommended items from the first model that the reactive party will dislike. A variety of combination methods for these models are discussed in [482].

The recommender systems $\mathcal{R}_1$ and $\mathcal{R}_2$ are often content-centric systems because of the cold-start problem. However, in some case, the ratings data can be augmented by treating past users and items as instantiations of similar users in the system and constructing pseudo-transactions between users and items. In such cases, collaborative filtering methods can also be used because one can use the additional data resulting from the pseudo-transactions.

13.8.3.2 Leveraging Link Prediction Methods

In cases where the cold-start problem is not a serious issue or the ratings data can be augmented with the data from similar users and items, link prediction methods can be adapted to this setting. Matrix factorization methods for directed and undirected link prediction are discussed in section 10.4.5 of Chapter 10. In these cases, one can construct a bipartite network in which the two reciprocal parties form the two partitions of the network. For example, one partition might be a set of employers and another partition might be a set of employees. In the dating application, one partition might correspond to men and the other partition might correspond to women. The edges in this network correspond to (previous) successful transactions between the nodes in these partitions (or their similar representatives). These scenarios are illustrated in Figure 13.5(a) and (b), respectively. However, in other applications, the underlying graph might not be bipartite. For example, in a same-gender dating application, the underlying preference graph might not be bipartite. In some cases, when the preferences are specified in an asymmetric way, the underlying graph might be directed. In all these cases, the asymmetric and symmetric matrix factorization methods discussed in section 10.4.5 can be very useful. This is not particularly surprising, considering the fact that the link-prediction problem is a special case of reciprocal recommender systems. In cases where the links are constructed in a noisy way using representatives, robust matrix factorization methods may be used to improve accuracy based on the ideas in Chapter 12.

13.9 Summary

This chapter reviews several advanced topics in recommender systems, such as group recommendations, multi-criteria recommendations, active learning, and privacy. In addition, some interesting applications of recommender systems have been covered.

Group recommendations are designed to provide recommendations to groups of users with possibly diverse interests. In general, straightforward averaging methods might not always work in these scenarios because of various social factors in the recommendation process. In multi-criteria recommender systems, diverse user interests are used to provide more robust recommendations. The basic idea is that the user behavior can be more accurately modeled when details for the user ratings of various criteria are available.

The problem of active learning, studies the issue of ratings acquisition in recommender systems. Ratings acquisition is sometimes expensive. Therefore, techniques need to be designed to judiciously query specific user-item combinations for ratings. The approach of active learning in recommender systems is very similar to that in classification.

Privacy remains a significant challenge for recommender systems, as in any other domain. Privacy-preserving methods can be applied either at data-collection time, or at data-publication time. Methods that preserve privacy at data-collection time generally provide better guarantees, but they are harder to implement from an infrastructure point of view.

Numerous applications have been proposed in recent years for recommender systems. Some examples include query recommendations, news personalization, computational advertising, and reciprocal recommendations. This chapter introduces some of the basic methods in these domains.

13.10 Bibliographic Notes

The problem of learning to rank is widely studied in the classification, internet search, and information retrieval [15, 115, 284, 370]. A tutorial on learning to rank from the perspective of recommender systems may be found in [323]. Ranking methods can be either pairwise methods or listwise methods [136]. Pairwise methods include the Bayesian personalized ranking model (BPR) [499], EigenRank model [367], pLPA [368], and CR [59]. Listwise methods include CoFiRank [624], CLiMF, xCLiMF and several other variations [545–548]. Some of these methods have also been generalized to the contextual scenario [549].

Multi-arm bandit methods can be viewed as a class of reinforcement learning algorithms [579]. A simple discussion of several bandit algorithms may be found in [628], although the book is written in the context of Website optimization. Bandit algorithms for recommender systems are discussed in [92, 348]. The work in [349] introduces the problem of evaluating bandit algorithms in the offline setting. The use of multi-armed bandits for computational advertising is discussed in [160, 290].

Group recommender systems are discussed in detail in [271, 272, 407, 408]. A review of social factors for group recommender systems may be found in [489]. Case-based methods for group recommendations are discussed in [413, 415]. Group recommendations have been used in a variety of domains, such as movies [168], television [653], music [412], and travel [52, 272, 413]. The limitations of the averaging strategy for group recommender systems are discussed in [409, 654]. A variety of aggregation strategies for group recommender systems, such as plurality voting, multiplicative aggregation, Borda count, Copeland rule, approval voting, and fairness are suggested in [407]. An experimental study comparing the various strategies is also included in the same work. In some cases, one is interested in recommending complex items containing sequences of items. An example is the case of a television program for a set of viewers, where the overall program may contain several components of various types. In such cases, the ordering of items is also important. Such systems are discussed in [407].

Surveys on multi-criteria recommender systems may be found in [11, 398, 604]. The multi-criteria recommendation problem was first defined in the seminal work of [12]. Neighborhood-based methods for multi-criteria recommendations are discussed in [12, 399, 596]. The work in [399] proposes three different methods to perform the aggregated similarity computation in neighborhood methods. However, the overall approach is not different in principle from that discussed in [12]. An ensemble-based method was also proposed in [12]. A number of model-based methods have also been proposed in the context of multi-criteria recommender systems. These include the flexible mixture model [514] and a multi-linear singular value decomposition (*MSVD*) approach [353]. Methods have also been proposed for cases in which overall ratings are not available. For example, the work in [328] proposes a method to combine the predicted ratings across various criteria with the use of a UTilities Additive method (UTA). The work in [276] uses a support vector regression model to determine the relative importance of different criteria. These are used to combine the user-based and item-based regression models with a weighted approach. A pareto-optimal approach with the use of skyline queries on a restaurant rating system is proposed in [340].

A detailed review of active learning methods is provided in [513]. However, this review is mostly based on the classification problem, since the available work on recommender systems is limited. Only a limited amount of work [192–194, 257, 295, 330, 578] has been proposed in recent years on this topic. The area of active learning is still quite open as far as the recommendation problem is concerned. An interesting class of algorithms, related to temporal collaborative filtering, is the multi-arm bandit class of algorithms in which the recommender trades off exploration vs exploitation in the recommendation space [92, 348].

Privacy-preserving techniques may include the use of perturbation techniques [35, 38, 484, 485], group-based anonymization methods [27, 352, 386, 521], or distributed methods [75, 133, 334, 551, 606]. Both perturbation methods and distributed techniques have a common aspect that they tend to preserve the privacy at data-collection time. This provides a greater level of privacy. On the other hand, these systems are generally harder to implement because of the greater infrastructural and customization issues involved in the final use of the stored data. These issues surface because the stored data is in a form that cannot be used by a traditional collaborative filtering algorithm. Group-based anonymization techniques are designed to publish the data, which is collected by a centralized entity. These techniques are more popular and the output can be used in conjunction with traditional collaborative filtering algorithms. All these methods are affected by the curse of dimensionality [30], which prevents effective privacy preservation for high-dimensional data. Some methods for anonymization of high-dimensional and sparse data sets are proposed in [657]. Recently, the notion of differential privacy has been proposed [189] that is theoretically very popular, although its practical and commercial use remains limited. A differentially private matrix factorization has recently been proposed in [372]. A privacy-preserving approach that treats the collecting system as a distrusted entity is proposed in [642].

Recommender systems have many specialized applications in the Web domain. Query recommendation methods attempt to recommend similar queries to those already issued by the user in a particular session. The work in [57] returns the most similar queries to the queries *current query*, which also have sufficient popularity (support). Support is measured in terms of the number of times the query was issued by other users, and the corresponding results were found relevant. The work in [137] uses not just the current query, but the current session of queries as the context for the query suggestion. An interesting idea in this area is that of *query flow graphs* [90], which uses a graphical representation of the user's latent querying behavior to make recommendations. The work in [429] performs query recommendations with the use of random walks on the query-URL graph. The use of Markov models for query recommendations is discussed in [244].

The dynamic profiler system is discussed in [636]. Methods for Web portal personalization are discussed in [34]. The use of semantic contextualization for news recommendation was discussed in [134]. This work is based on contextual recommendation ideas presented in Chapter 8. The Google news personalization engine is described in more detail in [175]. Mobile recommender systems are discussed in [504].

One of the earliest systems for computational advertising was discussed in [28]. However, this system was not based on modern models of computational advertising. More recent discussions of such systems may be found in [106, 107]. The slate method for computational advertising is discussed in [290]. In some cases, linear payoffs are associated with the features of Web pages and advertisements. A variant of the LinUCB algorithm is proposed in [160] to handle this setting. The problem of computational advertising is related to the problem of *reciprocal recommendations* [481]. The basic idea is that the task of recommendation changes when one needs to consider the utility of the recommendation to multiple stakeholders with asymmetric interests. Examples of such applications include online dating [480, 482], job matching [253], and mentor-mentee recommendations [103, 621].

Bibliography

[1] A. Abdul-Rahman and S. Hailes. Supporting trust in virtual communities. *Proceedings of the 33rd Annual Hawaii International Conference on System Sciences*, pp. 1769–1777, 2000.

[2] G. Abowd, C. Atkeson, J. Hong, S. Long, R. Kooper, and M. Pinkerton. Cyberguide: A mobile context-aware tour guide. *Wireless Networks*, 3(5), pp. 421–433, 1997.

[3] G. Abowd, A. Dey, P. Brown, N. Davies, M. Smith, and P. Steggles. Towards a better understanding of context and context-awareness. *Handheld and Ubiquitous Computing*, pp. 304–307, 1999.

[4] P. Adamopoulos, A. Bellogin, P. Castells, P. Cremonesi, and H. Steck. REDD 2014 – International Workshop on Recommender Systems Evaluation: Dimensions and Design. Held in conjunction with *ACM Conference on Recommender systems*, 2014.

[5] G. Adomavicius, and A. Tuzhilin. Toward the next generation of recommender systems: A survey of the state-of-the-art and possible extensions. *IEEE Transactions on Knowledge and Data Engineering*, 17(6), pp. 734–749, 2005.

[6] G. Adomavicius, R. Sankaranarayanan, S. Sen, and A. Tuzhilin. Incorporating contextual information in recommender systems using a multidimensional approach. *ACM Transactions on Information Systems*, 23(1), pp. 103–145, 2005.

[7] G. Adomavicius and A. Tuzhilin. Context-aware recommender systems. *Recommender Systems handbook*, pp. 217–253, Springer, NY, 2011.

[8] G. Adomavicius and A. Tuzhilin. Incorporating context into recommender systems using multidimensional rating estimation methods. *International Workshop on Web Personalization, Recommender Systems and Intelligent User Interfaces (WPRSIUI)*, 2005.

[9] G. Adomavicius and A. Tuzhilin. Multidimensional recommender systems: a data warehousing approach. *International Workshop on Electronic Commerce. Lecture Notes in Computer Science*, Springer, Vol. 2232, pp. 180–192, 2001.

© Springer International Publishing Switzerland 2016
C.C. Aggarwal, *Recommender Systems: The Textbook*,
DOI 10.1007/978-3-319-29659-3

[10] G. Adomavicius, A. Tuzhilin, and R. Zheng. REQUEST: A query language for customizing recommendations. *Information Systems Research*, 22(1), pp. 99–117, 2011.

[11] G. Adomavicius, N. Manouselis, and Y. Kwon. Multi-criteria recommender systems. *Recommender Systems Handbook*, Springer, pp. 769–803, 2011.

[12] G. Adomavicius and Y. Kwon. New recommendation techniques for multicriteria rating systems. *IEEE Intelligent Systems*, 22(3), pp. 48–55, 2007.

[13] D. Agarwal, and B. Chen. Regression-based latent factor models. *ACM KDD Conference*, pp. 19–28. 2009.

[14] D. Agarwal, B.-C. Chen, and B. Long. Localized factor models for multi-context recommendation. *ACM KDD Conference*, pp. 609–617, 2011.

[15] S. Agarwal. Ranking methods in machine learning. Tutorial at *SIAM Conference on Data Mining*, 2010. Slides available at: `http://www.siam.org/meetings/sdm10/tutorial1.pdf`

[16] E. Agichtein, C. Castillo, D. Donato, A. Gionis, and G. Mishne. Finding high-quality content in social media. *Web Search and Data Mining Conference*, pp. 183–194, 2008.

[17] C. Aggarwal. Social network data analytics. *Springer*, New York, 2011.

[18] C. Aggarwal. Data classification: algorithms and applications. *CRC Press*, 2014.

[19] C. Aggarwal. Data clustering: algorithms and applications. *CRC Press*, 2014.

[20] C. Aggarwal and P. Yu. Privacy-preserving data mining: models and algorithms, *Springer*, 2008.

[21] C. Aggarwal and C. Zhai. A survey of text classification algorithms. *Mining Text Data*, Springer, 2012.

[22] C. Aggarwal. Data mining: the textbook. *Springer*, New York, 2015.

[23] C. Aggarwal and J. Han. Frequent pattern mining. *Springer*, New York, 2014.

[24] C. Aggarwal and S. Parthasarathy. Mining massively incomplete data sets by conceptual reconstruction. *ACM KDD Conference*, pp. 227–232, 2001.

[25] C. Aggarwal, C. Procopiuc, and P. S. Yu. Finding localized associations in market basket data. *IEEE Transactions on Knowledge and Data Engineering*, 14(1), pp. 51–62, 2001.

[26] C. Aggarwal and T. Abdelzaher. Social sensing. *Managing and Mining Sensor Data*, Springer, New York, 2013.

[27] C. Aggarwal and P. Yu. On static and dynamic methods for condensation-based privacy-preserving data mining. *ACM Transactions on Database Systems (TODS)*, 33(1), 2, 2008.

[28] C. Aggarwal, J. Wolf, and P. Yu. A framework for the optimizing of WWW advertising. *Trends in Distributed Systems for Electronic Commerce*, pp. 1–10, 1998.

[29] C. Aggarwal, S. Gates, and P. Yu. On using partial supervision for text categorization. *IEEE Transactions on Knowledge and Data Engineering*, 16(2), pp. 245–255, 2004.

[30] C. Aggarwal. On k-anonymity and the curse of dimensionality, *Very Large Databases Conference*, pp. 901–909, 2005.

[31] C. Aggarwal, Z. Sun, and P. Yu. Online generation of profile association rules. *ACM KDD Conference*, pp. 129–133, 1998.

[32] C. Aggarwal, Z. Sun, and P. Yu. Online algorithms for finding profile association rules, *CIKM Conference*, pp. 86–95, 1998.

[33] C. Aggarwal, J. Wolf, K.-L. Wu, and P. Yu. Horting hatches an egg: a new graph-theoretic approach to collaborative filtering. *ACM KDD Conference*, pp. 201–212, 1999.

[34] C. Aggarwal and P. Yu. An automated system for Web portal personalization. *Very Large Data Bases Conference*, pp. 1031–1040, 2002.

[35] D. Agrawal and C. Aggarwal. On the design and quantification of privacy-preserving data mining algorithms. *ACM PODS Conference*, pp. 247–255, 2001.

[36] C. Aggarwal, Y. Xie, and P. Yu. On dynamic link inference in heterogeneous networks. *SIAM Conference on Data Mining*, pp. 415–426, 2012.

[37] R. Agrawal and R. Srikant. Mining sequential patterns. *International Conference on Data Engineering*, pp. 3–14, 1995.

[38] R. Agrawal, and R. Srikant. Privacy-preserving data mining. *ACM SIGMOD Conference*, pp. 439–450, 2000.

[39] R. Agrawal, R. Rantzau, and E. Terzi. Context-sensitive ranking. *ACM SIGMOD Conference*, pp. 383–394, 2006.

[40] H. Ahn, K. Kim, and I. Han. Mobile advertisement recommender system using collaborative filtering: MAR-CF. *Proceedings of the 2006 Conference of the Korea Society of Management Information Systems*, 2006.

[41] J. Ahn, P. Brusilovsky, J. Grady, D. He, and S. Syn. Open user profiles for adaptive news systems: help or harm? *World Wide Web Conference*, pp. 11–20, 2007.

[42] M. Al Hasan, and M. J. Zaki. A survey of link prediction in social networks. *Social network data analytics*, Springer, pp. 243–275, 2011.

[43] G. K. Al Mamunur Rashid, G. Karypis, and J. Riedl. Influence in ratings-based recommender systems: An algorithm-independent approach. *SIAM Conference on Data Mining*, 2005.

[44] X. Amatriain, J. Pujol, N. Tintarev, and N. Oliver. Rate it again: increasing recommendation accuracy by user re-rating. *ACM Conference on Recommender Systems*, pp. 173–180, 2009.

[45] S. Amer-Yahia, S. Roy, A. Chawlat, G. Das, and C. Yu. (2009). Group recommendation: semantics and efficiency. *Proceedings of the VLDB Endowment*, 2(1), pp. 754–765, 2009.

[46] S. Anand and B. Mobasher. Intelligent techniques for Web personalization. Lectures Notes in Computer Science, Vol. 3169, pp. 1–36, Springer, 2005.

[47] S. Anand and B. Mobasher. Contextual recommendation, *Lecture Notes in Artificial Intelligence*, Springer, 4737, pp. 142–160, 2007.

[48] R. Andersen, C. Borgs, J. Chayes, U. Feige, A. Flaxman, A. Kalai, V. Mirrokni, and M. Tennenholtz. Trust-based recommendation systems: An axiomatic approach. *World Wide Web Conference*, pp. 199–208, 2008.

[49] C. Anderson. The long tail: why the future of business is selling less of more. *Hyperion*, 2006.

[50] A. Ansari, S. Essegaier, and R. Kohli. Internet recommendation systems. *Journal of Marketing Research*, 37(3), pp. 363–375, 2000.

[51] F. Aiolli. Efficient top-n recommendation for very large scale binary rated datasets. *ACM conference on Recommender Systems*, pp. 273–280, 2013.

[52] L. Ardissono, A. Goy, G. Petrone, M. Segnan, and P. Torasso. INTRIGUE: personalized recommendation of tourist attractions for desktop and hand-held devices. *Applied Artificial Intelligence*, 17(8), pp. 687–714, 2003.

[53] W. G. Aref and H. Samet. Efficient processing of window queries in the pyramid data structure. *ACM PODS Conference*, pp. 265–272, 1990.

[54] D. Ashbrook and T. Starner. Using GPS to learn significant locations and predict movement across multiple users. *Personal and Ubiquitous Computing*, 7(5), pp. 275–286, 2003.

[55] F. Asnicar and C. Tasso. IfWeb: a prototype of user model-based intelligent agent for document filtering and navigation in the world wide web. *International Conference on User Modeling*, pp. 3–12, 1997.

[56] A. Azran. The rendezvous algorithm: Multiclass semi-supervised learning with markov random walks. *International Conference on Machine Learning*, pp. 49–56, 2007.

[57] R. Baeza-Yates, C. Hurtado, and M. Mendoza. Query recommendation using query logs in search engines. *EDBT 2004 Workshops on Current Trends in Database Technology*, pp. 588–596, 2004.

[58] R. Battiti. Accelerated backpropagation learning: Two optimization methods. *Complex Systems*, 3(4), pp. 331–342, 1989.

[59] S. Balakrishnan and S. Chopra. Collaborative ranking. *Web Search and Data Mining Conference*, pp. 143–152, 2012.

[60] M. Balabanovic, and Y. Shoham. Fab: content-based, collaborative recommendation. *Communications of the ACM*, 40(3), pp. 66–72, 1997.

[61] L. Baltrunas and X. Amatriain. Towards time-dependant recommendation based on implicit feedback. *RecSys Workshop on Context-Aware Recommender Systems*, 2009.

[62] L. Baltrunas and F. Ricci. Context-dependant items generation in collaborative filtering. *RecSys Workshop on Context-Aware Recommender Systems*, 2009.

[63] L. Baltrunas, B. Ludwig, and F. Ricci. Matrix factorization techniques for context aware recommendation. *ACM Conference on Recommender systems*, pp. 301–304, 2011.

[64] J. Bao, Y. Zheng, and M. Mokbel. Location-based and preference-aware recommendation using sparse geo-social networking data. *International Conference on Advances in Geographic Information Systems*, pp. 199–208, 2012.

[65] X. Bao. Applying machine learning for prediction, recommendation, and integration. *Ph.D dissertation*, Oregon State University, 2009. http://ir.library.oregonstate.edu/xmlui/bitstream/handle/1957/12549/Dissertation_XinlongBao.pdf?sequence=1

[66] X. Bao, L. Bergman, and R. Thompson. Stacking recommendation engines with additional meta-features. *ACM Conference on Recommender Systems*, pp. 109–116, 2009.

[67] A. Bar, L. Rokach, G. Shani, B. Shapira, and A. Schclar. Boosting simple collaborative filtering models using ensemble methods. *Arxiv Preprint*, arXiv:1211.2891, 2012. Also appears in *Multiple Classifier Systems*, Springer, pp. 1–12, 2013. http://arxiv.org/ftp/arxiv/papers/1211/1211.2891.pdf

[68] J. Basilico, and T. Hofmann. Unifying collaborative and content-based filtering. *International Conference on Machine Learning*, 2004.

[69] C. Basu, H. Hirsh, and W. Cohen. Recommendation as classification: using social and content-based information in recommendation. *AAAI*, pp. 714–720, 1998.

[70] G. Begelman, P. Keller, and F. Smadja. Automated tag clustering: Improving search and exploration in the tag space. *Collaborative Web Tagging Workshop* (colocated with *WWW Conference*), pp. 15–23, 2006.

[71] R. Bell, Y. Koren, and C. Volinsky. Modeling relationships at multiple scales to improve accuracy of large recommender systems. *ACM KDD Conference*, pp. 95–104, 2007.

[72] R. Bell and Y. Koren. Scalable collaborative filtering with jointly derived neighborhood interpolation weights. *IEEE International Conference on Data Mining*, pp. 43–52, 2007.

[73] R. Bell and Y. Koren. Lessons from the Netflix prize challenge. *ACM SIGKDD Explorations Newsletter*, 9(2), pp. 75–79, 2007.

[74] R. Bergmann, M. Richter, S. Schmitt, A. Stahl, and I. Vollrath. Utility-oriented matching: a new research direction for case-based reasoning. *German Workshop on Case-Based Reasoning*, pp. 264–274, 2001.

[75] S. Berkovsky, Y. Eytani, T. Kuflik, and F. Ricci. Enhancing privacy and preserving accuracy of a distributed collaborative filtering. *ACM Conference on Recommender Systems*, pp. 9–16, 2007.

[76] D. P. Bertsekas. Nonlinear programming. *Athena Scientific Publishers*, Belmont, 1999.

[77] S. Bhagat, G. Cormode, and S. Muthukrishnan. Node classification in social networks. *Social Network Data Analytics*, Springer, pp. 115–148. 2011.

[78] R. Bhaumik, C. Williams, B. Mobasher, and R. Burke. Securing collaborative filtering against malicious attacks through anomaly detection. *Workshop on Intelligent Techniques for Web Personalization (ITWP)*, 2006.

[79] R. Bhaumik, R. Burke, snd B. Mobasher. Crawling Attacks Against Web-based Recommender Systems. *International Conference on Data Mining (DMIN)*, pp. 183–189, 2007.

[80] B. Bi, Y. Tian, Y. Sismanis, A. Balmin, and J. Cho. Scalable topic-specific influence analysis on microblogs. *Web Search and Data Mining Conference*, pp. 513–522, 2014.

[81] J. Bian, Y. Liu, D. Zhou, E. Agichtein, and H. Zha. Learning to recognize reliable users and content in social media with coupled mutual reinforcement. *World Wide Web Conference*, pp. 51–60, 2009.

[82] D. Billsus and M. Pazzani. Learning collaborative information filters. *ICML Conference*, pp. 46–54, 1998.

[83] D. Billsus and M. Pazzani. Learning probabilistic user models. *International Conference on User Modeling, Workshop on Machine Learning for User Modeling*, 1997.

[84] D. Billsus and M. Pazzani. A hybrid user model for news story classification. *International Conference on User Modeling*, 1999.

[85] D. Billsus and M. Pazzani. User modeling for adaptive news access. *User Modeling and User-Adapted Interaction*, 10(2–3), pp. 147–180, 2000.

[86] C. M. Bishop. Pattern recognition and machine learning. *Springer*, 2007.

[87] C. M. Bishop. Neural networks for pattern recognition. *Oxford University Press*, 1995.

[88] J. Bobadilla, F. Ortega, A. Hernando, and A. Gutierrez. Recommender systems survey. *Knowledge-Based Systems*, 46, pp. 109–132, 2013.

[89] F. Bohnert, I. Zukerman, S. Berkovsky, T. Baldwin, and L. Sonenberg. Using interest and transition models to predict visitor locations in museums. *AI Communications*, 2(2), pp. 195–202, 2008.

[90] P. Boldi, F. Bonchi, C. Castillo, D. Donato, A. Gionis, and S. Vigna. The query-flow graph: model and applications. *ACM Conference on Information and Knowledge Management*, pp. 609–618, 2008.

[91] K. Bollacker, S. Lawrence, and C. L. Giles. CiteSeer: An autonomous web agent for automatic retrieval and identification of interesting publications. *International Conference on Autonomous Agents*, pp. 116–123, 1998.

[92] B. Bouneffouf, A. Bouzeghoub, and A. Gancarski. A contextual-bandit algorithm for mobile context-aware recommender system. *Neural Information Processing*, pp. 324–331, 2012.

[93] G. Box, W. Hunter, and J. Hunter. Statistics for experimenters, *Wiley*, New York, 1978.

[94] K. Bradley and B. Smyth. Improving recommendation diversity. *National Conference in Artificial Intelligence and Cognitive Science*, pp. 75–84, 2001.

[95] K. Bradley, R. Rafter, and B. Smyth. Case-based user profiling for content personalization. *International Conference on Adaptive Hypermedia and Adaptive Web-Based Systems*, pp. 62–72, 2000.

[96] M. Brand. Fast online SVD revisions for lightweight recommender systems. *SIAM Conference on Data Mining*, pp. 37–46, 2003.

[97] L. Branting. Acquiring customer preferences from return-set selections. *Case-Based Reasoning Research and Development*, pp. 59–73, 2001.

[98] J. Breese, D. Heckerman, and C. Kadie. Empirical analysis of predictive algorithms for collaborative filtering. *Conference on Uncertainty in Artificial Inetlligence*, 1998.

[99] L. Breiman. Bagging predictors. *Machine Learning*, 24(2), pp. 123–140, 1996.

[100] A. Brenner, B. Pradel, N. Usunier, and P. Gallinari. Predicting most rated items in weekly recommendation with temporal regression. *Workshop on Context-Aware Movie Recommendation*, pp. 24–27, 2010.

[101] D. Bridge. Diverse product recommendations using an expressive language for case retrieval. *European Conference on Case-Based Reasoning*, pp. 43–57. 2002.

[102] D. Bridge, M. Goker, L. McGinty, and B. Smyth. Case-based recommender systems. *The Knowledge Engineering Review*, 20(3), pp. 315–320, 2005.

[103] A. Brun, S. Castagnos, and A. Boyer. Social recommendations: mentor and leader detection to alleviate the cold-start problem in collaborative filtering. *Social Network Mining, Analysis, and Research Trends: Techniques and Applications: Techniques and Applications*, 270, 2011.

[104] S. Brin, and L. Page. The anatomy of a large-scale hypertextual web search engine. *Computer Networks*, 30(1–7), pp. 107–117, 1998.

[105] A. Broder, M. Fontoura, V. Josifovski, and L. Riedel. A semantic approach to contextual advertising. *SIGIR Conference*, pp. 559–566, 2007.

[106] A. Broder. Computational advertising and recommender systems. *ACM Conference on Recommender Systems*, pp. 1–2, 2008.

[107] A. Broder and V. Josifovski. Introduction to Computational Advertising. *Course Material*, Stanford University, 2010. http://www.stanford.edu/class/msande239/

[108] M. Brunato and R. Battiti. PILGRIM: A location broker and mobility-aware recommendation system. *International Conference on Pervasive Computing and Communications*, pp. 265–272, 2003.

[109] P. Brusilovsky, A. Kobsa, and W. Nejdl. The adaptive web: methods and strategies of web personalization, *Lecture Notes in Computer Sceince*, Vol. 4321, Springer, 2007.

[110] K. Bryan, M. O'Mahony, and P. Cunningham. Unsupervised retrieval of attack profiles in collaborative recommender systems. *ACM Conference on Recommender Systems*, pp. 155–162, 2008.

[111] P. Buhlmann. Bagging, subagging and bragging for improving some prediction algorithms, *Recent advances and trends in nonparametric statistics*, Elsivier, 2003.

[112] P. Buhlmann and B. Yu. Analyzing bagging. *Annals of statistics*, 20(4), pp. 927–961, 2002.

[113] L. Breiman. Bagging predictors. *Machine learning*, 24(2), pp. 123–140, 1996.

[114] C. Burges. A tutorial on support vector machines for pattern recognition. *Data mining and knowledge discovery*, 2(2), pp. 121–167, 1998.

[115] C. Burges, T. Shaked, E. Renshaw, A. Lazier, M. Deeds, N. Hamilton, and G. Hullender. Learning to rank using gradient descent. *International Conference on Machine Learning*, pp. 89–96, 2005.

[116] R. Burke. Knowledge-based recommender systems. *Encyclopedia of library and information systems*, pp. 175–186, 2000.

[117] R. Burke. Hybrid recommender systems: Survey and experiments. *User Modeling and User-adapted Interaction*, 12(4), pp. 331–370, 2002.

[118] R. Burke. Hybrid Web recommender systems. *The adaptive Web*, pp. 377–406, Springer, 2007.

[119] R. Burke, M. O'Mahony, and N. Hurley. Robust collaborative recommendation. *Recommender Systems Handbook*, Springer, pp. 805–835, 2011.

[120] R. Burke, K. Hammond, and B. Young. Knowledge-based navigation of complex information spaces. *National Conference on Artificial Intelligence*, pp. 462–468, 1996.

[121] R. Burke, K. Hammond, and B. Young. The FindMe approach to assisted browsing. *IEEE Expert*, 12(4), pp. 32–40, 1997.

[122] R. Burke, B. Mobasher, R. Zabicki, and R. Bhaumik. Identifying attack models for secure recommendation. *Beyond Personalization: A Workshop on the Next Generation of Recommender Systems*, 2005.

[123] R. Burke, B. Mobasher, and R. Bhaumik. Limited knowledge shilling attacks in collaborative filtering systems. *IJCAI Workshop in Intelligent Techniques for Personalization*, 2005.

[124] R. Burke, B. Mobasher, C. Williams, and R. Bhaumik. Classification features for attack detection in collaborative recommender systems. *ACM KDD Conference*, pp. 542–547, 2006.

[125] R. Burke. The Wasabi personal shopper: a case-based recommender system. *National Conference on Innovative Applications of Artificial Intelligence*, pp. 844–849, 1999.

[126] D. Cai, S. Yu, J. Wen, and W. Y. Ma. Extracting content structure for web pages based on visual representation. *Web Technologies and Applications*, pp. 406–417, 2003.

[127] J. Cai, E. Candes, and Z. Shen. A singular value thresholding algorithm for matrix completion. *SIAM Journal on Optimization*, 20(4), 1956–1982, 2010.

[128] Z. Cao, T. Qin, T. Liu, M. F. Tsai, and H. Li. Learning to rank: from pairwise approach to listwise approach. *International Conference on Machine Learning*, pp. 129–137, 2007.

[129] L. M. de Campos, J. Fernandez-Luna, J. Huete, and M. Rueda-Morales. Combining content-based and collaborative recommendations: A hybrid approach based on Bayesian networks. *International Journal of Approximate Reasoning*, 51(7), pp. 785–799, 2010.

[130] P. Campos, F. Diez, and I. Cantador. Time-aware recommender systems: a comprehensive survey and analysis of existing evaluation protocols. *User Modeling and User-Adapted Interaction*, 24(1–2), pp. 67–119, 2014.

[131] P. Campos, A. Bellogin, F. Diez, and J. Chavarriaga. Simple time-biased KNN-based recommendations. *Workshop on Context-Aware Movie Recommendation*, pp. 20–23, 2010.

[132] E. Candes, X. Li, Y. Ma, and J. Wright. Robust principal component analysis?. *Journal of the ACM (JACM)*, 58(3), 11, 2011.

[133] J. Canny. Collaborative filtering with privacy via factor analysis. *ACM SIGR Conference*, pp. 238–245, 2002.

[134] I. Cantador and P. Castells. Semantic contextualisation in a news recommender system. *Workshop on Context-Aware Recommender Systems*, 2009.

[135] I. Cantador, A. Bellogin, and D. Vallet. Content-based recommendation in social tagging systems. *ACM Conference on Recommender Systems*, pp. 237–240, 2010.

[136] H. Cao, E. Chen, J. Yang, and H. Xiong. Enhancing recommender systems under volatile user interest drifts. *ACM Conference on Information and Knowledge Management*, pp. 1257–1266, 2009.

[137] H. Cao, D. Jiang, J. Pei, Q. He, Z. Liao, E. Chen, and H. Li. Context-aware query suggestion by mining click-through and session data. *ACM KDD Conference*, pp. 875–883, 2008.

[138] O. Celma, M. Ramirez, and P. Herrera. Foafing the music: A music recommendation system based on RSS feeds and user preferences. *International Conference on Music Information Retrieval*, pp. 464–467, 2005.

[139] O. Celma, and X. Serra. FOAFing the music: Bridging the semantic gap in music recommendation. *Web Semantics: Science, Services and Agents on the World Wide Web*, 6(4), pp. 250–256, 2008.

[140] O. Celma and P. Herrera. A new approach to evaluating novel recommendations. *ACM Conference on Recommender Systems*, pp. 179–186, 2008.

[141] T. Chai and R. Draxler. Root mean square error (RMSE) or mean absolute error (MAE)?– Arguments against avoiding RMSE in the literature. *Geoscientific Model Development*, 7(3), pp. 1247–1250, 2004. ,

[142] D. Chakrabarti, D. Agarwal, and V. Josifovski. Contextual advertising by combining relevance with click feedback. *World Wide Web Conference*, 2008.

[143] S. Chakrabarti, B. Dom, and P. Indyk. Enhanced hypertext categorization using hyperlinks. *ACM SIGMOD Conference*, pp. 307–318, 1998.

[144] S. Chakrabarti. Mining the Web: Discovering knowledge from hypertext data. *Morgan Kaufmann*, 2003.

[145] S. Chaudhuri and U. Dayal. An overview of data warehousing and OLAP technology. *ACM SIGMOD Record*, 26(1), pp. 65–74, 1997.

[146] S. Chee, J. Han, and K. Wang. Rectree: An efficient collaborative filtering method. *Data Warehousing and Knowledge Discovery*, pp. 141–151, 2001.

[147] G. Chen and D. Kotz. A survey of context-aware mobile computing research. *Technical Report TR2000-381*, Department of Computer Science, Dartmouth College, 2000.

[148] L. Chen and P. Pu. Survey of preference elicitation methods *EPFL-REPORT-52659*, 2004. http://hci.epfl.ch/wp-content/uploads/publications/2004/IC_TECH_REPORT_200467.pdf

[149] L. Chen and P. Pu. Critiquing-based recommenders: survey and emerging trends. *User Modeling and User-Adapted Interaction*, 22(1–2), pp. 125–150, 2012.

[150] L. Chen, and K. Sycara. WebMate: a personal agent for browsing and searching. *International conference on Autonomous agents*, pp. 9–13, 1998.

[151] T. Chen, Z. Zheng, Q. Lu, W. Zhang, and Y. Yu. Feature-based matrix factorization. *arXiv preprint* arXiv:1109.2271, 2011.

[152] W. Chen, Y. Wang, and S. Yang. Efficient influence maximization in social networks. *ACM KDD Conference*, pp. 199–208, 2009.

[153] W. Chen, C. Wang, and Y. Wang. Scalable influence maximization for prevalent viral marketing in large-scale social networks. *ACM KDD Conference*, pp. 1029–1038, 2010.

[154] W. Chen, Y. Yuan, and L. Zhang. Scalable influence maximization in social networks under the linear threshold model. *IEEE International Conference on Data Mining*, pp. 88–97, 2010.

[155] Y. Chen, I. Hsu, and C. Lin. Website attributes that increase consumer purchase intention: a conjoint analysis. *Journal of Business Research*, 63(9), pp. 1007–1014, 2010.

[156] K. Cheverst, N. Davies, K. Mitchell, A. Friday, and C. Efstratiou. Developing a context-aware electronic tourist guide: some issues and experiences. *ACM SIGCHI Conference on Human Factors in Computing Systems*, pp. 17–24, 2000.

[157] K. Y. Chiang, C. J. Hsieh, N. Natarajan, I. S., Dhillon, and A. Tewari. Prediction and clustering in signed networks: a local to global perspective. *The Journal of Machine Learning Research*, 15(1), pp. 1177–1213, 2014.

[158] P. Chirita, W. Nejdl, and C. Zamfir. Preventing shilling attacks in online recommender systems. *ACM International Workshop on Web Information and Data Management*, pp. 67–74, 2005.

[159] E. Christakopoulou and G. Karypis. HOSLIM: Higher-order sparse linear method for top-n recommender systems. *Advances in Knowledge Discovery and Data Mining*, pp. 38–49, 2014.

[160] W. Chu, L. Li, L. Reyzin, and R. Schapire. Contextual bandits with linear payoff functions. *AISTATS Conference*, pp. 208–214, 2011.

[161] A. Cichocki and R. Zdunek. Regularized alternating least squares algorithms for non-negative matrix/tensor factorization. *International Symposium on Neural Networks*, pp. 793–802. 2007.

[162] M. Claypool, A. Gokhale, T. Miranda, P. Murnikov, D. Netes, and M. Sartin. Combining content-based and collaborative filters in an online newspaper. *Proceedings of the ACM SIGIR Workshop on Recommender Systems: Algorithms and Evaluation*, 1999.

[163] W. Cohen, R. Schapire and Y. Singer. Learning to order things. *Advances in Neural Information Processing Systems*, pp. 451–457, 2007.

[164] W. Cohen. Learning rules that classify e-mail. *AAAI symposium on machine learning in information access.* pp. 18–25, 1996.

[165] W. Cohen. Fast effective rule induction. *ICML Conference*, pp. 115–123, 1995.

[166] M. Condliff, D. Lewis, D. Madigan, and C. Posse. Bayesian mixed-effects models for recommender systems. *ACM SIGIR Workshop on Recommender Systems: Algorithms and Evaluation*, pp. 23–30, 1999.

[167] M. O'Connor and J. Herlocker. Clustering items for collaborative filtering. *Proceedings of the ACM SIGIR workshop on recommender systems*, Vol 128. 1999.

[168] M. O'Connor, D. Cosley, J. Konstan, and J. Riedl. PolyLens: a recommender system for groups of users. *European Conference on Computer Supported Cooperative Work*, pp. 199–218, 2001.

[169] R. Cooley, B. Mobasher, and J. Srivastava. Data preparation for mining World Wide Web browsing patterns. *Knowledge and Information Systems*, 1(1), pp. 5–32, 1999.

[170] L. Coyle and P. Cunningham. Improving recommendation ranking by learning personal feature weights. *European Conference on Case-Based Reasoning*, Springer, pp. 560–572, 2004.

[171] H. Cramer, V. Evers, S. Ramlal, M. Someren, L. Rutledge, N. Stash, L. Aroyo, and B. Wielinga. The effects of transparency on trust in and acceptance of a content-based art recommender. *User Modeling and User-Adapted Interaction*, 18(5), pp. 455–496, 2008.

[172] D. Crandall, D. Cosley, D. Huttenlocher, J. Kleinberg, and S. Suri. Feedback effects between similarity and social influence in online communities. *ACM KDD Conference*, pp. 160–168, 2008.

[173] P. Cremonesi, Y. Koren, and R. Turrin. Performance of recommender algorithms on top-n recommendation tasks. *RecSys*, pp. 39–46, 2010.

[174] A. Csomai and R. Mihalcea. Linking documents to encyclopedic knowledge. *IEEE Intelligent Systems*, 23(5), pp. 34–41, 2008.

[175] A. Das, M. Datar, A. Garg, and S. Rajaram. Google news personalization: scalable online collaborative filtering. *World Wide Web Conference*, pp. 271–280, 2007.

[176] P. Domingos and M. Richardson. Mining the network value of customers. *ACM KDD Conference*, pp. 57–66, 2001.

[177] B. De Carolis, I. Mazzotta, N. Novielli, and V. Silvestri. Using common sense in providing personalized recommendations in the tourism domain. *Workshop on Context-Aware Recommender Systems*, 2009.

[178] M. De Gemmis, P. Lops, and G. Semeraro. A content-collaborative recommender that exploits WordNet-based user profiles for neighborhood formation. *User Modeling and User-Adapted Interaction*, 17(3), pp. 217–255, 2007.

[179] M. De Gemmis, P. Lops, G. Semeraro and P. Basile. Integrating tags in a semantic content-based recommender. *Proceedings of the ACM Conference on Recommender Systems*, pp. 163–170, 2008.

[180] D. DeCoste. Collaborative prediction using ensembles of maximum margin matrix factorizations. *International Conference on Machine Learning*, pp. 249–256, 2006.

[181] M. Deshpande and G. Karypis. Item-based top-n recommendation algorithms. *ACM Transactions on Information Systems (TOIS)*, 22(1), pp. 143–177, 2004.

[182] M. Deshpande and G. Karypis. Selective Markov models for predicting Web page accesses. *ACM Transactions on Internet Technology (TOIT)*, 4(2), pp. 163–184, 2004.

[183] C. Desrosiers and G. Karypis. A comprehensive survey of neighborhood-based recommendation methods. *Recommender Systems Handbook*, pp. 107–144, 2011.

[184] R. Devooght, N. Kourtellis, and A. Mantrach. Dynamic matrix factorization with priors on unknown values. *ACM KDD Conference*, 2015.

[185] Y. Ding and X. Li. Time weight collaborative filtering. *ACM International Conference on Information and Knowledge Management*, pp. 485–492, 2005.

[186] Y. Ding, X. Li, and M. Orlowska. Recency-based collaborative filtering. *Australasian Database Conference*, pp. 99–107, 2009.

[187] J. O'Donovan and B. Smyth. Trust in recommender systems. *International Conference on Intelligent User Interfaces*, pp. 167–174, 2005.

[188] P. Dourish, What we talk about when we talk about context. *Personal and ubiquitous computing*, 8(1), pp. 19–30, 2004.

[189] C. Dwork. Differential privacy. *Encyclopedia of Cryptography and Security*, Springer, pp. 338–340, 2011.

[190] C. Dwork, R. Kumar, M. Naor, and D. Sivakumar. Rank aggregation methods for the web. *World Wide Web Conference*, pp. 613–622, 2010.

[191] D. Eck, P. Lamere, T. Bertin-Mahieux, and S. Green. Automatic generation of social tags for music recommendation. *Advances in Neural Information Processing Systems*, pp. 385–392, 2008.

[192] M. Elahi, V. Repsys, and F. Ricci. Rating elicitation strategies for collaborative filtering. *E-Commerce and Web Technologies*, pp. 160–171, 2011.

[193] M. Elahi, F. Ricci, and N. Rubens. Active learning strategies for rating elicitation in collaborative filtering: a system-wide perspective. *ACM Transactions on Intelligent Systems and Technology (TIST)*, 5(1), 13, 2013.

[194] M. Elahi, M. Braunhofer, F. Ricci, and M. Tkalcic. Personality-based active learning for collaborative filtering recommender systems. *Advances in Artificial Intelligence*, pp. 360–371, 2013.

[195] T. Fawcett. ROC Graphs: Notes and Practical Considerations for Researchers. *Technical Report HPL-2003-4*, Palo Alto, CA, HP Laboratories, 2003.

[196] A. Felfernig and R. Burke. Constraint-based recommender systems: technologies and research issues. *International conference on Electronic Commerce*, 2008. (p.

[197] A. Felfernig, G. Friedrich, D. Jannach, and M. Zanker. Developing constraint-based recommenders. *Recommender Systems Handbook*, Springer, pp. 187–216, 2011.

[198] A. Felfernig, G. Friedrich, D. Jannach, and M. Stumptner. Consistency-based diagnosis of configuration knowledge bases. *Artificial Intelligence*, 152(2), 213–234, 2004.

[199] A. Felfernig, G. Friedrich, M. Schubert, M. Mandl, M. Mairitsch, and E. Teppan. Plausible repairs for inconsistent requirements. *IJCAI Conference*, pp. 791–796, 2009.

[200] A. Felfernig, E. Teppan, E., and B. Gula. Knowledge-based recommender technologies for marketing and sales. *International Journal of Pattern Recognition and Artificial Intelligence*, 21(02), pp. 333–354, 2007.

[201] A. Felfernig, K. Isak, K. Szabo, and P. Zachar. The VITA financial services sales support environment. National conference on artificial intelligence, 22(2), pp. 1692–1699, 2007.

[202] R. A. Finkel and J. L. Bentley. Quad trees: A data structure for retrieval on composite keys. *Acta Informatica*, 4, pp. 1–9, 1974.

[203] D. M. Fleder and K. Hosanagar. Recommender systems and their impact on sales diversity. *ACM Conference on Electronic Commerce*, pp. 192–199, 2007.

[204] F. Fouss, A. Pirotte, J. Renders, and M. Saerens. Random-walk computation of similarities between nodes of a graph with application to collaborative recommendation. *IEEE Transactions on Knowledge and Data Engineering*, 19(3), pp. 355–369, 2007.

[205] F. Fouss, L. Yen, A. Pirotte, and M. Saerens. An experimental investigation of graph kernels on a collaborative recommendation task. *IEEE International Conference on Data Mining (ICDM)*, pp. 863–868, 2006.

[206] Y. Freund, and R. Schapire. A decision-theoretic generalization of online learning and application to boosting. *Computational Learning Theory*, pp. 23–37, 1995.

[207] Y. Freund and R. Schapire. Experiments with a new boosting algorithm. *ICML Conference*, pp. 148–156, 1996.

[208] X. Fu, J. Budzik, and K. J. Hammond. Mining navigation history for recommendation. *International Conference on Intelligent User Interfaces*, 2000.

[209] S. Funk. Netflix update: Try this at home, 2006. http://sifter.org/~simon/journal/20061211.html

[210] E. Gabrilovich and S. Markovitch. Computing semantic relatedness using wikipedia-based explicit semantic analysis. *IJCAI Conference*, pp. 1606–1611, 2007.

[211] E. Gabrilovich, and S. Markovitch. Overcoming the brittleness bottleneck using Wikipedia: Enhancing text categorization with encyclopedic knowledge. *AAAI Conference*, pp. 1301–1306, 2006.

[212] Z. Gantner, S. Rendle, and L. Schmidt-Thieme. Factorization models for context-/time-aware movie recommendations. *Workshop on Context-Aware Movie Recommendation*, pp. 14–19, 2010.

[213] A. Garcia-Crespo, J. Chamizo, I. Rivera, M. Mencke, R. Colomo-Palacios, and J. M. Gomez-Berbis. SPETA: Social pervasive e-Tourism advisor. *Telematics and Informatics* 26(3), pp. 306–315. 2009.

[214] M. Ge, C. Delgado-Battenfeld, and D. Jannach. Beyond accuracy: evaluating recommender systems by coverage and serendipity. *ACM Conference on Recommender Systems*, pp. 257–260, 2010.

[215] J. Gemmell, A. Shepitsen, B. Mobasher, and R. Burke. Personalization in folksonomies based on tag clustering. *Workshop on Intelligent Techniques for Web Personalization and Recommender Systems* , 2008. http://www.aaai.org/Papers/Workshops/2008/WS-08-06/WS08-06-005.pdf

[216] J. Gemmell, T. Schimoler, B. Mobasher, and R. Burke. Resource recommendation in social annotation systems: A linear-weighted hybrid approach. *Journal of Computer and System Sciences*, 78(4), pp. 1160–1174, 2012.

[217] R. Gemulla, E. Nijkamp, P. Haas, and Y. Sismanis. Large-scale matrix factorization with distributed stochastic gradient descent. *ACM KDD Conference*, pp. 69–77, 2011.

[218] M. Gery and H. Haddad. Evaluation of Web usage mining approaches for user's next request prediction. *ACM international workshop on Web information and data management*, pp. 74–81, 2003.

[219] L. Getoor and M. Sahami. Using probabilistic relational models for collaborative filtering. *Workshop on Web Usage Analysis and User Profiling*, 1999.

[220] F. Girosi, M. Jones, and T. Poggio. Regularization theory and neural networks architectures. *Neural Computation*, 2(2), pp. 219–269, 1995.

[221] J. Golbeck. Computing with social trust. *Springer*, 2008.

[222] J. Golbeck. Computing and applying trust in Web-based social networks, *Ph.D. Thesis*, 2005.

[223] J. Golbeck. Generating predictive movie recommendations from trust in social networks, *Lecture Notes in Computer Science*, Vol. 3986, pp. 93–104, 2006.

[224] J. Golbeck. Trust and nuanced profile similarity in online social networks. *ACM Transactions on the Web (TWEB)*, 3(4), 12, 2009.

[225] J. Golbeck and J. Hendler. Filmtrust: Movie recommendations using trust in Web-based social networks. *IEEE Consumer Communications and Networking Conference*, 96, pp. 282–286, 2006.

[226] J. Golbeck and J. Hendler. Inferring binary trust relationships in Web-based social networks. *ACM Transactions on Internet Technology (TOIT)*, 6(4), pp. 497–529, 2006.

[227] J. Golbeck and A. Mannes. Using Trust and Provenance for Content Filtering on the Semantic Web. *Models of Trust on the Web (WWW'06 Workshop)*, 2006.

[228] K. Goldberg, T. Roeder, D. Gupta, and C. Perkins. Eigentaste: A constant time collaborative filtering algorithm. *Information Retrieval*, 4(2), pp. 133–151, 2001.

[229] N. Good, J. Schafer, J. Konstan, A. Borchers, B. Sarwar, J. Herlocker, and J. Riedl. Combining collaborative filtering with personal agents for better recommendations. *National Conference on Artificial Intelligence (AAAI/IAAI)*, pp. 439–446, 1999.

[230] S. Gordea and M. Zanker. Time filtering for better recommendations with small and sparse rating matrices. *International Conference on Web Information Systems Engineering*, pp. 171–183, 2007.

[231] M. Gorgoglione and U. Panniello. Including context in a transactional recommender system using a pre- filtering approach: two real e-commerce applications. *International Conference on Advanced Information Networking and Applications Workshops*, pp. 667–672, 2009.

[232] M. Gori and A. Pucci. Itemrank: a random-walk based scoring algorithm for recommender engines. *IJCAI Conference*, pp. 2766–2771, 2007.

[233] A. Goyal, F. Bonchi, and L. V. S. Lakshmanan. A data-based approach to social influence maximization. *VLDB Conference*, pp. 73–84, 2011.

[234] A. Goyal, F. Bonchi, and L. V. S. Lakshmanan. Learning influence probabilities in social networks. *ACM WSDM Conference*, pp. 241–250, 2011.

[235] Q. Gu, J. Zhou, and C. Ding. Collaborative filtering: Weighted nonnegative matrix factorization incorporating user and item graphs. *SIAM Conference on Data Mining*, pp. 199–210, 2010.

[236] I. Gunes, C. Kaleli, A. Bilge, and H. Polat. Shilling attacks against recommender systems: a comprehensive survey. *Artificial Intelligence Review*, 42(4), 767–799, 2014.

[237] M. Gupta, R. Li, Z. Yin, and J. Han. A survey of social tagging techniques, *ACM SIGKDD Explorations*, 12(1), pp. 58–72, 2010.

[238] A. Gunawardana and C. Meek. A unified approach to building hybrid recommender systems. *ACM Conference on Recommender Systems*, pp. 117–124, 2009.

[239] R. Guttman, A. Moukas, and P. Maes. Agent-mediated electronic commerce: A survey, *Knowledge Engineering Review*, 13(2), pp. 147–159, 1998.

[240] R. Guha. Open rating systems. *Techical Report*, Stanford University, 2003. http://www.w3.org/2001/sw/Europe/events/foaf-galway/papers/fp/open_rating_systems/wot.pdf

[241] R. Guha, R. Kumar, P. Raghavan, and A. Tomkins. Propagation of trust and distrust. *World Wide Web Conference*, pp. 403–412, 2004.

[242] T. Hastie, R. Tibshirani, and J. Friedman. The elements of statistical learning. *Springer*, 2009.

[243] T. H. Haveliwala. Topic-sensitive pagerank. *World Wide Web Conference*, pp. 517–526, 2002.

[244] Q. He, D. Jiang, Z. Liao, S. Hoi, K. Chang, E. Lim, and H. Li. Web query recommendation via sequential query prediction. *IEEE International Conference on Data Engineering*, pp. 1443–1454, 2009.

[245] J. Herlocker, J. Konstan, A. Borchers, and J. Riedl. An algorithmic framework for performing collaborative filtering. *ACM SIGIR Conference*, pp. 230–237, 1999.

[246] J. Herlocker, J. Konstan, L. Terveen, and J. Riedl. Evaluating collaborative filtering recommender systems. *ACM Transactions on Information Systems (TOIS)*, 22(1), pp. 5–53, 2004.

[247] J. Herlocker, J. Konstan,, and J. Riedl. An empirical analysis of design choices in neighborhood-based collaborative filtering algorithms. *Information Retrieval*, 5(4), pp. 287–310, 2002.

[248] J. Herlocker, J. Konstan, and J. Riedl. Explaining collaborative filtering recommendations. *ACM Conference on Computer Supported Cooperative work*, pp. 241–250, 2000.

[249] C. Hermann. Time-based recommendations for lecture materials. *World Conference on Educational Multimedia, Hypermedia and Telecommunications*, pp. 1028–1033, 2010.

[250] P. Heymann, D. Ramage, and H. Garcia-Molina. Social tag prediction. *ACM SIGIR Conference*, pp. 531–538, 2008.

[251] W. Hill, L. Stead, M. Rosenstein, and G. Furnas. Recommending and evaluating choices in a virtual community of use. *ACNM SIGCHI Conference*, pp. 194–201, 1995.

[252] T. Hofmann. Latent semantic models for collaborative filtering. *ACM Transactions on Information Systems (TOIS)*, 22(1), pp. 89–114, 2004.

[253] W. Hong, S. Zheng, H. Wang, and J. Shi. A job recommender system based on user clustering. *Journal of Computers*, 8(8), 1960–1967, 2013.

[254] J. Hopcroft, T. Lou, and J. Tang. Who will follow you back?: reciprocal relationship prediction. *ACM International Conference on Information and Knowledge Management*, pp. 1137–1146, 2011.

[255] A. Hotho, R. Jaschke, C. Schmitz, and G. Stumme. Folkrank: A ranking algorithm for folksonomies. *Fachgruppe Informatik Ret. (FGIR)*, pp. 111–114, 2006.

[256] A. Hotho, R. Jaschke, C. Schmitz, and G. Stumme. BibSonomy: A social bookmark and publication sharing system. *Conceptual Structures Tool Interoperability Workshop*, pp. 87–102, 2006.

[257] N. Houlsby, J. M. Hernandez-Lobato, and Z. Ghahramani. Cold-start active learning with robust ordinal matrix factorization. *International Conference on Machine Learning (ICML)*, pp. 766–774, 2014.

[258] A. Howe, and R. Forbes. Re-considering neighborhood-based collaborative filtering parameters in the context of new data. *Proceedings of the 17th ACM Conference on Information and Knowledge Management*, pp. 1481–1482, 2008.

[259] C. Hsieh, N. Natarajan, and I. Dhillon. PU learning for matrix completion. *ICML Conference*, 2015.

[260] Y. Hu, Y. Koren, and C. Volinsky. Collaborative filtering for implicit feedback datasets. *IEEE International Conference on Data Mining*, pp. 263–272, 2008.

[261] Z. Huang, X. Li, and H. Chen. Link prediction approach to collaborative filtering. *ACM/IEEE-CS joint conference on Digital libraries*, pp. 141–142, 2005.

[262] Z. Huang, H. Chen, and D. Zheng. Applying associative retrieval techniques to alleviate the sparsity problem in collaborative filtering. *ACM Transactions on Information Systems*, 22(1), pp. 116–142, 2004.

[263] G. Hurley and D. Wilson. DubLet: An online CBR system for rental property accommodation. *International Conference on Case-Based Reasoning*, pp. 660–674, 2001.

[264] J. Illig, A. Hotho, R. Jaschke, and G. Stumme. A comparison of content-based tag recommendations in folksonomy systems. *Knowledge Processing and Data Analysis*, Springer, pp. 136–149, 2011.

[265] D. Isaacson and R. Madsen. Markov chains, theory and applications, *Wiley*, 1976.

[266] M. Jahrer, A. Toscher, and R. Legenstein. Combining predictions for accurate recommender systems. *ACM KDD Conference*, pp. 693–702, 2010.

[267] P. Jain and I. Dhillon. Provable inductive matrix completion. *arXiv preprint arXiv:1306.0626* http://arxiv.org/abs/1306.0626.

[268] P. Jain, P. Netrapalli, and S. Sanghavi. Low-rank matrix completion using alternating minimization. *ACM Symposium on Theory of Computing*, pp. 665–674, 2013.

[269] M. Jamali and M. Ester. TrustWalker: A random-walk model for combining trust-based and item-based recommendation. *ACM KDD Conference*, pp. 397–406, 2009.

[270] M. Jamali and M. Ester. A matrix factorization technique with trust propagation for recommendation in social networks. *ACM Internatonal Conference on Recommender Systems*, pp 135–142, 2010.

[271] A. Jameson and B. Smyth. Recommendation to groups. *The Adaptive Web*, pp. 596–627, 2007.

[272] A. Jameson. More than the sum of its members: challenges for group recommender systems. *Proceedings of the working conference on Advanced visual interfaces*, pp. 48–54, 2004.

[273] D. Jannach. Finding preferred query relaxations in content-based recommenders. *Intelligent Techniques and Tools for Novel System Architectures*, Springer, pp. 81–97, 2006.

[274] D. Jannach. Techniques for fast query relaxation in content-based recommender systems. *Advances in Artificial Intelligence*, Springer, pp. 49–63, 2006.

[275] D. Jannach, M. Zanker, A. Felfernig, and G. Friedrich. An introduction to recommender systems, *Cambridge University Press*, 2011.

[276] D. Jannach, Z. Karakaya, and F. Gedikli. Accuracy improvements for multi-criteria recommender systems. *ACM Conference on Electronic Commerce*, pp. 674–689, 2012.

[277] R. Jaschke, L. Marinho, A. Hotho, L. Schmidt-Thieme, and G. Stumme. Tag recommendations in folksonomies. *Knowledge Discovery in Databases (PKDD)*, pp. 506–514, 2007.

[278] G. Jeh, and J. Widom. SimRank: a measure of structural-context similarity. *ACM KDD Conference*, pp. 538–543, 2003.

[279] Z. Jiang, W. Wang, and I. Benbasat. Multimedia-based interactive advising technology for online consumer decision support. *Communications of the ACM*, 48(9), pp. 92–98, 2005.

[280] R. Jin, J. Chai, and L. Si. An automatic weighting scheme for collaborative filtering. *ACM SIGIR Conference*, pp. 337–344, 2004.

[281] R. Jin, L. Si, and C. Zhai. Preference-based graphic models for collaborative filtering. *Proceedings of the Nineteenth conference on Uncertainty in Artificial Intelligence*, pp. 329–336, 2003.

[282] R. Jin, L. Si, C. Zhai, and J. Callan. Collaborative filtering with decoupled models for preferences and ratings. *ACM CIKM Conference*, pp. 309–316, 2003.

[283] T. Joachims. Training linear SVMs in linear time. *ACM KDD Conference*, pp. 217–226, 2006.

[284] T. Joachims. Optimizing search engines using click-through data. *ACM KDD Conference*, pp. 133–142, 2002.

[285] I. Jolliffe. Principal component analysis, 2nd edition, *Springer*, 2002.

[286] N. Jones and P. Pu. User technology adoption issues in recommender systems. *Networking and Electronic Conference*, pp. 379–394, 2007.

[287] A. Josang, S. Marsh, and S. Pope. Exploring different types of trust propagation. In Trust management, *Lecture Notes in Computer Science*, Springer, 3986, pp. 179–192, 2006.

[288] P. Juell and P. Paulson. Using reinforcement learning for similarity assessment in case-based systems. *IEEE Intelligent Systems*, 18(4), pp. 60–67, 2003.

[289] U. Junker. QUICKXPLAIN: preferred explanations and relaxations for over-constrained problems. *AAAI Conference*, pp. 167–172, 2004.

[290] S. Kale, L. Reyzin, and R. Schapire. Non-stochastic bandit slate problems. *Advances in Neural Information Processing Systems*, pp. 1054–1062, 2010.

[291] M. Kaminskas and F. Ricci. Contextual music information retrieval and recommendation: State of the art and challenges. *Computer Science Review*, 6(2), pp. 89–119, 2012.

[292] S. Kamvar, M. Schlosser, and H. Garcia-Molina. The eigentrust algorithm for reputation management in P2P networks. *World Wide Web Conference*, pp. 640–651, 2003.

[293] A. Karatzoglou. Collaborative temporal order modeling. *ACM Conference on Recommender Systems*, pp. 313–316, 2011.

[294] A. Karatzoglou, X. Amatriain, L. Baltrunas, and N. Oliver. Multiverse recommendation: N-dimensional tensor factorization for context-aware collaborative filtering. *ACM Conference on Recommender Systems*, pp. 79–86, 2010.

[295] R. Karimi, C. Freudenthaler, A. Nanopoulos, L. Schmidt-Thieme. Exploiting the characteristics of matrix factorization for active learning in recommender systems. *ACM Conference on Recommender Systems*, pp. 317–320, 2012.

[296] J. Kemeny and J. Snell. Finite Markov chains. *Springer*, New York, 1983.

[297] D. Kempe, J. Kleinberg, and E. Tardos. Maximizing the spread of influence through a social network. *ACM KDD Conference*, pp. 137–146, 2003.

[298] M. Kendall. A new measure of rank correlation. *Biometrika*, pp. 81–93, 1938.

[299] M. Kendall and J. Gibbons. Rank correlation methods. *Charles Griffin*, 5th edition, 1990.

[300] D. Kim, and B. Yum. Collaborative filtering Based on iterative principal component analysis, *Expert Systems with Applications*, 28, pp. 623–830, 2005.

[301] H. Kim and H. Park. Nonnegative matrix factorization based on alternating nonnegativity constrained least squares and active set method. *SIAM Journal on Matrix Analysis and Applications*, 30(2), pp. 713–730, 2008.

[302] J. Kleinberg. Authoritative sources in a hyperlinked environment. *Journal of the ACM (JACM)*, 46(5), pp. 604–632, 1999.

[303] J. Kleinberg, C. Papadimitriou, and P. Raghavan. On the value of private information. *Proceedings of the 8th Conference on Theoretical Aspects of Rationality and Knowledge*, pp. 249–257, 2001.

[304] N. Koenigstein, G. Dror, and Y. Koren. Yahoo! Music recommendations: modeling music ratings with temporal dynamics and item taxonomy. *ACM Conference on Recommender Systems*, pp. 165–172, 2011.

[305] R. Kohavi, R. Longbotham, D. Sommerfield, R. Henne. Controlled experiments on the Web: survey and practical guide. *Data Mining and Knowledge Discovery*, 18(1), pp. 140–181, 2009.

[306] X. Kong, X. Shi, and P. S. Yu. Multi-Label collective classification. *SIAM Conference on Data Mining*, pp. 618–629, 2011.

[307] J. Konstan. Introduction to recommender systems: algorithms and evaluation. *ACM Transactions on Information Systems*, 22(1), pp. 1–4, 2004.

[308] J. Konstan, S. McNee, C. Ziegler, R. Torres, N. Kapoor, and J. Riedl. Lessons on applying automated recommender systems to information-seeking tasks. *AAAI Conference*, pp. 1630–1633, 2006.

[309] Y. Koren. Factorization meets the neighborhood: a multifaceted collaborative filtering model. *ACM KDD Conference*, pp. 426–434, 2008. Extended version of this paper appears as: "Y. Koren. Factor in the neighbors: Scalable and accurate collaborative filtering. *ACM Transactions on Knowledge Discovery from Data (TKDD)*, 4(1), 1, 2010."

[310] Y. Koren. Collaborative filtering with temporal dynamics. *ACM KDD Conference*, pp. 447–455, 2009. Another version also appears in the *Communications of the ACM*,, 53(4), pp. 89–97, 2010.

[311] Y. Koren. The Bellkor solution to the Netflix grand prize. *Netflix prize documentation*, 81, 2009. `http://www.netflixprize.com/assets/GrandPrize2009_BPC_BellKor.pdf`

[312] Y. Koren and R. Bell. Advances in collaborative filtering. *Recommender Systems Handbook*, Springer, pp. 145–186, 2011. (Extended version in 2015 edition of handbook).

[313] Y. Koren, R. Bell, and C. Volinsky. Matrix factorization techniques for recommender systems. *Computer*, 42(8), pp. 30–37, 2009.

[314] Y. Koren and J. Sill. Collaborative filtering on ordinal user feedback. *IJCAI Conference*, pp. 3022–3026, 2011.

[315] R. Krestel and P. Fankhauser. Personalized topic-based tag recommendation. *Neurocomputing*, 76(1), pp. 61–70, 2012.

[316] R. Krestel, P. Fankhauser, and W. Nejdl. Latent dirichlet allocation for tag recommendation. *ACM Conference on Recommender Systems*, pp. 61–68, 2009.

[317] V. Krishnan, P. Narayanashetty, M. Nathan, R. Davies, and J. Konstan. Who predicts better? Results from an online study comparing humans and an online recommender system. *ACM Conference on Recommender Systems*, pp. 211–218, 2008.

[318] J. Krosche, J. Baldzer, and S. Boll. MobiDENK -mobile multimedia in monument conservation. *IEEE MultiMedia*, 11(2), pp. 72–77, 2004.

[319] A. Krogh, M. Brown, I. Mian, K. Sjolander, and D. Haussler. Hidden Markov models in computational biology: Applications to protein modeling. *Journal of molecular biology*, 235(5), pp. 1501–1531, 1994.

[320] B. Krulwich. Lifestyle finder: Intelligent user profiling using large-scale demographic data. *AI Magazine*, 18(2), pp. 37–45, 1995.

[321] S. Kabbur, X. Ning, and G. Karypis. FISM: factored item similarity models for top-N recommender systems. *ACM KDD Conference*, pp. 659–667, 2013.

[322] S. Kabbur and G. Karypis. NLMF: NonLinear Matrix Factorization Methods for Top-N Recommender Systems. *IEEE Data Mining Workshop (ICDMW)*, pp. 167–174, 2014.

[323] A. Karatzoglou, L. Baltrunas, and Y. Shi. Learning to rank for recommender systems. *ACM Conference on Recommender Systems*, pp. 493–494, 2013. Slides available at http://www.slideshare.net/kerveros99/learning-to-rank-for-recommender-system-tutorial-acm-recsys-2013

[324] J. Kunegis, S. Schmidt, A. Lommatzsch, J. Lerner, E. De Luca, and S. Albayrak. Spectral analysis of signed graphs for clustering, prediction and visualization. *SIAM Conference on Data Mining*, pp. 559–559, 2010.

[325] J. Kunegis, E. De Luca, and S. Albayrak. The link prediction problem in bipartite networks. *Computational Intelligence for Knowledge-based Systems Design*, Springer, pp. 380–389, 2010.

[326] J. Kunegis and A. Lommatzsch. Learning spectral graph transformations for link prediction. *International Conference on Machine Learning*, pp. 562–568, 2009.

[327] A. Lacerda, M. Cristo, W. Fan, N. Ziviani, and B. Ribeiro-Neto. Learning to advertise. *ACM SIGIR Conference*, pp. 549–556, 2006.

[328] K. Lakiotaki, S. Tsafarakis, and N. Matsatsinis. UTA-Rec: a recommender system based on multiple criteria analysis. *ACM Conference on Recommender Systems*, pp. 219–226, 2008.

[329] S. Lam and J. Riedl. Shilling recommender systems for fun and profit. *World Wide Web Conference*, pp. 393–402, 2004.

[330] B. Lamche, U. Trottmann, and W. Worndl. Active learning strategies for exploratory mobile recommender systems. *Proceedings of the 4th Workshop on Context-Awareness in Retrieval and Recommendation*, pp. 10–17, 2014.

[331] A. Langville, C. Meyer, R. Albright, J. Cox, and D. Duling. Initializations for the nonnegative matrix factorization. *ACM KDD Conference*, pp. 23–26, 2006.

[332] L. Lathauwer, B. Moor, and J. Vandewalle. A multilinear singular value decomposition. *SIAM Journal on Matrix Analysis and Applications*, 21(4), pp. 1253–1278. 2000.

[333] N. Lathia, S. Hailes, and L. Capra. Temporal collaborative filtering with adaptive neighbourhoods. *ACM SIGIR Conference*, pp. 796–797, 2009.

[334] N. Lathia, S. Hailes, and L. Capra. Private distributed collaborative filtering using estimated concordance measures. *ACM Conference on Recommender Systems*, pp. 1–8, 2007.

[335] N. Lathia, S. Hailes, L. Capra, and X. Amatriain. Temporal diversity in recommender systems. *ACM SIGIR Conference*, pp. 210–217, 2010.

[336] S. Lawrence. Context in Web search. *IEEE Data Engineering Bulletin*, 23(3):25, 2000.

[337] D. Lee, S. Park, M. Kahng, S. Lee, and S. Lee. Exploiting contextual information from event logs for personalized recommendation. Chapter in *Computer and Information Science*, Springer, 2010.

[338] J.-S. Lee and S. Olafsson. Two-way cooperative prediction for collaborative filtering recommendations. *Expert Systems with Applications*, 36(3), pp. 5353–5361, 2009.

[339] B.-H. Lee, H. Kim, J. Jung, and G.-S. Jo. Location-based service with context data for a restaurant recommendation. *Database and Expert Systems Applications*, pp. 430–438, 2006.

[340] H. Lee and W. Teng. Incorporating multi-criteria ratings in recommendation systems. *IEEE International Conference on Information Reuse and Integration (IRI)*, pp. 273–278, 2007.

[341] J. Lees-Miller, F. Anderson, B. Hoehn, and R. Greiner. Does Wikipedia information help Netflix predictions?. *Machine Learning and Applications*, pp. 337–343, 2008.

[342] D. Lemire and A. Maclachlan. Slope one predictors for online rating-based collaborative filtering. *SIAM Conference on Data Mining*, 2005.

[343] J. Levandoski, M. Sarwat, A. Eldawy, and M. Mokbel. LARS: A location-aware recommender system. *IEEE ICDE Conference*, pp. 450–461, 2012.

[344] R. Levien. Attack-resistant trust metrics. *Computing with Social Trust*, Springer, pp. 121–132, 2009.

[345] M. Lesani and S. Bagheri. Applying and inferring fuzzy trust in semantic web social networks. *Canadian Semantic Web, Semantic Web and Beyond*, Springer, Vol 2, pp. 23–43, 2006.

[346] J. Leskovec, D. Huttenlocher, and J. Kleinberg. Predicting positive and negative links in online social networks. *World Wide Web Conference*, pp. 641–650, 2010.

[347] M. Levy and K. Jack. Efficient Top-N Recommendation by Linear Regression. *Large Scale Recommender Systems Workshop (LSRS) at RecSys*, 2013.

[348] L. Li, W. Chu, J. Langford, and R. Schapire. A contextual-bandit approach to personalized news article recommendation. *World Wide Web Conference*, pp. 661–670, 2010.

[349] L. Li, W. Chu, J. Langford, and X. Wang. Unbiased offline evaluation of contextual-bandit-based news article recommendation algorithms. *International Conference on Web Search and Data Mining*, pp. 297–306, 2011.

[350] M. Li, B. M. Dias, I. Jarman, W. El-Deredy, and P. J. Lisboa. Grocery shopping recommendations based on basket-sensitive random walk. *KDD Conference*, pp. 1215–1224, 2009.

[351] M. Li, T. Zhang, Y. Chen, and A. Smola. Efficient mini-batch training for stochastic optimization. *ACM KDD Conference*, pp. 661–670, 2014.

[352] N. Li, T. Li, and S. Venkatasubramanian. t-closeness: Privacy beyond k-anonymity and ℓ-diversity. *IEEE International Conference on Data Enginering*, pp. 106–115, 2007.

[353] Q. Li, C. Wang, and G. Geng. Improving personalized services in mobile commerce by a novel multicriteria rating approach. *World Wide Web Conference*, pp. 1235–1236, 2008.

[354] D. Liben-Nowell and J. Kleinberg. The link-prediction problem for social networks. *Journal of the American society for information science and technology*, 58(7), pp. 1019–1031, 2007.

[355] R. Lichtenwalter, J. Lussier, and N. Chawla. New perspectives and methods in link prediction. *ACM KDD Conference*, pp. 243–252, 2010.

[356] H. Lieberman. Letizia: An agent that assists Web browsing, *IJCAI*, pp. 924–929, 1995.

[357] C.-J. Lin. Projected gradient methods for nonnegative matrix factorization. *Neural Computation*, 19(10), pp. 2576–2779, 2007.

[358] W. Lin. Association rule mining for collaborative recommender systems. *Masters Thesis*, Worcester Polytechnic Institute, 2000.

[359] W. Lin, S. Alvarez, and C. Ruiz. Efficient adaptive-support association rule mining for recommender systems. *Data Mining and Knowledge Discovery*, 6(1), pp. 83–105, 2002.

[360] G. Linden, B. Smith, and J. York. Amazon.com recommendations: item-to-item collaborative filtering. *IEEE Internet Computing*, 7(1), pp. 76–80, 2003.

[361] C. Ling and C. Li. Data Mining for direct marketing: problems and solutions. *ACM KDD Conference*, pp. 73–79, 1998.

[362] R. Little and D. Rubin. Statistical analysis with missing data. *Wiley*, 2002.

[363] M. Littlestone and M. Warmuth. The weighted majority algorithm. *Information and computation*, 108(2), pp. 212–261, 1994.

[364] B. Liu. Web data mining: exploring hyperlinks, contents, and usage data. *Springer*, New York, 2007.

[365] B. Liu, W. Hsu, and Y. Ma. Mining association rules with multiple minimum supports. *ACM KDD Conference*, pp. 337–341, 1999.

[366] N. Liu, M. Zhao, E. Xiang, and Q Yang. Online evolutionary collaborative filtering. *ACM Conference on Recommender Systems*, pp. 95–102, 2010.

[367] N. Liu and Q. Yang. Eigenrank: a ranking-oriented approach to collaborative filtering. *ACM SIGIR Conference*, pp. 83–90, 2008.

[368] N. Liu, M. Zhao, and Q. Yang. Probabilistic latent preference analysis for collaborative filtering. *ACM Conference on Information and Knowledge Management*, pp. 759–766, 2009.

[369] L. Liu, J. Tang, J. Han, M. Jiang, and S. Yang. Mining topic-level influence in heterogeneous networks. *ACM CIKM Conference*, pp. 199–208, 2010.

[370] T. Y. Liu. Learning to rank for information retrieval. *Foundations and Trends in Information Retrieval*, 3(3), pp. 225–331, 2009.

[371] X. Liu, C. Aggarwal, Y.-F. Lee, X. Kong, X. Sun, and S. Sathe. Kernelized matrix factorization for collaborative filtering. *SIAM Conference on Data Mining*, 2016.

[372] Z. Liu, Y.-X. Wang, and A. Smola. Fast differentially private matrix factorization. *ACM Conference on Recommender Systems*, 2015.

[373] S. Lohr. A \$1 million research bargain for Netflix, and maybe a model for others, *The New York Times*, September 21, 2009. http://www.nytimes.com/2009/09/22/technology/internet/22netflix.html?_r=0

[374] S. Lombardi, S. Anand, and M. Gorgoglione. Context and customer behaviour in recommendation. *Workshop on Customer Aware Recommender Systems*, 2009.

[375] B. London, and L. Getoor. Collective classification of network data. *Data Classification: Algorithms and Applications*, CRC Press, pp. 399–416, 2014.

[376] P. Lops, M. de Gemmis, and G. Semeraro. Content-based recommender systems: state of the art and trends. *Recommender Systems Handbook*, Springer, pp. 73–105, 2011.

[377] F. Lorenzi and F. Ricci. Case-based recommender systems: a unifying view. *Intelligent Techniques for Web Personalization*, pp. 89–113, Springer, 2005.

[378] L. Lu, M. Medo, C. Yeung, Y. Zhang, Z. Zhang, and T. Zhou. Recommender systems. *Physics Reports*, 519(1), pp. 1–49, 2012. http://arxiv.org/pdf/1202.1112.pdf

[379] Q. Lu, and L. Getoor. Link-based classification. *ICML Conference*, pp. 496–503, 2003.

[380] H. Ma, I. King, and M. Lyu. Effective missing data prediction for collaborative filtering. *ACM SIGIR Conference*, pp. 39–46, 2007.

[381] H. Ma, H. Yang, M. Lyu, and I. King. SoRec: Social recommendation using probabilistic matrix factorization. *ACM Conference on Information and knowledge Management*, pp. 931–940, 2008.

[382] H. Ma, D. Zhou, C. Liu, M. Lyu, and I. King. Recommender systems with social regularization. *ACM International Conference on Web search and Data Mining*, pp. 287–296, 2011.

[383] H. Ma, M. Lyu, and I. King. Learning to recommend with trust and distrust relationships. *ACM International Conference on Recommender Systems*, pp. 189–196, 2009.

[384] H. Ma, M. Lyu, and I. King. Learning to recommend with social trust ensemble. *ACM SIGIR Conference*, pp. 203–210, 2009.

[385] Z. Ma, G. Pant, and O. Sheng. Interest-based personalized search. *ACM Transactions on Information Systems*, 25(1), 2007.

[386] A. Machanavajjhala, D. Kifer, J. Gehrke, and M. Venkitasubramaniam. ℓ-diversity: privacy beyond k-anonymity. *ACM Transactions on Knowledge Discovery from Data (TKDD)*, 1(3), 2007.

[387] S. Macskassy, and F. Provost. A simple relational classifier. *Second Workshop on Multi-Relational Data Mining (MRDM) at ACM KDD Conference*, 2003.

[388] S. A. Macskassy, and F. Provost. Classification in networked data: A toolkit and a univariate case study. *Joirnal of Machine Learning Research*, 8, pp. 935–983, 2007.

[389] T. Mahmood and F. Ricci. Learning and adaptivity in interactive recommender systems. *International Conference on Electronic Commerce*, pp. 75–84, 2007.

[390] T. Mahmood and F. Ricci. Improving recommender systems with adaptive conversational strategies. *ACM Conference on Hypertext and Hypermedia*, pp. 73–82, 2009.

[391] H. Mak, I. Koprinska, and J. Poon. Intimate: A web-based movie recommender using text categorization. *International Conference on Web Intelligence*, pp. 602–605, 2003.

[392] B. Magnini, and C. Strapparava. Improving user modelling with content-based techniques. *International Conference on User Modeling*, pp. 74–83, 2001.

[393] M. O'Mahony, N. Hurley, N. Kushmerick, and G. Silvestre. Collaborative recommendation: A robustness analysis. *ACM Transactions on Internet Technology*, 4(4), pp. 344–377, 2004.

[394] M. O'Mahony, N. Hurley, and G. Silvestre. Promoting recommendations: An attack on collaborative filtering. *Database and Expert Systems Applications*, pp. 494–503, 2002.

[395] M. O'Mahony, N. Hurley, G. Silvestre. An evaluation of the performance of collaborative filtering. *International Conference on Artificial Intelligence and Cognitive Science (AICS)*, pp. 164–168, 2003.

[396] M. O'Mahony, N. Hurley, G. Silvestre. Recommender systems: Attack types and strategies. *National Conference on Artificial Intelligence (AAAI)*, pp. 334–339, 2005.

[397] M. O'Mahony, N. Hurley, G. Silvestre. An evaluation of neighbourhood formation on the performance of collaborative filtering. *Artificial Intelligence Review*, 21(1), pp. 215–228, 2004.

[398] N. Manouselis and C. Costopoulou. Analysis and classification of multi-criteria recommender systems. *World Wide Web*, 10(4), pp. 415–441, 2007.

[399] N. Manouselis and Costopoulou. Experimental Analysis of Design Choices in a Multi-Criteria Recommender System. *International Journal of Pattern Recognition and AI*, 21(2), pp. 311–332, 2007.

[400] C. Manning, P. Raghavan, and H. Schutze. Introduction to information retrieval. *Cambridge University Press*, Cambridge, 2008.

[401] L. Marinho, A. Nanopoulos, L. Schmidt-Thieme, R. Jaschke, A. Hotho, G, Stumme, and P. Symeonidis. Social tagging recommender systems. *Recommender Systems Handbook*, Springer, pp. 615–644, 2011.

[402] B. Marlin and R. Zemel. Collaborative prediction and ranking with non-random missing data. *ACM Conference on Recommender Systems*, pp. 5–12, 2009.

[403] P. Massa and P. Avesani. Trust-aware collaborative filtering for recommender systems. *On the Move to Meaningful Internet Systems*, pp. 492–508, 2004.

[404] P. Massa and P. Avesani. Trust-aware recommender systems. *ACM Conference on Recommender Systems*, pp. 17–24, 2007.

[405] P. Massa and B. Bhattacharjee. Using trust in recommender systems: An experimental analysis. *Trust Management*, pp. 221–235, Springer, 2004.

[406] P. Massa and P. Avesani. Trust metrics on controversial users: balancing between tyranny of the majority. *International Journal on Semantic Web and Information Systems*, 3(1), pp. 39–64, 2007.

[407] J. Masthoff. Group recommender systems: combining individual models. *Recommender Systems Handbook*, Springer, pp. 677–702, 2011.

[408] J. Masthoff. Group modeling: Selecting a sequence of television items to suit a group of viewers. *Personalized Digital Television*, pp. 93–141, 2004.

[409] J. Masthoff and A. Gatt. In pursuit of satisfaction and the prevention of embarrassment: affective state in group recommender systems. *User Modeling and User-Adapted Interactio*, 16(3–4), pp. 281–319, 2006.

[410] J. Masthoff. Modeling the multiple people that are me. *International Conference on User Modeling*, Also appears in *Lecture Notes in Computer Science*, Springer, Vol. 2702, pp. 258–262, 2003.

[411] J. McAuley and J. Leskovec. Hidden factors and hidden topics: understanding rating dimensions with review text. *ACM Conference on Recommender systems*, pp. 165–172, 2013.

[412] J. McCarthy and T. Anagnost. MusicFX: An Arbiter of Group Preferences for Computer Supported Collaborative Workouts. *ACM Conference on Computer Supported Cooperative Work*, pp. 363–372, 1998.

[413] K. McCarthy, L. McGinty, B. Smyth, and M. Salamo. The needs of the many: a case-based group recommender system. *Advances in Case-Based Reasoning*, pp. 196–210, 2004.

[414] K. McCarthy, J. Reilly, L. McGinty, and B. Smyth. On the dynamic generation of compound critiques in conversational recommender systems. *Adaptive Hypermedia and Adaptive Web-Based Systems*, pp. 176–184, 2004.

[415] K. McCarthy, M. Salamo, L. McGinty, B. Smyth, and P. Nicon. Group recommender systems: a critiquing based approach. *International Conference on Intelligent User Interfaces*, pp. 267–269, 2006.

[416] K. McCarthy, L. McGinty, and B. Smyth. Dynamic critiquing: an analysis of cognitive load. *Irish Conference on Artificial Intelligence and Cognitive Science*, pp. 19–28, 2005.

[417] L. McGinty and J. Reilly. On the evolution of critiquing recommenders. *Recommender Systems Handbook*, pp. 419–453, 2011.

[418] S. McNee, J. Riedl, and J. Konstan. Being accurate is not enough: how accuracy metrics have hurt recommender systems. *SIGCHI Conference*, pp. 1097–1101, 2006.

[419] D. McSherry. Incremental relaxation of unsuccessful queries. *Advances in Case-Based Reasoning*, pp. 331–345, 2004.

[420] D. McSherry. Diversity-Conscious Retrieval. *European Conference on Case-Based Reasoning*, pp. 219–233, 2002.

[421] D. McSherry. Similarity and Compromise. *International Conference on Case-Based Reasoning*, pp. 291–305, 2003.

[422] D. McSherry and D. Aha. The ins and outs of critiquing. *IJCAI*, pp. 962–967, 2007.

[423] D. McSherry and D. Aha. Avoiding long and fruitless dialogues in critiquing. *Research and Development in Intelligent Systems*, pp. 173–186, 2007.

[424] B. Mehta, and T. Hofmann. A survey of attack-resistant collaborative filtering algorithms. *IEEE Data Enginerring Bulletin*, 31(2), pp. 14–22, 2008.

[425] B. Mehta, T. Hofmann, and P. Fankhauser. Lies and propaganda: detecting spam users in collaborative filtering. *International Conference on Intelligent User Interfaces*, pp. 14–21, 2007.

[426] B. Mehta, T. Hofmann, and W. Nejdl. Robust collaborative filtering. *ACM Conference on Recommender Systems*, pp. 49–56, 2007.

[427] B. Mehta and W. Nejdl. Unsupervised strategies for shilling detection and robust collaborative filtering. *User Modeling and User-Adapted Interaction*, 19(1–2), pp. 65–97, 2009.

[428] B. Mehta and W. Nejdl. Attack resistant collaborative filtering. *ACM SIGIR Conference*, pp. 75–82, 2008.

[429] Q. Mei, D. Zhou, and K. Church. Query suggestion using hitting time. *ACM Conference on Information and Knowledge Management*, pp. 469–478, 2009. .

[430] N. Meinshausen. Sign-constrained least squares estimation for high-dimensional regression. *Electronic Journal of Statistics*, 7, pp. 607–1631, 2013.

[431] P. Melville, R. Mooney, and R. Nagarajan. Content-boosted collaborative filtering for improved recommendations. *AAAI/IAAI*, pp. 187–192, 2002.

[432] A. K. Menon, and C. Elkan. Link prediction via matrix factorization. *Machine Learning and Knowledge Discovery in Databases*, pp. 437–452, 2011.

[433] S. Middleton, N. Shadbolt, and D. de Roure. Ontological user profiling in recommender systems. *ACM Transactions on Information Systems*, 22(1), pp. 54–88, 2004.

[434] A. Mild and M. Natter. Collaborative filtering or regression models for Internet recommendation systems?. *Journal of Targeting, Measurement and Analysis for Marketing*, 10(4), pp. 304–313, 2002.

[435] S. Min and I. Han. Detection of the customer time-variant pattern for improving recommender systems. *Expert Systems and Applications*, 28(2), pp. 189–199, 2005.

[436] T. M. Mitchell. Machine learning. *McGraw Hill International Edition*, 1997.

[437] K. Miyahara, and M. J. Pazzani. Collaborative filtering with the simple Bayesian classifier. *Pacific Rim International Conference on Artificial Intelligence*, 2000.

[438] D. Mladenic. Machine learning used by Personal WebWatcher. *Proceedings of the ACAI-99 Workshop on Machine Learning and Intelligent Agents*, 1999.

[439] D. Mladenic. Text learning and related intelligent agents: A survey. *IEEE Intelligent Systems*, 14(4), pp. 44–54, 1999.

[440] B. Mobasher, R. Cooley, and J. Srivastava. Automatic personalization based on Web usage mining. *Communications of the ACM*, 43(8), pp. 142–151, 2000.

[441] B. Mobasher, H. Dai, T. Luo, and M. Nakagawa. Effective personalization based on association rule discovery from Web usage data. *ACM Workshop on Web Information and Data Management*, pp. 9–15, 2001.

[442] B. Mobasher, H. Dai, T. Luo, and H. Nakagawa. Using sequential and non-sequential patterns in predictive web usage mining tasks. *International Conference on Data Mining*, pp. 669–672, 2002.

[443] B. Mobasher, H. Dai, M. Nakagawa, and T. Luo. Discovery and evaluation of aggregate usage profiles for web personalization. *Data Mining and Knowledge Discovery*, 6: pp. 61–82, 2002.

[444] B. Mobasher, R. Burke, R. Bhaumik, and C. Williams. Toward trustworthy recommender systems: an analysis of attack models and algorithm robustness. *ACM Transactions on Internet Technology (TOIT)*, 7(4), 23, 2007.

[445] B. Mobasher, R. Burke, R. Bhaumik, and C. Williams. Effective attack models for shilling item-based collaborative filtering systems. *WebKDD Workshop*, 2005.

[446] B. Mobasher, R. Burke, and J. Sandvig. Model-based collaborative filtering as a defense against profile injection attacks. *AAAI Conference*, Vol. 6, p. 1388, 2006.

[447] M. Mokbel and J. Levandoski. Toward context and preference-aware location-based services. *ACM International Workshop on Data Engineering for Wireless and Mobile Access*, pp. 25–32, 2009.

[448] R. J. Mooney and L. Roy. Content-based book recommending using learning for text categorization. *ACM Conference on Digital libraries*, pp. 195–204, 2000.

[449] L. Mui, M. Mohtashemi, and A. Halberstadt. A computational model of trust and reputation. *IEEE International Conference on System Sciences*, pp. 2413–2439, 2002.

[450] T. Murakami, K. Mori, and R. Orihara. Metrics for evaluating the serendipity of recommendation lists. *New Frontiers in Artificial Intelligence*, pp. 40–46, 2008.

[451] A. Narayanan and V. Shmatikov. How to break anonymity of the Netflix prize dataset. *arXiv preprint cs/0610105*, 2006. http://arxiv.org/abs/cs/0610105

[452] G. Nemhauser, and L. Wolsey. Integer and combinatorial optimization. *Wiley*, New York, 1988.

[453] J. Neville, and D. Jensen. Iterative classification in relational data. *AAAI Workshop on Learning Statistical Models from Relational Data*, pp. 13–20, 2000.

[454] Q. Nguyen and F. Ricci. User preferences initialization and integration in critique-based mobile recommender systems. *Artificial Intelligence in Mobile Systems*, pp. 71–78, 2004.

[455] X. Ning and G. Karypis. SLIM: Sparse linear methods for top-N recommender systems. *IEEE International Conference on Data Mining*, pp. 497–506, 2011.

[456] X. Ning and G. Karypis. Sparse linear methods with side information for top-n recommendations. *ACM Conference on Recommender Systems*, pp. 155–162, 2012.

[457] D. Oard and J. Kim. Implicit feedback for recommender systems. *Proceedings of the AAAI Workshop on Recommender Systems*, pp. 81–83, 1998.

[458] K. Oku, S. Nakajima, J. Miyazaki, and S. Uemura. Context-aware SVM for context-dependent information recommendation. *International Conference on Mobile Data Management*, pp. 109–109, 2006.

[459] F. Del Olmo and E. Gaudioso. Evaluation of recommender systems: A new approach. *Expert Systems with Applications*, 35(3), pp. 790–804, 2008.

[460] P. Paatero and U. Tapper. Positive matrix factorization: A non-negative factor model with optimal utilization of error estimates of data values. *Environmetrics*, 5(2), pp. 111–126, 1994.

[461] A. Paolo, P. Massa, and R. Tiella. A trust-enhanced recommender system application: Moleskiing. *ACM Symposium on Applied Computing*, pp. 1589–1593, 2005.

[462] D. Park, H. Kim, I. Choi, and J. Kim. A literature review and classification of recommender systems research. *Expert Systems with Applications*, 29(11), pp. 10059–10072, 2012.

[463] Y. Park and A. Tuzhilin. The long tail of recommender systems and how to leverage it. *Proceedings of the ACM Conference on Recommender Systems*, pp. 11–18, 2008.

[464] M. Park, J. Hong, and S. Cho. Location-based recommendation system using Bayesian user's preference model in mobile devices. *Ubiquitous Intelligence and Computing*, pp. 1130–1139, 2007.

[465] L. Page, S. Brin, R. Motwani, and T. Winograd. The PageRank citation engine: Bringing order to the web. *Technical Report*, 1999–0120, Computer Science Department, Stanford University, 1998.

[466] C. Palmisano, A. Tuzhilin, and M. Gorgoglione. Using context to improve predictive modeling of customers in personalization applications. *IEEE Transactions on Knowledge and Data Engineering*, 20(11), pp. 1535–1549, 2008.

[467] R. Pan, Y. Zhou, B. Cao, N. Liu, R. Lukose, M. Scholz, Q. Yang. One-class collaborative filtering. *IEEE International Conference on Data Mining*, pp. 502–511, 2008.

[468] R. Pan, and M. Scholz. Mind the gaps: weighting the unknown in large-scale one-class collaborative filtering. *ACM KDD Conference*, pp. 667–676, 2009.

[469] W. Pan and L. Chen. CoFiSet: Collaborative filtering via learning pairwise preferences over item-sets. *SIAM Conference on Data Mining*, 2013.

[470] U. Panniello, A. Tuzhilin, and M. Gorgoglione. Comparing context-aware recommender systems in terms of accuracy and diversity. *User Modeling and User-Adapted Interaction*, 24: pp. 35–65, 2014.

[471] U. Panniello, A. Tuzhilin, M. Gorgoglione, C. Palmisano, and A. Pedone. Experimental comparison of pre- vs. post-filtering approaches in context-aware recommender systems. *ACM Conference on Recommender Systems*, pp. 265–268, 2009.

[472] S. Parthasarathy and C. Aggarwal. On the use of conceptual reconstruction for mining massively incomplete data sets. *IEEE Transactions on Knowledge and Data Engineering*, 15(6), pp. 1512–1521, 2003.

[473] A. Paterek. Improving regularized singular value decomposition for collaborative filtering. *Proceedings of KDD Cup and Workshop*, 2007.

[474] V. Pauca, J. Piper, and R. Plemmons. Nonnegative matrix factorization for spectral data analysis. *Linear algebra and its applications*, 416(1), pp. 29–47, 2006.

[475] M. Pazzani. A framework for collaborative, content-based and demographic filtering. *Artificial Intelligence Review*, 13, (5–6), 1999.

[476] M. Pazzani and D. Billsus. Learning and revising user profiles: The identification of interesting Web sites. *Machine learning*, 27(3), pp. 313–331, 1997.

[477] M. Pazzani and D. Billsus. Content-based recommendation systems. *Lecture Notes in Computer Science*, Springer, 4321, pp. 325–341, 2007.

[478] M. Pazzani, J. Muramatsu, and D. Billsus. Syskill and Webert: Identifying interesting Web sites. *AAAI Conference*, pp. 54–61, 1996.

[479] J. Pitkow and P. Pirolli. Mining longest repeating subsequences to predict WWW surfing. *USENIX Annual Technical Conference*, 1999.

[480] L. Pizzato, T. Rej, T. Chung, I. Koprinska, and J. Kay. RECON: a reciprocal recommender for online dating. *ACM Conference on Recommender systems*, pp. 207–214, 2010.

[481] L. Pizzato, T. Rej, T. Chung, K. Yacef, I. Koprinska, and J. Kay. Reciprocal recommenders. *Workshop on Intelligent Techniques for Web Personalization and Recommender Systems*, pp. 20–24, 2010.

[482] L. Pizzato, T. Rej, K. Yacef, I. Koprinska, and J. Kay. Finding someone you will like and who won't reject you. *User Modeling, Adaption and Personalization*, Springer, pp. 269–280, 2011.

[483] B. Polak, A. Herrmann, M. Heitmann, and M. Einhorn. Die Macht des Defaults – Wirkung von Empfehlungen und Vorgaben auf das individuelle Entscheidungsverhalten. [English Translation: *The power of defaults: Effect on individual choice behavior.*] *Zeitschrift fur Betriebswirtschaft*, 78(10), pp. 1033–1060, 2008.

[484] H. Polat and W. Du. Privacy-preserving collaborative filtering using randomized perturbation techniques. *IEEE International Conference on Data Mining*, pp. 625–628, 2003.

[485] H. Polat and W. Du. SVD-based collaborative filtering with privacy. *ACM symposium on Applied Computing*, pp. 791–795, 2005.

[486] P. Pu and L. Chen. Trust building with explanation interfaces. *International conference on Intelligent User Interfaces*, pp. 93–100, 2006.

[487] G. Qi, C. Aggarwal, Q. Tian, H. Ji, and T. S. Huang. Exploring context and content links in social media: A latent space method. *IEEE Transactions on Pattern Analysis and Machine Intelligence*, 34(5), pp. 850–862, 2012.

[488] G. Qi, C. Aggarwal, and T. Huang. Link prediction across networks by biased cross-network sampling. *IEEE ICDE Conference*, pp. 793–804, 2013.

[489] L. Quijano-Sanchez, J. Recio-Garcia, B. Diaz-Agudo,and G. Jimenez-Diaz. Social factors in group recommender systems. *ACM Transactions on Intelligent Systems and Technology (TIST)*, 4(1), 8, 2013.

[490] C. Quoc and V. Le. Learning to rank with nonsmooth cost functions. *Advances in Neural Information Processing Systems*, 19, pp. 193–200, 2007.

[491] J. Reilly, B. Smyth, L. McGinty, and K. McCarthy. Critiquing with confidence. *Case-Based Reasoning Research and Development*, pp. 436–450, 2005.

[492] J. Reilly, K. McCarthy, L. McGinty, and B. Smyth. Explaining compound critiques. *Artificial Intelligence Review*, 24(2), pp. 199–220, 2005.

[493] S. Rendle. Factorization machines. *IEEE International Conference on Data Mining*, pp. 995–100, 2010.

[494] S. Rendle. Factorization machines with libfm. *ACM Transactions on Intelligent Systems and Technology (TIST)*, 3(3), 57, 2012.

[495] S. Rendle. Context-aware ranking with factorization models. *Studies in Computational Intelligence*, Chapter 9, Springer, 2011.

[496] S. Rendle, Z. Gantner, C. Freudenthaler, and L. Schmidt-Thieme. Fast context-aware recommendations with factorization machines. *ACM SIGIR Conference*, pp. 635–644, 2011.

[497] S. Rendle, L. Balby Marinho, A. Nanopoulos, and A. Schmidt-Thieme. Learning optimal ranking with tensor factorization for tag recommendation. *ACM KDD Conference*, pp. 727–736, 2009.

[498] S. Rendle and L. Schmidt-Thieme. Pairwise interaction tensor factorization for personalized tag recommendation. *ACM International Conference on Web Search and Data Mining*, pp. 81–90, 2010.

[499] S. Rendle, C. Freudenthaler, Z. Gantner, and L. Schmidt-Thieme. BPR: Bayesian personalized ranking from implicit feedback. *Uncertainty in Artificial Intelligence (UAI)*, pp. 452–451, 2009.

[500] J. Rennie and N. Srebro. Fast maximum margin matrix factorization for collaborative prediction. *ICML Conference*, pp. 713–718, 2005.

[501] P. Resnick, N. Iacovou, M. Suchak, P. Bergstrom, and J. Riedl. GroupLens: an open architecture for collaborative filtering of netnews. *Proceedings of the ACM Conference on Computer Supported Cooperative Work*, pp. 175–186, 1994.

[502] P. Resnick and R. Sami. The influence limiter: provably manipulation-resistant recommender systems. *ACM Conference on Recommender Systems*, pp. 25–32, 2007.

[503] P. Resnick and R. Sami. The information cost of manipulation resistance in recommender systems. *ACM Conference on Recommender Systems*, pp. 147–154, 2008.

[504] F. Ricci. Mobile recommender systems. *Information Technology and Tourism*, 12(3), pp. 205–213, 2010.

[505] F. Ricci, L. Rokach, B. Shapira, and P. Kantor. Recommender systems handbook. *Springer*, New York, 2011.

[506] F. Ricci and P. Avesani. Learning a local similarity metric for case-based reasoning. *International Conference on Case-Based Reasoning Research and Development*, pp. 301–312, 1995.

[507] F. Ricci, B. Arslan, N. Mirzadeh, and A. Venturini. LTR: A case-based travel advisory system. *European Conference on Case-Based Reasoning*, pp. 613–627, 2002.

[508] E. Rich. User modeling via stereotypes. *Cognitive Science*, 3(4), pp. 329–354, 1979.

[509] M. Richardson, R. Agrawal, and P. Domingos. Trust management for the semantic Web. *The Semantic Web*, Springer, pp. 351–368, 2003.

[510] M. Richardson and P. Domingos. Mining knowledge-sharing sites for viral marketing. *ACM KDD Conference*, pp. 61–70, 2002.

[511] J. Rocchio. Relevance feedback information retrieval. *The SMART retrieval system – experiments in automated document processing* , pp. 313–323, Prentice-Hall, Englewood Cliffs, NJ, 1971.

[512] P. Rousseeuw and A. Leroy. Robust regression and outlier detection *John Wiley and Sons*, 2005.

[513] N. Rubens, D. Kaplan, and M. Sugiyama. Active learning in recommender systems. *Recommender Systems Handbook*, Springer, pp. 735–767, 2011.

[514] N. Sahoo, R. Krishnan, G. Duncan, and J. Callan. Collaborative filtering with multi-component rating for recommender systems. *Proceedings of the sixteenth workshop on information technologies and systems*, 2006.

[515] A. Said, S. Berkovsky, and E. de Luca. Putting things in context: challenge on context-aware movie recommendation. *Proceedings of the Workshop on Context-Aware Movie Recommendation*, 2010.

[516] T. Sainath, B. Kingsbury, V. Sindhwani, E. Arisoy, and B. Ramabhadran. Low-rank matrix factorization for deep neural network training with high-dimensional output targets. *Acoustics, Speech and Signal Processing (ICASSP)*, pp. 6655–6659, 2013.

[517] R. Salakhutdinov, and A. Mnih. Probabilistic matrix factorization. *Advances in Neural and Information Processing Systems*, pp. 1257–1264, 2007.

[518] R. Salakhutdinov, and A. Mnih. Bayesian probabilistic matrix factorization using Markov chain Monte Carlo. *International Conference on Machine Learning*, pp. 880–887, 2008.

[519] R. Salakhutdinov, A. Mnih, and G. Hinton. Restricted Boltzmann machines for collaborative filtering. *International conference on Machine Learning*, pp. 791–798, 2007.

[520] J. Salter, and N. Antonopoulos. CinemaScreen recommender agent: combining collaborative and content-based filtering. *Intelligent Systems*, 21(1), pp. 35–41, 2006.

[521] P. Samarati. Protecting respondents identities in microdata release. *IEEE Transaction on Knowledge and Data Engineering*, 13(6), pp. 1010–1027, 2001.

[522] J. Sandvig, B. Mobasher, and R. Burke. Robustness of collaborative recommendation based on association rule mining. *ACM Conference on Recommender Systems*, pp. 105–12, 2007.

[523] J. Sandvig, B. Mobasher, and R. Burke. A survey of collaborative recommendation and the robustness of model-based algorithms. *IEEE Data Engineering Bulletin*, 31(2), pp. 3–13, 2008.

[524] B. Sarwar, G. Karypis, J. Konstan, and J. Riedl. Item-based collaborative filtering recommendation algorithms. *World Wide Web Conference*, pp. 285–295, 2001.

[525] B. Sarwar, G. Karypis, J. Konstan, and J. Riedl. Application of dimensionality reduction in recommender system – a case study. *WebKDD Workshop at ACM SIGKDD Conference, 2000*. Also appears at *Technical Report TR-00-043*, University of Minnesota, Minneapolis, 2000. https://wwws.cs.umn.edu/tech_reports_upload/tr2000/00-043.pdf

[526] B. Sarwar, J. Konstan, A. Borchers, J. Herlocker, B. Miller, and J. Riedl. Using filtering agents to improve prediction quality in the grouplens research collaborative filtering system. *ACM Conference on Computer Supported Cooperative Work*, pp. 345–354, 1998.

[527] B. Sarwar, G. Karypis, J. Konstan, and J. Riedl. Incremental singular value decomposition algorithms for highly scalable recommender systems. *International Conference on Computer and Information Science*, pp. 27–28, 2002.

[528] B. Sarwar, G. Karypis, J. Konstan, and J. Riedl. Recommender systems for large-scale e-commerce: Scalable neighborhood formation using clustering. *International Conference on Computer and Information Technology*, 2002.

[529] J. Schafer, D. Frankowski, J. Herlocker,and S. Sen. Collaborative filtering recommender systems. *Lecture Notes in Computer Science*, Vol. 4321, pp. 291–324, 2006.

[530] J. Schafer, J. Konstan, and J. Riedl. Recommender systems in e-commerce. *ACM Conference on Electronic Commerce*, pp. 158–166, 1999.

[531] L. Schaupp and F. Belanger. A conjoint analysis of online consumer satisfaction. *Journal of Electronic Commerce Research*, 6(2), pp. 95–111, 2005.

[532] S. Schechter, M. Krishnan, and M. D. Smith. Using path profiles to predict http requests. *World Wide Web Conference*, 1998.

[533] A. Schein, A. Popescul, L. Ungar, and D. Pennock. Methods and metrics for cold-start recommendations. *ACM SIGIR Conference*, 2002.

[534] I. Schwab, A. Kobsa, and I. Koychev. Learning user interests through positive examples using content analysis and collaborative filtering. Internal Memo, GMD, St. Augustin, Germany, 2001.

[535] S. Sen, J. Vig, and J. Riedl. Tagommenders: connecting users to items through tags. *World Wide Web Conference*, pp. 671–680, 2009.

[536] S. Sen, J. Vig, and J. Riedl. Learning to recognize valuable tags. *International Conference on Intelligent User Interfaces*, pp. 87–96, 2009.

[537] D. Seung, and L. Lee. Algorithms for non-negative matrix factorization. *Advances in Neural Information Processing Systems*, 13, pp. 556–562, 2001.

[538] G. Shani and A. Gunawardana. Evaluating recommendation systems. *Recommender Systems Handbook*, pp. 257–297, 2011.

[539] G. Shani, M. Chickering, and C. Meek. Mining recommendations from the Web. *ACM Conference on Recommender Systems*, pp. 35–42, 2008.

[540] U. Shardanand and P. Maes. Social information filtering: algorithms for automating word of mouth. *ACM Conference on Human Factors in Computing Systems*, 1995.

[541] H. Shen and J. Z. Huang. Sparse principal component analysis via regularized low rank matrix approximation. *Journal of multivariate analysis.* 99(6), pp. 1015–1034, 2008.

[542] A. Shepitsen, J. Gemmell, B. Mobasher, and R. Burke. Personalized recommendation in social tagging systems using hierarchical clustering. *ACM Conference on Recommender Systems*, pp. 259–266. 2008.

[543] B. Sheth and P. Maes. Evolving agents for personalized information filtering. *Ninth Conference on Artificial Intelligence for Applications*, pp. 345–352, 1993.

[544] Y. Shi, M. Larson, and A. Hanjalic. Collaborative filtering beyond the user-item matrix: A survey of the state of the art and future challenges. *ACM Computing Surveys (CSUR)*, 47(1), 3, 2014.

[545] Y. Shi, M. Larson, and A. Hanjalic. List-wise learning to rank with matrix factorization for collaborative filtering. *ACM Conference on Recommender Systems*, 2010.

[546] Y. Shi, A. Karatzoglou, L. Baltrunas, M. Larson, N. Oliver, and A. Hanjalic. CLiMF: Learning to maximize reciprocal rank with collaborative less-is-more collaborative filtering. *ACM Conference on Recommender Systems*, pp. 139–146, 2012.

[547] Y. Shi, A. Karatzoglou, L. Baltrunas, M. Larson, and A. Hanjalic. GAPfm: Optimal top-n recommendations for graded relevance domains. *ACM Conference on Information and Knowledge Management*, pp. 2261–2266, 2013.

[548] Y. Shi, A. Karatzoglou, L. Baltrunas, M. Larson, and A. Hanjalic. xCLiMF: optimizing expected reciprocal rank for data with multiple levels of relevance. *ACM Conference on Recommender Systems*, pp. 431–434, 2013.

[549] Y. Shi, A. Karatzoglou, L. Baltrunas, M. Larson, A. Hanjalic, and N. Oliver. TFMAP: Optimizing MAP for top-n context-aware recommendation. *ACM SIGIR Conference on Research and Development in Information Retrieval*, pp. 155–164, 2012.

[550] H. Shimazu, A. Shibata, and K. Nihei. ExpertGuide: A conversational case-based reasoning tool for developing mentors in knowledge spaces. *Applied Intelligence*, 14(1), pp. 33–48, 2002.

[551] R. Shokri, P. Pedarsani, G. Theodorakopoulos, and J. Hubaux. Preserving privacy in collaborative filtering through distributed aggregation of offline profiles. *ACM Conference on Recommender Systems*, pp. 157–164, 2009.

[552] M.-L. Shyu, C. Haruechaiyasak, S.-C. Chen, and N. Zhao. Collaborative filtering by mining association rules from user access sequences. *Workshop on Challenges in Web Information Retrieval and Integration*, pp. 128–135, 2005.

[553] B. Sigurbjornsson and R. Van Zwol. Flickr tag recommendation based on collective knowledge. *World Wide Web Conference*, pp. 327–336, 2008.

[554] J. Sill, G. Takacs, L. Mackey, and D. Lin. Feature-weighted linear stacking. *arXiv preprint*, arXiv:0911.0460, 2009. http://arxiv.org/pdf/0911.0460.pdf

[555] Y. Song, L. Zhang and C. L. Giles. Automatic tag recommendation algorithms for social recommender systems. *ACM Transactions on the Web (TWEB)*, 5(1), 4, 2011.

[556] Y. Song, Z. Zhuang, H. Li, Q. Zhao, J. Li, W. Lee, and C. L. Giles. Real-time automatic tag recommendation. *ACM SIGIR Conference*, pp. 515–522, 2008.

[557] A. P. Singh and G. J. Gordon. Relational learning via collective matrix factorization. *ACM KDD Conference*, pp. 650–658, 2008.

[558] B. Smyth. Case-based recommendation. *The Adaptive Web*, pp. 342–376, Springer, 2007.

[559] B. Smyth and P. Cotter. A personalized television listings service. *Communications of the ACM*, 43(8), pp. 107–111, 2000.

[560] B. Smyth and P. McClave. Similarity vs. diversity. *Case-Based Reasoning Research and Development*, pp. 347–361, 2001.

[561] H. Sorensen and M. McElligott. PSUN: a profiling system for Usenet news. *CIKM Intelligent Information Agents Workshop*, 1995.

[562] J. Srivastava, R. Cooley, M. Deshpande, and P.-N. Tan. Web usage mining: discovery and applications of usage patterns from Web data. *ACM SIGKDD Explorations*, 1(2), pp. 12–23, 2000.

[563] A. Stahl. Learning feature weights from case order feedback. *International Conference on Case-Based Reasoning*, pp. 502–516, 2001.

[564] H. Steck. Item popularity and recommendation accuracy. *ACM Conference on Recommender Systems*, pp. 125–132, 2011.

[565] H. Steck. Training and testing of recommender systems on data missing not at random. *ACM KDD Conference*, pp. 713–722, 2010.

[566] H. Steck. Evaluation of recommendations: rating-prediction and ranking. *ACM Conference on Recommender Systems*, pp. 213–220, 2013.

[567] H. Stormer. Improving e-commerce recommender systems by the identification of seasonal products. *Conference on Artificial Intelligence*, pp. 92–99, 2007.

[568] G. Strang. An introduction to linear algebra. *Wellesley Cambridge Press*, 2009.

[569] N. Srebro, J. Rennie, and T. Jaakkola. Maximum-margin matrix factorization. *Advances in neural information processing systems*, pp. 1329–1336, 2004.

[570] X. Su and T. Khoshgoftaar. A survey of collaborative filtering techniques. *Advances in artificial intelligence*, 4, 2009.

[571] X. Su, T. Khoshgoftaar, X. Zhu, and R. Greiner. Imputation-boosted collaborative filtering using machine learning classifiers. *ACM symposium on Applied computing*, pp. 949–950, 2008.

[572] X. Su, H. Zeng, and Z. Chen. Finding group shilling in recommendation system. *World Wide Web Conference*, pp. 960–961, 2005.

[573] K. Subbian, C. Aggarwal, and J. Srivasatava. Content-centric flow mining for influence analysis in social streams. *CIKM Conference*, pp. 841–846, 2013.

[574] B. O'Sullivan, A. Papadopoulos, B. Faltings, and P. Pu. Representative explanations for over-constrained problems. *AAAI Conference*, pp. 323–328, 2007.

[575] J. Sun and J. Tang. A survey of models and algorithms for social influence analysis. *Social Network Data Analytics*, Springer, pp. 177–214, 2011.

[576] Y. Sun, J. Han, C. Aggarwal, and N. Chawla. When will it happen?: relationship prediction in heterogeneous information networks. *ACM International Conference on Web Search and Data Mining*, pp. 663–672, 2012.

[577] Y. Sun, R. Barber, M. Gupta, C. Aggarwal, and J. Han. Co-author relationship prediction in heterogeneous bibliographic networks. *Advances in Social Networks Analysis and Mining (ASONAM)*, pp. 121–128, 2011.

[578] D. Sutherland, B. Poczos, and J. Schneider. Active learning and search on low-rank matrices. *ACM KDD Conference*, pp. 212–220, 2013.

[579] R. Sutton and A. Barto. Reinforcement learning: An introduction, *MIT Press*, Cambridge, 1998.

[580] P. Symeonidis, E. Tiakas, and Y. Manolopoulos. Transitive node similarity for link prediction in social networks with positive and negative links. *ACM Conference on Recommender Systems*, pp. 183–190, 2010.

[581] P. Symeonidis, E. Tiakas, and Y. Manolopoulos. Product recommendation and rating prediction based on multi-modal social networks. *ACM Conference on Recommender Systems*, pp. 61–68, 2011.

[582] P. Symeonidis, A. Nanopoulos, and Y. Manolopoulos. A unified framework for providing recommendations in social tagging systems based on ternary semantic analysis. *IEEE Transactions on Knowledge and Data Engineering*, 22(2), pp. 179–192, 2010.

[583] P. Symeonidis, A. Nanopoulos, and Y Manolopoulos. Tag recommendations based on tensor dimensionality reduction. *ACM Conference on Recommender Systems*, pp. 43–50, 2008.

[584] M. Szomszor, C. Cattuto, H. Alani, K. O'Hara, A. Baldassarri, V. Loreto, and V. Servedio. Folksonomies, the semantic web, and movie recommendation. *Bridging the Gap between the Semantic Web and Web 2.0*, pp. 71–84, 2007.

[585] N. Taghipour, A. Kardan, and S. Ghidary. Usage-based web recommendations: a reinforcement learning approach. *ACM Conference on Recommender Systems*, pp. 113–120, 2007.

[586] G. Takacs, I. Pilaszy, B. Nemeth, and D. Tikk. Matrix factorization and neighbor based algorithms for the Netflix prize problem. *ACM Conference on Recommender Systems*, pp. 267–274, 2008.

[587] G. Takacs, I. Pilaszy, B. Nemeth, and D. Tikk. Scalable collaborative filtering approaches for large recommender systems. *Journal of Machine Learning Research*, 10, pp. 623–656, 2009.

[588] J. Tang, X. Hu, and H. Liu. Social recommendation: a review. *Social Network Analysis and Mining*, 3(4), pp. 1113–1133, 2013.

[589] J. Tang, J. Sun, C. Wang, and Z. Yang. Social influence analysis in large-scale networks. *ACM KDD Conference*, pp. 807–816, 2009.

[590] J. Tang, C. Aggarwal, and H. Liu. Recommendations in signed social networks. *World Wide Web Conference*, 2016.

[591] J. Tang, S. Chang, C. Aggarwal, and H. Liu. Negative link prediction in social media. *Web Search and Data Mining Conference*, 2015.

[592] J. Tang, X. Hu, Y. Chang, and H. Liu. Predictability of distrust with interaction data. *ACM International Conference on Information and Knowledge Management (CIKM)*, pp. 181–190, 2014.

[593] J. Tang, X. Hu and H. Liu. Is distrust the negation of trust? The value of distrust in social media. *ACM Hypertext Conference (HT)*, pp. 148–157, 2014.

[594] J. Tang, H. Gao, X. Hu, and H. Liu. Exploiting homophily effect for trust prediction. *ACM International Conference on Web Search and Data Mining*, pp. 53–62, 2013.

[595] T. Tang, P. Winoto, and K. C. C. Chan. On the temporal analysis for improved hybrid recommendations. *International Conference on Web Intelligence*, pp. 214–220, 2003.

[596] T. Tang and G. McCalla. The pedagogical value of papers: a collaborative-filtering based paper recommender. *Journal of Digital Information*, 10(2), 2009.

[597] W. Tang, Y. Ma, and Z. Chen. Managing trust in peer-to-peer networks. *Journal of Digital Information Management*, 3(2), pp. 58–63, 2005.

[598] N. Tintarev and J. Masthoff. Designing and evaluating explanations for recommender systems. *Recommender Systems Handbook*, pp. 479–510, 2011.

[599] E. G. Toms. Serendipitous information retrieval. *DELOS Workshop: Information Seeking, Searching and Querying in Digital Libraries*, 2000.

[600] R. Torres, S. M. McNee, M. Abel, J. Konstan, and J. Riedl. Enhancing digital libraries with TechLens+. *ACM/IEEE-CS Joint Conference on Digital libraries*, pp. 228–234, 2004.

[601] T. Tran and R. Cohen. Hybrid recommender systems for electronic commerce. *Knowledge-Based Electronic Markets, Papers from the AAAI Workshop*, Technical Report WS-00-04, pp. 73–83, 2000.

[602] M.-H. Tsai, C. Aggarwal, and T. Huang. Ranking in heterogeneous social media. *Web Search and Data Mining Conference*, 2014.

[603] K. Tso-Sutter, L. Marinho, L. Schmidt-Thieme. Tag-aware recommender systems by fusion of collaborative filtering algorithms. *ACM Symposium on Applied Computing*, pp. 1995–1999, 2008.

[604] A. Tsoukias, N. Matsatsinis, and K. Lakiotaki. Multi-criteria user modeling in recommender systems. *IEEE Intelligent Systems*, 26(2), pp. 64–76, 2011.

[605] L. Tucker. Some mathematical notes on three-model factor analysis. *Psychometrika*, 31, pp. 279–311, 1966.

[606] A. Tveit. Peer-to-peer based recommendations for mobile commerce. *Proceedings of the International Workshop on Mobile Commerce*, pp. 26–29, 2001.

[607] A. Umyarov, and A. Tuzhilin. Using external aggregate ratings for improving individual recommendations. *ACM Transactions on the Web (TWEB)*, 5(1), 3, 2011.

[608] L. Ungar and D. Foster. Clustering methods for collaborative filtering. *AAAI Workshop on Recommendation Systems*. Vol. 1, 1998.

[609] B. van Roy and X. Yan. Manipulation-resistant collaborative filtering systems. *ACM Conference on Recommender Systems*, pp. 165–172, 2009.

[610] M. van Satten. Supporting people in finding information: Hybrid recommender systems and goal-based structuring. *Ph.D. Thesis*, Telemetica Instituut, University of Twente, Netherlands, 2005.

[611] M. van Setten, S. Pokraev, and J. Koolwaaij. Context-aware recommendations in the mobile tourist application compass. *Adaptive Hypermedia*, Springer, pp. 235–244, 2004.

[612] K. Verbert, N. Manouselis, X. Ochoa, M. Wolpers, H. Drachsler, I. Bosnic, and E. Duval. Context-aware recommender systems for learning: a survey and future challenges. *IEEE Transactions on Learning Technologies*, 5(4), pp. 318–335, 2012.

[613] K. Verstrepen and B. Goethals. Unifying nearest neighbors collaborative filtering. *ACM Conference on Recommender Systems*, pp. 177–184, 2014.

[614] P. Victor, C. Cornelis, M. De Cock, and P. Da Silva. Gradual trust and distrust in recommender systems. *Fuzzy Sets and Systems*, 160(10), pp. 1367–1382, 2009.

[615] P. Victor, C. Cornelis, M. De Cock, and E. Herrera-Viedma. Practical aggregation operators for gradual trust and distrust. *Fuzzy Sets and Systems*, 184(1), pp. 126–147, 2011.

[616] P. Victor, M. De Cock, and C. Cornelis. Trust and Recommendations. *Recommender Systems Handbook*, Springer, pp. 645–675, 2011.

[617] P. Victor, C. Cornelis, M. De Cock, and A. Teredesai. Trust-and distrust-based recommendations for controversial reviews. *Proceedings of the WebSci*, 2009. http://journal.webscience.org/161/2/websci09_submission_65.pdf

[618] V. Vlahakis, N. Ioannidis, J. Karigiannis, M. Tsotros, M. Gounaris, D. Stricker, T. Gleue, P. Daehne, and L. Almeida. Archeoguide: an augmented reality guide for archaeological sites. IEEE Computer Graphics and Applications, 22(5), pp. 52–60, 2002.

[619] L. von Ahn, M. Blum, N. Hopper, and J. Langford. CAPTCHA: Using hard AI problems for security. *Advances in Cryptology – EUROCRYPT*, pp. 294–311, 2003.

[620] S. Vucetic and Z. Obradovic. Collaborative filtering using a regression-based approach. *Knowledge and Information Systems*, 7(1), pp. 1–22, 2005.

[621] C. Wang, J. Han, Y. Jia, J. Tang, D. Zhang, Y. Yu, and J. Guo. Mining advisor-advisee relationships from research publication networks. *ACM KDD Conference*, pp. 203–212, 2010.

[622] J. Wang, A. de Vries, and M. Reinders. Unifying user-based and item-based similarity approaches by similarity fusion. *ACM SIGIR Conference*, pp. 501–508, 2006.

[623] A. M. Ahmad Wasfi. Collecting user access patterns for building user profiles and collaborative filtering. *International Conference on Intelligent User Interfaces*, pp. 57–64, 1998.

[624] M. Weimer, A. Karatzoglou, Q. Le, and A. Smola. CoFiRank: Maximum margin matrix factorization for collaborative ranking. *Advances in Neural Information Processing Systems*, 2007.

[625] M. Weimer, A. Karatzoglou, and A. Smola. Improving maximum margin matrix factorization. *Machine Learning*, 72(3), pp. 263–276, 2008.

[626] S.-S. Weng, L. Binshan, and W.-T. Chen. Using contextual information and multidimensional approach for recommendation. *Expert Systems and Applications*, 36, pp. 1268–1279, 2009.

[627] D. Wettschereck and D. Aha. Weighting features. *International Conference on Case-Based Reasoning*, pp. 347–358. 1995.

[628] J. White. Bandit algorithms for Website optimization. *O'Reilly Media, Inc*, 2012.

[629] S. Wild, J. Curry, and A. Dougherty. Improving non-negative matrix factorizations through structured initialization. *Pattern Recognition*, 37(11), pp. 2217–2232, 2004.

[630] C. Williams, B. Mobasher, and R. Burke. Defending recommender systems: detection of profile injection attacks. *Service Oriented Computing and Applications*, 1(3), pp. 157–170, 2007.

[631] C. Williams, B. Mobasher, R. Burke, J. Sandvig, and R. Bhaumik. Detection of obfuscated attacks in collaborative recommender systems. *ECAI Workshop on Recommender Systems*, 2006.

[632] C. Willmott and K. Matsuura. Advantages of the mean absolute error (MAE) over the root mean square error (RMSE) in assessing average model performance. *Climate Research*, 30(1), 79, 2005.

[633] W. Woerndl, C. Schueller, and R. Wojtech. A hybrid recommender system for context-aware recommendations of mobile applications. *IEEE International Conference on Data Engineering Workshop*, pp. 871–878, 2007.

[634] D. H. Wolpert. *Stacked generalization.* Neural Networks, 5(2), pp. 241–259, 1992.

[635] P. Wu, C. Yeung, W. Liu, C. Jin, and Y. Zhang. Time-aware collaborative filtering with the piecewise decay function. *arXiv preprint*, arXiv:1010.3988, 2010. http://arxiv.org/pdf/1010.3988.pdf

[636] K. L. Wu, C. C. Aggarwal, and P. S. Yu. Personalization with dynamic profiler. *International Workshop on Advanced Issues of E-Commerce and Web-Based Information Systems*, pp. 12–20, 2001. Also available online as *IBM Research Report*, RC22004, 2001. Search interface at http://domino.research.ibm.com/library/cyberdig.nsf/index.html

[637] M. Wu. Collaborative filtering via ensembles of matrix factorizations. *Proceedings of the KDD Cup and Workshop*, 2007.

[638] Z. Xia, Y. Dong, and G. Xing. Support vector machines for collaborative filtering. *Proceedings of the 44th Annual Southeast Regional Conference*, pp. 169–174, 2006.

[639] L. Xiang, Q. Yuan, S. Zhao, L. Chen, X. Zhang, Q. Yang, and J. Sun. Temporal recommendation on graphs via long-and short-term preference fusion. *ACM KDD Conference*, pp. 723–732, 2010.

[640] Z. Xiang and U. Gretzel. Role of social media in online travel information search. *Tourism Management*, 31(2), pp. 179–188, 2010.

[641] H. Xie, L. Chen, and F. Wang. Collaborative Compound Critiquing. *User Modeling, Adaptation, and Personalization*, Springer, pp. 254–265, 2014.

[642] Y. Xin and T. Jaakkola. Controlling privacy in recommender systems. *Advances in Neural Information Processing Systems*, pp. 2618–2626, 2014.

[643] B. Xu, J. Bu, C. Chen, and D. Cai. An exploration of improving collaborative recommender systems via user-item subgroups. *World Wide Web Conference*, pp. 21–30, 2012.

[644] G. Xue, C. Lin, Q. Yang, W. Xi, H. Zeng, Y. Yu, and Z. Chen. Scalable collaborative filtering using cluster-based smoothing. *ACM SIGIR Conference*, pp. 114–121, 2005.

[645] W. Yang, H. Cheng, and J. Dia. A location-aware recommender system for mobile shopping environments. *Expert Systems with Applications*, 34(1), pp. 437–445, 2008.

[646] X. Yang, Y. Guo, Y. Liu, and H. Steck. A survey of collaborative filtering based social recommender systems. *Computer Communications*, 41, pp. 1–10, 2014.

[647] H. Yildirim, and M. Krishnamoorthy. A random walk method for alleviating the sparsity problem in collaborative filtering. *ACM Conference on Recommender Systems*, pp. 131–138, 2008.

[648] H. Yin, B. Cui, J. Li, J. Yao, and C. Chen. Challenging the long tail recommendation. *Proceedings of the VLDB Endowment*, 5(9), pp. 896–907, 2012.

[649] H. Yin, Y. Sun, B. Cui, Z. Hu, and L. Chen. LCARS: A location-content-aware recommender system. *ACM KDD Conference*, pp. 221–229, 2013.

[650] H. F. Yu, C. Hsieh, S. Si, and I. S. Dhillon. Scalable coordinate descent approaches to parallel matrix factorization for recommender systems. *IEEE International Conference on Data Mining*, pp. 765–774, 2012.

[651] K. Yu, S. Zhu, J. Lafferty, and Y. Gong. Fast nonparametric matrix factorization for large-scale collaborative filtering. *ACM SIGIR Conference*, pp. 211–218, 2009.

[652] K. Yu, A. Shcwaighofer, V. Tresp, W.-Y. Ma, and H. Zhang. Collaborative ensemble learning. combining collaborative and content-based filtering via hierarchical Bayes, *Conference on Uncertainty in Artificial Intelligence*, pp. 616–623, 2003.

[653] Z. Yu, X. Zhou, Y. Hao, and J. Gu. TV program recommendation for multiple viewers based on user profile merging. *User Modeling and User-Adapted Interaction*, 16(1), pp. 63–82, 2006.

[654] Z. Yu, X. Zhou, D. Zhang, C. Y. Chin, and X. Wang. Supporting context-aware media recommendations for smart phones. *IEEE Pervasive Computing*, 5(3), pp. 68–75, 2006.

[655] Q. Yuan, G. Cong, Z. Ma, A. Sun, and N. Thalmann. Time-aware point-of-interest recommendation. *ACM SIGIR Conference*, pp. 363–372, 2013.

[656] R. Zafarani, M. A. Abbasi, and H. Liu. Social media mining: an introduction. *Cambridge University Press*, New York, 2014.

[657] H. Zakerzadeh, C. Aggarwal and K. Barker. Towards breaking the curse of dimensionality for high-dimensional privacy. *SIAM Conference on Data Mining*, pp. 731–739, 2014.

[658] F. Zaman and H. Hirose. Effect of subsampling rate on subbagging and related ensembles of stable classifiers. *Lecture Notes in Computer Science*, Springer, Volume 5909, pp. 44–49, 2009.

[659] M. Zanker and M. Jessenitschnig. Case studies on exploiting explicit customer requirements in recommender systems. *User Modeling and User-Adapted Interaction*, 19(1–2), pp. 133–166, 2009.

[660] M. Zanker, M. Aschinger, and M. Jessenitschnig. Development of a collaborative and constraint-based web configuration system for personalized bundling of products and services. *Web Information Systems Engineering–WISE*, pp. 273–284, 2007.

[661] M. Zanker, M. Aschinger, and M. Jessenitschnig. Constraint-based personalised configuring of product and service bundles. *International Journal of Mass Customisation*, 3(4), pp. 407–425, 2010.

[662] Y. Zhai, and B. Liu. Web data extraction based on partial tree alignment. *World Wide Web Conference*, pp. 76–85, 2005.

[663] J. Zhang, M. Ackerman, and L. Adamic. Expertise networks in online communities: structure and algorithms. *World Wide Web Conference*, pp. 221–230, 2007.

[664] J. Zhang and P. Pu. A comparative study of compound critique generation in conversational recommender systems. *Adaptive Hypermedia and Adaptive Web-Based Systems*, pp. 234–243, Springer, 2006.

[665] J. Zhang, N. Jones, and P. Pu. A visual interface for critiquing-based recommender systems. *Proceedings of the ACM conference on Electronic Commerce*, pp. 230–239, 2008.

[666] S. Zhang, W. Wang, J. Ford, and F. Makedon. Learning from incomplete ratings using nonnegative matrix factorization. *SIAM Conference on Data Mining*, pp. 549–553, 2006.

[667] S. Zhang, J. Ford, and F. Makedon Deriving Private Information from Randomly Perturbed Ratings. *SIAM Conference on Data Mining*, pp. 59–69, 2006. .

[668] S. Zhang, A. Chakrabarti, J. Ford, and F. Makedon. Attack detection in time series for recommender systems. *ACM KDD Conference*, pp. 809–814, 2006.

[669] T. Zhang and V. Iyengar. Recommender systems using linear classifiers. *Journal of Machine Learning Research*, 2, pp. 313–334, 2002.

[670] Y. Zhang, J. Callan, and T. Minka. Novelty and redundancy detection in adaptive filtering. *ACM SIGIR Conference*, pp. 81–88, 2002.

[671] Z. Zhang, T. Zhou, and Y. Zhang. Tag-aware recommender systems: A state-of-the-art survey. *Journal of Computer Science and Technology*, 26(5), pp. 767–777, 2011.

[672] Z. Zhang, C. Liu, and Y, Zhang. Solving the cold-start problem in recommender systems with social tags. *EPL (Europhysics Letters)*, 92(1), 2800, 2010.

[673] Y. Zhen, W. Li, and D. Yeung. TagiCoFi: tag informed collaborative filtering. *ACM Conference on Recommender Systems*, pp. 69–76, 2009.

[674] D. Zhou, O. Bousquet, T. Lal, J. Weston, and B. Scholkopf. Learning with local and global consistency. *Advances in Neural Information Processing Systems*, 16(16), pp. 321–328, 2004.

[675] D. Zhou, J. Huang, and B. Scholkopf. Learning from labeled and unlabeled data on a directed graph. *ICML Conference*, pp. 1036–1043, 2005.

[676] K. Zhou, S. Yang, and H. Zha. Functional matrix factorizations for cold-start recommendation. *ACM SIGIR Conference*, pp. 315–324, 2011.

[677] Y. Zhou, D. Wilkinson, R. Schreiber, and R. Pan. Large-scale parallel collaborative filtering for the Netflix prize. *Algorithmic Aspects in Information and Management*, pp. 337–348, 2008.

[678] X. Zhu, Z. Ghahramani, and J. Lafferty. Semi-supervised learning using gaussian fields and harmonic functions. *ICML Conference*, pp. 912–919, 2003.

[679] C. Ziegler. Applying feed-forward neural networks to collaborative filtering, Master's Thesis, Universitat Freiburg, 2006.

[680] C. Ziegler, S. McNee, J. Konstan, and G. Lausen. Improving recommendation lists through topic diversification. *World Wide Web Conference*, pp. 22–32, 2005.

[681] C. Ziegler and J. Golbeck. Investigating interactions of trust and interest similarity. *Decision Support Systems*, 43(2), pp. 460–475, 2007.

[682] C. Ziegler and G. Lausen. Propagation models for trust and distrust in social networks. *Information Systems Frontiers*, 7(4–5), pp. 337–358, 2005.

[683] C. Ziegler and G. Lausen. Spreading activation models for trust propagation. *IEEE International Conference on e-Technology, e-Commerce and e-Service*, pp. 83–97, 2004.

[684] A. Zimdars, D. Chickering, and C. Meek. Using temporal data for making recommendations. *Uncertainty in Artificial Intelligence*, pp. 580–588, 2001.

[685] A. Zimmermann, M. Specht, and A. Lorenz. Personalization and context management. *User Modeling and User-Adapted Interaction*, 15(3–4), pp. 275–302, 2005.

[686] http://www.foursquare.com

[687] http://grouplens.org

[688] http://grouplens.org/datasets/movielens/

[689] http://eigentaste.berkeley.edu/user/index.php

[690] http://www.netflix.com

[691] http://www.facebook.com

[692] http://www.last.fm

[693] http://www.pandora.com

[694] http://www.youtube.com

[695] http://www.tripadvisor.com

[696] http://www.google.com

[697] http://news.google.com

[698] http://www.amazon.com

[699] http://www.imdb.com

[700] http://www.flickr.com

[701] http://www.bibsonomy.org

[702] http://delicious.com

[703] http://www.pandora.com/about/mgp

[704] http://www.the-ensemble.com/

[705] http://www.epinions.com

[706] http://www.slashdot.org

[707] http://vanderwal.net/folksonomy.html

[708] http://www.bibsonomy.org

[709] http://www.amazon.com/gp/help/customer/display.html?nodeId=16238571

[710] http://opennlp.apache.org/index.html

[711] http://snowball.tartarus.org/

[712] https://code.google.com/p/ir-themis/

[713] http://www.netflixprize.com/community/viewtopic.php?id=828

[714] http://blog.netflix.com/2010/03/
      this-is-neil-hunt-chief-product-officer.html

[715] http://www.kddcup2012.org/workshop

Index

χ^2 Statistic, 148
ϵ-Greedy Algorithm, 420

A/B Testing, 225, 228
Accuracy Metrics, 240
Accuracy of Ratings Prediction, 240
Active Learning in Recommender Systems, 25, 412
AdaBoost.RT, 213
Adjusted Cosine Similarity, 40
Advagato Trust Metric, 383
Adversary, 385
Alternating Least Squares, 105
Amazon.com Recommender System, 5
Anomalous Attack Detection, 409
Appleseed, 352, 383
Applications of Recommender Systems, 435
Approval Voting, 447
Archeoguide, 307
ARHR, 247
Aspects in NMF, 122
Association Rules, 78
Association Rules for Attack Detection, 405
Asymmetric Factor Models, 110, 135
Attack Detection, 398
Attack Detection with H_v-Score, 403
Attack Detection with Time-Series Analysis, 409
Attack Resistance of Social Recommenders, 366
Attack-Resistant Recommender Systems, 385
Attacking Click-streams, 398, 409

Attacking Implicit Feedback, 398, 409
Average Attack, 393
Average Reciprocal Hit Rate, 247
Average without Misery, 425

Bagging, 209
Bandwagon Attack, 394
Baseline Estimators, 129
Bayes Classifier, 153
Bellkor's Pragmatic Chaos, 201
Bibsonomy, 366
Binary Ratings, 31
Bold Driver Algorithm, 100, 103
BookLens, 5
Boosting, 213
Bootstrapped Sampling, 209
Borda Count, 447
Bounded Random Selection Strategy, 187
Bucket-of-Models, 212

CAPTCHA, 404
CAPTCHAs for Attack Detection, 403
Cascade, 338
Catalog Coverage, 232
CATS, 423
Cinematch Recommender Algorithm, 6
Citeseer, 164
CLiMF, 417
Clustering for Neighborhood-based Methods, 45
CoFiRank, 416
Cold-Start Problem, 24, 136, 167, 212, 349, 366

Collaboration via Content, 217
Collaborative Advisory Travel System, 423
Collaborative Filtering, 8
Collaborative Tagging, 367
Collective Classification, 323
COMPASS, 307
Compatibility Conditions, 173
Compound Critiques, 188
Computational Advertising, 438
Condensation-Based Privacy, 434
Conditionally Factored RBMs, 276
Confidence, 77
Confidence in Evaluation, 232
Confidence-pruned Markov Model, 299
Conformity, 424
Conjoint Analysis, 179
Content-based Recommender Systems, 139
Context-based Recommender Systems, 20
Contextual Bandit Algorithms, 421
Contextual Filtering, 262, 289
Contextual Pre-Filtering, 289
Contextual Reduction, 262
Continuous Ratings, 31
Conversion Rate, 228
Coordinate Descent, 106
Copeland Rule, 447
Correlation Graph, 66, 319
Cosine Similarity, 36
Coverage, 231
Critiques in Case-Based Recommenders, 188
Critiquing Recommender Systems, 18, 170
Cross-Validation, 103, 239
Curse of Dimensionality, 434
Customer-Specific ROC Curves, 250
CWAdvisor, 196

Daily Learner System, 212
DCG, 245
DD (Attack Detection), 400
Decay-Based Temporal Methods, 286
Decision Trees, 160
Decision Trees for Collaborative Filtering, 74
Degree of Disagreement for Attack Detection, 400
Demographic Recommender Systems, 19
Diffusion Models, 338

Dimensionality Reduction for Neighborhood Methods, 47
Directional Critique, 188
Display Advertising, 439
Diversity, 234
Diversity in Evaluation, 234
Dynamic Critiques, 188
Dynamic Profiler, 436
Dynamic-Profiler, 164

Efficiency of Collaborative Filtering, 45
Efficient Attack, 386
EigenTrust, 383
Elastic-Net Regularizer, 59
Emotional Contagion, 424
Ensemble Recommender System, 199
EntreeC, 213
Error-pruned Markov Model, 299
Evaluating Recommender Systems, 225
Evaluation, 20

Factorization Machines, 270, 272, 280, 384, 416
Favorite Item Attack, 397
Feature Extraction in Content-based Systems, 142
Feature Weighted Linear Stacking, 219
Feature Weighting, 150
Filmtrust, 350
Filter Conditions, 173
Filtering in Neighborhood-based Methods, 36
FindMe Systems, 196
Fisher's Discrimination Index, 150
Folksonomy, 367
Forced Choice, 31
Forced Choice Rating System, 11
Frequent Itemsets, 77
Frequent Patterns, 77
Frequent Sequential Pattern, 300
Fusion Ensemble, 208

Gini Index, 74, 147
Global ROC Curves, 250
Google News Personalization, 436
Gradient Descent, 98
Graph-based Collaborative Filtering, 61
Group Attack Profile Detection, 402
Group Recommendation, 24, 423

Group Shilling Attacks, 409
GroupLens Recommender System, 4
Grundy, 19
GUIDE, 307

Handling Periodic Context, 288
Hidden Markov Models, 300
High-Knowledge Attack, 386
Hinge Loss, 127
Hit-Ratio, 391
Hold-Out, 238
Homophily, 349
Horting, 63
Hybrid Recommender System, 199
Hybrid Recommender Systems, 19

IDCG, 245
Ideal Discounted Cumulative Gain, 245
IfWeb System, 164
Implicit Feedback Data Sets, 11, 119
Implicit Feedback in Latent Factor Models,
    109
Implicit Ratings, 5
Impressions in Advertising, 439
Independent Cascade Model, 339
Individual Attack Profile Detection, 399
Inductive Models, 14, 111, 116
Influence Analysis, 337
Influence Limiter, 404
Interval-based Ratings, 31
INTIMATE, 165
INTRIGUE, 307
Intrigue, 423
Item-based Neighborhood Models, 40
Item-Item Graphs, 66
Item-Space Coverage, 232
Iterative Classification Algorithm, 324
ITR System, 165

Job Recommendations, 443
Joint Interpolation, 57

K-Means, 46
Katz Centrality, 329
Katz Measure, 61, 62, 328
Kendall Rank Correlation Coefficient, 243
Kernel Collaborative Filtering, 135
Knowledge-Based Recommender Systems,
    167

Knowledge-based Recommender Systems,
    15
Kronecker Product, 278

Label Propagation Algorithm, 325
LaboUr, 217
LARS, 303
Latent Dirichlet Allocation, 376
Latent Factor Models, 47, 90, 269
LDA, 376
Learning to Rank, 413
Least Misery Strategy, 424
Leave-One-Out Cross-Validation, 239
Left Eigenvector, 314
Letizia, 164
libFM, 275
Libra, 215, 223
Lift Index, 247
Linear Threshold Model, 339
Link Prediction, 326
Link Recommendation, 22
LinUCB Algorithm, 422
LISTEN, 307
Listwise Rank Learning, 415
LOCALBAL, 364
Location-Aware Recommender Systems,
    302
Location-based Recommender Systems, 21
Long-Tail Property, 32
Love/Hate Attack, 395
Low-Knowledge Attack, 386

MAE, 241
MAP, 246
Markov Decision Process, 181
Markovian Models, 295
Matrix Completion Problem, 3, 71
Matrix Factorization for Link Prediction,
    330
Maximum Margin Factorization, 127
Mean Absolute Error, 241
Mean Average Precision, 246
Mean Reciprocal Rank, 246
Mean Squared Error, 230, 240
Memory-based Collaborative Filtering, 9, 29
Mentor-Mentee Recommendation, 443
Minimum Support, 77
MINRELAX, 180
Missing Not at Random, 251

MNAR, 251
MobiDENK, 307
Model-based Collaborative Filtering, 71
Modified Degree of Similarity, 401
Moleskiing, 350, 383
MoleTrust, 356
MovieLens, 5
MRR, 246, 417
MSE, 230, 240
MSVD, 447
Multi-Arm Bandits, 228, 306, 418
Multi-Criteria Recommender Systems, 24, 426
Multi-Linear Singular Value Decomposition, 447
Multiplicative Aggregation, 447
Multiverse Recommendation, 269
Music Genome Project, 144
Music Recommendation, 144
MusicFX Group Recommender, 423
MyMap, 307

Naive Bayes Collaborative Filtering, 82
NDCG, 245
Neighborhood-based Collaborative Filtering, 9, 29, 33
Netflix Recommendation Challenge, 5
News Personalization, 436
Node Recommendation, 22
Non-negative Matrix Factorization, 119
Normalized Deviation, 149
Normalized Discounted Cumulative Gain, 245
Novelty, 161, 233
Novelty in Evaluation, 233
NPD, 399
Number of Prediction Differences, 399

Observed Ratings, 8
OLAP, 257
One class Collaborative Filtering, 120
Online Analytical Processing, 257
Online Attack Detection, 403
Online Dating Recommendations, 443
Online Evaluation, 227
Online Recruitment Recommendation, 196
Opinion Mining, 146
Ordinal Ratings, 31

OrdRec, 415
Out-of-Sample Recommendations, 14, 111, 116
Overfitting, 73
Overspecialization, 161

PageRank, 311
Pairwise Interaction Tensor Factorization, 270, 373
Pairwise Rank Learning, 415
Parallel Ensemble, 202
Pearson Correlation Coefficient, 35
Performance-Based Models, 432
Persistent Personalization in Knowledge-based Systems, 194
Personal WebWatcher, 164
Personalized PageRank, 314
Phrase Extraction, 145
Pipelined Ensemble, 202
PITF, 270, 373
PLSA, 127
Plurality Voting, 447
Pointwise Rank Learning, 415
PolyLens Group Recommender, 423
Popular Attack, 395
Popular Items, 32
Popular Page Attack, 398, 409
Portal Content Personalization, 435
Post-filtering, 266, 289
Power-Iteration Method, 314
Pre-filtering, 262, 289
Prediction Shift, 390, 391
Preference Locality, 302
Preprocessing in Content-based Systems, 142
Principal Component Analysis, 48
Privacy in Recommender Systems, 25, 432
Probabilistic Latent Semantic Analysis, 127
Probe Attack, 396
Product Recommendations with Social Cues, 23
Profile Association Rules, 81
Projected Gradient Descent, 116

Query Recommendation, 435
Query-Flow Graphs, 448
QUICKXPLAIN, 180
Qwikshop, 196

R-Score, 244
Random Attack, 393
Random Walks, 312
Randomness Injection, 211
Ranking Algorithms, 311
Rating Deviation from Mean Agreement,
        400
Raw Cosine, 36
RDMA (Attack Detection), 400
Recency-Based Collaborative Filtering, 286
Reciprocal Recommendations, 443, 448
Reciprocal Relationship Prediction, 443
Recommendation Query Language, 259, 280
Reduction-Based Multidimensional
        Approach, 262
Regression Trees for Collaborative
        Filtering, 74
Regression-based Models, 158
Regularization, 54, 100
Reinforcement Learning, 181, 228, 418
Relevance Feedback, 152
Repair Proposals, 179
Replacement Critique, 188
REQUEST, 280
Restaurant Recommender, 213
Reverse Bandwagon Attack, 396
Right Eigenvector, 314
RIPPER, 217
RMSE, 230, 240
Robust Matrix Factorization, 405
Robust Recommender Systems, 403
Robust Regression, 207
Robustness, 235
Rocchio Classification, 152
Root Mean Squared Error, 230, 240
RQL, 259, 280
Rule-based Collaborative Filtering, 77

Scalability, 235
Segment Attack, 396
Selective Markov Models, 298
Sentimental Analysis, 146
Sequential Ensemble, 202
Sequential Pattern Mining, 300
Serendipity, 161, 233
Serendipity in Evaluation, 233
Shilling Attacks, 385
Shills, 385
Short Memory Assumption, 297

Significance Weighting, 37
Simple Critiques, 188
SimRank, 321
Single Attack Profile Detection, 399
Singular Value Decomposition, 48, 113
Slashdot, 383
Slate Problem for Multi-Armed Bandits,
        442
SLIM, 58, 136, 218
SLIM with Side Information, 218
Slope-One Predictors, 68, 136
Smoothing Support Vector Machines, 86
Social Choice Theory, 424
Social Context, 345
Social Influence Analysis, 337
Social Recommender Systems, 22
Social Streams, 341
Social Tagging, 23, 366
Social Trust for Attack Detection, 404
SocialMF, 365
SocialRank, 317, 373
Sparse Linear Models, 58, 136, 218
Spearman Rank Correlation Coefficient, 243
Specified Ratings, 8
SPETA, 307
Sponsored Search, 439
Stability, 235
Standardization, 38
Stemming, 145
Stochastic Gradient Descent, 99
Stop-Words, 145
STREAM, 224
Subagging, 209
Subsampling, 209
Supervised Attack Detection, 399
Support, 77
Support-pruned Markov Model, 298
SVD++, 113
SVDFeature, 276
Symmetric Matrix Factorization, 333
Syskill & Webert, 164

Tag Informed Collaborative Filtering, 370
TagiCoFi, 380
Tagommenders, 370
Tags, 366
Temporal Recommender Systems, 283
Tensor Factorization, 269
The Ensemble, 201

TidalTrust, 353
Tie-Corrected Spearman Coefficient, 243
Tikhonov Regularization, 105, 158
Time-Periodic Biased K-NN, 288
Time-Sensitive Recommender Systems, 21, 283
Time-SVD++, 291
Top-k Recommendation Problem, 3
Topic-Sensitive PageRank, 314
Transductive Models, 14, 116
Travel Decision Forum, 423
Travel Locality, 302
Travel Recommendations, 196
Tree-Matching Algorithm, 144
Trust, 349
Trust Aggregation, 351
Trust in Evaluation, 232
Trust Metrics, 351
Trust Network, 350
Trust Propagation, 351
Trust Weighted Mean, 353
Trust-Enhanced Recommender Systems, 351
TrustWalker, 357
Trustworthy Recommender Systems, 23
Tucker Decomposition, 269

Unary Ratings, 32
Unbalanced Rating Scale, 10

Unconstrained Matrix Factorization, 96
UnRAP Algorithm, 402
Unsupervised Attack Detection, 399
Upper Bounding for Bandit Algorithms, 421
User Selection Bias, 251
User Studies, 227
User-based Neighborhood Models, 34
User-Item Graphs, 61
User-Space Coverage, 231
User-User Graphs, 63
Utility Matrix, 11
Utility-Based Recommender Systems, 18

Variable Selection in PCA, 402
Vector-Space Representation, 145
Viral Marketing, 23, 338
VITA Recommender, 196

Wasabi Personal Shopper, 196
WDA (Attack Detection), 401
WDMA (Attack Detection), 401
Web of Trust, 350
Web personalization, 295
WebMate System, 164
Weighted Degree of Agreement, 401
Weighted Deviation from Mean Agreement, 401
Window-Based Temporal Methods, 288